Photosynthesis under Biotic and Abiotic Environmental Stress

Photosynthesis under Biotic and Abiotic Environmental Stress

Editors

Suleyman Allakhverdiev
Alexander G. Ivanov
Marian Brestic

MDPI

Basel • Beijing • Wuhan • Barcelona • Belgrade • Novi Sad • Cluj • Manchester

Editors

Suleyman Allakhverdiev
Laboratory of Controlled Photobiosynthesis
Institute of Plant Physiology, RAS
Moscow
Russia

Alexander G. Ivanov
Department of Photoexcitable Membranes
Institute of Biophysics and Biomedical Engineering, Bulgarian Academy of Sciences
Sofia
Bulgaria

Marian Brestic
Department of Plant Physiology
The Slovak University of Agriculture
Nitra
Slovakia

Editorial Office
MDPI
St. Alban-Anlage 66
4052 Basel, Switzerland

This is a reprint of articles from the Special Issue published online in the open access journal *Cells* (ISSN 2073-4409) (available at: www.mdpi.com/journal/cells/special_issues/Photosynthesis_Biotic_Abiotic_Environmental_Stress).

For citation purposes, cite each article independently as indicated on the article page online and as indicated below:

Lastname, A.A.; Lastname, B.B. Article Title. *Journal Name* **Year**, *Volume Number*, Page Range.

ISBN 978-3-0365-9143-8 (Hbk)
ISBN 978-3-0365-9142-1 (PDF)
doi.org/10.3390/books978-3-0365-9142-1

Contents

About the Editors

Suleyman Allakhverdiev

Suleyman I. Allakhverdiev is the Head of the Controlled Photobiosynthesis Laboratory at the K.A. Timiryazev Institute of Plant Physiology of the Russian Academy of Sciences (RAS). He obtained his BS and MS in Physics from the Department of Physics, Azerbaijan State University, Baku. He obtained his Dr. Sci. degree in Plant Physiology and Biochemistry from the Institute of Plant Physiology, RAS (2002, Moscow). In 2018, he was recognized as the first in the ranking of the most highly cited scientists in Russia. In 2019, he was the winner of the K.A. Timiryazev Prize of the Russian Academy of Sciences. In 2021, he became the laureate of the "Global Energy Prize" in the "Non-conventional energy" category for his outstanding contribution to the development of alternative energy, particularly his design for a system of artificial photosynthesis and for his cycle of scientific works on bioenergy and hydrogen energy. In 2022, the Honored Scientist of the Russian Federation title was given to him as an assessment of his outstanding contribution to plant biology development and the study of the fundamental principles of photosynthesis. In 2022, he was elected a corresponding member of the Russian Academy of Sciences. In 2018–2023, he was recognized by Thomson Reuters-WoS (Clarivate Analytics) as one of the highly cited world-class researchers selected for their exceptional research performance and entered the list of the most highly cited scientists in the world (1% of the total number of researchers).

Alexander G. Ivanov

Alexander G. Ivanov received the MSc degree in ecology from the University of Sofia and the PhD degree in biophysics from the Institute of Biophysics, Bulgarian Academy of Sciences. He is currently a professor of biophysics at the Institute of Biophysics and Biomedical Engineering, Bulgarian Academy of Sciences. He is also an associate member of the Department of Biology at the University of Western Ontario. He has more than 30 years of experience in photosynthesis and plant physiology research. His research interests include the biophysics of photosynthesis, bioenergetics, membrane dynamics, and plant stress physiology.

Marian Brestic

Marian Brestic received the MS and PhD degrees at the Slovak University of Agriculture in Nitra. He completed postdoctoral stays at the University of Paris XI and CEA Cadarache in France. He is currently a professor of crop and tree physiology and head of the Institute of Plant and Environmental Studies at that university. He is also a professor at the Department of Botany and Plant Physiology at the Czech University of Life Sciences in Prague and a visiting professor at the Shandong Agricultural University in China. He led or participated in over 15 national and 20 bilateral and multilateral international projects. He published more than 350 research papers. His research is focused on crop photosynthesis under abiotic stress, plant drought and high-temperature tolerance, and the impact and mitigation of climate extremes on plants.

Preface

Photosynthesis, an extraordinary phenomenon fundamental to terrestrial ecosystems, has exerted a profound influence on the evolution and biodiversity of life forms on our planet. From the simplest microorganisms to the towering entities in the botanical realm, the process of photosynthesis assumes a central role. Across the course of evolutionary history, a myriad of plant species and photosynthetic mechanisms have arisen, each intricately adapted to function optimally under specific ecological conditions. However, the relentless march of change, driven by climatic fluctuations, environmental contamination, and the complexities of biological interactions, has introduced novel challenges. Amidst this swiftly changing landscape, the processes of plant growth, biomass production, and reproduction confront unprecedented constraints. It is in this context that the research presented in this reprint attains its significance.

The primary focus of this reprint is to delve into the intricacies of photosynthesis in the context of a mutable and often adverse environment and unravel the strategies employed by plants to respond and acclimatize to the challenges posed by abiotic and biotic stressors. The journey embarked upon within this reprint spans from the molecular intricacies of photosynthetic reactions to the comprehensive spectrum of whole-plant responses. Through original empirical investigations and erudite literature reviews, we investigate the ramifications of stress factors that can detrimentally affect the structural, photochemical, and biochemical facets that define the process of photosynthesis. At the core of our investigation in this reprint lie the regulatory mechanisms that govern the process of photosynthesis. We scrutinize not only the foundational scientific principles but also the pragmatic applications of this knowledge in diverse domains such as agriculture, forestry, and biotechnology. Armed with this knowledge, we can formulate strategies to augment plant adaptability, thereby ensuring the sustainability of natural and cultivated ecosystems. As we synthesize these multifarious strands of research, we aspire to stimulate interdisciplinary collaboration and ignite the curiosity for further exploration that promises to enrich our understanding of photosynthesis. The insights proffered herein afford a glimpse into the intricate interplay between life and light, a choreography that sustains the entirety of our biosphere. We extend our profound appreciation to the dedicated authors who have contributed their scholarly endeavors to this initiative, and we anticipate that this reprint will serve as an invaluable resource for scholars, educators, and all individuals impassioned by the unraveling of the profound processes governing life.

Suleyman Allakhverdiev, Alexander G. Ivanov, and Marian Brestic
Editors

cells

Editorial

Photosynthesis under Biotic and Abiotic Environmental Stress

Marian Brestic [1,*] and Suleyman I. Allakhverdiev [2,*]

[1] Institute of Plant and Environmental Sciences, Slovak University of Agriculture, A. Hlinku 2, 94976 Nitra, Slovakia.
[2] K.A. Timiryazev Institute of Plant Physiology, Russian Academy of Sciences, Botanicheskaya Street 35, 127276 Moscow, Russia
* Correspondence: marian.brestic@gmail.com (M.B.); suleyman.allakhverdiev@gmail.com (S.I.A.)

Photosynthesis is a unique process that has shaped life on our planet and created the conditions for all known life forms. During evolution, plant species and photosynthetic forms have emerged, and partial mechanisms have been optimized to work within a range of environmental conditions. However, variable environmental conditions caused by climate change, environmental pollution, and biotic factors significantly limit growth, biomass production, and plant reproduction.

By improving our understanding of the reactions and partial processes of photosynthesis in a changing or stressful environment, we can predict how plants will function in different climate-change scenarios when exposed to abiotic and biotic stressors, as well as how to improve their adaptability.

In this Special Issue of *Cells*, we present a collection of articles dedicated to various aspects of photosynthesis, representing research into the effects of abiotic and biotic stresses that can adversely affect structures and function, metabolism, and photochemical and biochemical processes from the molecular to the whole-plant level.

Cyanobacteria are prokaryotic photosynthetic organisms with considerable agro-biotechnological potential. They are used to produce biofertilizers, and contribute significantly to plant drought resistance and nitrogen enrichment in soil [1]. Sadvakasova et al. sought, isolated, and investigated nitrogen-fixing cyanobacterial strains in rice fields and evaluated the effect of Mo and Fe on photosynthetic and nitrogenase activities under nitrogen starvation. Cyanobacterial isolates isolated from rice paddies in Kazakhstan were identified as *Trichormus variabilis K-31 (MZ079356), Cylindrospermum badium J-8 (MZ079357), Nostoc sp. J-14 (MZ079360), Oscillatoria brevis SH-12 (MZ090011),* and *Tolypothrix tenuis J-1 (MZ079361).* The study of the influence of various concentrations of Mo and Fe on photosynthetic and nitrogenase activities under conditions of nitrogen starvation revealed the optimal concentrations of metals that stimulate the studied parameters [1].

The quality of light is an important factor in regulating plant growth and development during ontogenesis, including germination, photomorphogenesis, flowering induction, etc. Pashkovskiy et al. studied the influence of light of different qualities on the growth, gas exchange, fluorescence indices of Chl a, and expression of key light-dependent genes of Pinus sylvestris L. seedlings. In plants growing under red light (RL), the biomass of needles and root systems increased by more than two and three times, respectively, compared with those of the white fluorescent light (WFL) control. Meanwhile, the rates of photosynthesis and respiration in RL and blue light (BL) plants were lower than those of blue–red-light (BRL) plants, and the difference between the rates of photosynthesis and respiration, which characterizes the carbon balance, reached its maximum under RL [2].

According to the authors, RL influenced the number of xylem cells, activated the expression of genes involved in the transduction of cytokinin and auxin signals, and reduced the expression of the gene encoding the transcription factor phytochrome-interacting factor 3 (PIF3). It was suggested that RL-induced activation of key genes of cytokinin and

check for updates

Citation: Brestic, M.; Allakhverdiev, S.I. Photosynthesis under Biotic and Abiotic Environmental Stress. *Cells* **2022**, *11*, 3953. https://doi.org/10.3390/cells11243953

Received: 25 November 2022
Accepted: 3 December 2022
Published: 7 December 2022

auxin signaling might indicate a phytochrome-dependent change in cytokinin and auxin activity [2].

The effects of different nanoparticles and their combined effects on terrestrial plants have not been thoroughly investigated. This subject is a challenge for modern ecotoxicology. Cerium oxide nanoparticles (CeO2 NPs) and zinc oxide nanoparticles (ZnO NPs) are emerging pollutants [3]. In the study by Skiba et al., Pisum sativum L. plants were exposed to either CeO2 NPs or ZnO NPs alone or mixtures of these nano-oxides (at two concentrations: 100 and 200 mg/L) [3]. The authors concluded that CeO2 NPs moderate ZnO NP toxicity by protecting the photosynthetic apparatus in Pisum sativum leaves from oxidative stress triggered by Zn. Additionally, they observed that both nano-oxides affected nutrient uptake and transport at all concentrations applied. These results indicate that the free-radical scavenging properties of CeO2 NPs mitigate the toxicity induced by ZnO NPs [3].

Light plays an essential role in photosynthesis; however, its excess can cause damage to cellular components. Photosynthetic organisms thus developed a set of photoprotective mechanisms (e.g., nonphotochemical quenching, photoinhibition) that can be studied by classic biochemical and biophysical methods in cell suspensions. Here, we combined these bulk methods with single-cell identification of microdomains in the thylakoid membrane during high light (HL)-stress [4]. Canonico et al. used Synechocystis sp. PCC 6803 cells with yellow fluorescent protein (YFP)-tagged photosystem I and identified a multiphase response of cyanobacteria to HL stress with three main phases: fast, intermediate, and slow [4]. The authors observed the accumulation of myxoxanthophyll and more even spatial distribution of photosystems and phycobilisomes between microdomains. They suggest that the overall carotenoid increase during HL stress could be involved either in direct photoprotection (e.g., in ROS scavenging) and/or could play an additional role in maintaining the optimal distribution of photosystems in the thylakoid membrane to attain efficient photoprotection [4].

PSI photoinhibition is usually avoided through P700 oxidation. Without this protective mechanism, excess light represents a potentially lethal threat to plants. PGR5 is suggested to be a major component of cyclic electron transport around PSI and is important for P700 oxidation in angiosperms. The Arabidopsis PGR5-deficient mutant pgr5-1 is incapable of P700 oxidation regulation and has been used in numerous photosynthetic studies. However, Wada et al. revealed that pgr5-1 was a double mutant with exaggerated PSI photoinhibition [5]. pgr5-1 significantly reduced growth compared to the newly isolated PGR5-deficient mutant pgr5hope1. The introduction of PGR5 into pgr5-1 restored P700 oxidation regulation but maintained a pale-green phenotype, indicating that pgr5-1 had additional mutations. Both pgr5-1 and pgr5hope1 tended to cause PSI photoinhibition by excess light, but pgr5-1 exhibited an enhanced reduction in PSI activity [5].

In the work by Fatima et al., the combined response of exclusion of solar ultraviolet radiation (UV−A+B and UV−B) and static magnetic field (SMF) pretreatment was studied on soybean (Glycine max) leaves using synchrotron imaging [6]. The solar UV exclusion results suggested that ambient UV caused a reduction in leaf growth, which ultimately reduced photosynthesis in soybean seedlings, while SMF treatment enhanced leaf growth along with photosynthesis, even in the presence of ambient UV-B stress. The results suggested that SMF pretreatment of seeds diminishes the ambient UV-induced adverse effects on soybean [6].

Principal component analysis of foliar pigment composition revealed that Malva was similar to fast-growing annuals, while Lemna was similar to slow-growing evergreens [7]. Overall, Lemna exhibited traits reminiscent of those of its close relatives in the family Araceae, with a remarkable ability to acclimate to both deep shade and full sunlight. Overall, duckweed exhibits a combination of traits of fast-growing annuals and slow-growing evergreens with foliar pigment features that represent an exaggerated version of that of terrestrial perennials combined with an unusually high growth rate [7]. Duckweed's ability to thrive under a wide range of light intensities can support success in a dynamic light environment with periodic cycles of rapid expansion [7].

The efficacy of microbial endophytes in promoting plant growth and their comparison with exogenously applied hormones have not been investigated until now. The aim of the work by Ismail et al. was the isolation, identification, and characterization of bacterial and fungal endophytes from the roots of the *Phaseolus vulgaris* plant and the exploration of their potentiality compared to two common exogenously applied hormones on the growth and biochemical properties of *P. vulgaris* plants to explore the possibility of applying these microbial isolates as biofertilizers for the improvement of the growth performance and metabolites of crops. Their results indicated that the endophyte *Brevibacillus agri* (PB5) provides high potential as a stimulator for the growth and productivity of common bean plants [8].

Photosynthesis is an important target of action of numerous environmental factors; in particular, stressors can strongly affect photosynthetic light reactions. Considering the reactions of photosynthetic light to electron and proton transport, it can be supposed that an extremely low-frequency magnetic field (ELFMF) may influence these reactions; however, this problem has been weakly investigated [9]. Sukhov et al. experimentally tested a hypothesis about the potential influence of ELFMF on photosynthetic light reactions in wheat and pea seedlings [9]. The authors showed that ELFMF with Schumann resonance frequencies could influence photosynthetic light processes; however, this effect depends on the plant species (wheat or pea) and the type of treatment (short-term or chronic).

The effect of abiotic factors on the abundance and photosynthetic performance of airborne Cyanobacteria and microalgae isolated from the southern Baltic Sea region was studied by Wisniewska et al. [10]. Their experiments suggest that the adaptive abilities of microorganisms—particularly those producing toxins—may contribute to the spread, potentially increasing human exposure to their negative health effects. Any distinctive adaptations of the genera give them an additional competitive advantage and a greater chance for territorial expansion [10].

Black spot disease, caused by *Alternaria brassicicola* in *Brassica* species, is one of the most devastating diseases worldwide, especially since there is no known fully resistant *Brassica cultivar*. In this context, Macioszek et al. provides a report on the susceptible interaction between *B. oleracea var. capitata f. alba* (cultivar *'Glory of Enkhuizen'*) and *A. brassicicola*, both from the fungus and host plant perspectives [11]. The authors focused on the details of fungal development and colony formation and plant-cell reactions during infection at both light and transmission microscopy levels. Ultrastructural, molecular, physiological, and transcriptional analyses of infected leaves revealed photosynthesis as the most downregulated process from the onset of the infection [11]. This finding should be taken into consideration in further research when composing a strategy for the management of black spot disease.

One of our most striking observations concerns the significant difference between the physiological responses of different Synechococcus sp. phenotypes to changeable environmental conditions. The main aim of Śliwińska-Wilczewska's paper was to determine the acclimatization capacity of three Baltic phenotypes of *Synechococcus sp.* [12], an important link in forecasting future changes in the occurrence of these organisms in the context of global warming. Furthermore, the study focused on the effect of irradiance, temperature, and their mutual interactions on the content and proportions of cell-specific photosynthetic pigments of the examined cyanobacterial phenotypes. The authors reported that the detailed characterization of the quantitative and qualitative composition of pigments is important to determine the level of acclimatization of the examined phenotypes of cyanobacteria to specific environmental conditions. Awareness of the biology and physiology of these organisms obtained by capturing their reactions to various environmental factors is important for forecasting their possible expansion [12].

A review by Allakhverdiev et al. provides a comprehensive overview of Raman spectroscopy (RS) and its modifications applied to biological and medical research [13].

In this review, several existing studies on biological, medical, analytical, photosynthetic, and algal research using RS have been analyzed and presented. RS is used for a variety of studies in animals and human research. A greater focus on the application of RS in algal research will be beneficial for biotechnological purposes and general knowledge of the mechanisms of biomolecule interactions in algae under natural/environmental conditions. RS is a very attractive, advantageous, and promising approach for algal research [13].

Plant seeds are an essential input in agriculture; however, during their developmental stages, seeds can be negatively affected by environmental stresses, which can adversely affect seed vigor, seedling establishment, and crop production. Seeds resistant to high salinity, droughts, and climate change can result in higher crop yield [14]. The major findings suggested in this review by Shelar et al. refer to nano-priming as an emerging seed technology for sustainable food amid growing demand from the increasing global population [14]. This novel technology could influence the crop yield and ensure the quality and safety of seeds in a sustainable way. When nano-primed seeds are germinated, they undergo a series of synergistic events due to their enhanced metabolism: modulating biochemical signaling pathways, triggering hormone secretion, and reducing reactive oxygen species, leading to improved disease resistance. In addition to providing an overview of the challenges and limitations of seed nanopriming technology, this review also describes some of the emerging nanoseed priming methods for sustainable agriculture and other technological developments using cold plasma technology and machine learning [14].

Abiotic stresses, such as drought, salinity, heavy metals, variations in temperature, and ultraviolet (UV) radiation, are antagonistic to plant growth and development, resulting in an overall decrease in plant yield [15]. These stresses have direct effects on the rhizosphere, severely affecting root growth and thereby affecting plant growth, health, and productivity. However, the growth-promoting rhizobacteria that colonize the rhizosphere/endorhizosphere protect the roots from the adverse effects of abiotic stress and facilitate plant growth by various direct and indirect mechanisms [15]. In the rhizosphere, plants constantly interact with thousands of these microorganisms, yet it is unclear when and how these complex root, rhizosphere, and rhizobacteria interactions occur under abiotic stresses. A review by Khan et al. focuses on root–rhizosphere and rhizobacterial interactions during stress, how roots respond to these interactions, and the role of rhizobacteria under these stresses. Furthermore, the review focuses on the underlying mechanisms employed by rhizobacteria to improve root architecture and plant tolerance to abiotic stresses [15].

The original research and review articles collected in this Special Issue will be of interest to a broad audience of scientists studying photosynthesis under conditions of biotic and abiotic environmental stress, addressing all aspects of photosynthesis, including regulatory mechanisms, as well as their value in agriculture, forestry, ecology, and biotechnology. We hope that interdisciplinary applications of knowledge in this Special Issue will stimulate future research.

We would like to express our sincerest thanks to our readers, authors, anonymous peer reviewers, editors, and all the people working for the journal, all of whom have made substantial contributions to this Special Issue. This would not have been possible without your support. For the details of the Special Issue "Photosynthesis under Biotic and Abiotic Environmental Stress "please click here (https://www.mdpi.com/journal/cells/special_issues/Photosynthesis_Biotic_Abiotic_Environmental_Stress, accessed on 1 June 2020).

Author Contributions: S.I.A. and M.B. wrote the manuscript. All authors have read and agreed to the published version of the manuscript.

Funding: This work was carried out with the support of project (VEGA 1/0664/22) to MB, also supported by the Russian Science Foundation (RSF) grants (19-14-00118 and 22-44-08001) and the state contract of the Ministry of Science and Higher Education of the Russian Federation (Theme No. 122050400128-1) to SIA.

Institutional Review Board Statement: Not applicable.

Informed Consent Statement: Not applicable.

Data Availability Statement: The datasets generated during and/or analyzed during the current study are available from the corresponding author on reasonable request.

Conflicts of Interest: The authors declare no conflict of interest.

References

1. Sadvakasova, A.K.; Kossalbayev, B.D.; Token, A.I.; Bauenova, M.O.; Wang, J.; Zayadan, B.K.; Balouch, H.; Alwasel, S.; Leong, Y.K.; Chang, J.-S.; et al. Influence of Mo and Fe on Photosynthetic and Nitrogenase Activities of Nitrogen-Fixing Cyanobacteria under Nitrogen Starvation. *Cells* **2022**, *11*, 904. [CrossRef] [PubMed]
2. Pashkovskiy, P.; Kreslavski, V.D.; Ivanov, Y.; Ivanova, A.; Kartashov, A.; Shmarev, A.; Strokina, V.; Kuznetsov, V.V.; Allakhverdiev, S.I. Influence of Light of Different Spectral Compositions on the Growth, Photosynthesis, and Expression of Light-Dependent Genes of Scots Pine Seedlings. *Cells* **2021**, *10*, 3284. [CrossRef] [PubMed]
3. Skiba, E.; Pietrzak, M.; Glińska, S.; Wolf, W.M. The Combined Effect of ZnO and CeO2 Nanoparticles on Pisum sativum L.: A Photosynthesis and Nutrients Uptake Study. *Cells* **2021**, *10*, 3105. [CrossRef] [PubMed]
4. Canonico, M.; Konert, G.; Crepin, A.; Šedivá, B.; Kaňa, R. Gradual Response of Cyanobacterial Thylakoids to Acute High-Light Stress—Importance of Carotenoid Accumulation. *Cells* **2021**, *10*, 1916. [CrossRef] [PubMed]
5. Wada, S.; Amako, K.; Miyake, C. Identification of a Novel Mutation Exacerbated the PSI Photoinhibition in pgr5/pgrl1 Mutants; Caution for Overestimation of the Phenotypes in Arabidopsis pgr5-1 Mutant. *Cells* **2021**, *10*, 2884. [CrossRef] [PubMed]
6. Fatima, A.; Kataria, S.; Agrawal, A.K.; Singh, B.; Kashyap, Y.; Jain, M.; Brestic, M.; Allakhverdiev, S.I.; Rastogi, A. Use of Synchrotron Phase-Sensitive Imaging for the Investigation of Magnetopriming and Solar UV-Exclusion Impact on Soybean (Glycine max) Leaves. *Cells* **2021**, *10*, 1725. [CrossRef] [PubMed]
7. Stewart, J.J.; Adams, W.W., III; López-Pozo, M.; Doherty Garcia, N.; McNamara, M.; Escobar, C.M.; Demmig-Adams, B. Features of the Duckweed Lemna That Support Rapid Growth under Extremes of Light Intensity. *Cells* **2021**, *10*, 1481. [CrossRef] [PubMed]
8. Ismail, M.A.; Amin, M.A.; Eid, A.M.; Hassan, S.E.-D.; Mahgoub, H.A.M.; Lashin, I.; Abdelwahab, A.T.; Azab, E.; Gobouri, A.A.; Elkelish, A.; et al. Comparative Study between Exogenously Applied Plant Growth Hormones versus Metabolites of Microbial Endophytes as Plant Growth-Promoting for Phaseolus vulgaris L. *Cells* **2021**, *10*, 1059. [CrossRef] [PubMed]
9. Sukhov, V.; Sukhova, E.; Sinitsyna, Y.; Gromova, E.; Mshenskaya, N.; Ryabkova, A.; Ilin, N.; Vodeneev, V.; Mareev, E.; Price, C. Influence of Magnetic Field with Schumann Resonance Frequencies on Photosynthetic Light Reactions in Wheat and Pea. *Cells* **2021**, *10*, 149. [CrossRef] [PubMed]
10. Wiśniewska, K.; Śliwińska-Wilczewska, S.; Lewandowska, A.; Konik, M. The Effect of Abiotic Factors on Abundance and Photosynthetic Performance of Airborne Cyanobacteria and Microalgae Isolated from the Southern Baltic Sea Region. *Cells* **2021**, *10*, 103. [CrossRef] [PubMed]
11. Macioszek, V.K.; Gapińska, M.; Zmienko, A.; Sobczak, M.; Skoczowski, A.; Oliwa, J.; Kononowicz, A.K. Complexity of Brassica oleracea–Alternaria brassicicola Susceptible Interaction Reveals Downregulation of Photosynthesis at Ultrastructural, Transcriptional, and Physiological Levels. *Cells* **2020**, *9*, 2329. [CrossRef] [PubMed]
12. Śliwińska-Wilczewska, S.; Konarzewska, Z.; Wiśniewska, K.; Konik, M. Photosynthetic Pigments Changes of Three Phenotypes of *Picocyanobacteria Synechococcus* sp. under Different Light and Temperature Conditions. *Cells* **2020**, *9*, 2030. [CrossRef] [PubMed]
13. Allakhverdiev, E.S.; Khabatova, V.V.; Kossalbayev, B.D.; Zadneprovskaya, E.V.; Rodnenkov, O.V.; Martynyuk, T.V.; Maksimov, G.V.; Alwasel, S.; Tomo, T.; Allakhverdiev, S.I. Raman Spectroscopy and Its Modifications Applied to Biological and Medical Research. *Cells* **2022**, *11*, 386. [CrossRef] [PubMed]
14. Shelar, A.; Singh, A.V.; Maharjan, R.S.; Laux, P.; Luch, A.; Gemmati, D.; Tisato, V.; Singh, S.P.; Santilli, M.F.; Shelar, A.; et al. Sustainable Agriculture through Multidisciplinary Seed Nanopriming: Prospects of Opportunities and Challenges. *Cells* **2021**, *10*, 2428. [CrossRef] [PubMed]
15. Khan, N.; Ali, S.; Shahid, M.A.; Mustafa, A.; Sayyed, R.Z.; Curá, J.A. Insights into the Interactions among Roots, Rhizosphere, and Rhizobacteria for Improving Plant Growth and Tolerance to Abiotic Stresses: A Review. *Cells* **2021**, *10*, 1551. [CrossRef] [PubMed]

cells

MDPI

Article

Influence of Mo and Fe on Photosynthetic and Nitrogenase Activities of Nitrogen-Fixing Cyanobacteria under Nitrogen Starvation

Asemgul K. Sadvakasova [1], Bekzhan D. Kossalbayev [1,2,*], Aziza I. Token [1], Meruert O. Bauenova [1], Jingjing Wang [3], Bolatkhan K. Zayadan [1], Huma Balouch [1], Saleh Alwasel [4], Yoong Kit Leong [5], Jo-Shu Chang [5,6,7] and Suleyman I. Allakhverdiev [1,4,8,*]

1 Faculty of Biology and Biotechnology, Al-Farabi Kazakh National University, Al-Farabi Avenue 71, Almaty 050038, Kazakhstan; asem182010@gmail.com (A.K.S.); t.aziza_93@mail.ru (A.I.T.); bauyen.meruyert@gmail.com (M.O.B.); zbolatkhan@gmail.com (B.K.Z.); huma@comsats.org (H.B.)
2 Department of Chemical and Biochemical Engineering, Geology and Oil-Gas Business Institute Named after K. Turyssov, Satbayev University, Almaty 050043, Kazakhstan
3 Tianjin Institute of Industrial Biotechnology, Chinese Academy of Sciences, No. 32, West 7th Road, Tianjin Airport Economic Area, Tianjin 300308, China; wang_jj@tib.cas.cn
4 College of Science, King Saud University, Riyadh 12372, Saudi Arabia; salwasel@ksu.edu.sa
5 Department of Chemical and Materials Engineering, Tunghai University, Taichung 407, Taiwan; yoongkitleong1014@gmail.com (Y.K.L.); changjs@mail.ncku.edu.tw (J.-S.C.)
6 Research Center for Smart Sustainable Circular Economy, Tunghai University, Taichung 407, Taiwan
7 Department of Chemical Engineering, National Cheng Kung University, Tainan 701, Taiwan
8 Controlled Photobiosynthesis Laboratory, K.A. Timiryazev Institute of Plant Physiology, Russian Academy of Sciences, Botanicheskaya Street 35, 127276 Moscow, Russia
* Correspondence: kossalbayev.bekzhan@gmail.com (B.D.K.); suleyman.allakhverdiev@gmail.com (S.I.A.)

check for updates

Citation: Sadvakasova, A.K.; Kossalbayev, B.D.; Token, A.I.; Bauenova, M.O.; Wang, J.; Zayadan, B.K.; Balouch, H.; Alwasel, S.; Leong, Y.K.; Chang, J.-S.; et al. Influence of Mo and Fe on Photosynthetic and Nitrogenase Activities of Nitrogen-Fixing Cyanobacteria under Nitrogen Starvation. *Cells* **2022**, *11*, 904. https://doi.org/10.3390/cells11050904

Academic Editor: Stanislaw Karpinski

Received: 6 February 2022
Accepted: 2 March 2022
Published: 5 March 2022

Publisher's Note: MDPI stays neutral with regard to jurisdictional claims in published maps and institutional affiliations.

Abstract: The potential of cyanobacteria to perform a variety of distinct roles vital for the biosphere, including nutrient cycling and environmental detoxification, drives interest in studying their biodiversity. Increasing soil erosion and the overuse of chemical fertilizers are global problems in developed countries. The option might be to switch to organic farming, which entails largely the use of biofertilisers. Cyanobacteria are prokaryotic, photosynthetic organisms with considerable potential, within agrobiotechnology, to produce biofertilisers. They contribute significantly to plant drought resistance and nitrogen enrichment in the soil. This study sought, isolated, and investigated nitrogen-fixing cyanobacterial strains in rice fields, and evaluated the effect of Mo and Fe on photosynthetic and nitrogenase activities under nitrogen starvation. Cyanobacterial isolates, isolated from rice paddies in Kazakhstan, were identified as *Trichormus variabilis* K-31 (MZ079356), *Cylindrospermum badium* J-8 (MZ079357), *Nostoc* sp. J-14 (MZ079360), *Oscillatoria brevis* SH-12 (MZ090011), and *Tolypothrix tenuis* J-1 (MZ079361). The study of the influence of various concentrations of Mo and Fe on photosynthetic and nitrogenase activities under conditions of nitrogen starvation revealed the optimal concentrations of metals that have a stimulating effect on the studied parameters.

Keywords: cyanobacteria; heavy metals; photosynthesis; nitrogenase; heterocyst

1. Introduction

The most important direction in the development of organic agriculture is the creation of microbial biotechnologies that help maintain fertility and intensify agricultural production. Soil fertility is directly related to the diversity, quantity, and activity of the soil microbiota, which determine the transformation, migration, and accumulation of substances in the soil ecosystem, and provide plants with all necessary assailable nutrients. At the same time, it is known that the increase in crop yields largely depends on access to mineral nutrients, especially nitrogen. The natural sources of nitrogen in the soil are microorganisms, which can fix molecular nitrogen from the atmosphere [1–4].

In this context, one of the most interesting groups of microorganisms are cyanobacteria, which, in addition to the ability to fix nitrogen, have a wide range of adaptations to different soil and hydrothermal conditions [5–7]. Herewith, both types of cyanobacteria (with heterocysts and without them) have the ability to fix nitrogen. However, in heterocystous cyanobacteria, the processes of photosynthesis and nitrogen fixation occur simultaneously and independently of the time of day, while in non-heterocystous cyanobacteria, these processes occur separately: nitrogen fixation, at night; photosynthesis during the day [8]. In the cyanobacterial species that form heterocysts, the nitrogenase is located in the heterocyst and is, thus, protected from the inhibitory effects of oxygen [9–13]. They are more attractive in terms of productivity and can enrich the soil with 20–30 kg N ha^{-1} per year, which is a great economic advantage for agriculture. In addition, they are technologically advanced, can grow on cheap media without sources of organic compounds and mineral nitrogen, and do not require expensive equipment [14,15]. Considering the main factors affecting nitrogenase activity, it is worth outlining the role of biogenic metals in nitrogenase enzyme cofactors that determine the intensity of nitrogen fixation by free-living cyanobacteria [16,17].

Despite the large amount of data available in the literature indicating the high metabolic potential of cyanobacteria in nitrogen fixation, there are still some problems related to their productivity that limit their widespread use in agro-biotechnology. Therefore, the attention given to this group of organisms is focused on the study of the basic cultivation conditions that increase their nitrogen fixation abilities.

In this regard, scientific research aimed at potentially enhancing the nitrogen-fixing ability of cyanobacteria is currently very relevant. Primarily, research in this area should focus on finding new, more productive strains of nitrogen-fixing cyanobacteria and developing new approaches to increase the efficiency of their metabolic activity.

Therefore, this work aims to select and study new strains of nitrogen-fixing cyanobacteria. The article demonstrates the results of isolation, investigation, and identification of five cyanobacterial strains isolated from rice paddies in Kazakhstan and the influence of Mo and Fe on their photosynthetic and nitrogenase activity.

2. Materials and Methods

2.1. Research Objects

Five different strains of cyanobacteria, including *Nostoc* sp. (J-14), *Cylindrospermum* sp. (J-8), *Anabaena variabilis* (K-31), *Oscillatoria brevis* (SH-12), and *Tolypothrix tenuis* (J-1), were isolated from the Kyzylorda region. The strains were isolated from the rice fields and aquatic ecosystems of the Zhanakorgan and Shieli districts of the Kyzylorda region (rice paddies of Avangard, Akzhol, and Zhayilma villages, and water basins adjacent to rice fields, i.e., the tributaries of the Syr-Darya river: Kuday Kul, Zhayilma).

2.2. Isolation and Cultivation of Cyanobacterial Strains

Eighteen water samples and twenty-six composite samples of soil and cyanobacterial mats collected from the surface of plants using classical algological methods served as material for this study. Identifiers of freshwater microalgae and cyanobacteria were used to determine the species. In order to obtain enrichment cultures of cyanobacteria, 5 mL of aliquot samples were placed in 100 mL flasks containing 5 mL of BG-11 [18], Allen [19], Bold basal liquid [20] culture medium with the addition of cycloheximide (Biochem Chemopharma, Cosne-Cours-sur-Loire, France) at a concentration of 50 μg mL^{-1} to inhibit the growth of eukaryotes. Simultaneously, samples were inoculated into Petri dishes with solid medium (1% agar) BG-11. Cultivation was carried out in parallel at temperatures of 18–22 °C with a constant illumination of 50 μmol (photons) m^{-2} s^{-1}. Unialgal cultures of cyanobacteria were obtained through the periodic inoculation of the enrichment culture on solid nutrient medium BG-11, Allen, Gromov, and Bold, using capillary pipettes to separate the cyanobacterial filaments from the mixed microbial suspension. Several antibiotics (penicillin, gentamicin, tetracycline, neomycin, ampicillin, chloramphenicol, and kanamycin) at different concentrations from 1500 unit (U) mL^{-1} to 25,000 U mL^{-1} and their combinations with a total concentration of 20,000 U mL^{-1} were used to purify isolated cyanobacterial cultures from satellite bacteria. The calculated concentrations of antibiotics were added to the culture media after sterilization. The strains of isolated cyanobacteria were cultured on liquid and agar nutrient media of standard composition BG-11 and Allen in flasks in the luminostat. Preliminary morphological identification was performed under a light microscope (Micro Optix C600, Wiener Neudorf, Austria). Morphological identification was based on some characteristic morphological features, such as cell size, shape, location of cells, presence or absence of mucus, etc., using taxonomic literature and various photo galleries. The determination of the species composition of phototrophic eukaryotes was based on morphological and cultural characteristics using various determinants [21,22]. Before conducting an experiment, the cyanobacterial culture was subcultured from the liquid medium with the stored culture on Petri dishes with a freshly prepared nutrient medium of the same composition.

2.3. Molecular Identification of Cyanobacterial Strains

Cyanobacterial species were identified using molecular genetic research methods by sequencing the conserved DNA locus. To accomplish the tasks assigned, the following steps were performed: the isolation of genomic DNA, sample preparation and polymerase chain reaction, DNA sequencing for a fragment of the conserved locus, and an analysis of the nucleotide sequence of the conserved 16S rRNA locus [23].

2.3.1. Isolation of Genomic DNA of Cyanobacteria

Genomic DNA was isolated following the protocol proposed by Dale et al. [24]. Cyanobacterial cells were pelleted via centrifugation at 4000× *g* for 5 min, the supernatant was removed, and the cells were suspended in 1 mL of extraction buffer with the following composition: 200 mmol Tris-HCl (pH-8.0), 400 mmol LiCl, 25 mM EDTA, and 1% SDS. Acid-treated glass beads with a diameter of 425–600 µm (Sigma-Aldrich, Darmstadt, Germany) were added to the suspension. The resulting suspension was shaken vigorously for 30 s and then incubated on ice for 30 s. This procedure was repeated 5 times. The resulting mixture was centrifuged at 3000× *g* for 15 min at 4 °C. The supernatant was then transferred to a sterile microtube containing 600 µL of ice-cold isopropanol and mixed by pipetting, then centrifuged at 16,000× *g* for 20 min at 4 °C. The precipitate was dissolved in 100 µL TE buffer (2 mmol Tris-HCl (pH-8.0), 1 mmol EDTA). Then, 10 µL 3 M sodium acetate (pH-5.2) and 300 µL of 96% ethanol were added and incubated at −20 °C for 2 h. The precipitate was centrifuged. Another round of centrifugation was performed for 30 min with 16,000× *g* at 4 °C. The pellet was washed with 70% ethanol, oven-dried, and dissolved in 50 µL TE buffer. DNA concentration was determined using a Nanovue plus spectrophotometer (Biochrom, Boston, MA, USA). The concentration of isolated genomic DNA was adjusted to 100 ng µL^{-1}.

2.3.2. Sample Preparation and Polymerase Chain Reaction

The ribosomal DNA gene was selected as a conservative DNA locus for the cyanobacterial identification. The amplification of a 16S rRNA gene fragment from bacterial genomic DNA was performed using primers: 8f, 806r, M23f, M23′f, 27f, 1492r, 1525r [25–27]. The separation of the amplification products was carried out via DNA electrophoresis using 1% agarose gel in TAE buffer containing ethidium bromide (15 µg mL^{-1}). The resulting PCR products were treated with a mixture of ExoI exonuclease and FastAP alkaline phosphatase to remove the remaining primers and dNTPs. To 10 µL of each amplicon, 1 µL of ExoI and FastAP was added and incubated for 30 min at 37 °C. The inactivation of enzymes was carried out at 85 °C for 15 min [28].

2.3.3. DNA Sequencing Using the Fragment of the Conserved Locus

DNA sequencing using the Sanger method was used to determine the nucleotide sequence of the conserved locus [29]. The analysis of chromatograms and their comparison with the reference sequence was carried out using the VectorNTI version 11 software package and the NCBI database using the BLAST (Basic Local Alignment Search Tool) service [30].

2.4. Methods for Quantification of Microalgal Cells and Determination of Biomass Growth Productivity

The change in the concentration of cyanobacterial biomass was determined by measuring the optical density on a PD-303UV spectrophotometer (Apel, Saitama, Japan) at a wavelength of 540–590 nm. The biomass productivity was determined according to Tsarenko et al. [31]. The cells were precipitated using a 5810R centrifuge (Eppendorf, Hamburg, Germany) at 5000× *g* for 5 min at 25 °C. The growth rate coefficient of the cyanobacterial cultures was calculated based on the increase in the cell number in the experimental vessels, taking into account the initial cell number and the cell number after a certain time.

2.5. Determination of the Heterocyst Formation Frequency

The frequency of heterocyst formation was analysed on nitrogen-free BG_0-11 [32] medium, and the control was a BG-11 nutrient medium with a standard composition. Before the experiment, the cultures were kept in a stationary phase without nutrients for 3 days to minimize the effects of the previous BG-11 starting medium on the growth of the cyanobacteria studied. Then, the cyanobacterial cells were centrifuged in the 5810R centrifuge at 1000× *g* for 10 min and the pellets were washed twice in saline solution to remove residual N and P from the cell surface. Cells were then incubated in flasks containing two versions of fresh media—BG-11 and BG_0-11 without N—at the same temperature and light intensity as described above. The frequency of heterocyst appearance was recorded daily for 9 days under a light microscope (Micro Optix OPTIX C600, Wiener Neudorf, Austria) at magnifications of 200 and 400. For this purpose, 4 mL of the sample was taken with a micropipette, and a crushed droplet was prepared and analysed. Heterocysts were identified by their thickened cell wall, pale colour, and the formation of the poles compared to neighbouring cells [33]. The frequency of heterocysts was calculated as the percentage of heterocyst density to the total density of vegetative cells when at least 10 fields of view were counted per sample. Thus, the number of vegetative cells per heterocyst was determined for 10 heterocysts, and the data are expressed as mean ± standard deviation. If there were not enough cyanobacterial cells to count 10 heterocysts, the data were noted as too small to count.

2.6. Determination of Photosynthetic Activity

Chlorophyll fluorescence parameters were measured using a Multi-function Plant Efficiency Analyzer (M-PEA-2, Hansatech Instruments, King's Lynn, UK) after 24 h of cultivation. The following fluorescence parameters were recorded during the experiments: F_0 is the fluorescence value at open reaction centres, and F_m is the maximum fluorescence after a series of light flashes saturate the photosynthesis reaction centres. Before measurement, samples were kept in the dark for 15 min. These measured quantities were used to calculate the following parameters [34,35]:

$F_V = F_m - F_0$ is the maximum variable fluorescence;

F_V/F_m is the maximum quantum yield of PSII of the primary photochemical reaction in open PSII reaction centres: $F_V/F_m = \varphi Po$.

All measurements were carried out in at least five repetitions. The figures show data of average values.

2.7. Determination of Nitrogenase Activity Using the Acetylene Method

The cultures of cyanobacteria were cultivated in the light until the beginning of the stationary phase of the cells. Then, they were centrifuged at 16,000× g for 20 min at 4 °C. The obtained biomass was adjusted to OD_{720} = 1.5, then 15 mL thick biomass was transferred to 25 mL GC vial. Next, a gas mixture of 10% acetylene and 90% argon was introduced into the GC vial within 30 min [36]. The cells inside the vial were cultured for 1, 2, 4, 8, 24, and 32 h at a photosynthetic photon flux density of 70 µmol (photons) m^{-2} s^{-1}. After incubation, 500 µL gas samples were withdrawn and the concentration of ethylene in the gas mixture was determined. The redox activity of acetylene was determined on gas chromatograph GC-15A (Shimadzu, Kyoto, Japan) in the form of nmol ethylene mg^{-1} DW (dry weight) h^{-1}.

2.8. Effect of Different Concentrations of Mo and Fe on Nitrogenase Activity

The influence of these metals on the cultures was studied in Allen's medium. Of the metals studied, Mo (0.017 µmol) is present in this medium as a trace element in trace amounts and Fe (0.36 µmol) is present in an insignificant concentration. The study of the stimulatory effect of these metals on nitrogenase and photosynthetic activities was carried out with their additional introduction at different concentrations. Both metals were not excluded from the medium because they are of crucial importance, since Fe, in addition to the process of nitrogen fixation, is also necessary for cell growth, in particular for the processes of photosynthesis and respiration.

Salt ($FeCl_3 \times 6H_2O$) was added to the nutrient medium in concentrations of 1, 5, 10, and 15 µmol and Na_2MoO_4 in concentrations of 0.1, 0.5, and 5 µmol. The control for all variants of the experiment was the $Allen_0$ medium (without nitrogen source), which contained Mo and Fe in concentrations equal to the composition of the standard medium. The cyanobacterial cultures were first cultured in Allen's medium for 7 days under light at a temperature of 26 °C, then the cultures were transferred to $Allen_0$ medium, which also contained various concentrations of metals, and then the strains were cultured for an additional 7 days until anaerobic conditions were established and the amount of ethylene was recorded. Nitrogenase activity was determined after 1, 2, 4, 8, 24, and 32 h.

2.9. Pigment Content

To determine the quantitative composition of pigments, cyanobacterial isolates were cultured for 14 days in Allen and $Allen_0$ media, as well as in experimental variants with added metals, and then pigment content was analysed. From each variant analysed, 4 mL of a suspension of 14-day cyanobacterial culture was taken and then centrifuged at 15,000× g for 7 min, and the supernatant was discarded. Then, 4 mL of pre-chilled (+4 °C) methanol was added. The sample was homogenized by stirring (Silamat S6, 2 s), shaking (2000× g, 4 s), and pipetting. Then, the samples were covered with aluminium foil and incubated at +4 °C for 20 min to extract pigments from the cells. Next, the samples were centrifuged at 15,000× g, +4 °C for 7 min and the precipitate was discarded. The content of pigments (chlorophyll a, carotenoid, and phycocyanin) was determined using a spectrophotometer at wavelengths (A) of 470, 615, 650, 652, 665, and 720 nm using methanol as a control [37–39].

The concentration of the pigment content was calculated using Equations (1–3):

$$\text{Chlorophyll a } (\mu g\ m^{-1}) = 12.9447\ (A_{665} - A_{720}) \quad (1)$$

$$\text{Carotenoid } (\mu g\ mL^{-1}) = 1000\ (A_{470} - A_{720}) - 2.86 \quad (2)$$

$$\text{Phycocyanin } (\mu g\ mL^{-1}) = (A_{615} - 0.474A_{652})/5.34 \quad (3)$$

2.10. Statistical Analysis

The studies were conducted in 3–5 repetitions. The statistical data processing was conducted using the RStudio software (version 1.3.959, R Studio PBC, 2020). An ANOVA was used for the evaluation of significance. When the ANOVA proved a significant difference, Tukey's HSD test was performed to compare. Based on Tukey's HSD test, treatments were categorized (into letters in descending gradation). To create graphs on the evaluation of the nitrogenase and photosynthetic activities of cyanobacteria isolated from rice paddies, Microsoft Office Excel 2013 was used. A phylogenetic tree was constructed using the neighbour-joining method in Mega software 6.0 with 1000 bootstrap values [40].

3. Results

3.1. Isolation of Cyanobacteria Axenic Cultures from Rice Fields

To study the biodiversity of phototrophic microorganisms and isolate nitrogen-fixing cyanobacteria, water, soil, algal mat, and plant samples were collected twice from rice fields at the end of the rice seedling and tillering phases. The study of rice fields in the Kyzylorda region, Republic of Kazakhstan (GeoCoordinate: 43°54′16″ N 67°15′04″ E), revealed a rich biodiversity of algoflora consisting of 58 species, forms and varieties of microalgae belonging to 5 phyla, 10 classes, 19 orders, 26 families, and 29 genera. The taxonomic structure of the studied algoflora is as follows: *Cyanobacteria*—25 (45%), *Bacillariophyta*—5 (10%), *Euglenophyta*—5 (3%), *Chlorophyta*—18 (33%), and *Charophyta*—5 (8%).

At the same time, it must be noted that there is a certain seasonality in the occurrence of the species studied. Thus, if at the end of the rice seedling phase there was a high frequency of occurrence of green algal species of the *Chlorophyta* and *Charophyta* phyla, which form a bright green mass on the surface and in the water layer, at the end of the tillering phase, they were replaced by representatives of the *Cyanobacteria* phylum, especially *Microcystis pulverea*, *Anabaena variabilis*, *Cylindrospermum* sp., *Merismopedia elegans*, *Gloeocapsa minuta*, *Stranonostoc linkia*, *Oscillatoria limosa*, *Spirulina major*, *Phormidium ambiguum*, and *Nostoc pruniforme*. They are found in the water layer and on the surface, on the leaves and stems of rice and weeds, and on the thallus of dying macroalgae. Representatives of diatoms were more common at the rice seedling phase.

To isolate cyanobacteria, samples of the algal mat were taken towards the end of the rice tillering phase. From the collected water, soil and algal mat samples and 18 enrichment cultures of the dominant species of phototrophic microorganisms were obtained on enriched media. Thirteen cyanobacterial and microalgal isolates, characterised by stable growth under laboratory conditions, were obtained using the serial passage method, nine of which were unialgal cultures.

Since the purpose of this work was to search for and study nitrogen-fixing cyanobacterial strains, our attention was focused only on the filamentous forms of the isolated strains. Thus, five promising isolates, namely SH-12, J-14, J-8, K-31, and J-1, had a filamentous structure and probably belonged to cyanobacteria.

The microscopic examination of these isolates showed that the isolates J-14, K-31, J-8, and J-1 belonged to the order of the *Nostocales* according to the morphological characteristics and differed from isolate SH-12 in the form and function of cell differentiation. Thus, the single, straight, sometimes slightly curved trichomes of these isolates consisted of blue-green, spherical or cylindrical, isodiametric vegetative cells, narrowed or not at the ends, with pronounced or slightly pronounced constrictions at stigma sites. A distinctive feature of these isolates was the formation of the terminal and intercalary heterocysts when the medium was exhausted on days 8–15 of cultivation; their size was not significantly different from that of the vegetative cells. In addition, the formation of akinetes was observed in the three isolates, J-14, K-31, and J-8, which were significantly different in size and shape from the vegetative cells, especially isolate J-8. They consisted of cylindrical or oval cells, were 5.1–7.7 × 2.6–3.4 μm in size, green, light-brown, yellow, granular and bordered on being heterocystous. The formation of akinetes was not observed in isolate J-1. In addition, this isolate was characterised by pronounced false branching of the trichomes, which made the culture appear bushy. The morphological characteristics of the fifth isolate SH-12 resembled those of the genus *Oscillatoria*, order *Oscillatoriales*, whose main feature was the non-differentiation of the trichomes. The culture consisted of homocytic, single, straight, single-rowed trichomes derived from small, homogeneous, cylindrical vegetative cells lying close together.

A detailed description of each isolate studied is given below (Figure 1):

1. Culture J-14. Filamentous cyanobacteria. The trichomes are single, straight, not narrowed at the ends and have pronounced constrictions at the stigma sites and consist of blue-green spherical vegetative cells (length of 6–8 μm, width of 3–5 μm). The heterocysts are, in most cases, intercalary, solitary, and light-brown. Akinetes are rare, barely noticeably oval, their size does not differ from the size of the cells, and they are characterised by a granular content. They reproduce by hormogonia. Cultural characteristics: Grows well on Allen, Gromov, Bold, and BG-11 media at a temperature of 23–28 °C, with an initial pH of 7. At the same time, on a solid medium, prostrate growth is characteristic; on a liquid culture medium, growth is in the form of films both on the surface and at the bottom of the flask. In the old culture or on a medium with nitrogen deficiency, the development of heterocysts and akinetes is observed. According to the systematic position, it is classified as cyanobacteria, class *Hormogeneae*, order *Nostocales*, genus *Nostoc*, *Nostoc* sp.
2. Culture J-8. Filamentous cyanobacteria. Cells form straight, loose, immobile (weakly motile) trichomes up to 15 μm in length, comprising up to 10 cells. The cells are cylindrical, isodiametric, sometimes barrel-shaped, and 1.8–6.8 × 2.3–4.3 μm in size. The cells in the trichome are light, blue-green, uniform, and not granular. The heterocysts are terminal, unipolar, spherical or elongated, yellow-green, and 3.0–5.9 × 2.3–3.2 μm in size. The akinetes are mainly cylindrical, oval, large (5.1–7.7 × 2.6–3.4 μm), light-brown, yellow, granular, and adjacent to the heterocysts. Sometimes there are two akinetes at one end of the trichome. Reproduction is by binary cell division, in one plane only. Cultural characteristics: On a solid medium they form slimy colonies, first green, and then brownish in colour. They grow well at a temperature of 25–30 °C on Allen and Bold media in the light at an initial pH of 7. Old cultures (10–15 days of cultivation) contain both heterocysts and akinetes. Culture J-8 exhibited several morpho-characteristics identical to *Cylindrospermum badium* [41], including similar size dimensions of trichomes, vegetative cell, heterocysts, and akinete, and a flattened exospore. Therefore, the isolate J-8, after preliminary morphological identification, was assigned to cyanobacteria, class *Hormogeneae*, order *Nostocales*, genus *Cylindrospermum*, and species *Cylindrospermum badium*.
3. Culture K-31. The cells are cylindrical, barrel-shaped or spherical, pale- or light-blue- or olive-green, with gas bubbles (vacuoles) or without, but sometimes with granular contents. The terminal cells are slightly elongated, not vacuolated. The heterocysts are intercalary, solitary, spaced apart, oval, sometimes spherical, and usually slightly larger than the vegetative cells. The development of heterocysts was observed when the medium was exhausted (on the 8–15th day of cultivation). The akinetes are spherical, solitary, and located in the middle between two heterocysts. This cyanobacterium reproduces by hormogonia. Cultural characteristics: On solid culture medium, they form mucilaginous colonies, on liquid culture medium, they precipitate mucilaginous amorphous tangles, and the walls of the flasks are fouled by the culture. They were isolated on Allen's culture medium and grow well on Bold's, Chu-10, and BG -11 media. The development of heterocysts was observed when the medium was emptied at 8–15 days of cultivation. Optimal growth on this medium was observed at a temperature of 25–28 °C [42,43]. Morphological identification of the isolate shows that the isolate belongs to cyanobacteria, class *Hormogeneae*, order *Nostocales*, family *Nostocaceae*, genus *Trichormus*, species *Trichormus variabilis*.
4. Culture J-1. Filamentous cyanobacteria. The cells formed straight, sometimes slightly curved, immobile trichomes up to 10–15 μm in length. At the same time, bushiness (bushy growth) is observed in some places due to the false branching of the trichomes, which often occurs at the sites of heterocyst formation. The cells in the trichomes are often cylindrical, isodiametric, and 1.3–3.2 μm in size. The cells in the trichome are

light, green, uniform, and non-granular. The heterocysts are intercalary, spherical or elongated, yellow-green, and 3–3.2 × −5.9 μm in size. The presence of akinetes was not observed. Reproduction is by binary cell division, only in one plane. Cultural characteristics: Yellow-green colonies form on a solid medium. They grow well at a temperature of 25–30 °C on BG-11, Allen, and Bold media in the light, at an initial pH of 7. According to taxonomic affiliation, they are assigned to cyanobacteria, class *Hormogeneae*, order *Nostocales*, family *Scytonemataceae*, genus *Tolypothrix*, and species *Tolypothrix* sp. [44].

5. Culture SH-12. A filamentous cyanobacterium with a mucilaginous sheath. No active movement of the trichomes was observed, but some rocking of the trichome tips was. The width of the cell is 2.6–5 μm, i.e., it is two- to three-fold shorter than the length. They reproduce by binary cell division and in one plane only. Cultural characteristics: On a liquid medium, the suspension is blue-green in mass, slightly mucous-like, and precipitates into an amorphous precipitate that is easily stirred up. The strain develops regardless of the season and remains axenic during storage. This cyanobacterium is isolated on Gromov's culture medium and grows well on Allen's and BG-11 media. Blue-green trichome balls form on the surface of Gromov's agar medium. Optimal cultivation conditions are on Gromov's medium at an incubation temperature of 25–28 °C, medium pH was 7.5–8. Based on the morphological characteristics, which are congruent with the description reported in previous literature, the isolated strain was identified as *Oscillatoria brevis* [45,46]. According to the systematic position, they belong to the cyanobacteria, class *Hormogeneae*, order *Oscillatoriales*, genus *Oscillatoria*, and species *Oscillatoria brevis*.

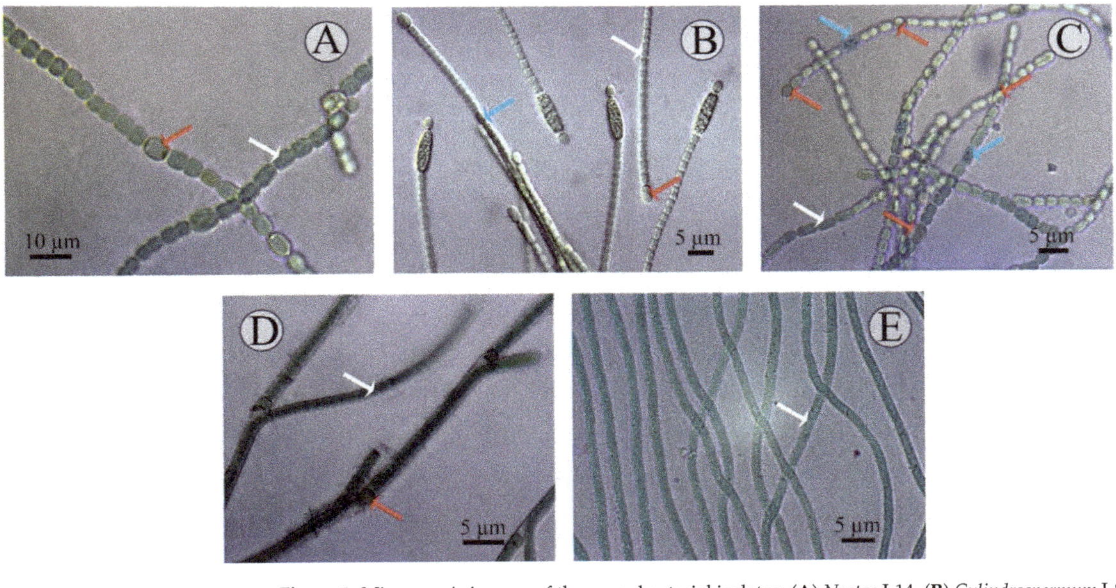

Figure 1. Microscopic images of the cyanobacterial isolates: (**A**) *Nostoc* J-14, (**B**) *Cylindrospermum* J-8, (**C**) *Trichormus* K-31, (**D**) *Tolypothrix* J-1, and (**E**) *Oscillatoria* SH-12. Formation of the different cell types are marked in arrow colours: white—vegetative cells; red—heterocysts; blue—akinetes.

The isolated pure cultures are stored in the biotechnology laboratory of the Department of Biotechnology of Al-Farabi Kazakh National University in the Collection of microalgae and cyanobacteria CCMKazNU (Culture Collection of microalgae, Al-Farabi Kazakh National University).

3.2. Genetic Identification of Isolated Strains and Phylogenetic Tree Construction

For the molecular identification of the isolated strains, DNA fragments of different lengths were obtained. These DNA fragments were amplified with 16 rRNA universal primers, and similar strands were identified in the NCBI BLAST online database.

Based on the data obtained, eight entries in the chain with high similarity were selected, and several alignments were performed using the MUSCLE program. The phylogenetic tree was constructed using MEGA 6 software based on the evolutionary distance calculated using the neighbour-joining method based on the Kimura-2-Parameter algorithm. The statistical analysis of the tree topology was carried out using a loading analysis of 1000 repeated samples [47].

The phylogenetic tree consists of four main clusters and one group, each represented mainly by different species of *Cylindrospermum, Anabaena, Tolypothrix,* and *Oscillatoria. Chlorella antarctica* strain CCAP 211/45 (KU291887.1) was used as an outgroup.

Figure 2 shows the phylogenetic relationships of the neighbouring strains and the topology of the tree, which is the basis for the sequential comparison of the taxa.

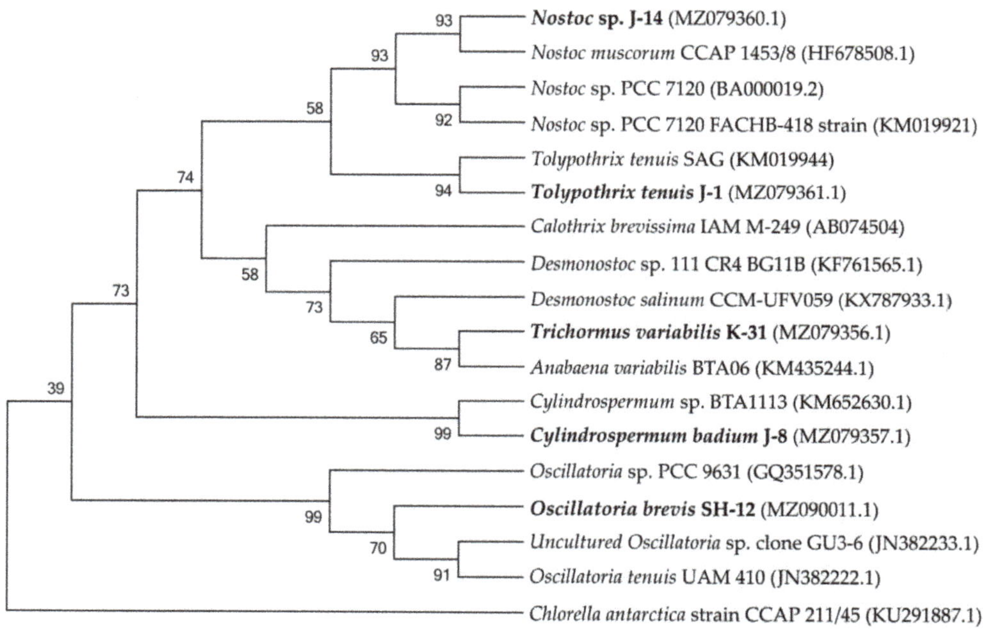

Figure 2. Isolated cyanobacterial strains and phylogenetic tree of neighbouring species.

The SH-12 strain (MH341423) was 99.76% close to *Oscillatoria tenuis* (JN382222.1) and Uncultured *Oscillatoria sp.* clone GU3-6 (JN382233.1). The phylogram showed that strain J-14 belongs to *Nostoc* genus, similar to *Nostoc muscorum* with accession number HF678508.1 (sc. = 2326; sim. = 99.38%), and *Nostoc sp.* with accession number KM019921.1 (sc. = 2298, sim. = 99.45%). Strain J-1 showed 99.33% affinity to *Tolypothrix tenuis* (KM019944.1) and 99.41% similarity to *Scytonema mirabile* (KM019943.1) showed. 1354 nucleotide sequences were obtained with a combination of two additional primers. The J-8 chains were identified with universal primers 27f and 1492r and showed a close affinity with *Cylindrospermum* sp. BTA1113 (KM652β630.1) (sc. = 785, sim. = 94.36%). The affinity of other neighbouring species was 90.47%, such as *Desmonostoc salinum* (sc. = 702), *Desmonostoc sp.* (90.36%). The results showed that strain K-31 was closely related to *Nostoc* sp. SAG 34.92, *Anabaena variabilis* BTA with 96.62% and *Trichormus variabilis* 94.21% similarity. Despite the fact that the percentage of similarity with *Anabaena variabilis* is greater, according to morphological features (the location of akinetes in relation to heterocysts), strain K-31 is assigned to the genus *Trihormus*. The genus *Trihormus* was isolated from the genus *Anabaena* within the subfamily Nostocoidae. According to the information available in the literature, the genus *Trichormus* includes *A. variabilis*, *A. azollae* Strasb., *A. doliolum* Bharadw, and a number of other species, and thus the names *A. variabilis* and *T. variabilis* (Kütz.) Kom. and Anagn. at one time were considered synonyms [48]. However, despite this, the genus *Trichormus* is characterized by the formation of akinetes from vegetative cells in the middle, between two heterocysts, while in *Anabaena* cells, akinetes are formed nearby with heterocysts, and their growth is directed away from the heterocysts. In the K-31 strain, akinetes were located in the middle between two heterocysts at an equal distance; therefore, this strain was assigned to the genus *Trichomus*. According to the research results, a genetic analysis of five cyanobacterial cultures was performed based on phylogenetic analysis, their close groups were identified, and phylogenetic trees were formed. The resulting nucleotide chains were uploaded to the GenBank database (NCBI) and accession numbers were assigned. Through phylogenetic and morphological analyses, the isolated new cyanobacterial isolates were identified as follows: *Trichormus variabilis* K-31 (MZ079356), *Cylindrospermum badium* J-8 (MZ079357), *Nostoc* sp. J-14 (MZ079360), *Oscillatoria brevis* SH-12 (MZ090011), and *Tolypothrix tenuis* J-1 (MZ079361).

3.3. Investigation of the Productivity of Isolated Cyanobacterial Strains on a Nitrogen-Free Medium

The next step was to investigate the ability of the isolated cyanobacteria to grow on a nitrogen-free nutrient medium (Allen$_0$) while determining the formation of heterocysts on the filament and the nitrogenase activity of the strains under these conditions. For this purpose, the cultures of four heterocystous cyanobacterial strains were grown on the nutrient medium Allen$_0$, and the number of heterocystous (HC) and vegetative cells (VC) within a filament, as well as the growth rate coefficient of the cultured cyanobacteria, were recorded daily. After 14 days, the biomass of the cyanobacteria was separated from the culture fluid and dried and, thereafter, the yield of dry biomass was determined. The results obtained are summarised in Table 1.

Table 1. The productivity results of isolated microalgal strains on a nitrogen-free medium (Allen$_0$). The data are presented as mean +/− SE (n = 3), $p < 0.05$.

Strain	Growth Coefficient, μ (per Day)	DW, g L^{-1}
Trichormus variabilis K-31	0.28 ± 0.02	1.06 ± 0.09
Nostoc sp. J-14	0.27 ± 0.02	1.01 ± 0.09
Cylindrospermum badium J-8	0.19 ± 0.02	0.87 ± 0.08
Tolypothrix tenuis J-1	0.24 ± 0.02	0.91 ± 0.07

In general, it can be stated that all strains examined were able to grow on a nitrogen-free medium. However, compared to the control, a significant prolongation of the lag phase of culture growth on a nitrogen-free medium was observed. In the course of cultivation, the growth rate coefficient of the examined strains reached different values, whereby the most active growth was observed in the representatives of the order *Nostocales*—the strains *Trihormus variabilis* K-31 and *Nostoc* sp. J-14—whose yield of dry biomass at the end of the experiment was 1.06 ± 0.09 and 1.01 ± 0.09 g L^{-1}, respectively.

The results obtained indicate a significant difference in the productivity of the representatives of the order *Nostocales* on a nitrogen-free medium: the highest productivity among the strains studied was observed in *Trihormus variabilis* K-31 and *Nostoc* sp. J-14. The ability to grow on a medium without nitrogen in all the cultures studied already indicates the presence of nitrogen-fixing activity in isolated cultures of cyanobacteria. The registration of the number of heterocysts showed that an increase in the number of heterocysts was observed in all cultures on a nitrogen-free medium, despite a slight decrease in the growth rate. After 9 days of culturing the strains on a nitrogen-free medium, it was found that there were no significant differences between the cultures grown on both Allen nutrient medium and $Allen_0$ nitrogen-free medium. In the experiment in Allen nutrient medium, three strains were characterised by the active formation of heterocysts. It was found that the cells of *Trihormus variabilis* K-31 formed 5 HC/filament, *Nostoc* sp. J-14, 3 HC/filament, and the strains *Cylindrospermum badium* J-8 and *Tolypothrix tenuis* J-1 released a maximum of 4 HC/filament. These values are about two- to three-fold higher than in the control, where the number of heterocysts correlates positively with cell density or biomass (Figure 3).

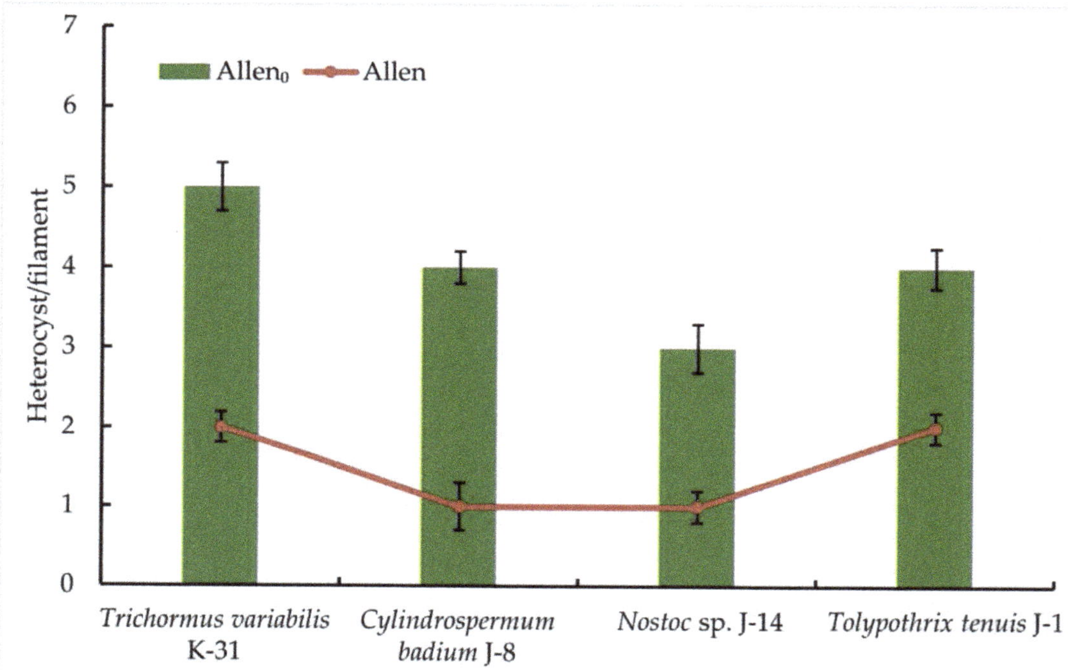

Figure 3. The frequency of heterocyst formation during nitrogen starvation in isolated cyanobacterial strains. The species were grown in the Allen medium for 14 days. The biomass was collected and transferred to the $Allen_0$ medium. Then, heterocyst formation was studied under a light microscope. Different letters on the bar and line chart indicate a significant difference at $p < 0.05$.

Then, five isolated cyanobacterial strains were tested for nitrogenase activity, including *Oscillatoria brevis* SH-12. A cyanobacterial strain that does not form heterocysts (*Synechocystis* sp. PCC 6803) was used as a control. The results of gas chromatography showed that all strains of nitrogen-fixing cyanobacteria exhibited nitrogen activity in the acetylene medium. However, their ability to reduce acetylene to ethylene was found to vary. According to the results, among the species studied, the strain *Oscillatoria brevis* SH-12 showed no nitrogenase activity, as expected, while the strain *Trichormus variabilis* K-31 had the highest ethylene content (3.57 ± 0.26 nmol ethylene mg^{-1} DW h^{-1}). In addition, the strains *Nostoc* sp. J-14 (2.82 ± 0.07 nmol ethylene mg^{-1} DW h^{-1}) and *Tolypothrix tenuis* J-1 (2.22 ± 0.06 nmol ethylene mg^{-1} DW h^{-1}) showed good results. In the strain *Cylindrospermum badium* J-8, this indicator was 1.36 ± 0.05 nmol ethylene mg^{-1} DW h^{-1} (Figure 4).

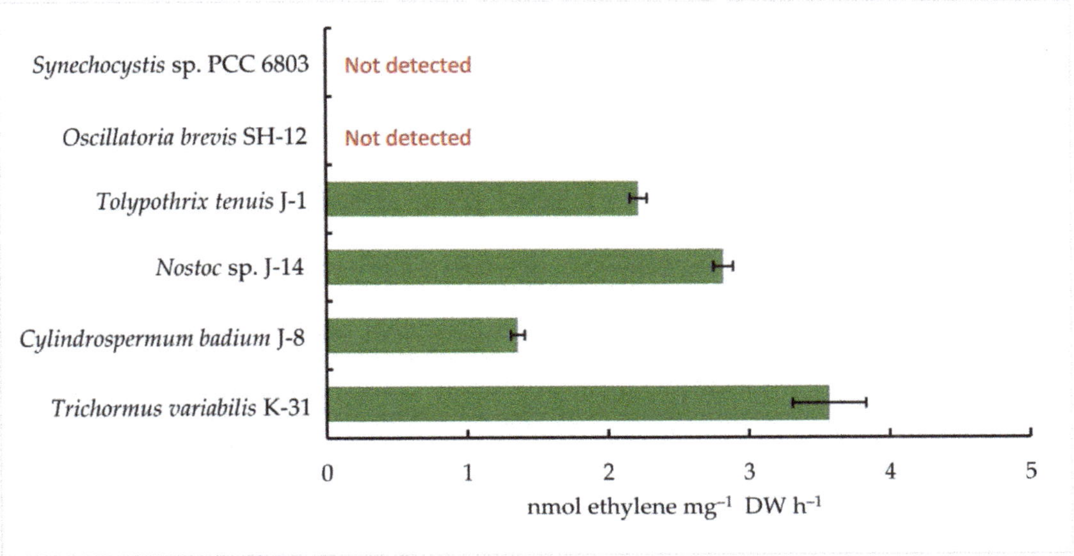

Figure 4. Nitrogenase activity of isolated cyanobacterial strains. The cells were cultured until the beginning of the stationary phase of the cells. Next, collected biomass was transferred to the $Allen_0$ nitrogen-free medium. After 24 h, the concentration of ethylene was measured. The data are presented as mean +/− SE (n = 3). Different letters on the bar indicate a significant difference at $p < 0.05$.

Thus, based on the experiments, it was found that four out of five isolated cyanobacterial strains have some ability to fix nitrogen. The best results were obtained for representatives of the order *Nostocales*. The strains *Trihormus variabilis* K-31 and *Nostoc* sp. J-14 had the highest nitrogen fixation activity.

3.4. Influence of Mo and Fe on Photosynthetic and Nitrogenase Activities of Isolated Cyanobacterial Strains

The effect of the investigated biogenic metals on the photosynthetic activity of the isolated strains was assessed using some fluorescence parameters and the biomass pigment composition. Using an M-PEA-2 fluorimeter, which allows for the simultaneous measurement of induction fluorescence curves and the redox conversions of photosystem components in the microsecond range, the effect of different concentrations of molybdenum (Mo) and iron (Fe) on the maximum quantum yield of PSII of the primary photochemical reactions Fv/Fm (φ_{Po}) was investigated as a first step. This indicator was determined at the time of the peak of nitrogenase activity in a nitrogen-free medium (i.e., according to preliminary data, this is about 24 h).

According to the data obtained, the maximum quantum yield of PSII of the primary photochemical reactions Fv/Fm correlated positively with an increase in the Fe concentration in the medium in both strains. In *Trihormus variabilis* K-31, a linear increase in the Fv/Fm parameter was observed with increasing Fe concentration, from values of 0.31 to 0.58 at 1 to 10 μmol Fe concentrations. With a further increase in the metal concentration to 15 μmol, a sharp decrease in the Fv/Fm value to 0.28 is observed (Figure 5).

The strain *Nostoc* sp. J-14 showed a similar pattern and differed from *Trichomus variabilis* K-31 with lower values. In general, a clear relationship between Fe concentration and PSII photochemical quantum yield was also observed. Thus, Fv/Fm values for *Nostoc* sp. J-14 were 50% lower under conditions of low metal concentration (1 μmol) and were 0.26 compared to conditions of medium enrichment with 10 μmol Fe (0.50).

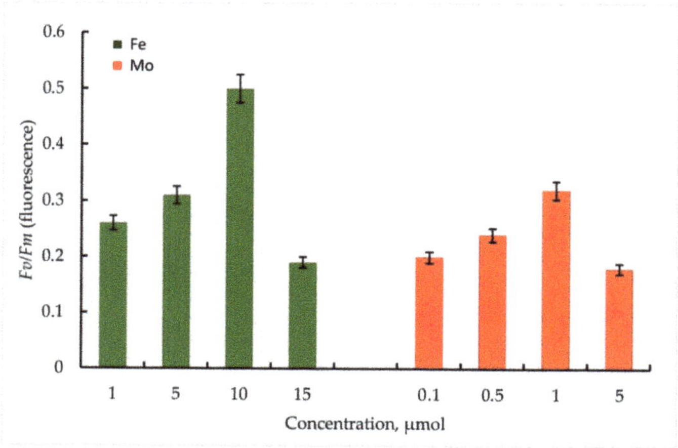

Figure 5. Maximum quantum yield of PSII (Fv/Fm) of primary photochemical reactions at different Mo and Fe concentrations for *Trihormus variabilis* K-31. The culture was grown on Allen medium for 7 days. The obtained biomass was transferred to the medium containing different concentrations of Fe (1, 5, 10, and 15 μmol) and Mo (0.1, 0.5, 1, and 5 μmol) metals. Then, the ratio of Fv/Fm values was measured. The data are presented as mean +/− SE ($n = 4$). Different letters on the bar indicate a significant difference at $p < 0.05$.

It should be noted that in the experiments with Mo, despite its stimulating effect on nitrogenase activity in cyanobacteria, the parameters of the maximum quantum yield of PSII of the primary photochemical reactions Fv/Fm were significantly lower compared to Fe.

Thus, within the investigated Mo concentrations, the maximum value of the Fv/Fm parameter was found at a Mo concentration of 1 μmol. For the strain *Trihormus variabilis* K-31, this value was 0.42, for *Nostoc* sp. J-14, 0.32. The ratio of Fv/Fm values for both strains at different Mo and Fe concentrations in comparison is shown in Figure 6.

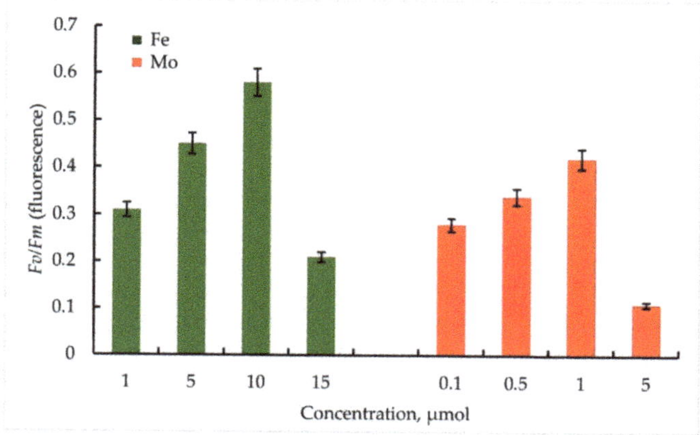

Figure 6. PSII maximum quantum yield Fv/Fm (φ_{Po}) of primary photochemical reactions at different Mo and Fe concentrations for *Nostoc* sp. J-14. The culture was grown on Allen medium for 7 days. The obtained biomass was transferred to the medium containing different concentrations of Fe (1, 5, 10, and 15 μmol) and Mo (0.1, 0.5, 1, and 5 μmol) metals. Then, the ratio of Fv/Fm values was measured. The data are presented as mean +/− SE ($n = 4$). Different letters on the bar indicate a significant difference at $p < 0.05$.

To investigate pigment composition, the strains were cultured for 14 days in Allen medium without nitrogen, after which the resulting biomass was analysed for chlorophyll *a*, carotenoid, and phycocyanin content. Metal concentrations were selected based on preliminary studies. Since the yield of dry biomass was different in the different variants of the experiment, the content of pigments was recalculated for 1 mg g^{-1} DW. The results obtained are summarised in Table 2.

Table 2. Pigment composition of cyanobacterial isolates under the influence of the metal. Different letters within one pigment indicate a significant difference at $p < 0.05$.

Strain	Metal	Concentration, µmol	Pigment Content, mg g^{-1} DW		
			Chlorophyll *a*	Carotenoid	Phycocyanin
Trichormus variabilis K-31	Mo	1	0.38 ± 0.024	0.18 ± 0.016	0.19 ± 0.011
	Fe	10	0.46 ± 0.018	0.23 ± 0.014	0.58 ± 0.018
	Control A [1]	NA [3]	0.29 ± 0.015	0.14 ± 0.012	0.15 ± 0.011
	Control B [2]	NA	0.45 ± 0.014	0.22 ± 0.011	0.18 ± 0.012
Nostoc sp. J-14	Mo	1	0.31 ± 0.021	0.23 ± 0.023	0.13 ± 0.013
	Fe	10	0.37 ± 0.019	0.25 ± 0.016	0.47 ± 0.015
	Control A [1]	NA	0.31 ± 0.011	0.22 ± 0.011	0.11 ± 0.012
	Control B [2]	NA	0.35 ± 0.011	0.24 ± 0.019	0.21 ± 0.015

[1] $Allen_0$ medium. [2] Allen medium. [3] NA— not added.

The obtained results show that the metals concentrations optimal for nitrogen fixa-tion affect the process of photosynthesis in different ways, especially the pigments content. As expected, the addition of Fe to the medium compared to Mo increases the photosynthetic activity of both cyanobacterial strains under nitrogen deficiency. Thus, considering Trichomus variabilis K-31, the content of chlorophyll a in control A (medium without nitrogen) was 0.29 ± 0.015 mg g^{-1} DW being significantly higher in the presence of 10 µmol Fe (0.46 ± 0.018 mg g^{-1} DW). In control B (Allen medium with standard composition) the content of chlorophyll a was 0.45 ± 0.014 mg g^{-1}. The content of carotenoid in the presence of Fe was amounted to 0.23 mg g^{-1}, which corresponded to control B—0.22 ± 0.011 mg g^{-1}, while this indicator on the medium without nitrogen was 0.14 ± 0.012 mg g^{-1}. A slightly different picture was obtained for phycocyanin. The best accumulation of phycocyanin was observed in the presence of Fe (0.58 ± 0.018 mg g^{-1}) compared to the controls and the Mo variant.

The influence of the selected Mo and Fe concentrations on the nitrogenase activity of *Trichormus variabilis* K-31 and *Nostoc* sp. J-14 was determined after 1, 2, 4, 8, 24, and 32 h.

The results obtained proved the stimulating effect of the investigated metals on the nitrogenase activity of the cyanobacteria.

Thus, in *Trichomus variabilis* K-31, when exposed to 1 µmol Mo, the maximum accumulation of C_2H_4 was observed after 24 h and 5.2 ± 0.6 nmol ethylene mg^{-1} DW h^{-1}, which is 48% higher than in the control (Figure 7). According to the data obtained, nitrogenase activity was high up to 24 h, and the amount of ethylene decreased in the following hours. This, in turn, is closely related to a decrease in glycogen stores. As you know, nitrogen fixation is an energy-rich process in which the nitrogenase releases three identical products simultaneously. Sufficiently good results were obtained with Fe addition, and the maximum oxidation of C_2H_4 showed 4.2 ± 0.8 nmol ethylene mg^{-1} DW h^{-1} at a concentration of 5 µmol after 24 h of the experiment.

It should be noted that nitrogenase activity was significantly lower in the control. The content of ethylene reached 1.6 ± 0.5 nmol ethylene mg^{-1} DW h^{-1} in the first hours of the experiment, which increased to a maximum point of 3.6 ± 0.5 nmol ethylene mg^{-1} DW h^{-1} after 24 h of the experiment.

In the experiment with the strain *Nostoc* sp. J-14, the isolate also showed ethylene activity when exposed to the investigated metals; however, the oxidation of C_2H_4 was comparatively lower than that of *Trichormus variabilis* K-31. At 1 µmol Mo, 1.5 ± 0.2 nmol ethylene mg^{-1} DW h^{-1} C_2H_4 was measured after 1 h. A high rate of oxidation of C_2H_2 to C_2H_4 was observed after 24 h (4.4 ± 2.6 nmol ethylene mg^{-1} DW h^{-1}), which is 57% more than in the control. Under the influence of Fe, the maximum oxidation of C_2H_4

at a concentration of 10 μmol after 24 h of the experiment was 3.8 ± 0.7 nmol ethylene mg^{-1} DW h^{-1} (35% more than in the control), in the following hours, a noticeable decrease in this indicator was observed (Figure 8).

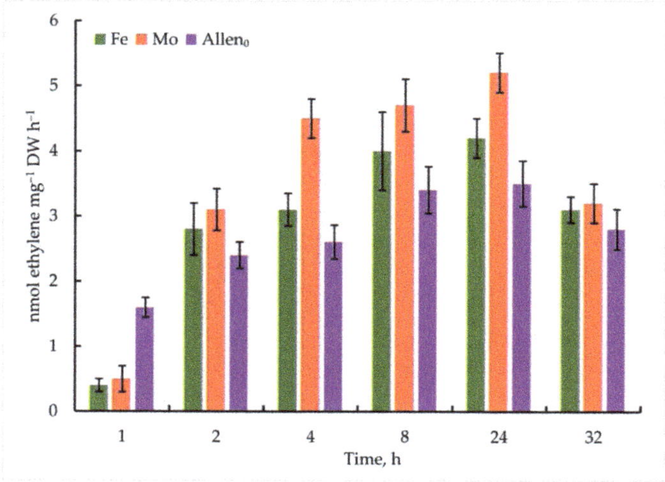

Figure 7. Metal-induced increase in nitrogenase activity in *Trichormus variabilis* K-31 cells. The culture was grown on Allen medium for 7 days. Then, the cells were transferred to media containing Mo (1 μmol) and Fe (10 μmol) at time 0. Acetylene yield was measured at 1, 2, 4, 8, 24, and 32 h after the addition of the Mo and Fe containing salts. The data are presented as mean +/− SE (n = 5). Different letters on the bar indicate a significant difference at $p < 0.05$.

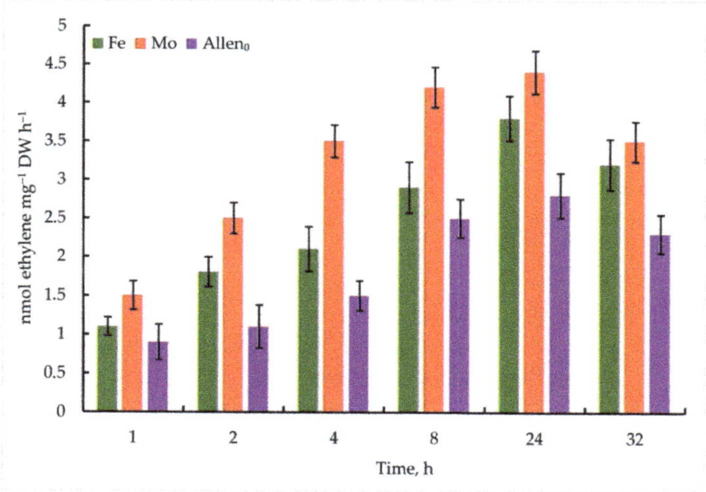

Figure 8. Metal-induced increase in nitrogenase activity in *Nostoc* sp. J-14 cells. The culture was grown on Allen medium for 7 days. Then, the cells were transferred to media containing Mo (1 μmol) and Fe (10 μmol) at time 0. Acetylene yield was measured at 1, 2, 4, 8, 24, and 32 h after the addition of the Mo and Fe containing salts. The data are presented as mean +/− SE (n = 5). Different letters on the bar indicate a significant difference at $p < 0.05$.

According to the results obtained, it was found that the examined concentrations of each metal have a stimulating effect on nitrogenase activity and operation.

4. Discussion

Soil erosion and the restoration of arable land after the excessive use of chemical fertilisers can be improved by switching to organic farming, and restoration is generally ensured by the use of biofertilisers as well as by restoring the biological components of the soil. Cyanobacteria, fulfilling many different functions in the soil ecosystem, can be considered to be as promising objects in agrobiotechnology to produce biofertilisers. It is known that the absorption activity of plant root systems leads to an increased outflow of nitrogenous metabolites from the soil. The replenishment of this pool can be maintained by the high activity of the cyanobacterial nitrogenase complex. The study of the phototrophic microorganism biodiversity in the rice fields of Kazakhstan and the isolation of nitrogen-fixing cyanobacteria from water, soil, algal mat, and plant samples was carried out to search for active strains of great practical interest.

The observed rich biodiversity of microalgae and cyanobacteria in the algoflora of the rice fields of the Kyzylorda region, with the characteristic seasonality of the species encountered, is certainly primarily related to the hydrological regime and soil properties, as well as the rice cultivation phase, the use of agrotechnical methods of rice cultivation, and seasonal temperature phenomena. Thus, during the period characterised by increased rice development, increasing shading of the water surface is observed, leading to a noticeable change in the ratio of the green species number, and they are replaced by shade-tolerant species of blue-green algae and diatoms, which form continuous films on the bottom of the rice fields.

The study of the morphological characteristics of the isolated strains J-14, K-31, J-8, J-1, and SH-12 allowed us to determine their affiliation to the orders *Nostocales* and *Oscillatoriales*, and through the analysis of phylogenetically similar species, they were identified as follows: *Trichormus variabilis* K-31 (MZ079356), *Cylindrospermum badium* J-8 (MZ079357), *Nostoc* sp. J-14 (MZ079360), *Oscillatoria brevis* SH-12 (MZ090011), and *Tolypothrix tenuis* J-1 (MZ079361).

An examination of the ability of the isolated heterocystous cyanobacteria to grow on a nitrogen-free nutrient medium, the heterocysts formation on a filament, and the nitrogenase activity of the strains under these conditions showed that all the strains examined could grow on a nitrogen-free medium. The results indicate a difference in the productivity of the representatives of the order *Nostocales* on a nitrogen-free medium. The cells of *Trichormus variabilis* K-31 and *Nostoc* sp. J-14 showed the highest productivity among the strains studied. The ability of all cultures examined to grow on a medium without a nitrogen source already indicates the presence of nitrogen-fixing activity in the isolated cyanobacterial cultures. However, compared to the control, a significant prolongation of the lag phase was observed on a nitrogen-free medium, which is apparently related to the adaptation of the cells to these conditions and the subsequent differentiation of some vegetative cells into heterocysts, with a corresponding shift in all phases of microbial population development. As is known, the formation of heterocysts begins after about 20 h of nitrogen starvation. This is because the genes for nitrogen fixation (*nif*) are expressed about 18–24 h after nitrogen deprivation [49].

Thus, the registration of the number of heterocysts in our studies showed that an increase in the number of heterocysts was observed in all cultures on a nitrogen-free medium, despite a slight decrease in the growth rate. Our results confirm the influence of nitrogen availability for cyanobacterial cells on the frequency of heterocyst formation and the species specificity of this factor. According to data in the literature, in the absence of a nitrogen source in the culture medium, 5–10% of the cells of the filaments of nitrogen-fixing cyanobacteria differentiate into heterocysts within about 24 h [50]. Heterocysts supply the vegetative cells with fixed nitrogen and receive carbohydrates from them in return [51,52]. This mutual exchange of balanced nutrients is regulated at the gene level to ensure their optimal growth and adaptation for long-term survival. At the same time, the genes responsible for heterocyst differentiation control their ratio to the number of vegetative cells on the filaments, ensuring an adequate balance between fixation by vegetative cells and nitrogen fixation by heterocysts. Thus, the NtcA-regulated *pat*D gene is reported to control the ratio of heterocysts relative to vegetative cells on the filaments of *Anabaena* sp. PCC 7120 [53]. It should be noted that according to our results, the frequency of heterocyst formation was even significantly higher under nitrogen deficiency in the medium in some strains. Thus, in the filamentous cyanobacterium *Trichormus variabilis* K-31, 15–20 cells differentiated into heterocysts with an average initial frequency of 8.2 $\pm$ 1.8 vegetative cells per heterocyst, indicating the species specificity of this factor.

Subsequently, five isolated cyanobacterial strains were tested for their nitrogenase activity, including *Oscillatoria brevis* SH-12. Although this strain does not form heterocysts, according to recent data, the ability to fix nitrogen is characteristic not only of heterocystous strains, but also of those in which the processes of nitrogen molecule reduction and photosynthesis are separated in time and regulated by a circadian rhythm [54]. Based on the experiments, it was found that four of the five isolated cyanobacterial strains have some ability to fix nitrogen. Consequently, the best results were obtained for representatives of the order *Nostocales*. The strains *Trihormus variabilis* K-31 and *Nostoc* sp. J-14 had the highest nitrogen fixation activity. In contrast to the existing data on the nitrogenase activity of species of the genus *Oscillatoria* [55], we could not detect nitrogenase activity in the cells of *Oscillatoria brevis* SH-12 in our study.

The activation of the process of nitrogen fixation is not only closely linked to the pool of nitrogen in the medium, but also depends on various external factors, including the content of various heavy metals. The process of nitrogen fixation is ensured by the work of a complex enzyme with several subunits and substrates called nitrogenase, which contains a metal in its active centre, and undoubtedly depends on their concentrations in the medium. Depending on the metal contained in the active centre, three types of nitrogenase are distinguished: Mo-nitrogenase (Mo), V-nitrogenase (V), and Fe-nitrogenase (Fe). In this context, it was advisable to investigate the effects of different Mo and Fe concentrations on the nitrogenase activity of the isolated strains and to select the optimal concentrations. For this purpose, two productive strains of the cyanobacteria *Trihormus variabilis* K-31 and *Nostoc* sp. J-14, which were proven to be the most active in previous studies, were selected.

The results obtained evidenced the stimulating effect of certain concentrations of the investigated metals on the cyanobacterial nitrogenase activity. The best results in the study of nitrogenase activity of *Trichormus variabilis* K-31 were obtained when it was exposed to Mo at a concentration of 1 μmol and 10 μmol Fe.

Thus, according to the research results, it was found that a certain amount of each metal studied has a stimulating effect on the operation of nitrogenase. According to the available literature data, nitrogenase responds to the availability of Fe and Mo in the media [56]. In our studies, the best results were obtained for both cyanobacterial cultures in the variants of the experiment with Mo at a concentration of 1 μmol and 10 μmol Fe. The results obtained confirm the presence of both types of nitrogenases in the cyanobacterial isolates studied. It is known that in such cases when the Mo concentration in the medium becomes limited, the work of alternative nitrogenases is expressed [57,58]. Thus, when Mo availability is higher, an increased efficiency of Mo nitrogenase is observed compared to V- and Fe-dependent enzymes [59]. In contrast to another metal studied, Mo is involved in nitrogen metabolism only in cyanobacterial cells and, although the cellular requirement for Mo is very low, this metal plays an essential role in the functioning of nitrogenase, the key element in nitrogen fixation [60]. A deficiency of Mo, which is part of the Fe-Mo cofactor of the enzyme, leads to a decrease in nitrogen fixation despite higher heterocyst and pigment concentrations [61]. When molybdate is excluded from the medium and vanadate is absent, the nitrogenase genes *nif*1 and *vnf* are expressed in heterocysts. However, since they cannot produce nitrogenase, these nitrogen-depleted cells produce a very high frequency of heterocysts and overexpress nitrogenase genes [62,63]. Our observations suggest the stimulatory effect of Mo on the activation of the enzyme. Increasing the Mo concentration to 1 μmol increases the activity of nitrogenase. The results obtained correlate with the data available in the literature. Thus, according to the data presented, the growth rate of *A. variabilis* was almost the same under the influence of V and Mo, but the catalytic efficiency of the alternative nitrogenase was significantly lower than that of the MoFe-nitrogenase [64]. It was also found that the activity of VFe-nitrogenase in *Azotobacter* was about 1.5-fold lower than that of MoFe-nitrogenase under the same cultivation conditions [65].

Iron plays an equally important role in the process of nitrogen fixation. As is known, nitrogen fixation is an Fe-dependent process, since the nitrogenase complex contains 38 Fe atoms per holoenzyme [66]. Among the strains examined, nitrogenase activity was significantly higher after Fe addition in the culture of *Trichormus variabilis* K-31 (4.2 nmol ethylene mg^{-1} DW h^{-1}) compared to *Nostoc* sp. J-14 (3.8 nmol ethylene mg^{-1} DW h^{-1}). At the same time, among the concentrations that were studied, 10 μmol was optimal, a lower concentration (1 μmol) was not effective enough, and a high concentration (15 μmol) had a toxic effect, leading to negative changes in culture growth and low nitrogen fixation. The need for Fe in nitrogen reduction processes is related to the fact that these processes require Fe both in enzymes and for the ferredoxin-mediated provision of the restorative energy of photosynthesis or respiration [67]. Nitrogen fixation metabolism requires the synthesis of Fe and Fe-Mo subunits. In nitrate reduction (either from nitrogen fixation or from an external source), Fe is directly involved in nitrite reductase and, as a ferredoxin cofactor, for nitrate reductase. An iron limitation has been reported to reduce nitrate reductase levels in algae [68]. The entire pathway of nitrogen reduction (from either dinitrogen or nitrate) requires a flux of reducing agents from photosynthesis, resulting in an increased cellular demand for these Fe compounds [69]. It has been reported that nitrogen-fixing cyanobacteria require up to 10-fold more Fe than the same species growing at the same rate on nitrates [70,71].

The chosen concentrations of these metals were optimal for both nitrogen fixation and photosynthesis. At the same time, as expected, the addition of Fe in a nitrogen-free medium compared to Mo increased the photosynthetic activity of both cyanobacterial strains compared to control A (Allen medium without N source). It is known that Fe is necessary as a redox catalyst in photosynthesis and nitrogen assimilation for electron transfer and the formation of NADPH. Iron deficiency can reduce the concentration of chlorophyll and carotenoid in the cells of phototrophic microorganisms [5,72].

5. Conclusions

The broad spectrum of activity of cyanobacteria and their importance for the biosphere and the economy of humankind are good reasons to study and understand their diversity, conservation, and use in the interest of our society. Nitrogen-fixing cyanobacteria, for example, which have the simplest nutrient requirements in nature, make an enormous contribution to the nitrogen cycle and, thus, to the greening of agriculture. Maintaining an equilibrium of nitrogen input into the soil as a result of the process of nitrogen fixation is crucial for maintaining soil fertility over long periods. Although a large amount of data has been collected today on the diversity of cyanobacteria in aquatic and soil ecosystems, the search for new strains of cyanobacteria that have great potential for agricultural biotechnology and the replenishment of microbial collections is very important and relevant. This paper presents the results of isolating new nitrogen-fixing cyanobacterial strains and studying their photosynthetic and nitrogenase activities, as well as the prospects for stimulating their nitrogenase activity with two metals important for the process of nitrogen fixation. The optimal concentrations for cyanobacterial growth and stimulation of nitrogenase activity were determined.

Author Contributions: Conceptualization, A.K.S. and B.D.K.; methodology, A.K.S.; software, S.I.A.; validation, S.A. and J.W; formal analysis, A.I.T. and M.O.B.; investigation, B.K.Z.; resources, B.D.K. and J.W.; data curation, H.B., A.I.T., and S.A.; writing—original draft preparation, A.K.S.; writing—review and editing, B.D.K. and S.I.A.; visualization, Y.K.L. and J.-S.C.; supervision, S.I.A.; project administration, A.K.S. and S.I.A.; funding acquisition, Y.K.L. and A.K.S. All authors have read and agreed to the published version of the manuscript.

Funding: This research has been funded by the Ministry of Education and Science of the Republic of Kazakhstan in the framework of the projects: AP08052402 and AP09260077. J.W. was funded by Science and Technology Partnership Program, Ministry of Science and Technology of China (KY202001017) and Tianjin Synthetic Biotechnology Innovation Capacity Improvement Project: TSBICIP-IJCP-001-03 S.A. would like to thank the Distinguished Scientists Fellowship Program, King Saud University, Saudi Arabia, for their support. Y.K.L., J.-S.C., and S.I.A. were supported by the joint grant from Russian Science Foundation and MOST (Taiwan) (no: 22-44-08001). The results of part (Results) were obtained within the state assignment of the Ministry of Science and Higher Education of the Russian Federation (project No. 121033000136-4).

Institutional Review Board Statement: Not applicable.

Informed Consent Statement: Not applicable.

Data Availability Statement: The data are available from the corresponding authors upon reasonable request.

Conflicts of Interest: The authors declare no competing financial interest.

References

1. Sundararaman, M.; Subramanian, G.; Averal, H.I.; Akbarsha, M.A. Evaluation of the Bio-Activity of Marine Cyanobacteria on Some Biochemical Parameters of Rat Serum. *Phytother. Res.* **1996**, *10*, 9–12. [CrossRef]
2. Kaushik, B.D.; Venkataraman, G.S. Effect of algal inoculation on the yield and vitamin c content of two varieties of tomato. *Plant Soil* **1979**, *52*, 135–137. [CrossRef]
3. Choudhary, K.K. Occurrence of Nitrogen-Fixing Cyanobacteria during Different Stages of Paddy Cultivation. *Bangladesh J. Plant Taxon.* **2011**, *18*, 73–76. [CrossRef]
4. Sadvakasova, A.K.; Kossalbayev, B.D.; Zayadan, B.K.; Kirbayeva, D.K.; Alwasel, S.; Allakhverdiev, S.I. Potential of Cyanobacteria in the Conversion of Wastewater to Biofuels. *World J Microbiol Biotechnol* **2021**, *37*, 140. [CrossRef] [PubMed]
5. Zayadan, B.; Usserbayeva, A.; Bolatkhan, K.; Akmukhanova, N.; Kossalbayev, B.; Baizhigitova, A.; Los, D. Screening of Isolated and Collection Strains Of Cyanobacteria On Productivity For Determining Their Biotechnological Potential. *Eur. J. Entomol.* **2018**, *55*, 121–130. [CrossRef]
6. Mahmud, K.; Makaju, S.; Ibrahim, R.; Missaoui, A. Current Progress in Nitrogen Fixing Plants and Microbiome Research. *Plants* **2020**, *9*, 97. [CrossRef]
7. Venkataraman, G.S. Blue-Green algae for rice production. *FAO Soil Bull.* **1981**, *16*, 33–42. Available online: https://www.fao.org/3/i5199e/I5199E.pdf (accessed on 1 January 2022).
8. Shashirekha, S.; Uma, L.; Subramanian, G. Phenol Degradation by the Marine Cyanobacterium *Phormidium Valderianum* BDU 30501. *J. Ind. Microbiol. Biotech.* **1997**, *19*, 130–133. [CrossRef]
9. Bolatkhan, K.; Kossalbayev, B.D.; Zayadan, B.K.; Tomo, T.; Veziroglu, T.N.; Allakhverdiev, S.I. Hydrogen Production from Phototrophic Microorganisms: Reality and Perspectives. *Int. J. Hydrog. Energy* **2019**, *44*, 5799–5811. [CrossRef]
10. Kossalbayev, B.D.; Tomo, T.; Zayadan, B.K.; Sadvakasova, A.K.; Bolatkhan, K.; Alwasel, S.; Allakhverdiev, S.I. Determination of the Potential of Cyanobacterial Strains for Hydrogen Production. *Int. J. Hydrog. Energy* **2020**, *45*, 2627–2639. [CrossRef]
11. Sadvakasova, A.K.; Kossalbayev, B.D.; Zayadan, B.K.; Bolatkhan, K.; Alwasel, S.; Najafpour, M.M.; Tomo, T.; Allakhverdiev, S.I. Bioprocesses of Hydrogen Production by Cyanobacteria Cells and Possible Ways to Increase Their Productivity. *Renew. Sustain. Energy Rev.* **2020**, *133*, 110054. [CrossRef]
12. Bolatkhan, K.; Sadvakasova, A.K.; Zayadan, B.K.; Kakimova, A.B.; Sarsekeyeva, F.K.; Kossalbayev, B.D.; Bozieva, A.M.; Alwasel, S.; Allakhverdiev, S.I. Prospects for the Creation of a Waste-Free Technology for Wastewater Treatment and Utilization of Carbon Dioxide Based on Cyanobacteria for Biodiesel Production. *J. Biotechnol.* **2020**, *324*, 162–170. [CrossRef] [PubMed]
13. Issa, A.A.; Abd-Alla, M.H.; Ohyama, T. *Nitrogen Fixing Cyanobacteria: Future Prospect*; IntechOpen: London, UK, 2014; ISBN 978-953-51-1216-7.
14. Singh, J.S.; Kumar, A.; Rai, A.N.; Singh, D.P. Cyanobacteria: A Precious Bio-Resource in Agriculture, Ecosystem, and Environmental Sustainability. *Front. Microbiol.* **2016**, *7*, 529. [CrossRef] [PubMed]
15. Durner, J.; Böhm, I.; Knörzer, O.C.; Böger, P. Proteolytic Degradation of Dinitrogenase Reductase from *Anabaena Variabilis* (ATCC 29413) as a Consequence of ATP Depletion and Impact of Oxygen. *J. Bacteriol.* **1996**, *178*, 610. [CrossRef] [PubMed]

16. Gallon, J.R. Reconciling the Incompatible: N_2 Fixation and O_2. *New Phytol.* **1992**, *122*, 571–609. Available online: https://www.jstor.org/stable/2557427 (accessed on 1 February 2022). [CrossRef]
17. Berman-Frank, I.; Lundgren, P.; Falkowski, P. Nitrogen Fixation and Photosynthetic Oxygen Evolution in Cyanobacteria. *Res. Microbiol.* **2003**, *154*, 157–164. [CrossRef]
18. UTEX—Culture Collection of Algae at The University of Texas at Austin. Available online: https://utex.org/products/bg-11-medium?variant=30991786868826 (accessed on 25 February 2022).
19. UTEX—Culture Collection of Algae at The University of Texas at Austin. Available online: https://utex.org/products/allen-medium?variant=30991788507226 (accessed on 25 February 2022).
20. UTEX—Culture Collection of Algae at The University of Texas at Austin. Available online: https://utex.org/products/bold-basal-medium (accessed on 25 February 2022).
21. Huynh, M.; Serediak, N. *Algae Identification Field Guide*; Agriculture and Agri-Food Canada: Ottawa, ON, Canada, 2006; p. 40. ISBN 978-1-100-18308-4.
22. Bellinger, E.G.; Sigee, D.C. *Freshwater Algae: Identification and Use as Bioindicators*; John Wiley & Sons: Hoboken, NJ, USA, 2011; p. 284. ISBN 978-1-119-96432-2.
23. Wilson, K.M.; Schembri, M.A.; Baker, P.D.; Saint, C.P. Molecular Characterization of the Toxic Cyanobacterium *Cylindrospermopsis Raciborskii* and Design of a Species-Specific PCR. *Appl. Environ. Microbiol.* **2000**, *66*, 332–338. [CrossRef]
24. Dale, R.K.; Pedersen, B.S.; Quinlan, A.R. Pybedtools: A Flexible Python Library for Manipulating Genomic Datasets and Annotations. *Bioinformatics* **2011**, *27*, 3423–3424. [CrossRef]
25. De Lillo, A.; Ashley, F.P.; Palmer, R.M.; Munson, M.A.; Kyriacou, L.; Weightman, A.J.; Wade, W.G. Novel Subgingival Bacterial Phylotypes Detected Using Multiple Universal Polymerase Chain Reaction Primer Sets. *Oral Microbiol. Immunol.* **2006**, *21*, 61–68. [CrossRef]
26. Walters, W.; Hyde, E.R.; Berg-Lyons, D.; Ackermann, G.; Humphrey, G.; Parada, A.; Gilbert, J.A.; Jansson, J.K.; Caporaso, J.G.; Fuhrman, J.A.; et al. Improved Bacterial 16S RRNA Gene (V4 and V4-5) and Fungal Internal Transcribed Spacer Marker Gene Primers for Microbial Community Surveys. *mSystems* **2015**, *1*, e00009-15. [CrossRef]
27. Hiraishi, A.; Kamagata, Y.; Nakamura, K. Polymerase Chain Reaction Amplification and Restriction Fragment Length Polymorphism Analysis of 16S RRNA Genes from Methanogens. *J. Ferment. Bioeng.* **1995**, *79*, 523–529. [CrossRef]
28. Johansen, J.R.; Mareš, J.; Pietrasiak, N.; Bohunická, M.; Zima, J.; Štenclová, L.; Hauer, T. Highly Divergent 16S RRNA Sequences in Ribosomal Operons of *Scytonema Hyalinum* (Cyanobacteria). *PLoS ONE* **2017**, *12*, e0186393. [CrossRef] [PubMed]
29. Sanger, F.; Nicklen, S.; Coulson, A.R. DNA Sequencing with Chain-Terminating Inhibitors. *Proc. Natl. Acad. Sci. USA* **1977**, *74*, 5463–5467. [CrossRef] [PubMed]
30. Lu, G.; Moriyama, E.; Lu, G.; Moriyama, E.N. Vector NTI, a Balanced All-in-One Sequence Analysis Suite. *Brief. Bioinform.* **2005**, *5*, 378–388. [CrossRef]
31. Tsarenko, P.M.; Borysova, O.V.; Korkhovyi, V.I.; Blume, Y.B. High-Efficiency Ukrainian Strains of Microalgae for Biodiesel Fuel Production (Overview). *Open Agric. J.* **2020**, *14*, 209–218. [CrossRef]
32. UTEX—Culture Collection of Algae at The University of Texas at Austin. Available online: https://utex.org/products/bg-11-n-medium (accessed on 12 December 2021).
33. Thajuddin, N.; Subramanian, G. Cyanobacterial Biodiversity and Potential Applications in Biotechnology. *Curr. Sci.* **2005**, *89*, 47–57.
34. Alekseev, A.A.; Yakovleva, O.V.; Matorin, D.N. The Fluorescence Methods for an Assessment Photosynthetic Capacity of Plants under the Anthropogenic Load. *AIP Conf. Proc.* **2018**, *2041*, 050013. [CrossRef]
35. Kumar, D.; Singh, H.; Raj, S.; Soni, V. Chlorophyll a Fluorescence Kinetics of Mung Bean (*Vigna Radiata* L.) Grown under Artificial Continuous Light. *Biochem. Biophys. Rep.* **2020**, *24*, 100813. [CrossRef]
36. Tsygankov, A.A.; Serebryakova, L.T.; Rao, K.K.; Hall, D.O. Acetylene Reduction and Hydrogen Photoproduction by Wild-Type and Mutant Strains of Anabaena at Different CO_2 and O_2 Concentrations. *FEMS Microbiol. Lett.* **1998**, *167*, 13–17. [CrossRef]
37. Ritchie, R.J. Consistent Sets of Spectrophotometric Chlorophyll Equations for Acetone, Methanol and Ethanol Solvents. *Photosynth. Res.* **2006**, *89*, 27–41. [CrossRef]
38. Wellburn, A.R. The Spectral Determination of Chlorophylls a and b, as Well as Total Carotenoids, Using Various Solvents with Spectrophotometers of Different Resolution. *J. Plant Physiol.* **1994**, *144*, 307–313. [CrossRef]
39. Bennett, A.; Bogorad, L. Complementary Chromatic Adaptation in a Filamentous Blue-Green Alga. *J. Cell. Biol.* **1973**, *58*, 419–435. [CrossRef] [PubMed]
40. Newman, L.; Duffus, A.L.J.; Lee, C. Using the Free Program MEGA to Build Phylogenetic Trees from Molecular Data. *Am. Biol. Teach.* **2016**, *78*, 608–612. [CrossRef]
41. Johansen, J.R.; Bohunická, M.; Lukešová, A.; Hrčková, K.; Vaccarino, M.A.; Chesarino, N.M. Morphological and Molecular Characterization within 26 Strains of the Genus *Cylindrospermum* (*Nostocaceae*, Cyanobacteria), with Descriptions of Three New Species. *J. Phycol.* **2014**, *50*, 187–202. [CrossRef]
42. Zulkefli, N.S.; Hwang, S.-J. Heterocyst Development and Diazotrophic Growth of Anabaena Variabilis under Different Nitrogen Availability. *Life* **2020**, *10*, 279. [CrossRef] [PubMed]
43. Gollerbakh, M.M.; Kosinskaya, E.K.; Polyanskii, V.I. *Opredelitel' Presnovodnykh Vodoroslei SSSR. Vyp. 2. Sinezelenye Vodorosli*; Identification Guide of Freshwater Algae of the USSR. Iss. 2. Blue-Green Algae; Sovetskaya Nauka: Moscow, Russia, 1953.

44. Hauer, T.; Bohunická, M.; Johansen, J.R.; Mareš, J.; Berrendero-Gomez, E. Reassessment of the Cyanobacterial Family Microchaetaceae and Establishment of New Families Tolypothrichaceae and Godleyaceae. *J. Phycol.* **2014**, *50*, 1089–1100. [CrossRef] [PubMed]
45. Munawar, S.; Khan, A. Occurrence of blooms of blue green algae from the coast of Buleji, Karachi, Pakistan. *Int. J. Biol. Res.* **2017**, *5*, 31–34.
46. Komárek, J.; Zapomělová, E. Planktic Morphospecies of the Cyanobacterial Genus *Anabaena* = Subg. Dolichospermum—1. Part: Coiled Types. *Fottea* **2007**, *7*, 1–31. [CrossRef]
47. Tamura, K.; Peterson, D.; Peterson, N.; Stecher, G.; Nei, M.; Kumar, S. MEGA5: Molecular Evolutionary Genetics Analysis Using Maximum Likelihood, Evolutionary Distance, and Maximum Parsimony Methods. *Mol. Biol. Evol.* **2011**, *28*, 2731–2739. [CrossRef] [PubMed]
48. Gladkikh, A.S.; Belykh, O.I.; Klimenkov, I.V.; Tikhonova, I.V. Nitrogen-Fixing Cyanobacterium Trichormus Variabilis of the Lake Baikal Phytoplankton. *Microbiology* **2008**, *77*, 726–733. [CrossRef]
49. Walsby, A.E. Cyanobacterial Heterocysts: Terminal Pores Proposed as Sites of Gas Exchange. *Trends Microbiol.* **2007**, *15*, 340–349. [CrossRef]
50. Kumar, K.; Mella-Herrera, R.A.; Golden, J.W. Cyanobacterial Heterocysts. *Cold Spring Harb. Perspect. Biol.* **2010**, *2*, a000315. [CrossRef] [PubMed]
51. Singh, P.; Khan, A.; Srivastava, A. Chapter 16—Heterocyst and Akinete Differentiation in Cyanobacteria: A View toward Cyanobacterial Symbiosis. In *Advances in Cyanobacterial Biology*; Singh, P.K., Kumar, A., Singh, V.K., Shrivastava, A.K., Eds.; Academic Press: Cambridge, MA, USA, 2020; pp. 235–248. ISBN 978-0-12-819311-2.
52. Allakhverdiev, E.S.; Khabatova, V.V.; Kossalbayev, B.D.; Zadneprovskaya, E.V.; Rodnenkov, O.V.; Martynyuk, T.V.; Maksimov, G.V.; Alwasel, S.; Tomo, T.; Allakhverdiev, S.I. Raman Spectroscopy and Its Modifications Applied to Biological and Medical Research. *Cells* **2022**, *11*, 386. [CrossRef] [PubMed]
53. Wang, L.; Lin, G.-M.; Niu, T.-C.; Zhang, S.-R.; Zhang, J.-Y.; Tang, G.-F.; Chen, W.; Zhang, C.-C. PatD, a Gene Regulated by NtcA, Is Involved in the Optimization of Heterocyst Frequency in the Cyanobacterium *Anabaena* sp. Strain PCC 7120. *J. Bacteriol.* **2019**, *201*, e00457-19. [CrossRef] [PubMed]
54. Magnuson, A. Heterocyst Thylakoid Bioenergetics. *Life* **2019**, *9*, 13. [CrossRef]
55. Bergman, B.; Gallon, J.R.; Rai, A.N.; Stal, L.J. N_2 Fixation by Non-Heterocystous Cyanobacteria. *FEMS Microbiol. Rev.* **1997**, *19*, 139–185. [CrossRef]
56. Seefeldt, L.C.; Hoffman, B.M.; Dean, D.R. Mechanism of Mo-Dependent Nitrogenase. *Annu. Rev. Biochem.* **2009**, *78*, 701. [CrossRef]
57. Betancourt, D.A.; Loveless, T.M.; Brown, J.W.; Bishop, P.E. Characterization of Diazotrophs Containing Mo-Independent Nitrogenases, Isolated from Diverse Natural Environments. *Appl. Environ. Microbiol.* **2008**, *74*, 3471–3480. [CrossRef]
58. Bothe, H.; Schmitz, O.; Yates, M.G.; Newton, W.E. Nitrogen Fixation and Hydrogen Metabolism in Cyanobacteria. *Microbiol. Mol. Biol. Rev.* **2010**, *74*, 529–551. [CrossRef]
59. Berrendero, E.; Valiente, E.F.; Perona, E.; Gómez, C.L.; Loza, V.; Muñoz-Martín, M.Á.; Mateo, P. Nitrogen Fixation in a Non-Heterocystous Cyanobacterial Mat from a Mountain River. *Sci. Rep.* **2016**, *6*, 30920. [CrossRef]
60. Raymond, J.; Siefert, J.L.; Staples, C.R.; Blankenship, R.E. The Natural History of Nitrogen Fixation. *Mol. Biol. Evol.* **2004**, *21*, 541–554. [CrossRef] [PubMed]
61. Esteves-Ferreira, A.A.; Cavalcanti, J.H.F.; Vaz, M.G.M.V.; Alvarenga, L.V.; Nunes-Nesi, A.; Araújo, W.L. Cyanobacterial Nitrogenases: Phylogenetic Diversity, Regulation and Functional Predictions. *Genet. Mol. Biol.* **2017**, *40*, 261–275. [CrossRef]
62. Thiel, T.; Pratte, B. Alternative Nitrogenases in Anabaena Variabilis: The Role of Molybdate and Vanadate in Nitrogenase Gene. *Adv. Microbiol.* **2013**, *3*, 87–95. [CrossRef]
63. Zahalak, M.; Pratte, B.; Werth, K.; Thiel, T. Molybdate Transport and Its Effect on Nitrogen Utilization in the Cyanobacterium Anabaena Variabilis ATCC 29413: Molybdate Transport in Anabaena. *Mol. Microbiol.* **2004**, *51*, 539–549. [CrossRef]
64. Kentemich, T.; Danneberg, G.; Hundeshagen, B.; Bothe, H. Evidence for the Occurrence of the Alternative, Vanadium-Containing Nitrogenase in the Cyanobacterium Anabaena Variabilis. *FEMS Microbiol. Lett.* **1988**, *51*, 19–24. [CrossRef]
65. Miller, R.W.; Eady, R.R. Molybdenum and Vanadium Nitrogenases of *Azotobacter Chroococcum*. Low Temperature Favours N2 Reduction by Vanadium Nitrogenase. *Biochem. J.* **1988**, *256*, 429–432. [CrossRef]
66. Seefeldt, L.C.; Yang, Z.-Y.; Lukoyanov, D.A.; Harris, D.F.; Dean, D.R.; Raugei, S.; Hoffman, B.M. Reduction of Substrates by Nitrogenases. *Chem. Rev.* **2020**, *120*, 5082–5106. [CrossRef]
67. Devadasu, E.; Chinthapalli, D.K.; Chouhan, N.; Madireddi, S.K.; Rasineni, G.K.; Sripadi, P.; Subramanyam, R. Changes in the Photosynthetic Apparatus and Lipid Droplet Formation in Chlamydomonas Reinhardtii under Iron Deficiency. *Photosynth. Res.* **2019**, *139*, 253–266. [CrossRef]
68. Rueter, J.G.; Petersen, R.R. Micronutrient Effects on Cyanobacterial Growth and Physiology. *N. Z. J. Mar. Freshw. Res.* **1987**, *21*, 435–445. [CrossRef]
69. Yadavalli, V.; Jolley, C.C.; Malleda, C.; Thangaraj, B.; Fromme, P.; Subramanyam, R. Alteration of Proteins and Pigments Influence the Function of Photosystem I under Iron Deficiency from *Chlamydomonas Reinhardtii*. *PLoS ONE* **2012**, *7*, e35084. [CrossRef]
70. Devadasu, E.; Pandey, J.; Dhokne, K.; Subramanyam, R. Restoration of Photosynthetic Activity and Supercomplexes from Severe Iron Starvation in *Chlamydomonas Reinhardtii*. *Biochim. Biophys. Acta (BBA) Bioenerg.* **2021**, *1862*, 148331. [CrossRef] [PubMed]

71. Huang, M.; Zhang, J.-Y.; Zeng, X.; Zhang, C.-C. C-Di-GMP Homeostasis Is Critical for Heterocyst Development in *Anabaena* sp. PCC 7120. *Front. Microbiol.* **2021**, *12*, 793336. [CrossRef] [PubMed]
72. Rochaix, J.-D. Reprint of: Regulation of Photosynthetic Electron Transport. *Biochim. Biophys. Acta (BBA) Bioenerg.* **2011**, *1807*, 878–886. [CrossRef] [PubMed]

cells

MDPI

Article

Influence of Light of Different Spectral Compositions on the Growth, Photosynthesis, and Expression of Light-Dependent Genes of Scots Pine Seedlings

Pavel Pashkovskiy [1,*], Vladimir D. Kreslavski [2], Yury Ivanov [1], Alexandra Ivanova [1], Alexander Kartashov [1], Alexander Shmarev [2], Valeriya Strokina [2], Vladimir V. Kuznetsov [1] and Suleyman I. Allakhverdiev [1,*]

1 K.A. Timiryazev Institute of Plant Physiology, Russian Academy of Sciences, Botanicheskaya Street 35, 127276 Moscow, Russia; ivanovinfo@mail.ru (Y.I.); aicheremisina@mail.ru (A.I.); botanius@yandex.ru (A.K.); vlkuzn@mail.ru (V.V.K.)

2 Institute of Basic Biological Problems, Russian Academy of Sciences, Institutskaya Street 2, 142290 Pushchino, Russia; vkreslav@rambler.ru (V.D.K.); shurik_bx_04@mail.ru (A.S.); strokina.93@mail.ru (V.S.)

* Correspondence: pashkovskiy.pavel@gmail.com (P.P.); suleyman.allakhverdiev@gmail.com (S.I.A.)

Abstract: Varying the spectral composition of light is one of the ways to accelerate the growth of conifers under artificial conditions for the development of technologies and to obtain sustainable seedlings required to preserve the existing areas of forests. We studied the influence of light of different quality on the growth, gas exchange, fluorescence indices of Chl *a*, and expression of key light-dependent genes of *Pinus sylvestris* L. seedlings. It was shown that in plants growing under red light (RL), the biomass of needles and root system increased by more than two and three times, respectively, compared with those of the white fluorescent light (WFL) control. At the same time, the rates of photosynthesis and respiration in RL and blue light (BL) plants were lower than those of blue red light (BRL) plants, and the difference between the rates of photosynthesis and respiration, which characterizes the carbon balance, was maximum under RL. RL influenced the number of xylem cells, activated the expression of genes involved in the transduction of cytokinin (Histidine-containing phosphotransfer 1, *HPT1*, Type-A Response Regulators, *RR-A*) and auxin (Auxin-induced protein 1, *Aux/IAA*) signals, and reduced the expression of the gene encoding the transcription factor phytochrome-interacting factor 3 (*PIF3*). It was suggested that RL-induced activation of key genes of cytokinin and auxin signaling might indicate a phytochrome-dependent change in cytokinins and auxins activity.

Keywords: photomorphogenesis; *Pinus sylvestris*; light of various spectral composition; photosynthesis; chlorophyll fluorescence; gene expression; pigment content

Citation: Pashkovskiy, P.; Kreslavski, V.D.; Ivanov, Y.; Ivanova, A.; Kartashov, A.; Shmarev, A.; Strokina, V.; Kuznetsov, V.V.; Allakhverdiev, S.I. Influence of Light of Different Spectral Compositions on the Growth, Photosynthesis, and Expression of Light-Dependent Genes of Scots Pine Seedlings. *Cells* **2021**, *10*, 3284. https://doi.org/10.3390/cells10123284

Academic Editor: Seiji Akimoto

Received: 22 October 2021
Accepted: 22 November 2021
Published: 24 November 2021

1. Introduction

The quality of light is an important factor in regulating plant growth and development during ontogenesis, including germination, photomorphogenesis, flowering induction, etc. At the beginning of ontogenesis, most plants are forced to vegetate under shading conditions while growing underneath taller plants, which also leads to a decrease in the quality of light. Green and far-red light (FRL) predominate under the forest canopy because light in the red and blue ranges of the spectrum is effectively absorbed by the chlorophyll of taller plants. This forces the seedlings of most woody plants to adapt to indifferent light qualities [1]. For example, plants growing under a forest canopy acclimatize to a low red:far red (R:FR) ratio, which causes shoot elongation [1]. It was previously shown that blue light (BL), on the contrary, inhibits shoot growth [2,3]. This leads to plants with a high BL level and a high R:FR ratio growing low but with an increased leaf surface, which in turn affects the intensity of photosynthesis [4].

The observed changes in the global climate under all forecast scenarios suggest a decline in coniferous species in the temperate zone of Europe [5]. In this regard, the development of technologies for obtaining sustainable seedlings is required to preserve the existing areas of coniferous forests. Varying the spectral composition of light can be one of the simplest ways to accelerate the growth of seedlings under artificial conditions [6]. However, to select the optimal spectral composition of light under artificial light conditions, it is necessary to know how the light of different spectral ranges affects the growth and photosynthetic parameters of seedlings of coniferous plants.

Plants have several types of photoreceptors that respond to changes in environmental light conditions. Under natural light, plants are simultaneously exposed to light of different wavelengths, resulting in cross-signaling of light between multiple photoreceptors. Phytochromes are among the most characterized photoreceptors. They are RL and FRL sensors and they regulate many developmental processes, including seed germination and hypocotyl growth [1]. Other well-known receptors, the cryptochromes, perceive light in the blue and UV ranges of the spectrum. They are involved in the growth processes and de-etiolation of seedlings and are also involved in circadian rhythms [7]. Photoreceptors have been studied in detail in *Arabidopsis thaliana*. They include five phytochromes (PHYA to PHYE) and two major cryptochromes (CRY1 and CRY2). In gymnosperms, PHYN is orthologous to PHYA of angiosperms, while PHYO is orthologous to PHYC [8]. In addition, gymnosperms have PHYP, which genetically occupies an intermediate position between the PHYB of *Oryza sativa* and the PHYE of *Arabidopsis thaliana* [9]. From the experiments carried out by Clapham et al., 2002, it was found that pine and spruce react to the RL/FRL ratio differently than angiosperms, which confirms the uniqueness of the phytochrome system of gymnosperms [10].

In addition to the quality of light, plant growth is also affected by hormonal balance [11]. The lighting conditions cause the level of hormones in plants to change, which leads to a change in photosensitivity. For example, exogenous hormones can stimulate plant growth by acting as mediators in the processes of light signal transduction [12]. In turn, light by photoreceptors regulates the metabolism of various hormonal signals. Thus, PHYA affects the metabolic pathways of gibberellins and indoleacetic acid [11], as well as key components of light signaling, such as the transcriptional factors (TFs): PIF3, PIF4, and HY5 [13]. The main phytohormones are associated with light-mediated growth regulation [14–16], while the stimulation of cell growth involves auxins and cytokinins [17–19]. The light of different spectral compositions affects the activity of endogenous hormones through the regulation of secondary metabolism; for example, blue light promotes the accumulation of flavonoids, which in turn affects the polar transport of auxin [20].

The aim of this work was to understand how the conditions of light of different spectral composition affect the growth, morphometric, and photosynthetic characteristics of *Pinus sylvestris* seedlings. At the same time, special attention was given to the possible relationships of the processes of growth, photosynthesis, and respiration with the intensity of expression of the main genes encoding proteins of photosystems, and the genes involved in light and hormonal signalling of *Pinus sylvestris* plants under light of various spectral quality. The obtained results can be used to create artificial lighting systems in forest nurseries to accelerate the cultivation of planting material, and they also have applications in biotechnology.

2. Materials and Methods

2.1. Plant Materials and Experimental Design

Seeds of Scots pine (*P. sylvestris* L.) were collected in the Bryansk region (Bryansk, Russia) from high-productivity pine stands in complex forest types. Seeds were germinated and grown in hydroculture on individual substrates in polypropylene cartridges filled with 1% agar bungs in individual boxes of the climatic chamber under red (maxima of 660 nm), blue (maxima of 450 nm), red + blue (maxima of 660 and 450 nm) LEDs, as well as under fluorescent lamps (58 W/33–640, white fluorescent lamps (Philips, Pila, Poland), 130 ± 10 μmol (photons) m^{-2} s^{-1}) for 6 weeks. (Figure 1).

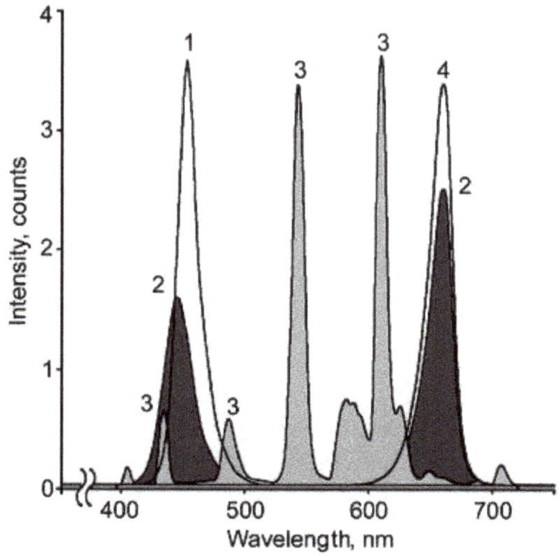

Figure 1. Emission spectra of the light sources used in the experiments. Spectra of blue light (BL) with a peak at 450 nm (1), blue + red (BRL) with two maxima at 445 nm and 660 nm (2), white fluorescent lamps (WFL) with a set of peaks in the visible spectral region (3), and red light (RL) with a peak at 660 nm (4).

After seed coat rupture and cotyledon expansion, the seedlings were transferred to a nutrient solution [21]. The seedlings were cultivated in 6 L plastic trays (171 seed beds per tray) in a growth chamber that provided a constant air temperature of 24 ± 2 °C and a 16 h photoperiod. The nutrient solutions were constantly aerated and renewed once a week. During the week, a constant volume of the nutrient solution was maintained by adding distilled water.

The fresh biomass of the roots and needles was determined using an analytical balance (Scout Pro SPU123, Ohaus Corporation, Parsippany, NJ, USA) with an accuracy of 1 mg, after which the samples were fixed in liquid nitrogen and stored at −70 °C until biochemical analyses. The fixation was carried out under the conditions of the light in which the plants grew without exposure to another light.

2.2. Pigment Contents

The contents of chlorophyll *a* (Chl *a*) and *b* (Chl *b*) and total carotenoids (Car) in pigment extracts of all studied needles were determined spectrophotometrically in 80% acetone [22].

2.3. Measurements of CO_2 Gas Exchange

The photosynthetic rate (P_n) was determined in a closed system under light conditions using an LCPro + portable infrared gas analyser from ADC BioScientific Ltd. (United Kingdom) connected to a leaf chamber. The CO_2 uptake per leaf area (μmol $m^{-2}s^{-1}$) was determined. The rate of photosynthesis of the leaves in the second layer from the top was determined at a saturating light intensity of 1000 μmol (photons) m^{-2} s^{-1}. Previously, we recorded the light curves in the interval of intensities from 0 to 1200 μmol (photons) m^{-2} s^{-1}. The light intensities in the region from 600 to 1200 μmol (photons) m^{-2} s^{-1} were saturated. After measuring the rate of photosynthesis, the light was turned off, and the rate of dark respiration was measured.

2.4. Determination of Photochemical Activity

Fluorescence parameters characterizing the state of the photosynthetic apparatus were calculated on the basis of induction fluorescence curves obtained using data from the JIP test, which is usually used to evaluate the state of PSII. Chlorophyll (Chl) fluorescence induction curves (OJIP curves) were recorded with the setup Plant Efficiency Analyser (Handy-PEA, Hansatech Instruments Ltd., London, UK). For the JIP test, OJIP curves were measured under illumination with blue light at an intensity of 3500 μmol (photons) m^{-2} s^{-1} for 1 s.

On the basis of induction fluorescence curves (OJIP curves), the following parameters, which characterize the PSII photochemical activity, were calculated: F_v/F_m, the PSII maximum quantum photochemical yield, and PI_{ABS}, the PSII performance index [23,24]. Here, F_v is the value of variable fluorescence, equal to the difference between F_m and F_0; F_0 is the minimum amplitude of fluorescence (F), and F_m is the maximum amplitude of fluorescence. For calculation of the PI_{ABS}, the following formula was used:

$$PI_{ABS} = (F_v/F_m)/(M_0/V_j) \times (F_v/F_0) \times (1 - V_j)/V_j);\ M_0 = 4 \times (F_{300\mu s} - F_0)/(F_m - F_0);\ \text{and}\ V_j = (F_{2ms} - F_0)/(F_m - F_0)$$

where M_0 is the average value of the initial slope of the relative variable fluorescence of Chl *a*, which reflects the closing rate of the PSII reaction centers, and V_j is the relative level of fluorescence in phase J after 2 ms.

PAM fluorimetry (Junior-PAM, Walz, Germany) was used to evaluate the photosynthetic apparatus state. The values F_0, F_v, F_m, F_m', and F', as well as the PSII maximum (F_v/F_m) and effective Y(II) $(F_m'-F_t)/F_m'$ photochemical quantum yields and non-photochemical quenching (NPQ) $(F_m/F_m'-1)$, were determined. Here, F_m and F_m' are the maximum Chl fluorescence levels under dark- and light-adapted conditions, respectively. F_v is the photoinduced change in fluorescence, and F_t is the level of fluorescence before a saturation impulse is applied. F_0 is the initial Chl fluorescence level. Actinic light was switched on for 10 min [I = 125 μmol (photons) m^{-2} s^{-1}].

2.5. RNA Extraction and qRT-PCR

RNA isolation was performed according to the method of Kolosova et al. (2004) [25] with some modifications suggested by Pashkovskiy et al. (2019) [26]. The quantity and quality of the total RNA were determined using a NanoDrop 2000 spectrophotometer (Thermo Fisher Scientific, Waltham, MA, USA). cDNA synthesis was performed using the M-MLV Reverse Transcriptase Kit (Fermentas, Canada) and the oligo (dT) 21 primer for genes of nuclear coding and random 6 (Evrogen, Moscow, Russia) for chloroplast coding genes expression. The expression patterns of the genes were assessed using the CFX96 Touch™ Real-Time PCR Detection System (Bio-Rad, Hercules, CA, USA). The transcript levels were normalized to the expression of the *Actin 1* gene. The mRNA levels were expressed as a ratio of the corresponding values for the WFL plants. The relative gene-expression signal intensity at the WFL plants was considered to have a value of 1. Gene-specific primers (Table 1) for photosystem II protein D1 (*psbA*, ABO77179.1), cryptochrome 1 (*Cry1*, K7R334), cryptochrome 2 (*Cry2*, T2FFB6), phytochrome P (*phyP*, AIY54822.1), phytochrome N (*phyN*, AFV79519.1), phytochrome O (*phyO*, A7Y6Q6), phytochrome-interacting factor 3 (*PIF3*, D5ABG4), chalcone synthase (*CHS*, AF543757.1), stilbene synthase (*STS*, S50350.1), histidine-containing phosphotransfer 1 (*HPT1*, ALN42232.1), type-A response regulators (*RR-A*, FJ717710.1), and auxin-induced protein 1 (*Aux/IAA*, AY289600.1) were selected using nucleotide sequences from the National Center for Biotechnology Information (NCBI) database (Available online: http://www.ncbi.nlm.nih.gov (accessed on 15 May 2021)) and database www.uniprot.org (Available online: http://www.uniprot.org (accessed on 20 May 2021)) with Vector NTI Suite 9 software (Invitrogen, Carlsbad, CA, USA).

Table 1. The primers for qRT-PCR analysis.

	Gene Bank ID	Gene Description	Plant	Gene	Primer 5′-3′ Forward	Reverse
1	ALN42232.1 (uniprot.org)	Histidine-containing phosphotransfer 1	*Pinus pinaster*	*HPT1*	GCTCAAGTAT-AGGAGCGCGG	CCAGCTTGTTT-TTCACGAGGT
2	FJ717710.1 (ncbi.nlm.nih.gov)	Type-A Response Regulators	*Pinus pinea*	*RR-A*	CAGAAGGCGC-TCAAGAGTTT	TTGTTGGTCCC-TGGATCTTC
3	AY289600.1 (ncbi.nlm.nih.gov)	Auxin-induced protein 1 (IAA1)	*Pinus taeda*	*Aux/IAA*	GCCACCTGTC-AAAGATTTCAG	TGAGGTCCACC-TTTCTGAGA
4	ABO77179.1 (uniprot.org)	Photosystem II protein D1	*Pinus sylvestris*	*psbA*	TGAAGGTTAC-AGATTCGGTCA	TGAATATGCAA-CAGCAATCCA
5	K7R334 (uniprot.org)	Cryptochrome 1	*Pinus sylvestris*	*Cry1*	TATGGTGCACA-GGGCAGATG	AAGCTGCAGAA-GCTGTTCCT
6	T2FFB6 (uniprot.org)	Cryptochrome 2	*Picea abies*	*Cry2*	TTCCCTGGCT-GCAACAGAAA	CCCAACATTGC-TAGGCAGGA
7	AIY54822.1 (uniprot.org)	Phytochrome P	*Pinus sylvestris*	*phyP*	GGCATGTCCC-TTGTTCAGGA	CTTCTGTGGGCC-AAAGGTCT
8	AFV79519.1 (uniprot.org)	Phytochrome N	*Pinus sylvestris*	*phyN*	GGCTCAGAGG-AGGACAAAGG	TTCTGCCCGGTC-ACATCTTG
9	A7Y6Q6 (uniprot.org)	Phytochrome O	*Pinus sylvestris*	*phyO*	AGATGTGACG-TGGCAAAGGA	TGCGGGATTCC-ACTCAGAAC
10	D5ABG4 (uniprot.org)	Phytochrome-interacting factor 3	*Picea sitchensis*	*PIF3*	ATCAGCACTT-CCTGGTTCCG	CAGGCTGAGTT-GTTCCAGGT
11	AF543757.1 (ncbi.nlm.nih.gov)	Chalcone synthase	*Pinus uliginosa*	*CHS*	ATGGCTGCAG-GAATGATGAAGG	AGTGCCAATAG-CGAGGATG
12	S50350.1 (ncbi.nlm.nih.gov)	Stilbene synthase	*Pinus sylvestris*	*STS*	TCCGACTGGA-ACAAGTTGTTC	GCTTGGCCTCC-ACCCGATCAAG
13	CBB44933.1 (uniprot.org)	Actin 1	*Pinus sylvestris*	*Act1*	TTAGCAACTGG-GATGACATGGA	CCTGAATGGCA-ACATACATAGCA

2.6. Histochemical Studies of the Hypocotyls

Anatomical studies were performed using light microscopy methods on live preparations of cross-sections of hypocotyls of *P. sylvestris* seedlings prepared immediately before the study. Sections 10–30 μm thick were prepared using an HM650 V vibrating blade microtome (Thermo Fisher Scientific, Waltham, MA, USA) through the central part of the organ. The obtained sections were stained with 3% phloroglucinol/HCl reagent (Sigma, P3502) for 2 min and then washed with phosphate buffered saline (70 mM, pH = 7.4) [27]. The sections were photographed under an Imager D1 light microscope (Carl Zeiss, Oberkochen, Germany) using a Levenhuk M800 PLUS digital photo attachment with a resolution of 8.0 MPI (Levenhuk, Tampa, FL, USA). The resulting images were processed with the public domain software ImageJ v.1.49 (NIH; http://rsb.info.nih.gov/ij). The diameter of the hypocotyl, the number of lignified xylem cells, the sectional area of the hypocotyl, the area of the xylem, and the average sectional area of one xylem cell were determined on the sections of the hypocotyls.

2.7. Statistics

The number of biological replicates for the determination of the fresh biomass of the roots and needles ranged from 27 to 39; 13 biological replicates were performed for the histochemical studies of the hypocotyls, and 6 biological replicates were used for photosynthetic and respiration rates as well as photochemical activity. Each plant sample fixed in liquid nitrogen was treated as a biological replicate; therefore, there were six biological replicates for the pigment contents and gene expression analyses.

The data were statistically analyzed using SigmaPlot 12.3 (Systat Software, San Jose, USA) with one-way analysis of variance (ANOVA) followed by Duncan's method for normally distributed data (in the figures, significant differences are denoted by different capital letters). Kruskal–Wallis one-way ANOVA was performed on ranks followed by a Student-Newman–Keuls post hoc test or by a Dunn's post hoc test for non-normally distributed data and data with unequal variance (in the figures, significant differences are denoted by different italic letters for the Student-Newman–Keuls post hoc test or by different boldface italic type for the Dunn's post hoc test). Different letters were used to indicate significance at $p < 0.05$. The values presented in the tables and figures are the arithmetic means ± standard errors.

3. Results

3.1. Growth and Morphological Parameters

Narrowband light caused marked changes in plant morphology and growth (Figure 2).

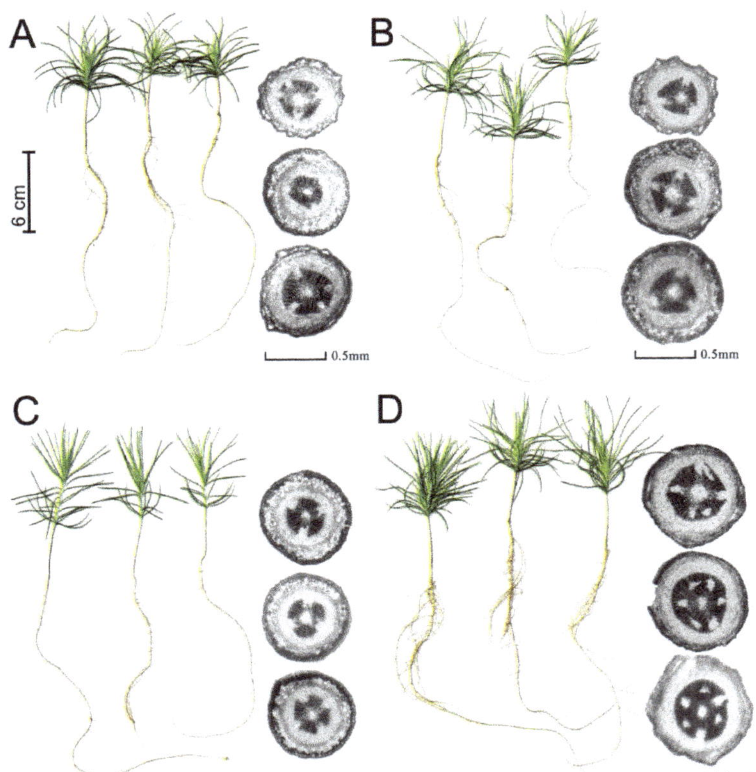

Figure 2. The morphological changes of *P. sylvestris* seedlings with dependence on light quality. WFL (**A**), BRL (**B**), BL (**C**), and RL (**D**).

Thus, RL increased the weight of the needles by 2.6 times, and BL increased the weight of the needles by 2.1 times relative to the WFL taken as a control (Figure 3A). BL caused hypocotyl elongation, and plants with RL formed short, larger plants. In addition, plant root weights in the RL variant were 3.8 times higher than the root weight in the control (WFL) and, on average, 2.7 times higher than in the other experimental variants

(Figure 3B). Moreover, there was an increase in both the fresh and dry weight of the root in RL (Figure 3B,E) and dry weight of needles in BL and RL in relation to WFL control (Figure 3D). In general, RL led to a more intensive development of the root system of the seedlings and an increase in the number of lateral roots of the first and second orders (Figure 2). Under the influence of RL, in the hypocotyls of the seedlings, 1.7 times more xylem cells were formed compared to the other variants of the experiment (Figure 3C), in the absence of any differences in the total area of the hypocotyls (Figure 3F). It is important to note that the growth of xylem under RL was not accompanied by thickening of the hypocotyls (Figure 3F), but the area of the xylem increased relative to other tissues (Figure 3G,H), while the diameter of xylem cells did not differ significantly (Figure 3I).

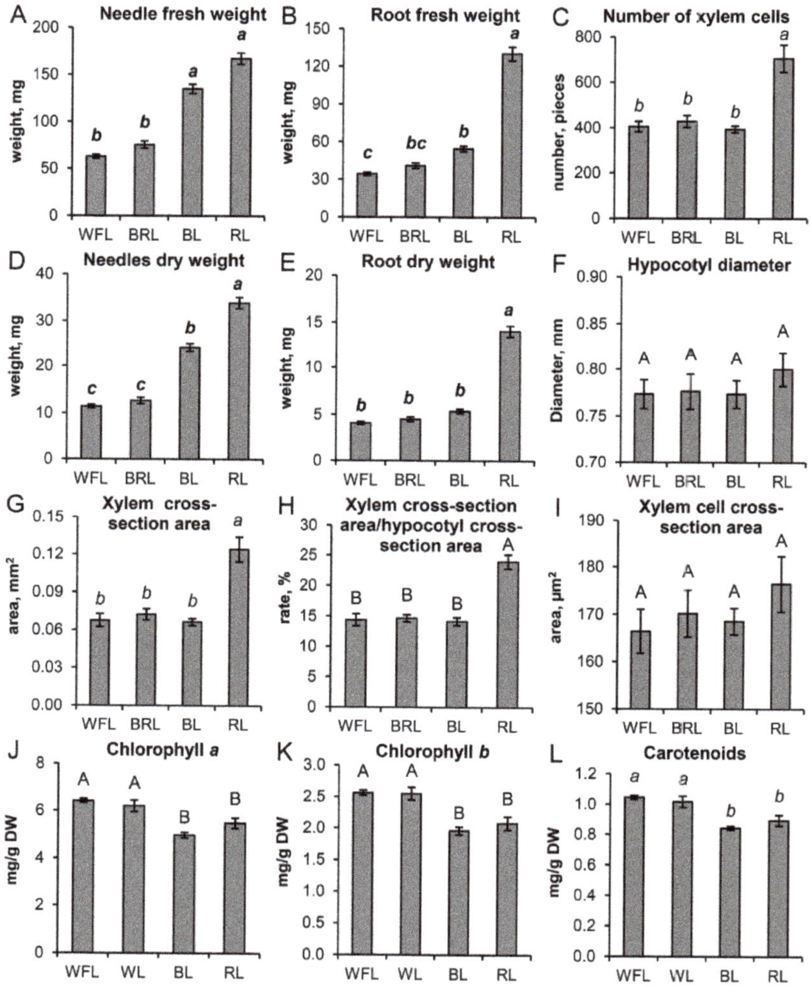

Figure 3. Effect of light quality on the needle (**A**) and root (**B**) fresh weight (mg); number of xylem cells (**C**) (pieces); needle (**D**) and root (**E**) dry weight (mg, DW); hypocotyl diameter (**F**) (mm); xylem cross-section area (**G**) (mm); xylem cross-section area and hypocotyls cross-section area ratio (**H**) (%); xylem cell cross-section area (**I**) (µm^2); and content of Chl *a* (**J**) mg/g DW, Chl *b* (**K**) mg/g DW, and total carotenoids (**L**) mg/g DW in *P. sylvestris* seedlings. Values are the mean ± SE. Different capital

letters denote statistically significant differences in the means at $p < 0.05$ (ANOVA followed by Duncan's method). Different italic letters denote statistically significant differences in the means at $p < 0.05$ (Kruskal–Wallis ANOVA of the ranks followed by the Student-Newman-Keuls post hoc test). Different boldface italic letters denote statistically significant differences in the means at $p < 0.05$ (Kruskal–Wallis ANOVA of the ranks followed by the Dunn's method).

3.2. Contents of Photosynthetic Pigments

When grown on a narrowband RL and BL, a reduced content of chlorophylls and carotenoids was observed (by an average of 10–20%), while the BRL and WFL variants were comparable (Figure 3J–L).

3.3. Fluorescence Parameters and CO_2 Gas Exchange

The photosystem II (PSII) maximum quantum yield (F_v/F_m) was approximately 0.81 and did not depend on the light quality used in the experiments (Table 2).

Table 2. Effect of light quality on the net photosynthetic and respiration rates, R/P_n ratio, (P_n–R) difference, PSII maximum quantum yield (F_v/F_m), effective quantum yield Y(II), performance index PSII (PI_{ABS}), and nonphotochemical fluorescence quenching (NPQ) in 6-week-old *P. sylvestris* seedlings. Values are the mean ± SE. Different normal-type letters denote statistically significant differences in the means at $p < 0.05$ (ANOVA followed by Duncan's method).

Parameter	WFL	BRL	BL	RL
P_n, µmol CO_2 m^{-2} s^{-1}	14.4 ± 2.1^{ab}	16.3 ± 0.2^{a}	13.7 ± 0.3^{b}	12.1 ± 0.3^{c}
R, µmol CO_2 m^{-2} s^{-1}	8.7 ± 0.8^{a}	8.8 ± 0.1^{a}	5.1 ± 0.2^{b}	2.1 ± 0.1^{c}
R/P_n	0.62 ± 0.12^{a}	0.54 ± 0.06^{a}	0.37 ± 0.02^{b}	0.17 ± 0.01^{c}
(P_n − R), µmol CO_2 m^{-2} s^{-1}	5.7 ± 0.5^{d}	7.5 ± 0.2^{c}	8.6 ± 0.1^{b}	10.0 ± 0.2^{a}
F_v/F_m	0.814 ± 0.005^{a}	0.803 ± 0.001^{a}	0.815 ± 0.007^{a}	0.812 ± 0.006^{a}
PI_{ABS}	7.73 ± 0.39^{b}	7.85 ± 0.64^{b}	6.45 ± 0.82^{b}	10.21 ± 0.59^{a}
Y_{II}	0.52 ± 0.01^{b}	0.44 ± 0.03^{c}	0.50 ± 0.01^{b}	0.57 ± 0.01^{a}
NPQ	0.86 ± 0.10^{b}	1.67 ± 0.21^{a}	0.95 ± 0.12^{ab}	0.66 ± 0.17^{b}

The PSII effective quantum yield (Y(II)) was the lowest in BRL plants (0.44); at the same time, in BL and WFL plants, this indicator remained at a comparable level and amounted to approximately 0.51. The highest value of the Y(II) parameter was observed when growing seedlings under RL (0.57) (Table 2). The non-photochemical quenching (NPQ) parameter was the smallest in RL plants. When plants were grown on BL and WFL, NPQ was comparable and amounted to approximately 0.90; the highest NPQ value (1.67) was observed in the BRL plants (Table 2).

The intensity of CO_2 gas exchange did not differ noticeably between the samples. The largest value of the parameter P_n (16.3 µmol CO_2 m^{-2} s^{-1}) was observed in the BRL variant, and the smallest value was 12.1 µmol CO_2 m^{-2} s^{-1} in the RL variant, while the photosynthesis rate in the seedlings under the WFL and BL conditions was intermediate (Table 2).

The respiration rate (R) in the WFL and BRL variants was the highest (approximately 8.8 µmol CO_2 m^{-2} s^{-1}), while in BL and RL plants, this value was reduced to 5.1 and 2.1 µmol CO_2 m^{-2} s^{-1}, respectively. As a consequence, the calculated parameters of the R/P_n ratio and the P_n–R carbon balance changed accordingly. The respiration/photosynthesis ratio was the highest in the WFL and BRL variants, lower in the BL variant, and the lowest in the RL plants (0.17). At the same time, the highest value of the carbon balance, assessed by the difference between the rates of photosynthesis and respiration, was observed in the RL variant (Table 2).

The PSII performance index (PI_{ABS}) was on average 1.5 times higher in the RL variant than in the other variants. At the same time, all other options did not differ significantly among themselves (Table 2).

3.4. Gene Expression

The transcript level of genes involved in hormonal signaling of cytokinins and auxins significantly changed in the variants of narrow-band light. Expression of the *HPT1* gene in the needles of RL plants was increased almost two times, and in the roots, it was 2.7 times higher than that in the WFL control (Figure 4A,D).

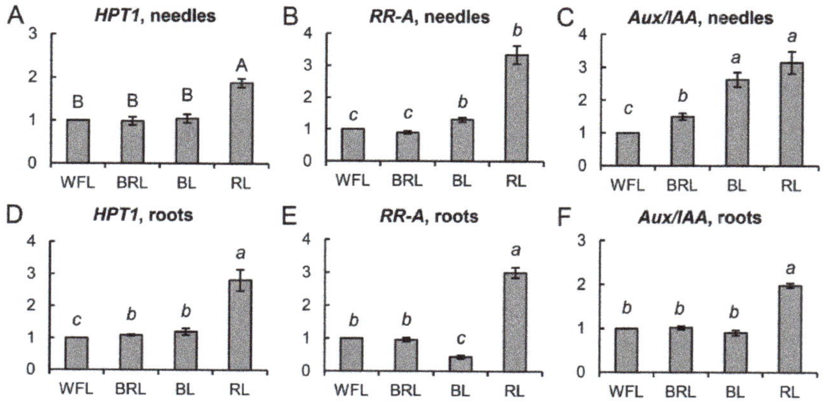

Figure 4. Effect of light quality on the transcript levels of different groups of genes in *P. sylvestris* seedlings. *HPT1*, *RR-A*, and *Aux/IAA* in needles (**A–C**) and roots (**D–F**). The mRNA levels of the genes were expressed as the BRL, BL, RL/WFL ratio (fold change BRL, BL, RL/WFL). Values are the mean ± SE. Different capital letters denote statistically significant differences in the means at $p < 0.05$ (ANOVA followed by Duncan's method). Different italic letters denote statistically significant differences in the means at $p < 0.05$ (Kruskal–Wallis ANOVA of the ranks followed by the Student-Newman-Keuls post hoc test).

At the same time, the level of *RR-A* transcription in the RL variant was 3.3 and 3.0 times higher than that in WFL needles (Figure 4B) and roots (Figure 4E), respectively. In contrast, in the roots of the BL variant, *RR-A* expression was more than 2-fold lower than that in the WFL control (Figure 4E). In the needles of the seedlings, the expression of the *Aux/IAA* gene in the RL and BL plants was, on average, 2.8 times higher than that in the WFL control (Figure 4C). At the same time, in the roots, the level of *Aux/IAA* transcripts was higher only in the RL variant (Figure 4F).

We also studied the transcription of genes responsible for light signaling and the synthesis of secondary metabolites, such as the transcription factors *PIF3*, *CHS*, and *STS*. The expression level of the *PIF3* gene in the RL variant compared to that of WFL was reduced in needles by 30% (Figure 5A), while that of the *CHS* gene, on the contrary, increased 1.5 times compared to those of BRL and WFL (Figure 5B).

An increase in the transcript level of the *STS* gene was also observed more than 11 times in needles (Figure 5C) and more than 23 times in roots (Figure 5F) in the RL variant, and approximately 4 times in needles and 6 times in roots in the BL variant (Figure 5C,F).

Over the course of the experiment, the gene expression of the main proteins of photosystems I and II was studied. Among the large number of analyzed genes (*psbA,B,C,D,S; petA,C,D,E; psaA,B; flvA,B; Lhc1,2*), only the transcription levels of *psbA* encoding the PSII key protein D1 changed significantly and reliably. Thus, in variants BRL and BL, a twofold increase in the level of *psbA* transcripts was observed, while RL did not cause changes in the expression of this gene (Figure 6A).

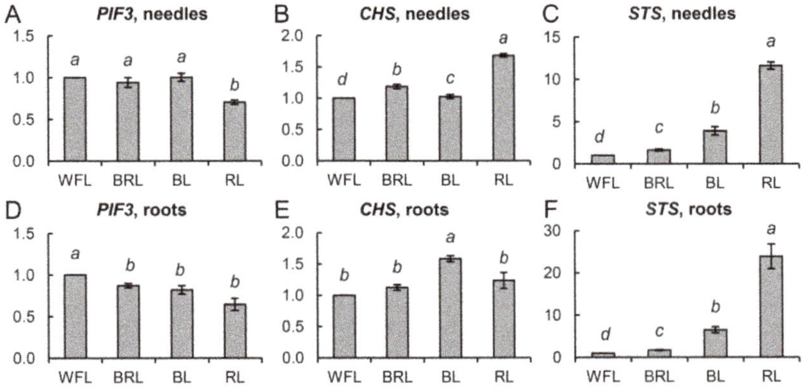

Figure 5. Effect of light quality on the transcript levels of different groups of genes in *P. sylvestris* seedlings. *PIF3, CHS,* and *STS* in needles(**A–C**), and roots (**D–E**). The mRNA levels of the genes were expressed as the BRL, BL, RL/WFL ratio (fold change BRL, BL, RL/WFL). Different italic letters denote statistically significant differences in the means at $p < 0.05$ (Kruskal–Wallis ANOVA of the ranks followed by the Student-Newman–Keuls post hoc test).

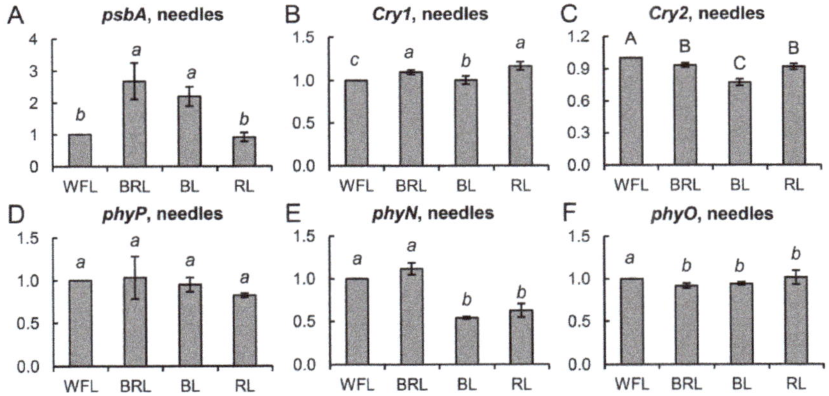

Figure 6. Effect of light quality on the transcript levels of different groups of genes in *P. sylvestris* needles. *psbA* (**A**), *Cry1* (**B**), and *Cry2* (**C**); and *phyP* (**D**), *phyN* (**E**), and *phyO* (**F**). The mRNA levels of the genes were expressed as the BRL, BL, RL/WFL ratio (fold change BRL, BL, RL/WFL). Values are the mean ± SE. Different capital letters denote statistically significant differences in the means at $p < 0.05$ (ANOVA followed by Duncan's method). Different italic letters denote statistically significant differences in the means at $p < 0.05$ (Kruskal–Wallis ANOVA of the ranks followed by the Student-Newman–Keuls post hoc test).

In addition, the gene expression of apoproteins of the main blue and red light photoreceptors was studied (Figure 6B–D,F). Significant differences were observed in the *phyN* gene transcripts, the level of which was reduced by almost 2-fold in BL and RL relative to the WFL control (Figure 6E). Significant but negligible differences were observed in the transcript levels of *Cry1* and *Cry2* genes (Figure 6B,C).

4. Discussion

Seedlings of *P. sylvestris* are photophilous; however, in the first few seasons of their life, they can only be higher than the surrounding herbaceous plants for a short period of time, which causes a lack of light, primarily in the red and blue spectral ranges. It is known that RL, in contrast to BL, has a significant effect on the stem growth, stem diameter, and size and dry weight of *Picea abies* needles [20]. Thus, in contrast to growing under BL and WFL, RL increased the biomass and stem diameter of *Brassica oleracea* plants [28]. At the same time, photosynthesis in RL plants was noticeably higher than in other variants, which probably led to an increasing biomass [28]. In our work, RL influenced the morphology of *P. sylvestris* seedlings, which was manifested in an increase in the mass of the root system by more than 3.8 times, complications of root branching, and an increase in the mass of the needles by 2.6 times in comparison with WFL. At the same time, RL caused a slight decrease in CO_2 gas exchange, as well as a significant (more than four times) decrease in respiration intensity relative to those of the WFL control (Table 2). This indicates a greater assimilation of carbon in RL plants. The slight difference in the pigment content is consistent with the fact that their photosynthetic apparatus is not dependent on light.

Regarding photochemical processes, the effective PSII quantum yield Y(II) was the highest in RL plants but the value of NPQ was small. We suppose that the increased Y(II) and complementary decreased NPQ are due to preferential excitation of photosystem I (PSI) by RL with a wavelength >685 nm. The contribution of the long-wave light is quite significant in red LEDs spectrum (Figure 1). Such light can lead to faster re-oxidization of the plastoquinone pool and reopening of PSII reaction centers [29]. As a result, reaction centers use absorbed light more efficiently.

Gymnosperms have the ability to synthesize chlorophyll in the dark due to the presence of the three genes of light-independent protochlorophyllide oxidoreductase L, N, and B (*ChlL*, *ChlN*, and *ChlB*) involved in the light-independent reduction of protochlorophyllide to chlorophyllide [30]. This also confirms the presence of special light regulation in conifers, which makes them a unique object of research.

The features of development and metabolism in plants induced by light of different spectral composition are primarily mediated by changes in the expression of light-dependent genes [20,30], including those encoding chloroplast proteins, photoreceptor apoproteins, transcription factors, and enzymes involved in the biosynthesis of secondary metabolites [31] and phytohormone signaling. In our work, we showed that several genes involved in the response to RL or BL are expressed in different ways. For example, the level of transcription of the *psbA* gene of the PSII main protein D1 increased in BL and BRL plants but decreased in RL plants (Figure 6A). This response to RL, as well as the accumulation of *psbA* transcripts under BL and RL conditions, is a typical response for most flowering plants [32]. Unfortunately, we did not observe noticeable changes in the expression of photoreceptor genes, with the exception of *phyN* (Figure 6B–F); on the other hand, under conditions of prolonged exposure to light of different spectra, it is impossible to exclude the presence of a sufficient number of active forms of photoreceptors, as well as the presence of regulation at the level of light signaling. In another work, it was shown that *phyN* is able to respond to different ratios of BL and RL in Scots pine seedlings of different growing regions [6].

The quality of light influences photomorphogenesis, photosynthesis, and plant growth through appropriate photoreceptors [33,34]. However, no significant difference in the expression level of photoreceptor genes was found between the light variants BL and RL, which is consistent with previous studies performed on *A. thaliana* seedlings [35]. The gene expression profiles of *A. thaliana* plants grown under BRL, RL, and BL were similar in all variants, and a significant proportion of differentially expressed genes under BL were also induced under RL. This indicates that the expression of light-regulated genes in *P. sylvestris* is not a unique response to BL or RL and that light of different spectral composition is able to regulate metabolic patterns in a similar way through the regulation of light signaling genes.

Transcription factors play an important role in the regulation of photosynthetic apparatus sensitivity to light of different spectral composition, since they, together with photoreceptors, are involved in the transduction of both light and hormonal signals. Phytochrome signaling transcription factors (PIFs) are important negative regulatory proteins that can alter the expression of a number of associated genes [6,36]. We observed a 30% decrease in the expression of the *PIF3* gene in RL plants, which may be associated with the activation of light signaling at the level of transcription of the corresponding genes (Figure 5A,D). These results demonstrate that the mechanisms by which light of different spectral composition controls the growth of *P. sylvestris* may involve angiosperm light signaling pathways. Transcription factors, together with photoreceptors, can influence the expression of hormonal signaling genes. Thus, a direct link between cytokinin signaling and light was found in a study demonstrating the key role of A-type RR regulators of cytokinin signaling [37–39]. RL induces *RR-A* expression in a PHYB-dependent manner. Thus, the overexpressing *RR-A* leads to hypersensitivity to RL [37]. Later research showed an important role for RR-A in photomorphogenesis [40,41]. RR-As are able to interact with PHYB via the cytokinin receptor (AHK). In support of this, we observed an increased expression of one of the main proteins of cytokinin signaling transduction (*HPT1*) by more than two times in roots and needles in the RL variant (Figure 4A,D). In addition, the PIFs are also involved in phytochrome-mediated regulation of auxin signaling under RL conditions, since these TFs are able to bind to the promoter regions of the *Aux/IAA* genes. PIFs modulate plant growth by directly controlling the expression of auxin signaling genes [11]. We assume that this, to a certain extent, explains the greater number of xylem cells in hypocotyls in the RL variant and, as a consequence, a greater accumulation of plant biomass (Figure 4C,F).

Along with the growth and development of plants, changes in the spectral composition of light also affect secondary metabolism. Chalcone synthase, the first enzyme in the biosynthesis of flavonoids, is expressed in needles of RL plants and in roots of BL plants (Figure 5B,E). In our study, RL stimulated gene families associated with the biosynthesis of the flavonoids (*CHS, STS*). Stilbenes are a family of polyphenolic secondary metabolites that act as phytoalexins [42,43]. Previously, it was shown that treatment with RL suspension culture of grape cells increased the biosynthesis of stilbenes [44,45]. In our study, we observed an increase in *STS* expression in roots and needles by more than 3.5 times in BL plants and more than 10 times in RL plants (Figure 5C,F). It can be assumed that, as in grape plants, stilbenes are involved in the photoadaptation of *P. sylvestris* plants to narrow-band RL and BL.

5. Conclusions

In this work, we tried to answer the question of what spectral range of light can be most favorable for the growth of *P. sylvestris* seedlings. It was shown that the RL spectral range is most favorable for growing *P. sylvestris* seedlings in a hydroculture, which manifested itself in both a greater mass of aboveground and underground organs and in an increase in the number of xylem cells. We found an increase in the level of transcripts of genes for auxin and cytokinin signaling (*HPT1, RR-A*, and *Aux/IAA*) and a decrease in the expression of TF *PIF3*, which in turn could activate the expression of a number of genes associated with the synthesis of secondary metabolites (*CHS, STS*). Based on the data obtained, we assumed that the large biomass of *P. sylvestris* plants under the RL might be due to a large accumulation of carbon in the needles, which corresponds to a better balance between photosynthesis and respiration, as well as to the increased activity of cytokinins and auxins in seedlings. Although *P. sylvestris* seedlings are able to grow in conditions of low RL content at the beginning of ontogenesis, better RL radiation can significantly improve their growth and development. The obtained results can serve as a basis for the development of a technology for the accelerated cultivation of planting material during reforestation when growing seedlings under artificial lighting, and they can also be used in biotechnology.

Author Contributions: Investigation, formal analysis, data curation, and responsible for the experimental part of the manuscript: P.P., V.D.K., Y.I., A.I., and A.K.; obtained main results: P.P., Y.I., A.I., A.S., V.S.; designed the experiments, interpreted the main results, drew the main conclusions, and prepared the first version of the manuscript: V.D.K., P.P., S.I.A. and V.V.K. All authors have read and agreed to the published version of the manuscript.

Funding: The research was carried out within the state assignment of Ministry of Science and Higher Education of the Russian Federation (theme No. 121040800153-1).

Institutional Review Board Statement: Not applicable.

Informed Consent Statement: Not applicable.

Data Availability Statement: The data presented in this study are available on request from the corresponding authors. The data are not public.

Conflicts of Interest: The authors declare no conflict of interest.

References

1. Casal, J.J. Shade avoidance. *Arab. Book* **2012**, *10*, e0157. [CrossRef]
2. Taulavuori, K.; Sarala, M.; Karhu, J.; Taulavuori, E.; Kubin, E.; Laine, K.; Poikolainen, J.; Pesonen, E. Elongation of Scots pine seedlings under blue light depletion. *Silva Fenn.* **2005**, *39*, 131–136. [CrossRef]
3. Hernández, R.; Kubota, C. Physiological responses of cucumber seedlings under different blue and red photon flux ratios using LEDs. *Environ. Exp. Bot.* **2016**, *121*, 66–74. [CrossRef]
4. Franklin, K.A. Photomorphogenesis: Plants feel blue in the shade. *Curr. Biol.* **2016**, *26*, R1275–R1276. [CrossRef] [PubMed]
5. Dyderski, M.K.; Paź, S.; Frelich, L.E.; Jagodziński, A.M. How much does climate change threaten European forest tree species distributions? *Glob. Chang. Biol.* **2018**, *24*, 1150–1163. [CrossRef]
6. Alakärppä, E.; Taulavuori, E.; Valledor, L.; Marttila, T.; Jokipii-Lukkari, S.; Karppinen, K.; Nguyen, N.; Taulavuori, K.; Häggman, H. Early growth of Scots pine seedlings is affected by seed origin and light quality. *J. Plant Physiol.* **2019**, *237*, 120–128. [CrossRef] [PubMed]
7. Ahmad, M. Photocycle and signaling mechanisms of plant cryptochromes. *Curr. Opin. Plant Biol.* **2016**, *33*, 108–115. [CrossRef]
8. Mathews, S. Evolutionary studies illuminate the structural-functional model of plant phytochromes. *Plant Cell.* **2010**, *22*, 4–16. [CrossRef] [PubMed]
9. Clapham, D.H.; Kolukisaoglu, H.Ü.; Larsson, C.T.; Qamaruddin, M.; Ekberg, I.; Wiegmann-Eirund, C.; Schneider-Poetsch, H.A.W.; von Arnold, S. Phytochrome types in *Picea* and *Pinus*. Expression patterns of PHYA-related types. *Plant Mol. Biol.* **1999**, *40*, 669–678. [CrossRef] [PubMed]
10. Clapham, D.H.; Ekberg, I.; Eriksson, G.; Norell, L.; Vince-Prue, D. Requirement for far-red light to maintain secondary needle extension growth in northern but not southern populations of *Pinus sylvestris* (Scots pine). *Physiol. Plantarum.* **2002**, *114*, 207–212. [CrossRef] [PubMed]
11. Zdarska, M.; Dobisová, T.; Gelová, Z.; Pernisová, M.; Dabravolski, S.; Hejátko, J. Illuminating light, cytokinin, and ethylene signalling crosstalk in plant development. *J. Exp. Bot.* **2015**, *66*, 4913–4931. [CrossRef] [PubMed]
12. Hornitschek, P.; Kohnen, M.V.; Lorrain, S.; Rougemont, J.; Ljung, K.; López-Vidriero, I.; Franco-Zorrilla, J.M.; Solano, R.; Trevisan, M.; Pradervand, S.; et al. Phytochrome interacting factors 4 and 5 control seedling growth in changing light conditions by directly controlling auxin signaling. *Plant J.* **2012**, *71*, 699–711. [CrossRef] [PubMed]
13. Lau, O.S.; Deng, X.W. Plant hormone signaling lightens up: Integrators of light and hormones. *Curr. Opin. Plant Biol.* **2010**, *13*, 571–577. [CrossRef] [PubMed]
14. Kurepin, L.V.; Emery, R.N.; Pharis, R.P.; Reid, D.M. The interaction of light quality and irradiance with gibberellins, cytokinins and auxin in regulating growth of *Helianthus annuus* hypocotyls. *Plant Cell Environ.* **2007**, *30*, 147–155. [CrossRef] [PubMed]
15. Pierik, R.; Keuskamp, D.H.; Sasidharan, R.; Djakovic-Petrovic, T.; de Wit, M.; Voesenek, L.A. Light quality controls shoot elongation through regulation of multiple hormones. *Plant Signal. Behav.* **2009**, *4*, 755–756. [CrossRef] [PubMed]
16. Kurepin, L.V.; Walton, L.J.; Hayward, A.; Emery, R.N.; Pharis, R.P.; Reid, D.M. Interactions between plant hormones and light quality signaling in regulating the shoot growth of *Arabidopsis thaliana* seedlings. *Botany* **2012**, *90*, 237–246. [CrossRef]
17. Gubler, F.; Hughes, T.; Waterhouse, P.; Jacobsen, J. Regulation of dormancy in barley by blue light and after-ripening: Effects on abscisic acid and gibberellin metabolism. *Plant Physiol.* **2008**, *147*, 886–896. [CrossRef]
18. Reddy, S.K.; Finlayson, S.A. Phytochrome B promotes branching in Arabidopsis by suppressing auxin signaling. *Plant Physiol.* **2014**, *164*, 1542–1550. [CrossRef]
19. Zhang, Z.; Ji, R.; Li, H.; Zhao, T.; Liu, J.; Lin, C.; Liu, B. CONSTANS-LIKE 7 (COL7) is involved in Phytochrome B (phyB)-mediated light-quality regulation of auxin homeostasis. *Mol. Plant* **2014**, *7*, 1429–1440. [CrossRef]
20. OuYang, F.; Mao, J.F.; Wang, J.; Zhang, S.; Li, Y. Transcriptome analysis reveals that red and blue light regulate growth and phytohormone metabolism in Norway spruce [*Picea abies* (L.) Karst.]. *PLoS ONE* **2015**, *10*, e0127896. [CrossRef] [PubMed]
21. Ivanov, Y.V.; Kartashov, A.V.; Ivanova, A.I.; Savochkin, Y.V.; Kuznetsov, V.V. Effects of zinc on Scots pine (*Pinus sylvestris* L.) seedlings grown in hydroculture. *Plant Physiol. Bioch.* **2016**, *102*, 1–9. [CrossRef]
22. Lichtenthaler, H.K. [34] Chlorophylls and carotenoids: Pigments of photosynthetic biomembranes. In *Methods in Enzymology*; Packer, L., Douce, R., Eds.; Academic Press: Cambridge, MA, USA, 1987; Volume 148, pp. 350–382. [CrossRef]

23. Lankin, A.V.; Kreslavski, V.D.; Khudyakova, A.Y.; Zharmukhamedov, S.K.; Allakhverdiev, S.I. Effect of naphthalene on photosystem 2 photochemical activity of pea plants. *Biochemistry* **2014**, *79*, 1216–1225. [CrossRef]
24. Goltsev, V.N.; Kalaji, H.M.; Paunov, M.; Bąba, W.; Horaczek, T.; Mojski, J.; Kociel, H.; Allakhverdiev, S.I. Variable chlorophyll fluorescence and its use for assessing physiological condition of plant photosynthetic apparatus. *Russ. J. Plant Physiol.* **2016**, *63*, 869–893. [CrossRef]
25. Kolosova, N.; Miller, B.; Ralph, S.; Ellis, B.E.; Douglas, C.; Ritland, K.; Bohlmann, J. Isolation of high-quality RNA from gymnosperm and angiosperm trees. *Biotechniques* **2004**, *36*, 821–824. [CrossRef]
26. Pashkovskiy, P.P.; Vankova, R.; Zlobin, I.E.; Dobrev, P.; Ivanov, Y.V.; Kartashov, A.V.; Kuznetsov, V.V. Comparative analysis of abscisic acid levels and expression of abscisic acid-related genes in Scots pine and Norway spruce seedlings under water deficit. *Plant Physiol. Bioch.* **2019**, *140*, 105–112. [CrossRef] [PubMed]
27. Mitra, P.P.; Loqué, D. Histochemical staining of *Arabidopsis thaliana* secondary cell wall elements. *J. Vis. Exp.* **2014**, *87*, 51381. [CrossRef]
28. Chen, X.W.; Liu, S.Q.; Wang, Y.; Liu, J.K.; Feng, L. Effects of different LED light qualities on growth, photosynthetic characteristics and nutritional quality of savoy. *Ying Yong Sheng Tai Xue Bao* **2014**, *25*, 1955–1962.
29. Zhen, S.; van Iersel, M.W. Far-red light is needed for efficient photochemistry and photosynthesis. *J. Plant Physiol.* **2017**, *209*, 115–122. [CrossRef] [PubMed]
30. Ranade, S.S.; Delhomme, N.; García-Gil, M.R. Global gene expression analysis in etiolated and de-etiolated seedlings in conifers. *PLoS ONE* **2019**, *14*, e0219272. [CrossRef] [PubMed]
31. Gyula, P.; Schäfer, E.; Nagy, F. Light perception and signalling in higher plants. *Curr. Opin. Plant Biol.* **2003**, *6*, 446–452. [CrossRef]
32. Zheng, J.; Hu, M.J.; Guo, Y.P. Regulation of photosynthesis by light quality and its mechanism in plants. *Ying Yong Sheng Tai Xue Bao* **2008**, *19*, 1619–1624. [PubMed]
33. Kreslavski, V.D.; Strokina, V.V.; Pashkovskiy, P.P.; Balakhnina, T.I.; Voloshin, R.A.; Alwasel, S.; Kosobryukhov, A.A.; Allakhverdiev, S.I. Deficiencies in phytochromes A and B and cryptochrome 1 affect the resistance of the photosynthetic apparatus to high-intensity light in *Solanum lycopersicum*. *J. Photoch. Photobio. B* **2020**, *210*, 111976. [CrossRef]
34. Kreslavski, V.D.; Strokina, V.V.; Khudyakova, A.Y.; Shirshikova, G.N.; Kosobryukhov, A.A.; Pashkovskiy, P.P.; Alwasel, S.; Allakhverdiev, S.I. Effect of high-intensity light and UV-B on photosynthetic activity and the expression of certain light-responsive genes in *A. thaliana phyA* and *phyB* mutants. *BBA-Bioenergetics* **2021**, *1862*, 148445. [CrossRef] [PubMed]
35. Ma, L.; Li, J.; Qu, L.; Hager, J.; Chen, Z.; Zhao, H.; Deng, X.W. Light control of Arabidopsis development entails coordinated regulation of genome expression and cellular pathways. *Plant Cell* **2001**, *13*, 2589–2607. [CrossRef] [PubMed]
36. Rohde, A.; Ruttink, T.; Hostyn, V.; Sterck, L.; Van Driessche, K.; Boerjan, W. Gene expression during the induction, maintenance, and release of dormancy in apical buds of poplar. *J. Exp. Bot.* **2007**, *58*, 4047–4060. [CrossRef] [PubMed]
37. Sweere, U.; Eichenberg, K.; Lohrmann, J.; Mira-Rodado, V.; Bäurle, I.; Kudla, J.; Nagy, F.; Schäfer, E.; Harter, K. Interaction of the response regulator ARR4 with phytochrome B in modulating red light signaling. *Science* **2001**, *294*, 1108–1111. [CrossRef]
38. Fankhauser, C. Light perception in plants: Cytokinins and red light join forces to keep phytochrome B active. *Trends Plant Sci.* **2002**, *7*, 143–145. [CrossRef]
39. Al-Sady, B.; Ni, W.; Kircher, S.; Schäfer, E.; Quail, P.H. Photoactivated phytochrome induces rapid PIF3 phosphorylation prior to proteasome-mediated degradation. *Mol. Cell* **2006**, *23*, 439–446. [CrossRef]
40. Thomas, T.H.; Hare, P.D.; van Staden, J. Phytochrome and cytokinin responses. *Plant Growth Regul.* **1997**, *23*, 105–122. [CrossRef]
41. Mira-Rodado, V.; Sweere, U.; Grefen, C.; Kunkel, T.; Fejes, E.; Nagy, F.; Schäfer, E.; Harter, K. Functional cross-talk between two-component and phytochrome B signal transduction in Arabidopsis. *J. Exp. Bot.* **2007**, *58*, 2595–2607. [CrossRef] [PubMed]
42. Dubrovina, A.S.; Kiselev, K.V. Regulation of stilbene biosynthesis in plants. *Planta* **2017**, *246*, 597–623. [CrossRef]
43. Valletta, A.; Iozia, L.M.; Leonelli, F. Impact of environmental factors on stilbene biosynthesis. *Plants* **2021**, *10*, 90. [CrossRef] [PubMed]
44. Tassoni, A.; Durante, L.; Ferri, M. Combined elicitation of methyl-jasmonate and red light on stilbene and anthocyanin biosynthesis. *J. Plant Physiol.* **2012**, *169*, 775–781. [CrossRef] [PubMed]
45. Ahn, S.Y.; Kim, S.A.; Choi, S.J.; Yun, H.K. Comparison of accumulation of stilbene compounds and stilbene related gene expression in two grape berries irradiated with different light sources. *Hortic. Environ. Biotechnol.* **2015**, *56*, 36–43. [CrossRef]

cells

MDPI

Article

The Combined Effect of ZnO and CeO_2 Nanoparticles on *Pisum sativum* L.: A Photosynthesis and Nutrients Uptake Study

Elżbieta Skiba [1,*], Monika Pietrzak [1], Sława Glińska [2] and Wojciech M. Wolf [1]

[1] Institute of General and Ecological Chemistry, Lodz University of Technology, 90-924 Lodz, Poland; monika.pietrzak@dokt.p.lodz.pl (M.P.); wojciech.wolf@p.lodz.pl (W.M.W.)

[2] Laboratory of Microscopic Imaging and Specialized Biological Techniques, Faculty of Biology and Environmental Protection, University of Lodz, 90-237 Lodz, Poland; slawa.glinska@biol.uni.lodz.pl

* Correspondence: elzbieta.skiba@p.lodz.pl; Tel.: +48-(42)-6313123

Abstract: Cerium oxide nanoparticles (CeO_2 NPs) and zinc oxide nanoparticles (ZnO NPs) are emerging pollutants that are likely to occur in the contemporary environment. So far, their combined effects on terrestrial plants have not been thoroughly investigated. Obviously, this subject is a challenge for modern ecotoxicology. In this study, *Pisum sativum* L. plants were exposed to either CeO_2 NPs or ZnO NPs alone, or mixtures of these nano-oxides (at two concentrations: 100 and 200 mg/L). The plants were cultivated in hydroponic system for twelve days. The combined effect of NPs was proved by 1D ANOVA augmented by Tukey's post hoc test at $p = 0.95$. It affected all major plant growth and photosynthesis parameters. Additionally, HR-CS AAS and ICP-OES were used to determine concentrations of Cu, Mn, Fe, Mg, Ca, K, Zn, and Ce in roots and shoots. Treatment of the pea plants with the NPs, either alone or in combination affected the homeostasis of these metals in the plants. CeO_2 NPs stimulated the photosynthesis rate, while ZnO NPs prompted stomatal and biochemical limitations. In the mixed ZnO and CeO_2 treatments, the latter effects were decreased by CeO_2 NPs. These results indicate that free radicals scavenging properties of CeO_2 NPs mitigate the toxicity symptoms induced in the plants by ZnO NPs.

Keywords: combined effect; *Pisum sativum* L.; nanoparticles; cerium oxide; zinc oxide; metal uptake; photosynthesis; hydroponic culture

check for updates

Citation: Skiba, E.; Pietrzak, M.; Glińska, S.; Wolf, W.M. The Combined Effect of ZnO and CeO_2 Nanoparticles on *Pisum sativum* L.: A Photosynthesis and Nutrients Uptake Study. *Cells* **2021**, *10*, 3105. https://doi.org/10.3390/cells10113105

Academic Editors: Suleyman Allakhverdiev, Alexander G. Ivanov and Marian Brestic

Received: 20 September 2021
Accepted: 5 November 2021
Published: 10 November 2021

1. Introduction

Nowadays, synthetic nanomaterials (NMs) are finding applications in almost all material aspects of human life and are particularly vital for contemporary technology and medicine [1–5]. The most important are sustainable energy production and storage, electronic devices, catalysts, sensors and adhesives, high-quality petro- and agrochemicals [6–14]. Their widespread use raises fundamental questions related to the environment, pollution, and safety [15–17].

In the diverse world of nanomaterials metal oxides firmly occupy a unique position. Their global market in 2020 was worth almost USD 5.3 billion with several forecasts indicating its steady rise to USD 9.3 billion in 2027 [18]. Obviously, this growing stream of NMs cannot be completely isolated from soil, air, or water and finally will find its way to terrestrial living organisms. Therefore, complex interactions of nanoparticles (NPs) with the biota and their further environmental fate deserve additional studies.

In a number of comprehensive papers on interactions of nanoparticles with living organisms, the authors investigated the role of chemical composition, particle shape, size, and mechanisms of aggregation of NPs [19–22]. The beneficial or harmful effects of NPs were also examined. Owing to their unique properties, nanomaterials can be successfully used in fertilizers, pesticides, or as dedicated chemical carriers or sensors [23–25]. This increasing flux of diverse nanomaterials approaching soil and plant environments should not be left without comprehensive studies on their migrations, uptake, and toxicities.

The emerging picture suggests dynamic processes which act in complicated matrices. Identification of interactions among various substances poses a real challenge.

Obviously, nanoparticles affect plant metabolism in a number of ways and finally introduce changes to plant physiology at cellular, organ, and individual plant levels [26–29]. In particular, photosynthesis—the essential, bioenergy-generating process [30]—may be either facilitated or hampered by nanoparticles [31–36]. As pointed out by Du et al. [37] and Tighe-Neira et al. [35], the latter may be conveniently examined by gas exchange parameters augmented with the contents of photosynthetic pigments. Unfortunately, results are rarely completely consistent and a wide range of plant responses to nanoparticles could be expected. Notably, metal oxide NPs usually alter the photosynthesis rate, photochemical fluorescence, and quantum yield in plants [37] with ZnO and CeO_2 being among the most active. Mukherjee et al. [38] described the negative effect of ZnO NPs at soil concentrations 125–500 mg/L on chlorophyll activity in *Pisum sativum* L. It results from the substitution of the Mg atoms at chlorophyll centers by Zn which finally hampers the photosynthesis process. The opposite effect was observed by Reddy Pullagurala et al. [39] during cilantro (*Coriandrum sativum*) cultivation. Supplementation with 100 or 200 mg/L ZnO NPs induced photosynthetic pigments production and boosted photosynthesis.

The ambient character of nanoparticulate cerium oxide combined with its increasing abundance in the environment makes this substance a useful agent to study plant metabolism effects triggered by anthropogenic nanomaterials. In particular, Wu et al. [32] demonstrated that CeO_2 NPs (nanoceria) are a reactive oxygen species (ROS) scavenger in leaf mesophyll cells and defend the chloroplast photosynthetic machinery from abiotic stresses. On the other hand, the negative effect of nanoceria on photosynthesis in soybean plants was investigated by Li et al. [40]. They pointed out several aspects which are at stake, namely: inhibited conversion efficiency of C5 to C3 in the Calvin–Benson cycle, destruction of thylakoid membranes, and reduced chlorophyll synthesis and activity. All these factors participate in the final plant destruction.

Regrettably, the most relevant works on the subject are aimed at particular types of nanoparticles. Investigations of combined, mutual effects induced by either mixture of NPs or their additives and stabilizers are quite scarce. In real ecosystems, nanoparticles rarely play a solo performance, which is especially important in modern efficient agriculture. Increasing pressure on massive food production for the rapidly growing population has prompted the development of new, smart, and efficient agrochemicals [41–44].

The fertilizing effect of CeO_2 and ZnO NPs attracted the attention of several research groups. It is quite well documented that cerium oxide NPs can alleviate plant salinity stress, act as a catalyst in chlorophyll production, and in scavenging reactive oxygen species which stabilize the chloroplast structure and cell walls [45–47]. On the other hand, nanoparticulate zinc oxide may be used to counteract cadmium toxicity in wheat and elevate zinc concentrations in plants. Thus, it can be a useful agent for Zn biofortification in cereals plantations and help to overcome the well recognized hidden hunger in humans resulting from the deficiency of Zn in cereals [25]. In this respect, optimization of photosynthetic efficiency by nanoparticles is a highly promising approach for a smart increase of crop production [48].

Mixtures of nanoparticles that are present in the growth environment affect plant development in a different way than single species [49–51]. In particular, the combined, mutual interactions between ZnO and CeO_2 are quite likely indeed. Both substances often coexist in all compartments of the environment and their simultaneous presence in the environment is more than likely. Our previous works on *Pisum sativum* L. were related to metal migration strategies as induced by single stressors such as CeO_2 or ZnO nanoparticles [52,53]. We have shown that zinc species alter Cu, Mn, and Fe uptake and their further migration in green pea. On the other hand, low concentrations of cerium oxide NPs increased the photosynthesis rate. Those investigations had a model character and did not account for combined effects as triggered by ZnO and CeO_2 together in real plant matrices. This study enhances this approach significantly. Here, we examine the mixture of nanometric cerium and zinc oxides. The methodology has been based on hydroponic pot experiments [54]. Pea is frequently applied in system biology experiments and is treated as a non-model plant with a roughly complete genome structure [55,56]. It is cultivated worldwide [57] and additive interactions which affect nutrient uptake and enhance plant growth yield are of practical relevance.

2. Materials and Methods

2.1. Nanoparticles Characteristic

CeO_2 NPs and ZnO NPs were purchased from the Byk (Byk-Chemie GmbH, Wesel, Germany) as commercially available products. The properties of nanoparticles, including average particle size, transmission electron microscopy (TEM) images, and zeta potential are given in Table S1 [52,54].

2.2. Plants Growth Conditions and Treatments

Pisum sativum L. plants were cultivated under hydroponic conditions in Hoagland's nutrient solution. The composition of the growing medium and the detailed setup of the experiment were described previously [53,54]. The seeds, ("Iłówiecki" sugar pea, "PNOS" Co., Ltd., Ożarów Mazowiecki, Poland), sterilized in 70% ethanol were germinated in dark for 3 days. Next, the seedlings (at BBCH 09 phenological stage [58]) were transferred to perforated plastic plates and placed in containers with the nutrient solution. Each growing vessel contained 26 pea seedlings. *Pisum sativum* L. was grown in Hoagland's nutrient solution for 4 days in a controlled environment: light of 170 $\mu mol/m^2$ intensity, average temperature 21 $\pm$ 3 °C, and 16/8 h day/night photoperiod. On the 5th day, the nutrient solutions were supplemented with nanoparticles. The treatments were as follows, CeO_2 NPs: 100 mg (Ce)/L; CeO_2 NPs: 200 mg (Ce)/L; ZnO NPs: 100 mg (Zn)/L; ZnO NPs: 200 mg (Zn)/L; two mixtures of CeO_2 and ZnO NPs: 100 mg (Ce)/L + 100 mg (Zn)/L and 200 mg (Ce)/L + 200 mg (Zn)/L. Plants grown in Hoagland's nutrient solution served as the control group. In each variant of the experiment, six growing vessels were used. Fresh liquid media were supplied every 48 h. *Pisum sativum* L. was harvested after 12 days of exposure when plants reached the BBCH 15 phenological stage.

2.3. Growth Parameters

Root and stem lengths were measured at the end of cultivation (Figure 1). Next, the roots were thoroughly rinsed with deionized water and separated from the shoots. The fresh weight of roots and shoots was measured and calculated per single plant (mg/plant). Prior to the chemical analysis, the collected fresh plants material was dried to a constant weight at 55 °C.

2.4. Elements Content

The dried samples of roots and shoots were digested in the mixture of concentrated HNO_3 and HCl (6:1, *v*/*v*) using a Multiwave 3000 Anton Paar microwave reaction system (Anton Paar GmbH, Graz, Austria). The concentrations of Cu, Mn, Zn, Fe, and Mg were determined by a High-Resolution Continuum Source Atomic Absorption spectrometer HR-CS AAS (contrAA300, Analytik Jena, Jena, Germany). Additionally, the digested solutions were analyzed for Ce, Ca, and K concentrations by an Inductively Coupled Plasma–Optical Emission spectrometer ICP-OES (PlasmaQuant PQ 9000, Analytik Jena, Jena, Germany). Certified reference material of plant leaves of Oriental Basma Tobacco Leaves (INCT-OBTL-5) obtained from the Institute of Nuclear Chemistry and Technology, Warsaw [59] was used to check the reliability of the applied analytical procedures. Recoveries ranging from 96% to 108% were obtained. The detailed numerical data are given in Table S2.

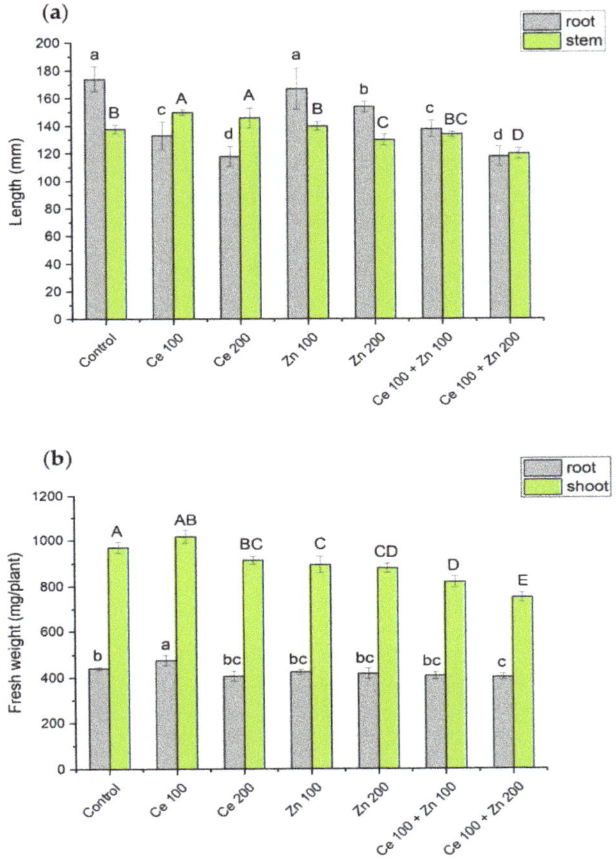

Figure 1. The influence of nanoparticulate CeO_2 and ZnO on: (**a**) root and stem length and (**b**) fresh weight of root and shoot as determined for a single pea plant. Concentrations of nanoparticles are given in mg/L of elemental cerium or zinc, the cultivation time was 12 days. Roots are represented by grey while above-ground parts are in green. Vertical bars represent standard deviations ($n = 6$). Distinct letters show the statistically significant differences among treatments as calculated with Tukey's HSD post hoc test. The probability level $p = 0.95$ was applied.

2.5. Tolerance Index (TI) and Translocation Factor (TF)

Tolerance indices (TI) and translocation factors (TF) were calculated for plants treated with nanoparticulate oxides. TI is defined as the ratio between the root length of treated plants and that of plants in the control group [60,61]. The effectiveness of metal translocation from roots to shoots was assessed using TF, defined as the ratio of average element content in shoots to that in roots [62,63].

2.6. Photosynthetic Pigments

Contents of photosynthetic pigments, i.e., chlorophyll a and b and carotenoids in *Pisum sativum* L. leaves were determined following the method developed by Oren et al. [64]. All pigments were extracted from leaves with 80% acetone and kept in the dark at 4 °C. The absorbance of extracts was measured at wavelengths 470, 646, and 663 nm by SpectraMaxi3x (Molecular Devices, San Jose, CA, USA) on samples centrifugated at 4000 rpm for 10 min. Contents of pigments were calculated by the formula proposed by Lichtenthaler and Wellburn [65]. For comparison, the nondestructive measurements of relative chlorophyll content in fully expanded leaves at the fourth node were performed by the Soil Plant Analysis Development SPAD-502Plus chlorophyll meter (Konica–Minolta, Inc., Osaka, Japan).

2.7. Photosynthetic Parameters

Photosynthetic parameters, such as: leaf net photosynthesis (A), sub-stomatal CO_2 concentration (C_i), transpiration (E), stomatal conductance (gs), photosynthetic water use efficiency (WUE), and photosynthetic CO_2 response curve (A/C_i) were determined with a portable gas exchange analyzer (CIRAS-3; PP Systems, Amesbury, MA, USA). Measurements were performed 12 days after the administration of nanoparticles on fully expanded leaves at the fourth node. PARi levels at 1000 μmol/m^2s were obtained from an LED Light Unit (RGBW) connected to the gas analyzer. The temperature within the chamber was kept at 25 °C, relative humidity at 80%, and reference CO_2 concentration at 390 μmol/mol. Photosynthetic CO_2 response curves were collected at a CO_2 concentration gradient ranging from 0 to 1500 μmol/mol. All measurements were performed on fully expanded leaves at the fourth node. The acclimatization time between measurements was 120 s. The results from each CO_2 level were recorded three times. Two biochemical parameters: maximum carboxylation rate (V_{cmax}) and maximum electron transport rate (J_{max}) were calculated in Rstudio (v.3.4.2; R Foundation for Statistical Computing, Vienna, Austria) [66] using the "plantecophys" package developed by Duursma [67].

2.8. Statistical Analysis

All treatments were replicated six times, numerical results are accompanied by their ± SD (standard deviation). Statistical analysis was performed with the OriginPro 2016 (OriginLab Corporation, Northampton, MA, USA) software. The normality of the sample distributions was proved by the Shapiro–Wilk test [68]. The initial hypothesis on equal variances of investigated populations was validated with the Bartlett test [69]. A one-way ANOVA with Tukey's post hoc approach was applied to validate differences between means. The probability level $p = 0.95$ was applied.

3. Results and Discussion

3.1. Growth Parameters

Several plant growth parameters were determined after twelve days of exposure time to illustrate the pea plant behavior under the combined treatments of nanoparticulate CeO_2 and ZnO. Roots and stem lengths and their fresh weights are shown in Figure 1.

Almost all supplementations decreased roots length of pea as compared to the control treatment. Nevertheless, this effect is less pronounced for plants cultivated with the addition of ZnO (100 and 200 mg/L) than for the CeO_2 (100 and 200 mg/L) treatments. It is directly correlated with the nanoparticles concentration applied. Notably, supplementation with mixed NPs resulted in roots shortening similar to that observed for relevant sole nanoceria concentrations. The order of tolerance indices (TI) supports above observations: (Zn 100) ≈ (Zn 200) > (Ce 100 + Zn 100) ≈ (Ce 100) > (Ce 200 + Zn 200) ≈ (Ce 200). Numbers in brackets represent supplementations of nanoparticles given in mg/L of elemental cerium or zinc, numerical values are presented in Table S3. Stem elongation was triggered by (Ce 100) and (Ce 200). A reverse effect was observed for (Zn 200) and (Ce 200 + Zn 200) administrations. Notably, the majority of treatments left the root mass unchanged with (Ce 100) and (Ce 200 + Zn 200) being the chief exceptions (Figure 1b). The biomass of the above-ground parts decreased upon all supplementations. The only exception was (Ce 100) with a biomass similar to that of a control sample.

3.2. Cerium and Zinc Concentrations

Cerium and zinc concentrations in roots and shoots along with their translocation factors are summarized in Table 1. The highest cerium content was determined for (Ce 200) treatment while the lowest levels were observed for combined (Ce 100 + Zn 100) and

(Ce 200 + Zn 200) administrations. Respective translocation factors were remarkably low, which indicates that plants accumulate cerium primarily in roots. Cerium was not detected in plants grown in Hoagland's nutrient solutions which were not supplemented with this element. Zinc is an essential element necessary for proper plant development [70,71]. The amounts of zinc accumulated by pea cultivated in solutions supplemented with the ZnO NPs alone were substantially larger than those observed for control samples. Combined treatments, namely (Ce 100 + Zn 100) and (Ce 200 + Zn 200) decreased Zn levels in both roots and shoots as compared to respective sole ZnO NPs administrations. However, all relevant concentrations highly exceeded critical toxic zinc level 300 mg/kg as suggested by Broadley et al. [70,71]. Similar to nanoceria administrations, the vast majority of zinc was immobilized in roots.

Table 1. Zinc and cerium concentrations in root and shoot of *Pisum sativum* L. grown under the sole or combined CeO_2 and ZnO NPs treatments accompanied by translocation factors (TF). Concentrations of nanoparticles are given in mg/L of elemental cerium or zinc, respectively. The cultivation time was 12 days. Data are the means ± SD ($n = 6$). Distinct letters show the statistically significant differences among treatments as calculated with Tukey's HSD post hoc test ($p = 0.95$). nd: concentration below the detection limit (18 µg/L).

Treatment	Zinc Concentration (mg/kg DW)		TF	Cerium Concentrations (mg/kg DW)		TF
	Root	Shoot		Root	Shoot	
Control	97.8 ± 4.5 d	61.4 ± 1.1 d	0.63	nd	nd	nd
Ce 100	74.3 ± 5.5 d	52.4 ± 2.3 d	0.71	14,874 ± 495 b	110 ± 11 c	0.01
Ce 200	68.3 ± 6.2 d	51.5 ± 2.2 d	0.75	16,977 ± 1067 a	206 ± 14 a	0.01
Zn 100	25,921 ± 2307 b	1160 ± 61 bc	0.04	nd	nd	nd
Zn 200	30,522 ± 2140 a	1455 ± 54 a	0.05	nd	nd	nd
Ce 100 + Zn 100	18,192 ± 703 c	1079 ± 113 c	0.06	6058 ± 824 d	120 ± 7 bc	0.02
Ce 200 + Zn 200	28,097 ± 3773 ab	1236 ± 88 b	0.04	10,301 ± 1484 c	142 ± 10 b	0.01

3.3. Photosynthetic Pigments

Measurements of photosynthetic pigments provide useful information on the physiological status of plants [72]. Concentrations of chlorophylls and carotenoids which are involved in the absorption and further transfer of light energy are prone to changes induced by inorganic chemical stressors [73]. Contents of chlorophyll a (Chl a), chlorophyll b (Chl b), and carotenoids (Car) in leaves of pea treated with nano-oxides are summarized in Figure 2a.

Significant differences among those parameters were observed. Almost all treatments prompted a decrease in photosynthetic pigments. The highest reduction was observed under (Zn 200) administration. Similar results were obtained by Hu et al. [74] who found that the lowest level of photosynthetic pigments in *Salvinia natans* grown in hydroponic conditions was reached in cultures supplemented with nanometric ZnO at 50 mg/L. Pea plants exposed to (Zn 100) and (Zn 200) showed initial symptoms of leaf chlorosis (Figure 2b). The chlorophyll loss associated with chlorosis is one of the typical symptoms of Zn excess in plants [75–77]. In this study, Zn levels in pea shoots under (Zn 100) and (Zn 200) administrations (Table 1) were significantly higher than the critical Zn toxicity level (>300 mg/kg) as quoted in the highly respected Marschner's Mineral Nutrition of Higher Plants [71].

Moreover, the lowest Chl a, Chl b, and Car concentrations were observed for the sole (Zn 200) treatment. Acute effects of nanoparticulate (Zn 200) were reduced by the (Ce 200) addition. It prompted the mutual, combined interactions which elevated photosynthetic pigments levels and significantly relieved leaf chlorosis symptoms. Following Wang et al. [75] and Broadley et al. [70], Zn may induce chlorosis of leaves by stimulating Mg, Fe, and Mn deficiency. These elements are crucial in the synthesis and stability of chlorophyll. A decline of green photosynthetic pigments in plants under ZnO NPs treatment was also reported by Zoufan et al. [78] who explained it by peroxidation of the chloroplast membrane due to exacerbation of oxidative stress. Oxidative damage triggered by ZnO NPs was also reported by Salehi et al. [79]. In turn, CeO_2 NPs have the ability to quench ROS, mainly produced in chloroplasts [80–82], and presumably mitigate stress induced by nanoparticulate ZnO.

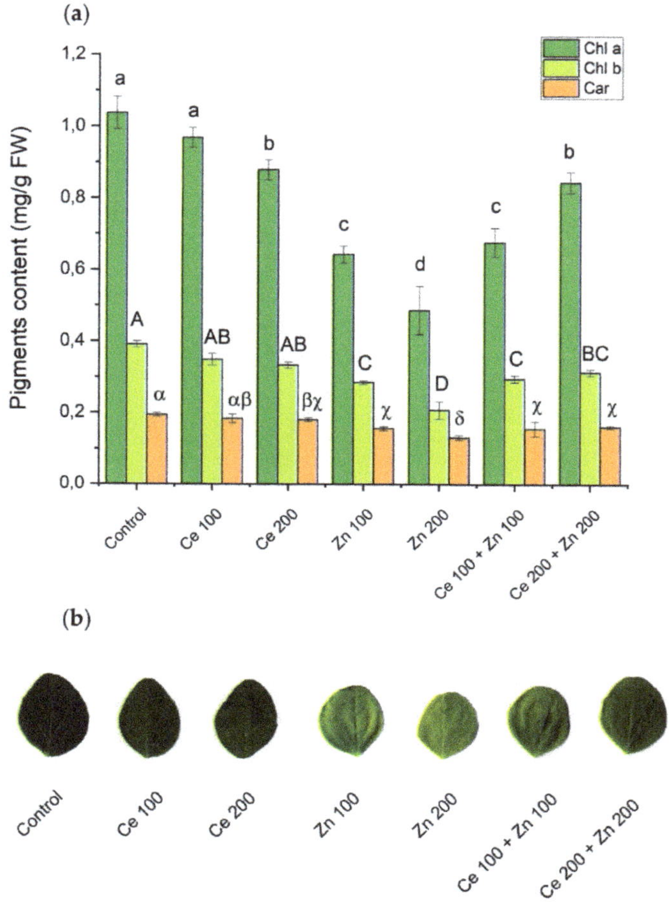

Figure 2. The influence of nanoparticulate CeO_2 and ZnO on: (**a**) photosynthetic pigments: chlorophyll a (Chl a), chlorophyll b (Chl b), and carotenoids (Car) in *Pisum sativum* L. The pigments were extracted from mature leaves after 12 days of contact with a particular nano-oxide. Vertical bars represent standard deviations ($n = 6$). Treatments with the same letter are not significantly different according to Tukey's post hoc test ($p = 0.95$) (**b**) appearance of pea leaves from each treatment.

Apart from the conventional wet chemical methods based on extraction of the chlorophyll pigments and their further spectrophotometric determination, the nondestructive measurements using the Soil Plant Analysis Development SPAD-502Plus chlorophyll meter were also performed (Figure 3). The results obtained by all those methods are highly correlated. The reduction in chlorophyll content expressed in SPAD units in *Pisum sativum* L., cultivated in soil supplemented by nanoparticulate ZnO, was also observed by Mukherjee et al. [38], and following Küpper et al. [83] was attributed to the Mg substitution by Zn.

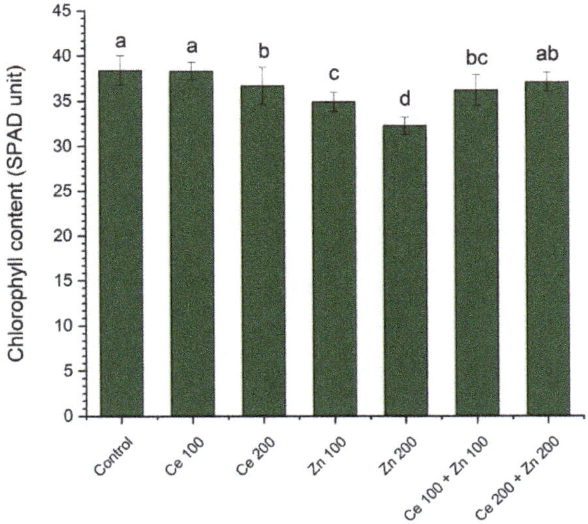

Figure 3. The influence of nanoparticulate CeO_2 and ZnO on chlorophyll content in *Pisum sativum* L. expressed in SPAD units. All measurements were performed on mature leaves on the 12th day from administration of the nanoparticles. Vertical bars represent standard deviations (n = 10). Treatments with the same letter are not significantly different according to Tukey's post hoc test (p = 0.95).

3.4. Photosynthesis Parameters

Photosynthesis efficiency is usually assessed with the gas exchange analysis of plant leaves [84–86]. Leaf net photosynthesis (A), transpiration rate (E), stomatal conductance (gs), sub-stomatal CO_2 concentration (C_i), and water use efficiency (WUE) determined under either sole or combined nanoparticles treatments are collected in Figure 4. Net photosynthesis was encouraged by CeO_2 supplementations and significantly reduced by ZnO additions, Figure 4a. The latter follows reduced chlorophyll content as induced by elevated zinc concentrations [75,87]. This effect is partially released by the nanoceria addition in combined (Ce 100 + Zn 100) and (Ce 200 + Zn 200) treatments. Zinc is an essential metal necessary for proper plant development [70]. However, above certain concentrations, it affects chlorophyll synthesis and photosystem II efficiency [88,89]. Transpiration and stomatal conductance behave in a similar way with the highest values observed for (Ce 100) and (Ce 200) treatments and their substantial decline induced by either (Zn 100) and (Zn 200) supplementations (Figure 4b,c). The sub-stomatal CO_2 concentrations exhibit a more uniform pattern and are significantly less prone to ZnO supplementations (Figure 4d). The largest water use efficiency was observed for (Ce 100) and (Zn 100) treatments, Figure 4e. High concentrations of (Ce 200) and (Zn 200) nanoparticles prompted a decrease in WUE. Maximum rates of ribulose 1,5-bisphosphate (RuBP) carboxylation as catalyzed by Rubisco (V_{cmax}), electron transfer driving the regeneration of RuBP (J_{max}), and CO_2 compensation points for pea plants cultivated in Hoagland's solution supplemented with either sole or combined CeO_2 and ZnO NPs are in Table 2 and Supplementary Figure S1. The highest V_{cmax} values were observed for the sole nanoceria augmentations [(Ce 100) and (Ce 200)] while the strongest decreases were induced by the sole (Zn 100) and (Zn 200) NPs additions. The electron transfer rates J_{max} followed patterns observed for the RuBP regeneration. The lowest carbon dioxide compensation point was determined for the (Ce 100) treatment, the largest was obtained for (Zn 200).

Nanoceria is an effective free radicals scavenger and mimics the activity of several enzymes, namely superoxide dismutase (SOD), catalase (CAT), and oxidase (OXD) [32,90–92]. In this respect, it was classified by Korsvik et al. [93] as the first antioxidant nanoenzyme [90]. Stimulation of photosynthesis under abiotic stress mitigated by nanoceria in

Arabidopsis thaliana was investigated by Wu et al. [32,94]. They pointed out that nanoparticulate CeO_2 reduced the stress induced by reactive oxygen species, namely •OH, which are not tackled by the usual plant enzymatic scavenging pathways. Those NPs had entered chloroplasts through the nonendocytic pathways, reduced the ROS concentrations, and finally increased the quantum yield of photosystem II, carbon assimilation rates, and chlorophyll content. The latter processes are governed by the reversible redox reactions between the Ce^{4+} and Ce^{3+} species which are followed by the oxygen vacancy generation or elimination [95–97]. Moreover, Cao et al. [22,98] identified a strong correlation between photosynthesis parameters and the applied doses of CeO_2 NPs in soybean. The resulting photosynthesis enhancement was explained by the elevated Rubisco activity (V_{cmax}) and promotion of the NADPH regeneration rate which prompted the RuBP synthesis.

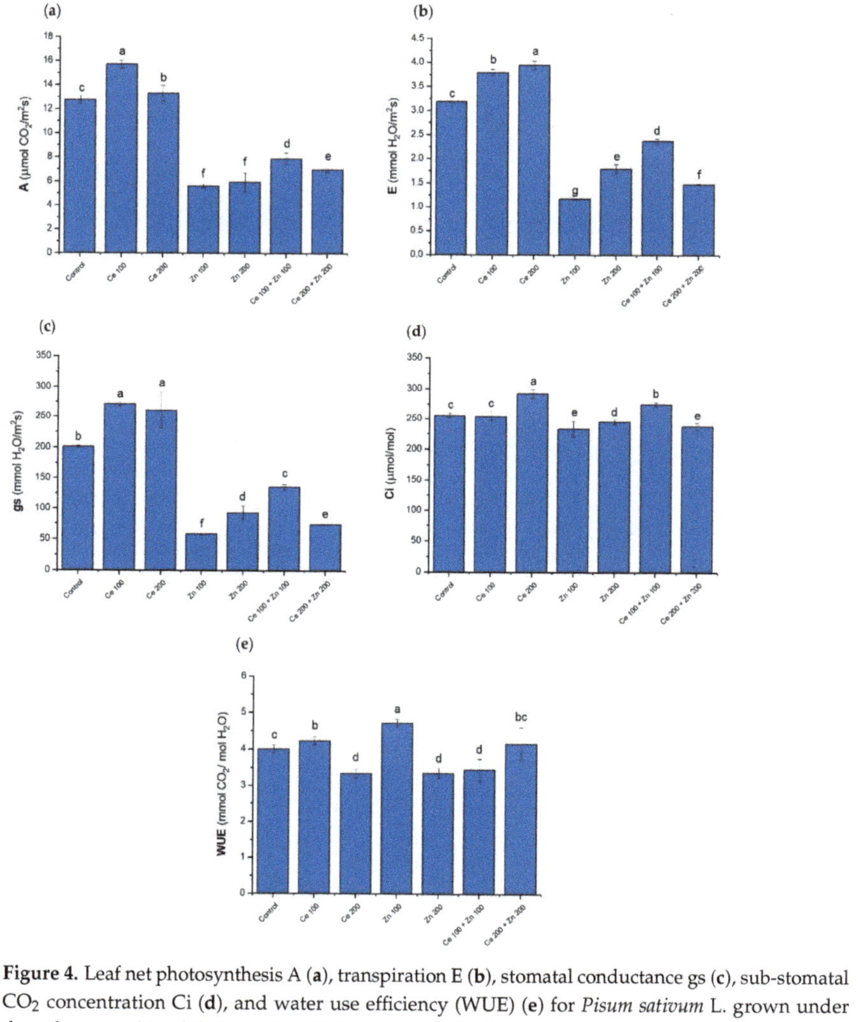

Figure 4. Leaf net photosynthesis A (**a**), transpiration E (**b**), stomatal conductance gs (**c**), sub-stomatal CO_2 concentration Ci (**d**), and water use efficiency (WUE) (**e**) for *Pisum sativum* L. grown under the sole or combined CeO_2 and ZnO NPs treatments. Concentrations of nanoparticles are given in mg/L of elemental cerium or zinc, respectively. The cultivation time was 12 days. Vertical bars represent standard deviations (n = 6). Distinct letters show the statistically significant differences among treatments as calculated with Tukey's HSD post hoc test (p = 0.95).

Table 2. Maximum ribulose 1,5-bisphosphate carboxylation rates (V_{cmax}), maximum electron transport rates (J_{max}) and CO_2 compensation points determined for *Pisum sativum* L. plants grown under the sole or combined CeO_2 and ZnO NPs treatments. Distinct letters show the statistically significant differences among treatments as calculated with Tukey's HSD post hoc test ($p = 0.95$), $n = 6$.

Treatment	V_{cmax} (μmol/m^2 s)	J_{max} (μmol/m^2 s)	CO_2 Compensation Point (μmol/mol)
Control	62.22 ± 1.22 c	80.31 ± 0.93 b	57.40 ± 1.12 b
Ce 100	71.07 ± 1.30 a	87.84 ± 0.69 a	43.00 ± 0.62 c
Ce 200	66.39 ± 0.58 b	85.02 ± 1.10 a	58.84 ± 1.00 b
Zn 100	50.29 ± 1.07 d	73.96 ± 1.27 c	97.51 ± 0.91 a
Zn 200	50.21 ± 0.82 d	67.82 ± 1.00 c	97.95 ± 1.28 a
Ce 100 + Zn 100	60.75 ± 1.41 c	79.09 ± 1.01 b	96.91 ± 1.34 a
Ce 200 + Zn 200	53.81 ± 2.58 d	72.78 ± 1.80 c	57.39 ± 3.27 b

Nanometric ZnO affected the CO_2 assimilation process in a roughly opposite way than CeO_2 NPs. At low concentrations, zinc is a micronutrient necessary for proper plant development. At elevated levels (above 300 μg/g, plant dry weight), zinc becomes a toxic pollutant responsible for the generation of ROS [99,100]. It alters stomata morphology and formation. The decreased gs and E indicated an increased stomatal closure and restriction of the transpiration rate. Carbon dioxide enters pea leaves by diffusion through stomatal pores. The major photosynthesis limitations result either from diffusion-controlled restrictions in CO_2 supply to the carboxylation sites or reduction in its consumption through the mitigated Rubisco activity and RuBP regeneration [101]. The decrease in the net photosynthesis A was correlated with the intercellular CO_2 concentration C_i and accompanied by the simultaneous reduction in V_{cmax} and J_{max}. These results suggest that ZnO NPs induced either stomatal or biochemical photosynthesis limitations. In the mixed ZnO and CeO_2 treatments, the latter effects are attenuated by the free radical scavenging properties of nanoceria.

3.5. Elements Content

Copper, manganese, iron, magnesium, calcium, and potassium contents were determined in plants from all treatments. The results are presented in Figure 5 and Table S4. Roots and shoots were treated separately. A one-way ANOVA was initially used to evaluate concentrations of macro- and micronutrients in pea plants. The 0.95 probability level was applied. Both nano-oxides altered the uptake of elements and their further translocation to the green parts of pea (Table S5).

Nanoceria in both tested concentrations [(Ce 100) and (Ce 200)] significantly reduced the uptake of all three investigated heavy metals (Cu, Mn, and Fe) by the roots. On the other hand, nanoparticulate ZnO [(Zn 100) and (Zn 200)] behaved in a less obvious way. Roots uptake was increased for copper and decreased for manganese. The uptake of iron by the roots was not affected by (Zn 100) and was reduced by high zinc supplementation (Zn 200). The combined treatments, either (Ce 100 + Zn 100) or (Ce 200 + Zn 200), prompted more complex root responses. For Cu, the resulting uptakes were between those induced by the administration of either CeO_2 or ZnO alone. The manganese levels were as low as those observed for the sole (Zn 100) and (Zn 200) supplementations. In the combined (Ce 200 + Zn 200) treatment Fe root level was as high as that in the control sample. However, (Ce 100 + Zn 100) treatment prompted much lower Fe concentrations. In green parts of the pea plants, the levels of Cu, Mn, and Fe were lower than those in roots. The only exception was manganese whose concentrations in shoots were higher for all samples supplemented with nanometric ZnO. Zinc hampers manganese uptake by roots. The latter element is essential for the water-splitting process during the light-dependent phase of photosynthesis. These circumstances facilitate Mn migration from roots to shoots. A quite similar situation was observed for magnesium and calcium. The former is an important chlorophyll cofactor while the latter is a structural component of photosystem II. Both ions are crucial for the overall photosynthesis yield. The additions of (Ce 100) and (Ce 200) NPs alone decreased potassium levels in roots and shoots. Both were inversely proportional to the Ce supplementations. The nanometric ZnO in either sole or combined administrations further restrained potassium contents in pea roots. However, unlike in the case of the nanoceria additions, higher K levels were observed in shoots.

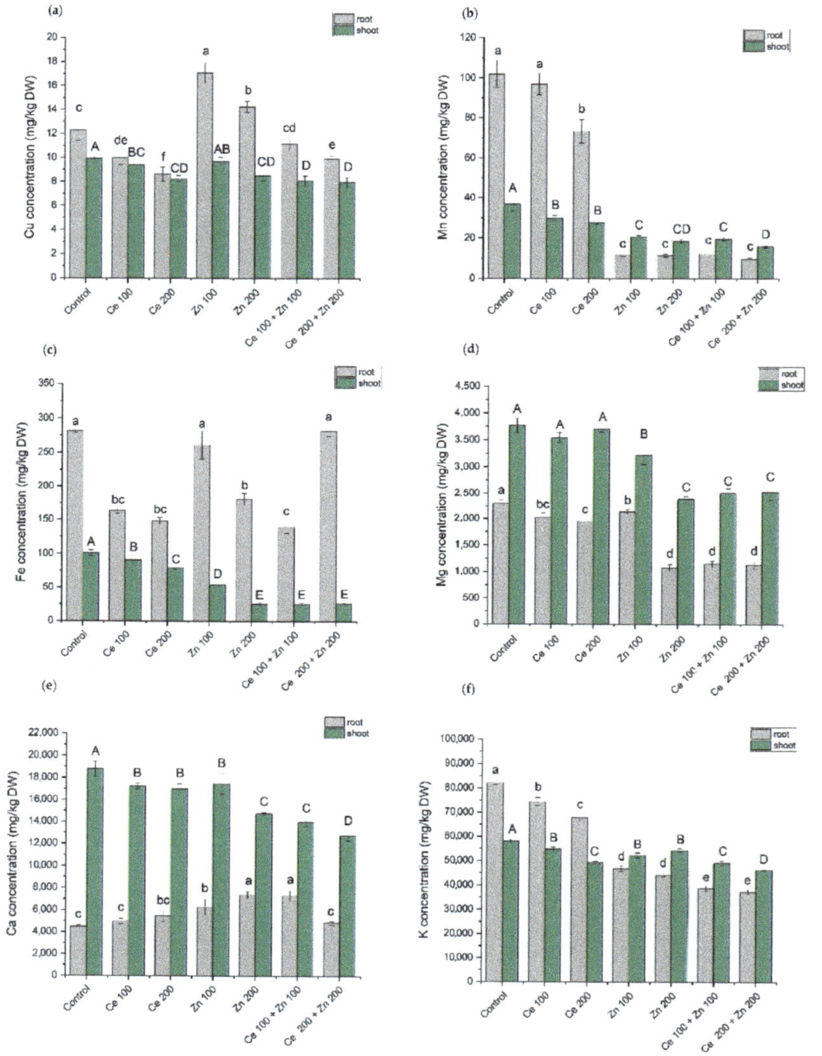

Figure 5. Copper (**a**), manganese (**b**), iron (**c**), magnesium (**d**), calcium (**e**), and potassium (**f**) concentrations in roots and shoots of *Pisum sativum* L. grown under the sole or combined CeO_2 and ZnO NPs treatment. Concentrations of nanoparticles are given in mg/L of elemental cerium and zinc, respectively. The cultivation time was 12 days. Vertical bars represent standard deviations ($n = 6$). Distinct letters show the statistically significant differences among treatments as calculated with Tukey's HSD post hoc test ($p = 0.95$).

4. Conclusions

Nanomaterials alter plant metabolism in a number of ways with photosystems I and II being their important targets. This paper continues our investigations on plant metabolism and uptake of nutrients in a model hydroponic environment subjected to nanoparticles pollution. Steadily increasing usage of nanomaterials in smart agriculture prompts thorough studies on their interactions with plant organisms. Nanoparticles rarely exist in the environment alone and their combined interactions can hardly be neglected.

Regrettably, relevant studies on their impact on plant photosynthesis are scarce. In this study, pea plants were cultivated in Hoagland's solutions supplemented with either sole or mixed cerium and zinc nano-oxides at 100 mg/L or 200 mg/L Ce or Zn levels. Despite relatively high sole CeO_2 administration (200 mg/L), no morphological symptoms of phytotoxicity were detected in *Pisum sativum* L. Leaf net photosynthesis, water use efficiency, and fresh biomass production were stimulated at the 100 mg/L Ce concentration and only slightly suppressed at its higher level. Contrarily, ZnO NPs applied alone caused serious impairment of metal homeostasis, decreased the level of photosynthetic pigments, induced leaf chlorosis, and finally hampered photosynthetic efficiency. We proved that ZnO NPs induced stomatal and biochemical limitations of photosynthesis. Such dysfunctions could lead to the overproduction of ROS in chloroplast and induce oxidative stress. In mixed CeO_2–ZnO NPs treatments, the beneficial effect of nanoceria was observed. In particular, pigments concentrations, leaf net photosynthesis, and maximum electron transport rate (J_{max}) depressed by ZnO NPs were significantly alleviated when CeO_2 NPs were present in the growing medium. It is well recognized that nanoceria has the potential to quench ROS. Therefore, we conclude that CeO_2 NPs moderate ZnO NPs toxicity by protecting the photosynthetic apparatus in *Pisum sativum* leaves from oxidative stress trigged by Zn. Additionally, we observed that both nano-oxides affected the nutrients uptake and transport at all concentrations applied.

Reactive oxygen species are by-products of aerobic metabolic processes in plants [102,103]. They usually increase membrane permeability and initiate stress signals often leading to programmed cell death [104]. At certain concentrations, the presence of NPs in growing media elevates ROS levels and induces oxidative damage in plant species. On the other hand, plant organisms have developed advanced antioxidant systems which involve either enzymatic or non-enzymatic pathways stabilizing ROS levels [105]. Those systems are enhanced under exposure to NPs, perhaps as an adaptive response to alleviate stress. We speculate that either CeO_2 or ZnO nanoparticles trigger oxidative stress in pea but only cerium dioxide acts as an antioxidant and reduces the stress symptoms while zinc oxide is mainly a prooxidant.

Our future studies will be aimed at the binary activity of nanoceria in agricultural plants. Namely, at high concentrations, it is a plant stressor that triggers ROS production while at certain, low levels nanoceria exhibits a ROS scavenging ability and supports the plant's defense mechanisms. The latter effect may find applications in agriculture and deserves further investigation.

Supplementary Materials: The following are available online at https://www.mdpi.com/article/10.3390/cells10113105/s1, Table S1: Characterization of CeO_2 NPs and ZnO NPs, Table S2: Elements content in certified reference materials Oriental Basma Tobacco Leaves—INCT-OBTL-5 (p = 0.95, n = 8), Table S3: Root length and tolerance index (TI) of *Pisum sativum* L. treated with nanoparticulate CeO_2 and ZnO, Table S4: ANOVA parameters for elements concentrations in roots and shoots of *Pisum sativum* L. plants across all treatments., Table S5: Translocation factors (TF) of nutrients from root to shoot in *Pisum sativum* L. plants grown under the sole or combined CeO_2 and ZnO NPs treatments, Figure S1: Comparison of CO_2 response curves (A/Ci) between control and treated plants.

Author Contributions: Conceptualization, E.S. and W.M.W.; methodology, E.S., M.P. and S.G.; validation, E.S. and W.M.W.; formal analysis, E.S. and M.P.; investigation, E.S. and M.P.; writing—original draft preparation, E.S., M.P. and W.M.W.; writing—review and editing, E.S., M.P., S.G. and W.M.W.; visualization, E.S. and M.P.; supervision, E.S. and W.M.W. All authors have read and agreed to the published version of the manuscript.

Funding: This work received support from the Regional Fund for Environmental Protection and Water Management in Lodz, Poland (Grant Number: 58/BN/D2018); additional funding from the Institute of General and Ecological Chemistry of Lodz University of Technology is also acknowledged.

Institutional Review Board Statement: Not applicable.

Informed Consent Statement: Not applicable.

Data Availability Statement: The data presented in this study are available on request from the corresponding author.

Acknowledgments: The authors wish to thank Lilianna Chęcińska for consultations on the solid structures of nano-oxides and their influence on properties of nanomaterials in micellar forms. Jakub Kubicki is kindly acknowledged for his support in heavy metals determination; Sylwia Michlewska and Sławomir Kadłubowski for their help in zeta potential measurements. The European University Foundation is acknowledged for advising on the legal and social dimensions of this study.

Conflicts of Interest: The authors declare no conflict of interest.

References

1. Verma, A.; Yadav, M. Application of nanomaterials in architecture—An overview. *Mater. Today Proc.* **2021**, *43*, 2921–2925. [CrossRef]
2. Hoang, A.T. Combustion behavior, performance and emission characteristics of diesel engine fuelled with biodiesel containing cerium oxide nanoparticles: A review. *Fuel Process. Technol.* **2021**, *218*, 106840. [CrossRef]
3. Yu, J.; Luo, M.; Lv, Z.; Huang, S.; Hsu, H.H.; Kuo, C.C.; Han, S.T.; Zhou, Y. Recent advances in optical and optoelectronic data storage based on luminescent nanomaterials. *Nanoscale* **2020**, *12*, 23391–23423. [CrossRef]
4. Caro, C.; Gámez, F.; Quaresma, P.; Páez-Muñoz, J.M.; Domínguez, A.; Pearson, J.R.; Pernía Leal, M.; Beltrán, A.M.; Fernandez-Afonso, Y.; De La Fuente, J.M.; et al. Fe_3O_4-Au core-shell nanoparticles as a multimodal platform for in vivo imaging and focused photothermal therapy. *Pharmaceutics* **2021**, *13*, 416. [CrossRef]
5. Jandt, K.D.; Watts, D.C. Nanotechnology in dentistry: Present and future perspectives on dental nanomaterials. *Dent. Mater.* **2020**, *36*, 1365–1378. [CrossRef]
6. Chen, L.; Liu, Z.; Guo, Z.; Huang, X.J. Regulation of intrinsic physicochemical properties of metal oxide nanomaterials for energy conversion and environmental detection applications. *J. Mater. Chem. A* **2020**, *8*, 17326–17359. [CrossRef]
7. Khoo, K.S.; Chia, W.Y.; Tang, D.Y.Y.; Show, P.L.; Chew, K.W.; Chen, W.H. Nanomaterials utilization in biomass for biofuel and bioenergy production. *Energies* **2020**, *13*, 892. [CrossRef]
8. Sengwa, R.J.; Dhatarwal, P.; Choudhary, S. A comparative study of different metal oxide nanoparticles dispersed PVDF/PEO blend matrix-based advanced multifunctional nanodielectrics for flexible electronic devices. *Mater. Today Commun.* **2020**, *25*, 101380. [CrossRef]
9. Degfie, T.A.; Mamo, T.T.; Mekonnen, Y.S. Optimized biodiesel production from waste cooking oil (WCO) using calcium oxide (CaO) nano-catalyst. *Sci. Rep.* **2019**, *9*, 18982. [CrossRef] [PubMed]
10. Sui, X.; Downing, J.R.; Hersam, M.C.; Chen, J. Additive manufacturing and applications of nanomaterial-based sensors. *Mater. Today* **2021**. [CrossRef]
11. Hou, Q.; Qi, X.; Zhen, M.; Qian, H.; Nie, Y.; Bai, C.; Zhang, S.; Bai, X.; Ju, M. Biorefinery roadmap based on catalytic production and upgrading 5-hydroxymethylfurfural. *Green Chem.* **2021**, *23*, 119–231. [CrossRef]
12. Shi, Y.; Xu, H.; Liu, T.; Zeb, S.; Nie, Y.; Zhao, Y.; Qin, C.; Jiang, X. Advanced development of metal oxide nanomaterials for H_2 gas sensing applications. *Mater. Adv.* **2021**, *2*, 1530–1569. [CrossRef]
13. Singh, R.P.; Handa, R.; Manchanda, G. Nanoparticles in sustainable agriculture: An emerging opportunity. *J. Control. Release* **2021**, *329*, 1234–1248. [CrossRef]
14. Singh, H.; Sharma, A.; Bhardwaj, S.K.; Arya, S.K.; Bhardwaj, N.; Khatri, M. Recent advances in the applications of nano-agrochemicals for sustainable agricultural development. *Environ. Sci. Process. Impacts* **2021**, *23*, 213–239. [CrossRef]
15. Tarrahi, R.; Mahjouri, S.; Khataee, A. A review on in vivo and in vitro nanotoxicological studies in plants: A headlight for future targets. *Ecotoxicol. Environ. Saf.* **2021**, *208*, 111697. [CrossRef]
16. Wang, M.; Li, S.; Chen, Z.; Zhu, J.; Hao, W.; Jia, G.; Chen, W.; Zheng, Y.; Qu, W.; Liu, Y. Safety assessment of nanoparticles in food: Current status and prospective. *Nano Today* **2021**, *39*, 101169. [CrossRef]
17. Ameen, F.; Alsamhary, K.; Alabdullatif, J.A.; ALNadhari, S. A review on metal-based nanoparticles and their toxicity to beneficial soil bacteria and fungi. *Ecotoxicol. Environ. Saf.* **2021**, *213*, 112027. [CrossRef] [PubMed]
18. Metal Oxide Nanoparticles Market—Growth, Trends, COVID-19 Impact, and Forecasts (2021–2026). Available online: https://www.mordorintelligence.com/industry-reports/metal-oxide-nanoparticles-market (accessed on 5 May 2021).
19. Choi, S.J.; Choy, J.H. Effect of physico-chemical parameters on the toxicity of inorganic nanoparticles. *J. Mater. Chem.* **2011**, *21*, 5547–5554. [CrossRef]

20. Chen, C.; Li, Y.F.; Qu, Y.; Chai, Z.; Zhao, Y. Advanced nuclear analytical and related techniques for the growing challenges in nanotoxicology. *Chem. Soc. Rev.* **2013**, *42*, 8266–8303. [CrossRef]
21. Bundschuh, M.; Filser, J.; Lüderwald, S.; McKee, M.S.; Metreveli, G.; Schaumann, G.E.; Schulz, R.; Wagner, S. Nanoparticles in the environment: Where do we come from, where do we go to? *Environ. Sci. Eur.* **2018**, *30*, 6. [CrossRef]
22. Cao, Z.; Stowers, C.; Rossi, L.; Zhang, W.; Lombardini, L.; Ma, X. Physiological effects of cerium oxide nanoparticles on the photosynthesis and water use efficiency of soybean (*Glycine max* (L.) Merr.). *Environ. Sci. Nano* **2017**, *4*, 1086–1094. [CrossRef]
23. Bindra, H.S.; Singh, B. Nanofertilizers and nanopesticides: Future of plant protection. In *Advances in Nano-Fertilizers and Nano-Pesticides in Agriculture*, 1st ed.; Jogaiah, S., Singh, H.B., Fraceto, L.F., de Lima, R., Eds.; Woodhead Publishing: Cambridge, UK, 2021; pp. 57–84.
24. Verma, S.K.; Das, A.K.; Patel, M.K.; Shah, A.; Kumar, V.; Gantait, S. Engineered nanomaterials for plant growth and development: A perspective analysis. *Sci. Total Environ.* **2018**, *630*, 1413–1435. [CrossRef] [PubMed]
25. Hussain, A.; Ali, S.; Rizwan, M.; Zia ur Rehman, M.; Javed, M.R.; Imran, M.; Chatha, S.A.S.; Nazir, R. Zinc oxide nanoparticles alter the wheat physiological response and reduce the cadmium uptake by plants. *Environ. Pollut.* **2018**, *242*, 1518–1526. [CrossRef] [PubMed]
26. Rizwan, M.; Ali, S.; Qayyum, M.F.; Ok, Y.S.; Adrees, M.; Ibrahim, M.; Zia-ur-Rehman, M.; Farid, M.; Abbas, F. Effect of metal and metal oxide nanoparticles on growth and physiology of globally important food crops: A critical review. *J. Hazard. Mater.* **2017**, *322*, 2–16. [CrossRef]
27. Marchiol, L.; Filippi, A.; Adamiano, A.; Esposti, L.D.; Iafisco, M.; Mattiello, A.; Petrussa, E.; Braidot, E. Influence of hydroxyapatite nanoparticles on germination and plant metabolism of tomato (*Solanum lycopersicum* L.): Preliminary evidence. *Agronomy* **2019**, *9*, 161. [CrossRef]
28. Marslin, G.; Sheeba, C.J.; Franklin, G. Nanoparticles alter secondary metabolism in plants via ROS burst. *Front. Plant Sci.* **2017**, *8*, 832. [CrossRef] [PubMed]
29. Dhiman, S.; Bakshi, P.; Kapoor, N.; Sharma, P.; Kohli, S.K.; Mir, B.A.; Bhardwaj, R. Nanoparticle-induced oxidative stress. In *Plant Responses to Nanomaterials. Recent Interventions and Physiological and Biochemical Responses*; Singh, V.P., Singh, S., Prasad, S.M., Chauhan, D.K., Tripathi, D.K., Eds.; Springer: New York, NY, USA, 2021; pp. 269–313.
30. Kalaji, M.H.; Goltsev, V.N.; Żuk-Golaszewska, K.; Zivcak, M.; Brestic, M. *Chlorophyll Fluorescence: Understanding Crop Performance—Basics and Applications*; CRC Press Taylor & Francis Group: New York, NY, USA, 2017.
31. Rodea-Palomares, I.; Gonzalo, S.; Santiago-Morales, J.; Leganés, F.; García-Calvo, E.; Rosal, R.; Fernández-Piñas, F. An insight into the mechanisms of nanoceria toxicity in aquatic photosynthetic organisms. *Aquat. Toxicol.* **2012**, *122–123*, 133–143. [CrossRef]
32. Wu, H.; Tito, N.; Giraldo, J.P. Anionic cerium oxide nanoparticles protect plant photosynthesis from abiotic stress by scavenging reactive oxygen species. *ACS Nano* **2017**, *11*, 11283–11297. [CrossRef]
33. Wang, X.P.; Li, Q.Q.; Pei, Z.M.; Wang, S.C. Effects of zinc oxide nanoparticles on the growth, photosynthetic traits, and antioxidative enzymes in tomato plants. *Biol. Plant.* **2018**, *62*, 801–808. [CrossRef]
34. Faizan, M.; Faraz, A.; Hayat, S. Effective use of zinc oxide nanoparticles through root dipping on the performance of growth, quality, photosynthesis and antioxidant system in tomato. *J. Plant Biochem. Biotechnol.* **2020**, *29*, 553–567. [CrossRef]
35. Tighe-Neira, R.; Carmora, E.; Recio, G.; Nunes-Nesi, A.; Reyes-Diaz, M.; Alberdi, M.; Rengel, Z.; Inostroza-Blancheteau, C. Metallic nanoparticles influence the structure and function of the photosynthetic apparatus in plants. *Plant Physiol. Biochem.* **2018**, *130*, 408–417. [CrossRef]
36. Zhao, L.; Peralta-Videa, J.R.; Rico, C.M.; Hernandez-Viezcas, J.A.; Sun, Y.; Niu, G.; Servin, A.; Nunez, J.E.; Duarte-Gardea, M.; Gardea-Torresdey, J.L. CeO_2 and ZnO nanoparticles change the nutritional qualities of cucumber (*Cucumis sativus*). *J. Agric. Food Chem.* **2014**, *62*, 2752–2759. [CrossRef] [PubMed]
37. Du, W.; Tan, W.; Peralta-Videa, J.R.; Gardea-Torresdey, J.L.; Ji, R.; Yin, Y.; Guo, H. Interaction of metal oxide nanoparticles with higher terrestrial plants: Physiological and biochemical aspects. *Plant Physiol. Biochem.* **2017**, *110*, 210–225. [CrossRef] [PubMed]
38. Mukherjee, A.; Peralta-Videa, J.R.; Bandyopadhyay, S.; Rico, C.M.; Zhao, L.; Gardea-Torresdey, J.L. Physiological effects of nanoparticulate ZnO in green peas (*Pisum sativum* L.) cultivated in soil. *Metallomics* **2014**, *6*, 132–138. [CrossRef] [PubMed]
39. Reddy Pullagurala, V.L.; Adisa, I.O.; Rawat, S.; Kalagara, S.; Hernandez-Viezcas, J.A.; Peralta-Videa, J.R.; Gardea-Torresdey, J.L. ZnO nanoparticles increase photosynthetic pigments and decrease lipid peroxidation in soil grown cilantro (*Coriandrum sativum*). *Plant Physiol. Biochem.* **2018**, *132*, 120–127. [CrossRef] [PubMed]
40. Li, J.; Mu, Q.; Du, Y.; Luo, J.; Liu, Y.; Li, T. Growth and photosynthetic inhibition of cerium oxide nanoparticles on soybean (*Glycine max*). *Bull. Environ. Contam. Toxicol.* **2020**, *105*, 119–126. [CrossRef]
41. Zulfiqar, F.; Navarro, M.; Ashraf, M.; Akram, N.A.; Munné-Bosch, S. Nanofertilizer use for sustainable agriculture: Advantages and limitations. *Plant Sci.* **2019**, *289*, 110270. [CrossRef]
42. Kumar, S.; Nehra, M.; Dilbaghi, N.; Marrazza, G.; Hassan, A.A.; Kim, K.H. Nano-based smart pesticide formulations: Emerging opportunities for agriculture. *J. Control. Release* **2019**, *294*, 131–153. [CrossRef]
43. Sun, Y.; Liang, J.; Tang, L.; Li, H.; Zhu, Y.; Jiang, D.; Song, B.; Chen, M.; Zeng, G. Nano-pesticides: A great challenge for biodiversity? *Nano Today* **2019**, *28*, 100757. [CrossRef]
44. Silva, L.P.; Bonatto, C.C. Green nanotechnology for sustained release of eco-friendly agrochemicals. In *Sustainable Agrochemistry. A Compendium of Technologies*; Vaz, S., Jr., Ed.; Springer: New York, NY, USA, 2019; pp. 113–129.

45. Mohammadi, M.H.Z.; Panahirad, S.; Navai, A.; Bahrami, M.K.; Kulak, M.; Gohari, G. Cerium oxide nanoparticles (CeO_2-NPs) improve growth parameters and antioxidant defense system in Moldavian Balm (*Dracocephalummoldavica*, L.) under salinity stress. *Plant Stress* **2021**, *1*, 100006. [CrossRef]
46. Jahani, S.; Saadatmand, S.; Mahmoodzadeh, H.; Khavari-Nejad, R.A. Effect of foliar application of cerium oxide nanoparticles on growth, photosynthetic pigments, electrolyte leakage, compatible osmolytes and antioxidant enzymes activities of *Calendula officinalis* L. *Biologia* **2019**, *74*, 1063–1075. [CrossRef]
47. Jurkow, R.; Pokluda, R.; Sękara, A.; Kalisz, A. Impact of foliar application of some metal nanoparticles on antioxidant system in oakleaf lettuce seedlings. *BMC Plant Biol.* **2020**, *20*, 290. [CrossRef] [PubMed]
48. Liu, Y.; Yue, L.; Wang, C.; Zhu, X.; Wang, Z.; Xing, B. Photosynthetic response mechanisms in typical C3 and C4 plants upon La_2O_3 nanoparticle exposure. *Environ. Sci. Nano* **2020**, *7*, 81–92. [CrossRef]
49. Naasz, S.; Altenburger, R.; Kühnel, D. Environmental mixtures of nanomaterials and chemicals: The Trojan-horse phenomenon and its relevance for ecotoxicity. *Sci. Total Environ.* **2018**, *635*, 1170–1181. [CrossRef]
50. Singh, D.; Kumar, A. Binary mixture of nanoparticles in sewage sludge: Impact on spinach growth. *Chemosphere* **2020**, *254*, 126794. [CrossRef] [PubMed]
51. Singh, D.; Kumar, A. Quantification of metal uptake in *Spinacia oleracea* irrigated with water containing a mixture of CuO and ZnO nanoparticles. *Chemosphere* **2020**, *243*, 125239. [CrossRef]
52. Skiba, E.; Michlewska, S.; Pietrzak, M.; Wolf, W.M. Additive interactions of nanoparticulate ZnO with copper, manganese and iron in *Pisum sativum* L., a hydroponic study. *Sci. Rep.* **2020**, *10*, 13574. [CrossRef]
53. Skiba, E.; Pietrzak, M.; Gapińska, M.; Wolf, W.M. Metal homeostasis and gas exchange dynamics in *Pisum sativum* L. exposed to cerium oxide nanoparticles. *Int. J. Mol. Sci.* **2020**, *21*, 8497. [CrossRef]
54. Skiba, E.; Wolf, W.M. Cerium oxide nanoparticles affect heavy metals uptake by pea in a divergent way than their ionic and bulk counterparts. *Water Air Soil Pollut.* **2019**, *230*, 248. [CrossRef]
55. Kreplak, J.; Madoui, M.A.; Cápal, P.; Novák, P.; Labadie, K.; Aubert, G.; Bayer, P.E.; Gali, K.K.; Syme, R.A.; Main, D.; et al. A reference genome for pea provides insight into legume genome evolution. *Nat. Genet.* **2019**, *51*, 1411–1422. [CrossRef]
56. Burstin, J.; Kreplak, J.; Macas, J.; Lichtenzveig, J. *Pisum sativum* (Pea). *Trends Genet.* **2020**, *36*, 312–313. [CrossRef] [PubMed]
57. Zhang, X.; Wan, S.; Hao, J.; Hu, J.; Yang, T.; Zong, X. Large-scale evaluation of pea (*Pisum sativum* L.) germplasm for cold tolerance in the field during winter in Qingdao. *Crop. J.* **2016**, *4*, 377–383. [CrossRef]
58. Meier, U. *Growth Stages of Mono- and Dicotyledonous Plants: BBCH Monograph*; Blackwell Wissenschafts: Berlin, Germany, 1997.
59. Samczyński, Z.; Dybczyński, R.S.; Polkowska-Motrenko, H.; Chajduk, E.; Pyszynska, M.; Danko, B.; Czerska, E.; Kulisa, K.; Doner, K.; Kalbarczyk, P. Two new reference materials based on tobacco leaves: Certification for over a dozen of toxic and essential elements. *Sci. World J.* **2012**, *2012*, 216380. [CrossRef]
60. Turner, R.G.; Marshal, C. Accumulation of zinc by subcellular root of *Agrostis tannis* Sibth. in relation of zinc tolerance. *New Phytol.* **1972**, *71*, 671–676. [CrossRef]
61. KaramiMehrian, S.; Heidari, R.; Rahmani, F.; Najafi, S. Effect of chemical synthesis silver nanoparticles on germination indices and seedlings growth in seven varieties of *Lycopersiconesculentum* Mill (tomato) plants. *J. Clust. Sci.* **2016**, *27*, 327–340. [CrossRef]
62. Liu, Z.; He, X.; Chen, W.; Yuan, F.; Yan, K.; Tao, D. Accumulation and tolerance characteristics of cadmium in a potential hyperaccumulator-*Lonicera japonica* Thunb. *J. Hazard. Mater.* **2009**, *169*, 170–175. [CrossRef]
63. Mayerová, M.; Petrová, Š.; Madaras, M.; Lipavský, J.; Šimon, T.; Vaněk, T. Non-enhanced phytoextraction of cadmium, zinc, and lead by high-yielding crops. *Environ. Sci. Pollut. Res.* **2017**, *24*, 14706–14716. [CrossRef]
64. Oren, R.; Werk, K.S.; Buchmann, N.; Zimmermann, R. Chlorophyll-nutrient relationships identify nutritionally caused decline in *Piceaabies* stands. *Can. J. For. Res.* **1993**, *23*, 1187–1195. [CrossRef]
65. Lichtenthaler, H.K.; Wellburn, A.R. Determinations of total carotenoids and chlorophylls a and b of leaf extracts in different solvents. *Biochem. Soc. Trans.* **1983**, *11*, 591–592. [CrossRef]
66. R Core Team. *R: A Language and Environment for Statistical Computing*; R Foundation for Statistical Computing: Vienna, Austria, 2019. Available online: https://www.R-project.org (accessed on 10 January 2021).
67. Duursma, R.A. Plantecophys—An R package for analysing and modelling leaf gas exchange data. *PLoS ONE* **2015**, *10*, e0143346. [CrossRef]
68. Marchiol, L.; Mattiello, A.; Pošćić, F.; Fellet, G.; Zavalloni, C.; Carlino, E.; Musetti, R. Changes in physiological and agronomical parameters of barley (*Hordeum vulgare*) exposed to cerium and titanium dioxide nanoparticles. *Int. J. Environ. Res. Public Health* **2016**, *13*, 332. [CrossRef] [PubMed]
69. Herber, R.F.M.; Sallé, H.J.A. Statistics and data evaluation. *Tech. Instrum. Anal. Chem.* **1994**, *15*, 257–271.
70. Broadley, M.R.; White, P.J.; Hammond, J.P.; Zelko, I.; Lux, A. Zinc in plants. *New Phytol.* **2007**, *173*, 677–702. [CrossRef] [PubMed]
71. Broadley, M.; Brown, P.; Cakmak, I.; Rengel, Z.; Zhao, F. Function of nutrients: Micronutrients. In *Marschner's Mineral Nutrition of Higher Plants*, 3rd ed.; Marschner, P., Ed.; Academic Press: Cambridge, MA, USA; Elsevier Ltd.: Amsterdam, The Netherlands, 2011; pp. 191–248.
72. Hu, X.; Tanaka, A.; Tanaka, R. Simple extraction methods that prevent the artifactual conversion of chlorophyll to chlorophyllide during pigment isolation from leaf samples. *Plant Methods* **2013**, *9*, 1–13. [CrossRef]
73. Genter, R.B. Ecotoxicology of inorganic chemical stress to algae. In *Algal Ecology Freshwater Benthic Ecosystems*, 1st ed.; Stevenson, R.J., Bothwell, M.L., Lowe, R.L., Thorp, J., Eds.; Academic Press: San Diego, CA, USA, 1996; pp. 403–468.

74. Hu, C.; Liu, X.; Li, X.; Zhao, Y. Evaluation of growth and biochemical indicators of *Salvinia natans* exposed to zinc oxide nanoparticles and zinc accumulation in plants. *Environ. Sci. Pollut. Res.* **2014**, *21*, 732–739. [CrossRef] [PubMed]
75. Wang, C.; Zhang, S.H.; Wang, P.F.; Hou, J.; Zhang, W.J.; Li, W.; Lin, Z.P. The effect of excess Zn on mineral nutrition and antioxidative response in rapeseed seedlings. *Chemosphere* **2009**, *75*, 1468–1476. [CrossRef] [PubMed]
76. Ebbs, S.; Uchil, S. Cadmium and zinc induced chlorosis in Indian mustard [*Brassica juncea* (L.) Czern] involves preferential loss of chlorophyll b. *Photosynthetica* **2008**, *46*, 49–55. [CrossRef]
77. Andrejić, G.; Gajić, G.; Prica, M.; Dželetović, Ž.; Rakić, T. Zinc accumulation, photosynthetic gas exchange, and chlorophyll a fluorescence in Zn-stressed *Miscanthus x giganteus* plants. *Photosynthetica* **2018**, *56*, 1249–1258. [CrossRef]
78. Zoufan, P.; Baroonian, M.; Zargar, B. ZnO nanoparticles-induced oxidative stress in *Chenopodium murale* L., Zn uptake, and accumulation under hydroponic culture. *Environ. Sci. Pollut. Res.* **2020**, *27*, 11066–11078. [CrossRef]
79. Salehi, H.; De Diego, N.; Chehregani Rad, A.; Benjamin, J.J.; Trevisan, M.; Lucini, L. Exogenous application of ZnO nanoparticles and $ZnSO_4$ distinctly influence the metabolic response in *Phaseolus vulgaris* L. *Sci. Total Environ.* **2021**, *778*, 146331. [CrossRef]
80. Choudhury, F.K.; Rivero, R.M.; Blumwald, E.; Mittler, R. Reactive oxygen species, abiotic stress and stress combination. *Plant J.* **2017**, *90*, 856–867. [CrossRef]
81. Mhamdi, A.; Van Breusegem, F. Reactive oxygen species in plant development. *Development* **2018**, *145*, 15. [CrossRef]
82. Kim, C. ROS-driven oxidative modification: Its impact on chloroplasts-nucleus communication. *Front. Plant Sci.* **2020**, *10*, 1729. [CrossRef] [PubMed]
83. Küpper, H.; Küpper, F.; Spiller, M. Environmental relevance of heavy metal-substituted chlorophylls using the example of water plants. *J. Exp. Bot.* **1996**, *47*, 259–266. [CrossRef]
84. Sharkey, T.D. What gas exchange data can tell us about photosynthesis. *Plant Cell Environ.* **2016**, *39*, 1161–1163. [CrossRef] [PubMed]
85. Long, S.P.; Bernacchi, C.J. Gas exchange measurements, what can they tell us about the underlying limitations to photosynthesis? Procedures and sources of error. *J. Exp. Bot.* **2003**, *54*, 2393–2401. [CrossRef] [PubMed]
86. Haworth, M.; Marino, G.; Centritto, M. An introductory guide to gas exchange analysis of photosynthesis and its application to plant phenotyping and precision irrigation to enhance water use efficiency. *J. Water Clim. Chang.* **2018**, *9*, 786–808. [CrossRef]
87. Vassilev, A.; Nikolova, A.; Koleva, L.; Lidon, F. Effects of excess Zn on growth and photosynthetic performance of young bean plants. *J. Phytol.* **2011**, *3*, 58–62.
88. Balafrej, H.; Bogusz, D.; Triqui, Z.-E.A.; Guedira, A.; Bendaou, N.; Smouni, A.; Fahr, M. Zinc hyperaccumulation in plants: A review. *Plants* **2020**, *9*, 562. [CrossRef]
89. Anwaar, S.A.; Ali, S.; Ali, S.; Ishaque, W.; Farid, M.; Farooq, M.A.; Najeeb, U.; Abbas, F.; Sharif, M. Silicon (Si) alleviates cotton (*Gossypium hirsutum* L.) from zinc (Zn) toxicity stress by limiting Zn uptake and oxidative damage. *Environ. Sci. Pollut. Res.* **2015**, *22*, 3441–3450. [CrossRef]
90. Liu, Y.; Xiao, Z.; Chen, F.; Yue, L.; Zou, H.; Lyu, J.; Wang, Z. Metallic oxide nanomaterials act as antioxidant nanozymes in higher plants: Trends, meta-analysis, and prospect. *Sci. Total Environ.* **2021**, *780*, 146578. [CrossRef]
91. Djanaguiraman, M.; Nair, R.; Giraldo, J.P.; Prasad, P.V.V. Cerium oxide nanoparticles decrease drought-induced oxidative damage in *Sorghum Leading* to higher photosynthesis and grain yield. *ACS Omega* **2018**, *3*, 14406–14416. [CrossRef]
92. Baldim, V.; Bedioui, F.; Mignet, N.; Margaill, I.; Berret, J.F. The enzyme-like catalytic activity of cerium oxide nanoparticles and its dependency on Ce^{3+} surface area concentration. *Nanoscale* **2018**, *10*, 6971–6980. [CrossRef]
93. Korsvik, C.; Patil, S.; Seal, S.; Self, W.T. Superoxide dismutase mimetic properties exhibited by vacancy engineered ceria nanoparticles. *Chem. Commun.* **2007**, 1056–1058. [CrossRef] [PubMed]
94. Wu, H.; Shabala, L.; Shabala, S.; Giraldo, J.P. Hydroxyl radical scavenging by cerium oxide nanoparticles improves *Arabidopsis* salinity tolerance by enhancing leaf mesophyll potassium retention. *Environ. Sci. Nano* **2018**, *5*, 1567–1583. [CrossRef]
95. Nelson, B.C.; Johnson, M.E.; Walker, M.L.; Riley, K.R.; Sims, C.M. Antioxidant cerium oxide nanoparticles in biology and medicine. *Antioxidants* **2016**, *5*, 15. [CrossRef] [PubMed]
96. Zhou, P.; Adeel, M.; Shakoor, N.; Guo, M.; Hao, Y.; Azeem, I.; Li, M.; Liu, M.; Rui, Y. Application of nanoparticles alleviates heavy metals stress and promotes plant growth: An overview. *Nanomaterials* **2021**, *11*, 26. [CrossRef]
97. Xue, Y.; Luan, Q.; Yang, D.; Yao, X.; Zhou, K. Direct evidence for hydroxyl radical scavenging activity of cerium oxide nanoparticles. *J. Phys. Chem. C* **2011**, *115*, 4433–4438. [CrossRef]
98. Cao, Z.; Rossi, L.; Stowers, C.; Zhang, W.; Lombardini, L.; Ma, X. The impact of cerium oxide nanoparticles on the physiology of soybean (*Glycine max* (L.) Merr.) under different soil moisture conditions. *Environ. Sci. Pollut. Res.* **2018**, *25*, 930–939. [CrossRef]
99. Khan, M.I.R.; Jahan, B.; Alajmi, M.F.; Rehman, M.T.; Khan, N.A. Exogenously-sourced ethylene modulates defense mechanisms and promotes tolerance to zinc stress in mustard (*Brassica juncea* L.). *Plants* **2019**, *8*, 540. [CrossRef]
100. Feigl, G.; Lehotai, N.; Molnár, Á.; Ördög, A.; Rodríguez-Ruiz, M.; Palma, J.M.; Corpas, F.J.; Erdei, L.; Kolbert, Z. Zinc induces distinct changes in the metabolism of reactive oxygen and nitrogen species (ROS and RNS) in the roots of two *Brassica* species with different sensitivity to zinc stress. *Ann. Bot.* **2015**, *116*, 613–625. [CrossRef]
101. Deans, R.M.; Farquhar, G.D.; Busch, F.A. Estimating stomatal and biochemical limitations during photosynthetic induction. *Plant Cell Environ.* **2019**, *42*, 3227–3240. [CrossRef]
102. Sharma, P.; Jha, A.B.; Dubey, R.S.; Pessarakli, M. Reactive Oxygen Species, Oxidative Damage, and Antioxidative Defense Mechanism in Plants under Stressful Conditions. *J. Bot.* **2012**, 217037. [CrossRef]

103. Apel, K.; Hirt, H. Reactive Oxygen Species: Metabolism, Oxidative Stress, and Signal Transduction. *Annu. Rev. Plant Biol.* **2004**, *55*, 373–399. [CrossRef]
104. Yanik, F.; Vardar, F. Mechanism and interaction of nanoparticle-induced programmed cel death in plants. In *Phytotoxicity of Nanoparticles*, 1st ed.; Faisal, M., Saquib, Q., Alator, A.A., Al-Khedhairy, A.A., Eds.; Springer International Publishing AG: Cham, Switzerland, 2018; pp. 175–196.
105. Yang, J.; Cao, W.; Rui, Y. Interactions between nanoparticles and plants: Phytotoxicity and defense mechanisms. *J. Plant Interact.* **2017**, *12*, 158–169. [CrossRef]

cells

MDPI

Article

Identification of a Novel Mutation Exacerbated the PSI Photoinhibition in *pgr5/pgrl1* Mutants; Caution for Overestimation of the Phenotypes in Arabidopsis *pgr5-1* Mutant

Shinya Wada [1,*], Katsumi Amako [2] and Chikahiro Miyake [1,3]

1 Graduate School of Agricultural Science, Kobe University, 1-1, Rokkodai, Nada-ku, Kobe 657-8501, Japan; cmiyake@hawk.kobe-u.ac.jp
2 Department of Health and Nutrition, Faculty of Human Life Studies, Jin-ai University, 3-1-1, Ohde-cho, Fukui, Echizen 915-8586, Japan; amakok@jindai.ac.jp
3 Core Research for Environmental Science and Technology (CREST), Japan Science and Technology Agency (JST), 7, Gobancho, Chiyoda-ku, Tokyo 102-0076, Japan
* Correspondence: swada@penguin.kobe-u.ac.jp; Tel.: +81-78-803-5850

Abstract: PSI photoinhibition is usually avoided through P700 oxidation. Without this protective mechanism, excess light represents a potentially lethal threat to plants. PGR5 is suggested to be a major component of cyclic electron transport around PSI and is important for P700 oxidation in angiosperms. The known Arabidopsis PGR5 deficient mutant, *pgr5-1*, is incapable of P700 oxidation regulation and has been used in numerous photosynthetic studies. However, here it was revealed that *pgr5-1* was a double mutant with exaggerated PSI photoinhibition. *pgr5-1* significantly reduced growth compared to the newly isolated PGR5 deficient mutant, $pgr5^{hope1}$. The introduction of PGR5 into *pgr5-1* restored P700 oxidation regulation, but remained a pale-green phenotype, indicating that *pgr5-1* had additional mutations. Both *pgr5-1* and $pgr5^{hope1}$ tended to cause PSI photoinhibition by excess light, but *pgr5-1* exhibited an enhanced reduction in PSI activity. Introducing AT2G17240, a candidate gene for the second mutation into *pgr5-1* restored the pale-green phenotype and partially restored PSI activity. Furthermore, a deficient mutant of PGRL1 complexing with PGR5 significantly reduced PSI activity in the double-deficient mutant with AT2G17240. From these results, we concluded that AT2G17240, named PSI photoprotection 1 (PTP1), played a role in PSI photoprotection, especially in PGR5/PGRL1 deficient mutants.

Keywords: proton gradient regulation 5 (PGR5); PGR5-like photosynthetic phenotype 1 (PGRL1); photosynthetic electron transport; PSI photoinhibition; oxidation of P700; oxidative stress

check for updates

Citation: Wada, S.; Amako, K.; Miyake, C. Identification of a Novel Mutation Exacerbated the PSI Photoinhibition in *pgr5/pgrl1* Mutants; Caution for Overestimation of the Phenotypes in Arabidopsis *pgr5-1* Mutant. *Cells* **2021**, *10*, 2884. https://doi.org/10.3390/cells10112884

Academic Editors: Suleyman Allakhverdiev, Alexander G. Ivanov and Marian Brestic

Received: 10 September 2021
Accepted: 22 October 2021
Published: 26 October 2021

Publisher's Note: MDPI stays neutral with regard to jurisdictional claims in published maps and institutional affiliations.

1. Introduction

Photosynthesis consists of two steps: the electron transport reaction and the carbon fixing reactions. The electron transport reaction converts light energy absorbed in chloroplast thylakoid membrane to chemical energy (NADPH and ATP), while the subsequent carbon fixing reaction (Calvin–Benson cycle) consumes NADPH and ATP to fix CO_2. These reactions are regarded as an electron source-sink relationship. The electron transport reaction consists of photophysical and biochemical processes, while the carbon fixing reaction is biochemical; therefore, the impacts of environmental stresses (such as strong light, temperature, drought, etc.) are expected to be different between these reactions, despite the activities of the two reactions being tightly linked [1].

An imbalance between electron source and sink can cause fatal damage to the photosynthetic apparatus, especially on photosystem I (PSI). When the electron transport chain is full of electrons, O_2 can be easily reduced into the superoxide radical (O_2^-) on the components with the lowest redox potential, in other words, the acceptor side of PSI [2].

Sejima et al. (2014), [3], devised an experimental method called repetitive short-pulse (rSP) illumination that promoted the electron accumulation in the electron transport chain without activation of the carbon fixing reaction. This rSP-illumination drives the electron source while suppressing the drive of the electron sink. That is, rSP-illumination induces the imbalance of the electron accumulation between them. Consequently, rSP-illumination induced O_2^-dependent PSI-specific inactivation. The acceptor side of PSI is the main generation site for reactive oxygen species (ROS) which cause oxidative damage to PSI (PSI photoinhibition) [2,4,5]. Photo-inactivated PSI has been shown to take days or weeks for full recovery [6–9], causing severe reductions in CO_2 fixation rate [3,10,11]. Thus, PSI photoinhibition can severely affect growth and may even be lethal for plants.

However, in nature, PSI photoinhibition rarely occurs in wild plants [12]. Under unsuitable environments for photosynthesis, coordinated linkage of the electron source and sink promotes oxidation of the PSI reaction center, P700, avoiding PSI photoinhibition [1,3,4]. P700 transports electrons as part of a redox cycle with three energy states. The oxidized form, $P700^+$, receives electrons from PSII via plastocyanin (PC) and is reduced into P700. The reduced form P700 is photo excited to P700*. P700* then passes the electron to the next electron acceptor and is re-oxidized to $P700^+$. The increase of oxidized P700 suggests that the rate-determining step in the P700 redox cycle is the reduction of $P700^+$ [1]. Under stress conditions, where CO_2 fixation (electron sink) was suppressed, P700 was generally more oxidized [13–17]. Theoretically, a decrease in the electron sink would predict a significant reduction of P700 (electron source). However, at this time, the trans-thylakoid ΔpH is increased, suppressing the electron transport activity of Cyt *b6f* [18]. Additionally, the redox balance of the plastoquinone (PQ) pool also slows the electron flow at the Cyt b6f [19]. The extent of the suppression of electron flow to PSI was larger than that of the decrease in the electron sink, resulting in oxidized P700 more [1,20]. Conversely, under the specific conditions (chilling and rSP-illumination) where PSI photoinhibition was observed, P700 oxidation was not induced [3,9]. PSI photoinhibition has also been observed in some mutants that could not promote P700 oxidation [21,22]. Oxidation of P700 avoids the accumulation of electrons in the acceptor side of PSI where ROS are generated, while oxidized P700 cannot use the light energy absorbed in PSI for electron transport but can directly dissipate it as heat [23–25]. In plants, promoting P700 oxidation can be regarded as a robust mechanism of avoiding ROS generation and protecting PSI.

Proton gradient regulation 5 (PGR5), a protein tethered in the thylakoid membrane, was identified in an Arabidopsis deficient mutant, *pgr5* [21], renamed as *pgr5-1* [26]; hereafter, also referred to as "*pgr5-1*" in this study. *pgr5-1* reduced proton gradient formation and P700 oxidation under high light [21,27]. PGR5 complexing with PGRL1 has been proposed to be a major component in cyclic electron flow around PSI [28]; however, the molecular function of PGR5 remains unclear. Under high light, preferential PSI photoinhibition was often observed in *pgr5-1* [21,26,29–31], and it was assumed that ROS production was accelerated by the accumulation of reduced iron-sulfur centers due to the loss of P700 oxidation. Moreover, *pgr5-1* significantly reduced survivability and growth under natural light conditions [32], indicating that PSI photoprotection by P700 oxidation is necessary for acclimation to natural light. Therefore, *pgr5-1* has been used as a valuable tool to investigate PSI photoinhibition, possible lethal damage caused by natural light stress [31–33].

In this study, we newly isolated a PGR5 deficient mutant named $pgr5^{hope1}$. Coincidentally, $pgr5^{hope1}$ and the known *pgr5-1* mutant were revealed as having the same point mutation in *PGR5* gene, although, *pgr5-1* grew smaller and had lower photo-oxidizable PSI (*Pm*) than $pgr5^{hope1}$ under both constant and natural light conditions. The phenotypic differences suggested the presence of additional factors besides PGR5, which was assumed to be related to the PSI photoprotection. So far, little is known about the factors involved in PSI photoprotection. Therefore, we aimed to find the novel factor, which we named PSI photoprotection 1 (PTP1). Whole-genome sequencing and complemented transformants revealed that PTP1 was encoded in AT2G17240 gene. AT2G17240 gene was identified previously as cgl20a, which interacted with plastidial ribosomes and affected plastome translation [34]. A mutation of the *PTP1* gene in *pgr5-1* enhanced PSI photoinhibition and exacerbated the growth reduction. Here, we introduce the novel factor involved in PSI photoprotection in Arabidopsis.

2. Materials and Methods

2.1. Plant Materials and Growth Condition

Arabidopsis thaliana (L.) Heynh. ecotype Columbia (Col) gl-1 and Col-0 were used as wild-type. In mutants used in this study, *pgr5*^{hope1} and *pgr5-1* were gl-1 backgrounds, and *ptp1-1*, *npq4* and *pgrl1ab* were Col-0 backgrounds. *pgr5*^{hope1} was screened from an ethyl methanesulfonate (EMS) mutant population, which was gl-1 background, by chlorophyll fluorescence measurement under low O_2 and CO_2 free conditions (Supplementary Figure S1) [35]. *pgr5-1*, *ptp1-1*, *npq4* and *pgrl1ab* were originated from [21], [34], [36], and [22], respectively. For the constant light condition, plants were cultivated in a growth chamber (BioTRON LPH-241, NKsystem, Osaka, Japan) with a light/dark regime of 10/14 h (light intensity; 250 μmol photons m^{-2} s^{-1}) with the temperature of 24/22 °C. For natural light conditions, plants were cultivated in a greenhouse with supplemental halogen lamps used in the morning and evening to extend the day length to 14 h and adjusted temperature at 23/20 °C, during March–May in 2020. The greenhouse is located in Kobe University (34°43′ north latitude, 135°14′ east longitude). Based on the climate data in Kobe, the daytime light intensity varied up to around 2000 μmol photons m^{-2} s^{-1}. The culture soil was a mixture of vermiculite and horticultural soil (Tanemaki-baido, Takii, Kyoto, Japan) at a ratio of 1:1.

2.2. Transformation

To produce complementation lines, *pgr5-1 PTP1*, *pgr5-1 PGR5* and *pgr5*^{hope1} *PGR5* constructs expressing *PGR5* or *PTP1* cDNA under the control of 35S promoter were introduced into *pgr5-1* or *pgr5*^{hope1} by agrobacterium-mediated transformation (*Agrobacterium tumefaciens* strain GV3101). mRNA was isolated from leaves of wild-type gl-1 using RNeasy plant mini kit (QUIAGEN, Hilden, Germany), which was used for cDNA synthesis by PrimeScript RT master mix (Takara, Shiga, Japan). Full-length CDS of *PGR5* and *PTP1* was amplified from the cDNA with specific primers described in Supplementary Table S1. Amplified CDSs were primarily inserted into the pENTR/D-topo entry vector (Thermo Fisher Scientific, Waltham, MA, USA) and finally introduced into pBI DAVL-GWR1 destination vectors (Inplanta Innovations, Yokohama, Japan) by the Gateway cloning system. Transformation into plants was performed by floral dip methods [37].

2.3. Determination of Phenotypic Recovery in Complementation Lines

The phenotypic recovery was assessed by four different parameters, the shoot fresh weight, SPAD value, the maximum quantum yield of PSII (*Fv/Fm*), and the maximum amount of photo-oxidizable P700 (PSI) (*Pm*). Sampled plants were weighed and the SPAD values were measured using a SPAD-502 (Konica Minolta, Tokyo, Japan). *Fv/Fm* and *Pm* were measured by Dual-Pam/F (Walz, Effertrich, Germany). The calculations of *Fv/Fm* and *Pm* were described below.

2.4. Simultaneous Measurement of Gas Exchange, Chlorophyll Fluorescence, and Absorbance Change Due to P700 Oxidation

Analysis of photosynthesis was conducted on rosette leaves of Arabidopsis about 30 days after sowing. CO_2 and H_2O gas exchanges were measured by a GFS-3000 system equipped with a Dual-PAM gas-exchange cuvette (Walz, Effertrich, Germany). For the measurement, ambient air CO_2 concentration (Ca), relative humidity, and leaf temperature were controlled at 400 ppm, 60%, and 25 °C, respectively. CO_2 fixation rate (A) was calculated by the system software based on the method of [38].

Chlorophyll fluorescence and absorbance change were measured using Dual-PAM 100 (Walz) simultaneously with gas exchange. Chlorophyll fluorescence parameters were calculated as follows based on [39]; *Fo*, minimal fluorescence in dark-adapted leaf; *Fm*, maximal fluorescence in dark-adapted leaf; Fm', maximal fluorescence in light-adapted leaf; Fs, fluorescence in steady-state; the effective PSII quantum yield (Y(II)), Y(II) = $(Fm' - Fs)/Fm'$; non-photochemical quenching (NPQ), NPQ = $(Fm - Fm')/Fm'$. And P700 absorbance parameters were calculated as follows based on [15]; Pm, the maximal signal of photo-oxidizable P700; Pm', the maximal signal of P700 photo-oxidized by saturating pulse flash under actinic light; P, the signal of P700 photo-oxidized under actinic light; the effective PSI quantum yield (Y(I)), Y(I) = $(Pm' - P)/Pm$; the quantum yield of non-photochemical energy dissipation due to donor-side limitation of PSI (Y(ND)), Y(ND) = P/Pm; the quantum yield of non-photochemical energy dissipation due to acceptor-side limitation of PSI (Y(NA)), Y(NA) = $(Pm - Pm')/Pm$. The sum of these quantum yields is 1. (Y(I) + Y(ND) + Y(NA) = 1) To determine the Fo and Fm, fully dark-adapted leaves (<30 min) were irradiated with a saturated light pulse (20,000 μmol-photons m^{-2} s^{-1}, for 300 ms). The maximal quantum yield of PSII, Fv/Fm, was calculated as; $Fv/Fm = (Fm - Fo)/Fm$. After Fo and Fm determination, Pm was determined by a saturating pulse in the presence of far-red light. Then, leaves were irradiated with actinic light. The light intensity of actinic light was increased stepwise (8.2, 26.1, 54.8, 110.8, 200.3, 434.6, 709.3, 903.0, 1150.3, 1447.6 μmol-photons m^{-2} s^{-1}), and the respective parameters were measured at each light intensity with application of a saturating pulse. After the photosynthesis measurement, the leaves were excised, frozen with liquid nitrogen, and stored at −80 °C before biochemical analysis.

2.5. Biochemical Analyses

The leaf N, chlorophyll and rubisco protein content were determined according to [40]. Leaf samples were ground with homogenization buffer (50 mM sodium phosphate, 5%(v/v) glycerol and 1 mM sodium iodoacetate). The aliquot was used for leaf N and chlorophyll determination based on the Kjeldahl method and [41]. For rubisco protein quantification, leaf soluble fractions were applied to sodium dodecyl sulfate polyacrylamide gel electrophoresis (SDS-PAGE). After Coomassie brilliant blue (CBB) staining, the bands corresponding to the large and small subunits of rubisco were excised, and the dye extracted with formamide was colorimetrically quantified. Bovine serum albumin (BSA) protein (Bovine Serum Albumin Standard, Thermo Fisher Scientific, Waltham, MA, USA) was used as a standard sample, and a standard curve was prepared.

2.6. Immunoblotting

Aliquots of leaf homogenate for biochemical analyses were applied for immunoblotting. The total leaf homogenate was combined with an equal volume of SDS-sample buffer (200 mM Tris-HCl (pH 8.5), 2%(w/v) SDS, 20%(v/v) glycerol, 5%(v/v) 2-ME), boiled for 2 min and stored at −30 °C until analysis. Immunoblotting was carried out with a 12.5%(w/v) acrylamide gel, a semi-dry blotting apparatus (Trans-Blot Turbo Transfer System; Bio-Rad, Hercules, CA, USA), a polyvinyldifluoridene (PVDF) membrane (Trans-Blot Turbo RTA transfer kit, mini, PVDF; Bio-Rad), a chemiluminescence detection kit (SuperSignal West Dura Extended Duration Substrate; Thermo Fisher Scientific), and an image analyzer (Ex-Capture MG; ATTO, Tokyo, Japan). All antibodies used in this study were purchased from Agrisera (Vännäs, Sweden). The product code of PsaA, Lhca1, PsbA, Lhcb5, PETB, NDHB and NDHH specific antibodies were AS06 172, AS01 005, AS05 084A, AS01 009, AS18 4169, AS16 4064 and AS16 4065, respectively. The Goat Anti-Rabbit IgG Horseradish Peroxidase Conjugated (Thermo Fisher Scientific) was used for the secondary antibody.

2.7. Photoinhibition Experiment

For the photoinhibition in PSII and PSI, attached leaves of plants were exposed to actinic light (900 μmol-photons m^{-2} s^{-1}) for 2 h in a Dual-PAM cuvette and Dual-PAM 100 (Walz, Effertrich, Germany). The experimental condition was the same as in the measurement of photosynthesis parameters described above. Fv/Fm and Pm were determined before tuning on the actinic light and after 30 min dark treatment after turning off the light.

2.8. Whole-genome Sequences Analysis

The whole-genome sequence analysis was performed with NovaSeq 6000 by Macrogen (Kyoto, Japan). The TAIR 10.1 was used as a reference genome sequence of Arabidopsis.

3. Results

3.1. Identification of the PSI Photoprotection 1, PTP1; a Recessive Mutation in AT2G17240 Gene Enhanced the Growth Phenotype of pgr5-1, a Well-known PGR5 Deficient Mutant

We previously performed screening for Arabidopsis mutants in ΔpH formation across thylakoid membrane by monitoring chlorophyll fluorescence under hypoxic conditions (*hope* mutant screening; hunger for oxygen in photosynthetic electron transport reaction) [34]. *hope1* was isolated as a high chlorophyll fluorescent mutant in hypoxic conditions (Supplementary Figure S1A,B) and was identified a mutation in proton gradient regulation 5 (*PGR5)* gene by genome mapping. Coincidently, the mutation was in the same position (388 G to A) as *pgr5-1*, a known PGR5 deficient mutant (Supplementary Figure S1C [21]). The mutation resulted in PGR5 protein deficiency in both *pgr5-1* and *hope1* (Supplementary Figure S1D and Supplementary Table S2). As *pgr5-1* and *hope1* were genetically the same alleles as *pgr5* mutants, hereafter, *hope1* was referred to as $pgr5^{hope1}$ in this study.

Figure 1 showed the growth phenotypes of *pgr5-1* and $pgr5^{hope1}$ grown under relatively high (250 μmol-photons m^{-2} s^{-1}) constant light (Figure 1A,B), and natural light condition (varying below 2000 μmol-photons m^{-2} s^{-1}; Figure 1C,D). Despite the identical mutation, under both light conditions, *pgr5-1* was significantly smaller than $pgr5^{hope1}$. Although the growth of *pgr5-1* was severely affected by light intensity, the growth declines of *pgr5-1* under constant and natural light conditions were consistent with [32,42]. Whereas $pgr5^{hope1}$ showed almost the same growth as wild-type under constant light conditions, although it tended to be slightly smaller (Figure 1A,B). Under natural light, $pgr5^{hope1}$ exhibited a reduction to around 35% of the fresh weight compared to the wild-type, while *pgr5-1* achieved masses around 11% of wild-type (Figure 1C,D). These differences suggested that the influence of genetic factors other than *PGR5* were involved in these growth phenotypes. All F1 hybrids of *pgr5-1* and $pgr5^{hope1}$ grew the same level as $pgr5^{hope1}$ (Figure 1A,B). Based on the fresh weight and the SPAD value, an indicator for chlorophyll amount, the growth phenotype of the F2 progeny exhibited Mendelian segregation of approximately 1:3 (16:62) in 78 plants, indicating the existence of a second genetic factor affecting *pgr5-1* or $pgr5^{hope1}$ growth. The growth reduction in *pgr5-1* compared to the wild-type was reported to be mainly due to the PSI photoinhibition [21,43]. We considered that the second mutation could act positively or negatively on the PSI photoinhibition in *pgr5* mutant, resulting in different phenotypes. Thus, we named this factor PSI photoprotection 1, *PTP1*.

For genetic identification, we performed whole-genome sequencing on *pgr5-1* and $pgr5^{hope1}$. So far, *pgr5-1* and $pgr5^{hope1}$ have no phenotypic segregation in hybrid strain production or backcrossing. Therefore, it was considered that the *PTP1* was positioned near the *PGR5* locus on chromosome 2 and was difficult to segregate from PGR5 mutation. The candidate mutations on chromosome 2 that existed independently in each of these mutants and that had a severe effect (e.g., amino acid substitution) were listed in Supplementary Table S3. We selected several candidate mutations from the list and analyzed their T-DNA insertion mutants. One of them, *salk_133989*, a T-DNA insertion allele for AT2G17240, exhibited a pale green phenotype which partially resembled *pgr5-1* phenotype (Supplementary Figure S2). AT2G17240 was reported to encode CGL20A protein and the *salk_133989* was its defective mutant [34]. A point mutation in AT2G17240 was found in *pgr5-1* (Supplementary Table S3) which altered C to T at base 278 of the coding DNA, predicting an amino acid substitution at position 93 from Pro to Leu (Supplementary Figure S3). Here, we tentatively termed AT2G17240 gene *PTP1*, while the mutant alleles, T-DNA insertion mutant alleles (*salk_133989*) and one base substitution in *pgr5-1* were named *ptp1-1* and *ptp1-2*, respectively (Supplementary Figure S3).

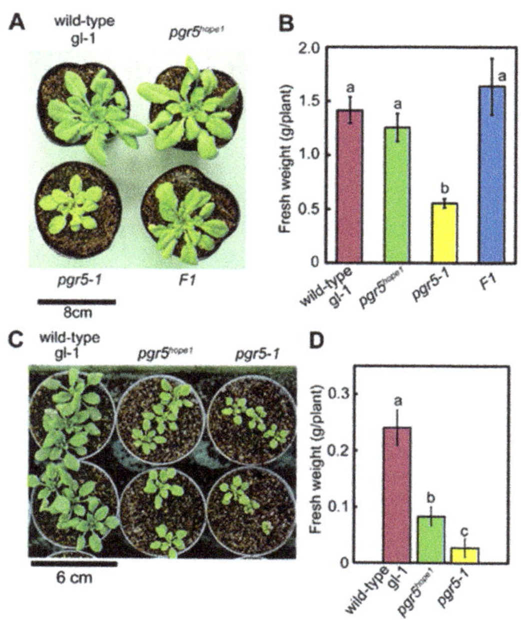

Figure 1. The difference of the growth phenotype between *pgr5-1* and *pgr5^{hope1}*. (**A**) is a picture of representative plants (wild-type (gl-1), *pgr5^{hope1}*, *pgr5-1*, and F1 hybrid of *pgr5-1* and *pgr5^{hope1}*) grown under constant light conditions (250 µmol-photons m^{-2} s^{-1}) for 30 days. (**B**) shows the fresh weight of plants in (**A**) Data are means ± sd. (n = 4–5) (**C**) is a picture of representative plants (wild-type (gl-1), *pgr5^{hope1}* and *pgr5-1*) grown under natural light condition for 20 days. (**D**) shows the fresh weight of plants in (**C**). Data is means ± sd. (n = 9) In (**B**,**D**), Different alphabets indicate significant differences analyzed by Tukey's HSD-test. ($p < 0.05$) Experiments were independently repeated at least 3 times and showed similar results. Figures showed the representatives.

To determine whether PTP1 caused the phenotypic difference between *pgr5^{hope1}* and *pgr5-1*, we produced several complementation lines and observed phenotypic recoveries under constant (Figure 2A–C) and natural light condition (Figure 2D–F). We assessed the phenotypic recovery as fresh weight and SPAD value (Figure 2B,E), and the maximum yield of PSII (*Fv*/*Fm*) and the maximum amount of photo-oxidizable PSI (*Pm*; Figure 2C,F). The genetic background is different between *pgr5-1*, *pgr5^{hope1}* (gl-1) and *ptp1-1* (Col-0). However, the wild-type gl-1 and Col-0 didn't show any significant differences in their growth (Figure 1A,D) and measured parameters (Figure 1B,C,E,F). Under the constant light condition, *pgr5^{hope1}* slightly reduced the fresh weight, SPAD value, and *Pm* compared to wild-type (Figure 2B,C). The complementation line, *pgr5^{hope1} PGR5*, which induced *PGR5* in the *pgr5^{hope1}* background under the control of 35S promoter, restored these reductions to the wild-type level (Figure 2B,C). *pgr5^{hope1}* significantly decreased in growth and *Pm* under natural light, but *pgr5^{hope1} PGR5* restored these decreases to the wild-type level (Figure 2E,F). These results strongly indicated that *pgr5^{hope1}* was a *pgr5* single mutant. In contrast, *pgr5-1* significantly reduced fresh weight, SPAD value, and *Pm* under both constant and natural light conditions (Figure 2B,C,E,F). *pgr5-1 PGR5*, which induced *PGR5* in the *pgr5-1* background, largely restored growth rate and *Pm*. However, *pgr5-1 PGR5* remained low SPAD value and showed pale green leaves similar to *ptp1-1*. Partial phenotypic recovery of *pgr5-1 PGR5* suggested the existence of other mutations causing the pale green leaves. Whereas *pgr5-1 PTP1*, which expressed *PTP1* in *pgr5-1* background, restored the fresh weight, the SPAD value and *Pm* to almost the same level as *pgr5^{hope1}* under constant light condition (Figure 2B,C). In addition to constant light, *pgr5-1 PTP1* restored these parameters to the *pgr5^{hope1}* level under natural light (Figure 2E,F). These recoveries indicated that the introduction of *PTP1* in *pgr5-1* was largely compensated for the phenotypic differences between *pgr5-1* and *pgr5^{hope1}*.

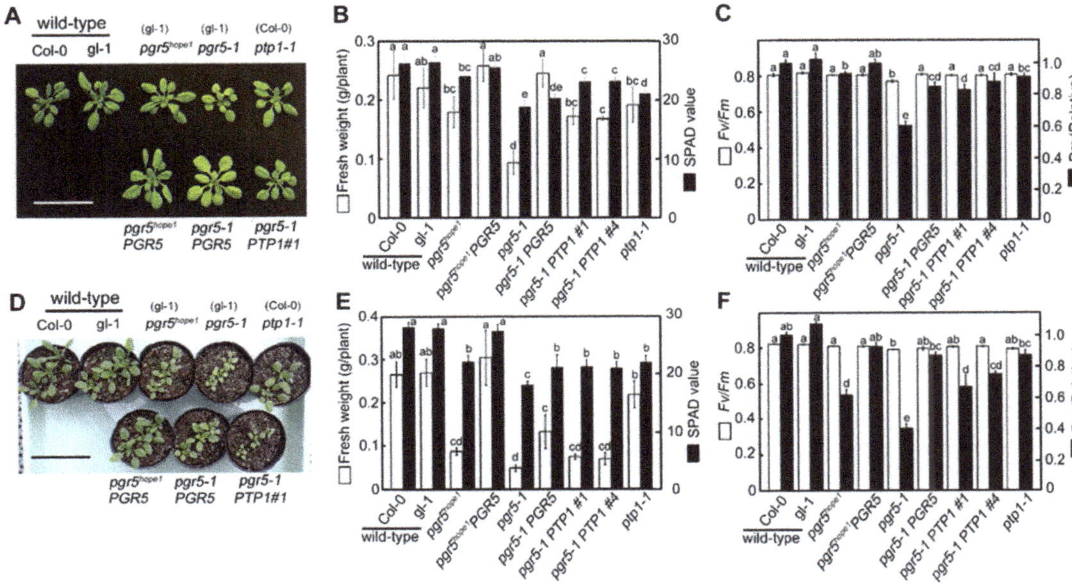

Figure 2. The phenotypic recovery in complementation lines. (**A–C**) and (**D–F**) were the data of plants grown under constant (250 μmol-photons m^{-2} s^{-1}) and natural light conditions, respectively. (**A**,**D**) are pictures of representative plants grown for 22 days and 20 days, respectively. The parentheses indicate the background genotypes of each mutant. The scale bars in (**A**,**D**) are 5 and 8 cm, respectively. (**B**,**E**) show fresh weight (white bars) and SPAD values (black bars). Data are means ± sd. ((**B**), $n = 6$; (**E**), $n = 4–5$) (**C**,**F**) show the maximum quantum yield in PSII (*Fv/Fm*; white bars) and the maximum amount of photo-oxidizable PSI (*Pm*; black bars). Data are means ± sd. (C, $n = 5–6$; F, $n = 4–5$) *Pm* is a relative value with the value wild-type Col-0 being 1. In (**B**,**C**,**E**,**F**), the different alphabets indicate significant differences analyzed by Tukey's HSD test. ($p < 0.05$).

To confirm the effect of double mutation of *PTP1* and *PGR5*, we attempted to cross-breed *pgr5^{hope1}* and *ptp1-1*. However, the double mutant could not be obtained from the cross-bred strains, probably due to the proximity of their loci on chromosome 2. Alternatively, we produced a double-deficient mutant of *PGRL1* and *PTP1* (*pgrl1ab ptp1-1*) and analyzed the growth phenotypes under constant light conditions (Figure 3). PGRL1 forms a protein complex with PGR5, and the deficient mutant, *pgrl1ab*, is incapable of P700 oxidation like PGR5 deficient mutant [22]. *pgrl1ab ptp1-1* double mutant significantly reduced the growth and *Pm* like *pgr5-1*, while the *pgrl1ab* grew similar to wild-type like *pgr5^{hope1}* (Figure 3). In addition, we produced a double mutant *npq4 ptp1-1* to compare the effect of PTP1 deficiency in another photodamage susceptible mutant. *npq4* mutant is deficient in PsbS protein, reducing its ability of photoprotective thermal energy dissipation [36]. Consequently, the growth, *Fv/Fm* and *Pm* did not change between *npq4* and *npq4 ptp1-1* (Figure 3), indicated that PTP1 deficiency did not exacerbate the growth inhibition of *npq4*. From the results of complementation lines and *pgrl1ab ptp1-1*, we concluded that the second genetic factor *PTP1* affecting the growth phenotype in *pgr5-1* was AT2G17240 gene. As a result, *pgr5-1* was determined to be a double mutant of *pgr5-1 ptp1-2*. We tried to but have been so far unsuccessful in detecting the PTP1 protein, even in wild-type plants. Therefore, the detailed effect of the *ptp1-2* mutation on PTP1 protein in *pgr5-1* remained unclear. However, under constant light conditions, *pgr5-1 PTP1* and *ptp1-1* deficient mutant reduced the SPAD value to a similar level (Figure 2B,E) and reduced chlorophyll content to the same level (Table 1). Furthermore, in comparison with wild-type, the growth reduction in *pgr5-1* was relatively larger than in the *pgrl1ab ptp1-1* (Figures 2B and 3B). Thus, it was estimated that the *ptp1-2* mutation in *pgr5-1* caused a similar situation to protein deficiency or further impairment of its function.

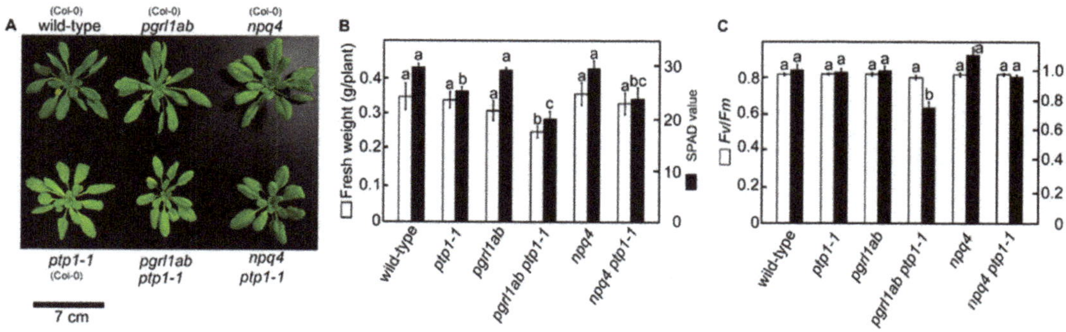

Figure 3. The effect of PTP1 mutation on the growth of PGRL1 deficient mutants. (**A–C**) were the data of plants grown under constant light conditions (250 μmol-photons m^{-2} s^{-1}). **A** is a picture of representative plants grown for 28 days. The parentheses indicate the background genotypes of each mutant. The scale bar is 7 cm. (**B**) shows the fresh weight (white bars) and SPAD values (black bars). Data are means ± sd. (n = 4). (**C**) shows the maximum quantum yield in PSII (*Fv/Fm*; white bars) and the maximum amount of photo-oxidizable PSI (*Pm*; black bars). Data are means ± sd. (n = 4) *Pm* is a relative value with the value wild-type Col-0 being 1. In (**B**,**C**), the different alphabets indicate significant differences analyzed by Tukey's HSD test. ($p < 0.05$).

Table 1. The chlorophyll, leaf total N and Rubisco amount in *pgr5-1*, *pgr5^{hope1}* and *ptp1-1*. After photosynthesis measurement (Figure 5), leaves were used for the determination of leaf components. Dara are means ± sd. (n = 4–5) Different alphabets beside the numbers indicate the significant difference analyzed by Tukey's HSD test ($p < 0.05$).

Genotypes (Background)	Chlorophyll mmol m^{-2}	Chlorophyll a/b	Leaf N mmol m^{-2}	Rubisco g m^{-2}
wild-type (Col-0)	0.377 ± 0.041 a	3.08 ± 0.20 a	80.6 ± 6.9 a	1.45 ± 0.17 a
wild-type (gl-1)	0.366 ± 0.024 a	3.09 ± 0.40 a	85.8 ± 4.6 a	1.42 ± 0.13 a
pgr5^{hope1} (gl-1)	0.385 ± 0.009 a	2.96 ± 0.30 a	82.4 ± 5.6 a	1.48 ± 0.15 a
pgr5-1 (gl-1)	0.287 ± 0.030 b	3.30 ± 0.20 a	78.4 ± 4.0 a	1.27 ± 0.09 a
ptp1-1 (Col-0)	0.293 ± 0.019 b	3.29 ± 0.19 a	80.2 ± 5.6 a	1.51 ± 0.11 a

3.2. The Biochemical and Physiological Damages on Photosynthetic Apparatus Caused by the ptp1 Mutation

pgr allele possessed the second recessive mutation, *ptp1-2*, and was regarded as a double mutant, *pgr5-1 ptp1-2*, whereas *pgr5^{hope1}* was regarded as a single PGR5 deficient allele, based on the phenotypic recovery. To reveal the molecular function of PTP1, we compared the difference between *pgr5^{hope1}* and *pgr5-1*. First, we determined the total leaf N, chlorophyll and rubisco content (Table 1), and thylakoid protein levels, mainly in electron transport chain components of plants grown under constant light conditions (Figure 4). No significant differences in leaf total N and rubisco content were noted between genotypes. On the other hand, *pgr5-1* and *ptp1-1* deficient mutants exhibited significantly reduced chlorophyll contents compared to wild-type and *pgr5^{hope1}*; consistent with the SPAD value and the pale green leaf color (Figures 1A and 2A). But the chlorophyll a/b ratio was not significantly changed in *pgr5-1* and *ptp1-1* (Table 1). Moreover, only *pgr5-1* largely reduced the concentration of the core subunit of PSI (PsaA) to around 66% of wild-type levels, but not other tested proteins in the electron transport chain (Lhca1, PsbA, Lhcb5, and PETB). Previous studies have shown that the core subunit of PSI, PsaA, is specifically degraded in the process of PSI photoinhibition [9,43]. In *pgr5-1*, the specific decrease of the PSI core protein was observed with PSI inactivation [32,44], suggesting severe PSI photoinhibition. However, the PsaA level in *pgr5^{hope1}* was only marginally reduced, to around 90% of wild-type levels (Figure 4). Thus, the PSI photoinhibition in *pgr5-1* was suggested not only due to PGR5 deficiency but was enhanced by the *ptp1-2* mutation. In addition, *ptp1-1* deficient mutant was reported a decrease in proteins of NDH complex [34]. We also detected the decrease of NDHB and NDHH proteins in the complex to around 70% of wild-type in *ptp1-1* and *pgr5-1* (Figure 4). The growth reduction of *pgr5-1* was largely enhanced by crossing with NDH complex deficient mutants (*crr2* and *crr4*) [45]. The reduction of NDH complex by PTP1 mutation may be associated with the growth phenotype of *pgr5-1* and *pgr5^{hope1}*.

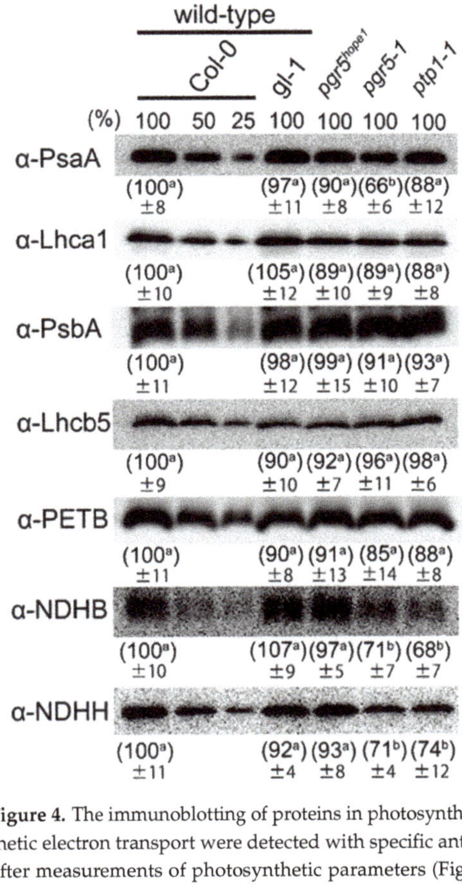

Figure 4. The immunoblotting of proteins in photosynthetic electron transport. Proteins in photosynthetic electron transport were detected with specific antibodies in each plant. Leaves were samples after measurements of photosynthetic parameters (Figure 5). Samples were loaded based on the same leaf area. The experiment was repeated at least three times with similar results. The values in parentheses show relative values of band intensities when the wild-type Col-0 is 100. Data are means ± sd. (n = 3–4) Different alphabets beside the numbers indicate the significant difference analyzed by Tukey's HSD test ($p < 0.05$). The images were representative.

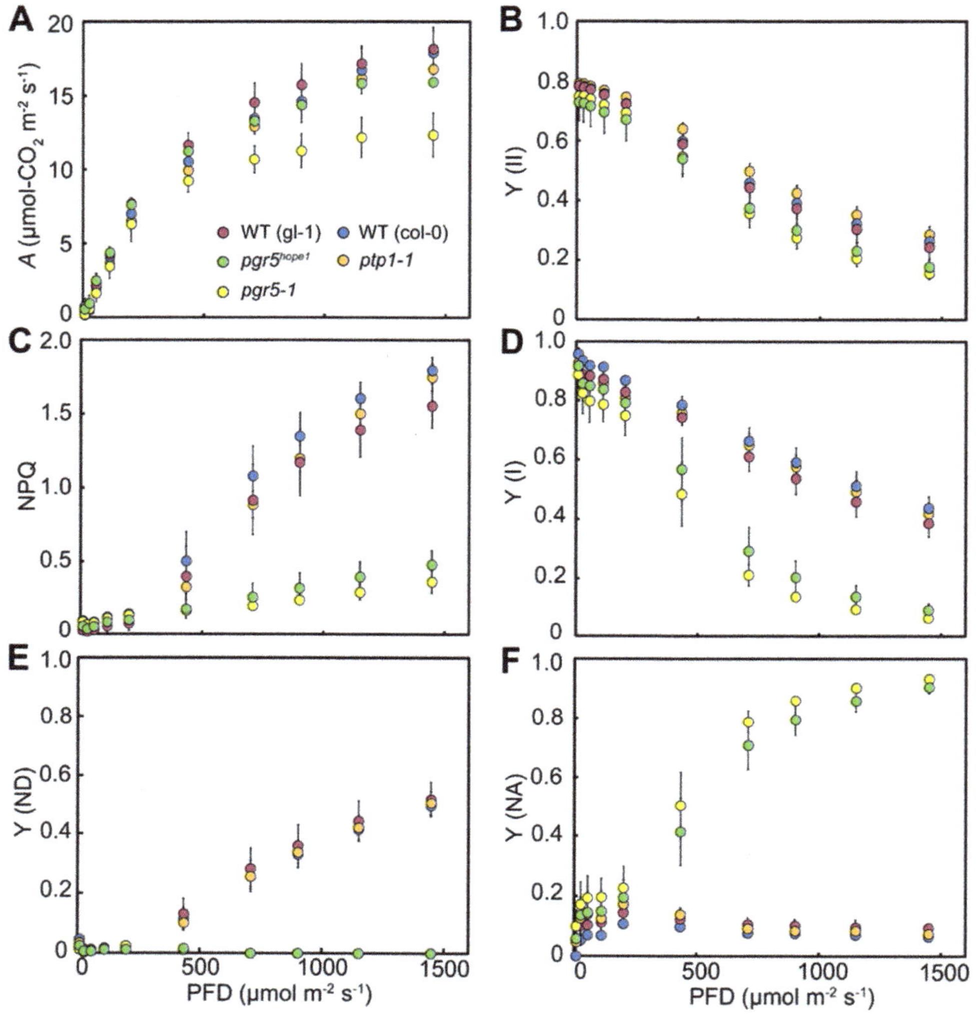

Figure 5. The photosynthetic capacity in *pgr5-1*, *pgr5^{hope1}* and *ptp1-1*. Light intensity-dependent changes in photosynthetic parameters; (**A**), CO_2 fixation rate, (**B**), the quantum yield in PSII (Y(II)), (**C**), non-photochemical quenching, (**D**), the quantum yield in PSI (Y(I)), (**E**), the ratio of oxidized P700 (P700$^+$) in PSI (Y(ND)), (**F**), the ratio of excited P700 (P700*) in PSI (Y(NA)). Magenta, blue, green, yellow and orange are wild-type gl-1, wild-type Col-0, *pgr5^{hope1}*, *pgr5-1* and *ptp1-1*, respectively. Data are means ± sd (n = 5). Experiments were independently repeated at least 3 times with similar results. Graphs show representative results. The results of statistical analysis of these data are summarized in the Supplementary Table S4.

PSI photoinhibition has been shown to cause severe reductions in CO_2 fixation rate [3,10,11]. We next determined CO_2 fixation rate (A), chlorophyll fluorescence and absorbance change simultaneously, using plants grown under constant light conditions (Figure 5). A in *pgr5-1* was significantly reduced, especially under high light (<500 μmol-photons m^{-2} s^{-1}), to about 60% of wild-type (Figure 5A). A large reduction of A in *pgr5-1* was previously observed in [42], mainly due to PSI photoinhibition caused by PGR5 deficiency. *pgr5^{hope1}* also reduced A, but not as much as *pgr5-1*, to about 90% of wild-type levels under high light (<500 μmol-photons m^{-2} s^{-1}) condition. This difference in *A* between *pgr5*^{hope1} and *pgr5-1* suggested that the *PTP1* mutation largely enhanced the reduction in *A* under PGR5 deficient condition. A similar decrease in *A* was confirmed between *pgrl1ab ptp1-1* and *pgrl1ab* (Supplementary Figure S4). In contrast, both *pgr5-1* and *pgr5*^{hope1} exhibited similarly reduced quantum yields in PSII and PSI, Y(II) and Y(I), compared to wild-type (Figure 5B,D). *pgr5-1* was originally isolated as a mutant incapable of qE-dependent non-photochemical quenching (NPQ), an indicator for trans-thylakoid ΔpH formation, and P700 oxidation [Y(ND)] under high light ([21]; Figure 5C,E in this study). *pgr5*^{hope1} also showed the same low levels of NPQ and Y(ND) as *pgr5-1*. These results indicated that changes in electron transport capacities, including the loss of P700 oxidation, were due solely to PGR5 deficiency. In contrast, *ptp1-1* deficient mutants showed almost the same photosynthetic capacities as wild-type (Figure 5). These results indicated that PTP1 was neither necessary in photosynthesis nor directly involved in P700 oxidation.

3.3. PSI Photoinhibition Was Triggered by PGR5/PGRL1 Deficiency

The specific core protein degradation in PSI and the reduction in CO_2 fixation rate suggested that the mutation of PTP1 enhanced PSI photoinhibition in *pgr5-1* (Figures 4 and 5A). Next, we analyzed the extent of photoinhibition in PSII and PSI caused by constant intense light (900 µmol-photons m^{-2} s^{-1}), using plants grown under constant light conditions (Figure 6). In wild-type and *ptp1-1*, no changes in the maximum quantum yield of PSII (Fv/Fm) and the maximum amount of photo-oxidizable PSI (*Pm*) were found before or after high light irradiation (Figure 6B,C). It suggested that PTP1 deficiency was not a trigger for PSI photoinhibition. On the contrary, *pgr5*^{hope1} and *pgr5-1* significantly decreased both Fv/Fm and *Pm*, especially *Pm*, after irradiation. Similar results were observed with the *pgrl1ab* mutant (Supplementary Figure S5). These results showed that PSI photoinhibition, whether PTP1 was present or absent, was primarily induced by PGR5/PGRL1 deficiency. The reduction rate of *Pm* was significantly larger in *pgr5-1* (53%) than in *pgr5*^{hope1} (40%). However, *pgr5-1* exhibited significantly lower *Pm* before irradiation, suggesting that the actual difference in PSI damage was even smaller. The low *Pm* before high light irradiation also suggested that the damage of PSI photoinhibition was accumulated constantly in *pgr5-1*. These results imply that PTP1 did not avoid the occurrence of PSI photoinhibition but suppressed the damages to PSI.

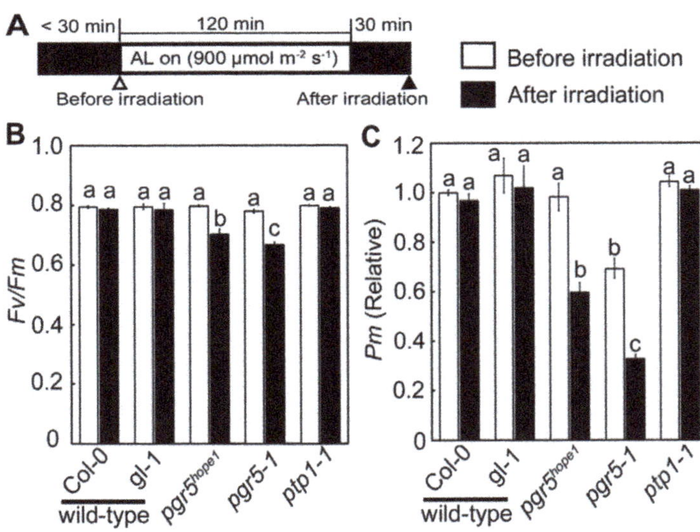

Figure 6. The extent of photoinhibition by constant intense light. (**A**) is an illustration of an experimental scheme for photoinhibition by constant intense light. The leaves that had been fully darkened

(<30 min) were exposed to intense constant light (AL; 900 μmol-photons m^{-2} s^{-1}) for 2 h under atmospheric conditions, and then darkened for 30 min. Saturated pulse flush was exposed before AL on (before irradiation) and after darkness for 30 min with AL off (after irradiation). (**B**,**C**) show the maximum quantum yield in PSII (*Fv*/*Fm*) and the maximum amount of photo-oxidizable PSI (*Pm*) before and after irradiation, respectively. Black and white bars are before and after irradiation. In (**C**), *Pm* is a relative value with the value wild-type Col-0 before irradiation being 1. Data are means ± sd. (n = 3) Experiments were independently repeated at least 3 times with similar results. Graphs show representative results.

4. Discussion

In this study, we identified the AT2G17240 gene, named *PTP1*, as a novel factor for PSI photoprotection. Mutations in PTP1 exacerbated PSI photoinhibition and reduced the growth of *pgr5* and *pgrl1* mutants. PGR5 and PGRL1 that form a protein complex are suggested to be the main components of cyclic electron transport around PSI (CEF-PSI) [21,22], but its actual molecular function has not yet been specified. *ptp1-1* deficient mutant performed the same photosynthetic activity including ΔpH formation (NPQ) as wild-type plants (Figure 5). In addition, PTP1 was reported as CGL20A to function in ribosome biogenesis in plastids [34]. Thus, PTP1 was unlikely to be directly involved in the linear electron transport or CEF-PSI. The physiological roles of CEF-PSI were suggested to protect PSI from excess light energy by sustaining high ΔpH and produce ATP for CO_2 assimilation [28]. However, *pgr5*^{hope1} showed significantly larger *Pm* than *pgr5-1* under both constant and natural light conditions (Figure 2C,F) and showed nearly the same CO_2 assimilation level as wild-type under saturated light conditions (Figure 5A). These results indicated that the physiological roles of CEF-PSI based on the analysis of *pgr5-1* could have been overestimated.

In addition, it was suggested that the existence of CEF-PSI itself needed to be reconsidered in other experiments. As for the CEF-PSI pathway carried by PGR5 and PGRL1, ferredoxin-quinone reductase dependent (FQR) pathway in which electrons transfer from ferredoxin (Fd) to plastoquinone (PQ) in the electron transport chain is proposed. However, the electron transport rates at Fd and PSII showed a positive linear relationship without intercept, suggesting that electron transports other than the linear electron transport were negligible in abundance [46]. Furthermore, a direct measurement system of CEF-PSI has not been established so far, and the difference in quantum yield between PSI and PSII (ΔY(I) = Y(I) − Y(II)), which has been used as an alternative index, is also doubtful as an evaluation of CEF-PSI. The quantum yield of PSI (Y(I)) at steady state measured by saturation pulse irradiation could be overestimated depending on the redox state of plastocyanin (PC) [20]. Assuming that ΔY(I) was properly assessing the CEF-PSI, ΔY(I) was not induced by PGR5/PGRL1 deficient mutants not only under steady-state but also under fluctuating light [26]. However, the introduction of moss flavodiiron protein (FLV) into *pgr5-1* mutant restored ΔY(I) along with the oxidation regulation of P700 [26]. This recovery was further confirmed by the introduction of FLV in *PGR5-RNAi* transformant in rice (*Oryza sativa*) [47]. FLV, theoretically, played a role as an alternative electron sink for linear electron transport by transporting electrons to oxygen to generate water (pseudo-cyclic electron transport). However, FLV stimulated the photosynthetic linear electron transport and oxidized P700 in PSI, which induced the extra electron flux in PSI, ΔY(I), in both wild-type and *pgr5-1* plants [26]. These studies indicated that ΔY(I) was not induced by PGR5/PGRL1. In this study, ΔY(I) was also observed in wild-type, and not in *pgr5*^{hope1} and *pgr5-1* (Figure 5B,C; Supplementary Figure S6). However, the reduction of CO_2 fixation rate in *pgr5-1* was caused by *ptp1-2* mutation, and *pgr5*^{hope1} and wild-type exhibited no significant differences in CO_2 fixation rate (Figure 5A). Therefore, it was shown that PGR5-dependent ΔY(I) contributed little to the CO_2 fixation reaction even when evaluated functionally. On the other hand, the light stress sensitivity in *pgr5*^{hope1} indicated that PGR5 was necessary for P700 oxidation regulation, and important for the avoidance of PSI photoinhibition (Figure 6).

However, considering the growth of *pgr5^hope1^* compared to *pgr5-1* under constant light and natural light of some intensity, it was implied that there was a mechanism that suppressed PSI photoinhibition other than P700 oxidation regulation.

In the Arabidopsis genome, two highly homologous genes encode *CGL20*, *CGL20A* (PTP1; AT2G17240) and *CGL20B* (AT3G24506). It is interesting to see if these two genes duplicate the role of PSI photoprotection. The double deficient mutant, *cgl20ab*, exhibited significantly reduced plastome-encoded protein production, such as PnsL1-4 and PnSB1-2 in NDH complex, PsbA in PSII, PETB and PETD in Cyt *b6f*, PsaN in PSI [34]. Not only in the *cgl20ab* double knockout mutant, but *ptp1-1* (*cgl20a* single mutant) also showed a slight reduction in some subunits of NDH complex [34]. We also detected the decrease of subunits of NDH complex, NDHB and NDHH (Figure 4). NDH complex was Fd-dependent plastoquinone reductase that was proposed to be responsible for the minor pathway of CEF-PSI [48]. Although the mechanism was still unclear, *pgr5-1* was further reduced in growth by double deficiency with NDH complex [44]. The reduced NDH complex by the mutation of PTP1 may reduce the growth of *pgr5-1*. However, NDH deficient mutants, such as *crr2*, did not exhibit pale green phenotypes like *ptp1-1* [49]. The effect of *PTP1* mutation which enhanced PSI photoinhibition in PGR5/PGRL1 deficient mutants may have another cause within the chloroplast. Further investigation of the relationship between NDH complex and PSI photoinhibition under PGR5 deficiency may help to understand the molecular mechanisms of PSI photoprotection. On the other hand, *pgr5-1* significantly reduced PsaA level while *pgr5^hope1^* was nearly the same as wild-type (Figure 4). PSI protein turnover was relatively slow and full recovery is known to take several days [6,44,50]. In PSI, however, the core protein PsaA showed relatively faster turnover rate than other PSI proteins [51]. If PTP1 was responsible for PsaA turnover, it might alleviate PSI photoinhibition in its repair or reassembly process.

PSI photoinhibition is caused by ROS generated at the acceptor side of PSI [3,52,53]. ROS scavenging was considered as a photoprotective mechanism for photosystems [2]. In the acceptor side of PSI, some ROS scavenging enzymes, such as APX and SOD, function in the well-known alternative electron flow, the water–water cycle [5]. Arabidopsis has several APX and SOD genes all of which are encoded in the nuclear genome. Thus, PTP1 was unlikely to be involved in the expression of these proteins encoded in the nuclear genome. However, it was reported that chloroplastic Fe-SOD deficient mutant in Arabidopsis, *fsd2* and *fsd3*, showed sensitivity to oxidative stress [54,55], while in cyanobacteria, the lack of Fe-SOD strain caused the PSI photoinhibition [56]. A reduction in ROS scavenging capacities may enhance PSI photoinhibition.

Supplementary Materials: The following are available online at https://www.mdpi.com/article/10.3390/cells10112884/s1, Supplementary Figure S1; *pgr5^hope1^* screening method and mutation identification; Figure S2; The growth phenotype of *ptp1-1* (*salk_133989*) mutant; Figure S3; The *PTP1* mutations in *ptp1-1* and *pgr5-1*; Figure S4; The photosynthesis capacity of *pgrl1ab* and *pgrl1ab ptp1* mutants; Figure S5; The extent of photoinhibition in PSII and PSI by constant intense light using *pgrl1ab* and *pgrl1ab ptp1* mutants; Figure S6; The relationship between Y(II) and Y(I) in wild-type, *pgr5^hope1^* and *pgr5-1*; Table S1; Primer information used in this study; Table S2; The relative amount of PGR5 protein in each mutant (per chlorophyll); Table S3; The mutation list on the chromosome 2 in *pgr5^hope1^* and *pgr5-1*; Table S4; The result of statistical analysis of photosynthetic parameters (Figure 5).

Author Contributions: S.W. performed experiments. K.A. provided technical support. S.W. and C.M. designed experiments analyzed results and wrote the manuscript. S.W. serves as the author responsible for contact and communication. All authors have read and agreed to the published version of the manuscript.

Funding: This work was supported by grants from Japan Society for the Promotion of Science (JSPS) KAKENHI grant number JP20K05769 (S.W.) and from JST CREST grant number JPMJCR1503 (C.M.).

Institutional Review Board Statement: Not applicable.

Informed Consent Statement: Not applicable.

Acknowledgments: The authors would like to thank D. Takagi for the isolation of Arabidopsis EMS mutant. The authors also would like to thank K. Noguchi, K. Ifuku, and L. Irving for the exciting discussion and critical reading of the manuscript.

Conflicts of Interest: The authors declare no conflict of interest.

References

1. Miyake, C. Molecular Mechanism of Oxidation of P700 and Suppression of ROS Production in Photosystem I in Response to Electron-Sink Limitations in C3 Plants. *Antioxidants* **2020**, *9*, 230. [CrossRef]
2. Asada, K. The water–water cycle as alternative photon and electron sinks. *Philos. Trans. R. Soc. B Biol. Sci.* **2000**, *355*, 1419–1431. [CrossRef]
3. Sejima, T.; Takagi, D.; Fukayama, H.; Makino, A.; Miyake, C. Repetitive Short-Pulse Light Mainly Inactivates Photosystem I in Sunflower Leaves. *Plant Cell Physiol.* **2014**, *55*, 1184–1193. [CrossRef]
4. Shimakawa, G.; Miyake, C. Changing frequency of fluctuating light reveals the molecular mechanism for P700 oxidation in plant leaves. *Plant Direct* **2018**, *2*, e00073. [CrossRef]
5. Asada, K. The water-water cycle in chloroplasts: Scavenging of Active Oxygens and Dissipation of Excess Photons. *Annu. Rev. Plant Biol.* **1999**, *50*, 601–639. [CrossRef]
6. Kudoh, H.; Sonoike, K. Irreversible damage to photosystem I by chilling in the light: Cause of the degradation of chlorophyll after returning to normal growth temperature. *Planta* **2002**, *215*, 541–548. [CrossRef]
7. Li, X.-G.; Wang, X.-M.; Meng, Q.-W.; Zou, Q. Factors Limiting Photosynthetic Recovery in Sweet Pepper Leaves After Short-Term Chilling Stress Under Low Irradiance. *Photosynthetica* **2004**, *42*, 257–262. [CrossRef]
8. Zhou, Y.H.; Yu, J.Q.; Huang, L.F.; Nogues, S. The relationship between CO2 assimilation, photosynthetic electron transport and water-water cycle in chill-exposed cucumber leaves under low light and subsequent recovery. *Plant Cell Environ.* **2004**, *27*, 1503–1514. [CrossRef]
9. Zhang, S.; Scheller, H. Photoinhibition of Photosystem I at Chilling Temperature and Subsequent Recovery in Arabidopsis thaliana. *Plant Cell Physiol.* **2004**, *45*, 1595–1602. [CrossRef]
10. Zivcak, M.; Brestic, M.; Kunderlikova, K.; Sytar, O.; Allakhverdiev, S.I. Repetitive light pulse-induced photoinhibition of photosystem I severely affects CO2 assimilation and photoprotection in wheat leaves. *Photosynth. Res.* **2015**, *126*, 449–463. [CrossRef]
11. Shimakawa, G.; Miyake, C. What Quantity of Photosystem I Is Optimum for Safe Photosynthesis? *Plant Physiol.* **2019**, *179*, 1479–1485. [CrossRef] [PubMed]
12. Sonoike, K. Photoinhibition of photosystem I. *Physiol. Plant* **2011**, *142*, 56–64. [CrossRef]
13. Foyer, C.; Furbank, R.; Harbinson, J.; Horton, P. The mechanisms contributing to photosynthetic control of electron transport by carbon assimilation in leaves. *Photosynth. Res.* **1990**, *25*, 83–100. [CrossRef] [PubMed]
14. Harbinson, J.; Hedley, C.L. Changes in P-700 Oxidation during the Early Stages of the Induction of Photosynthesis. *Plant Physiol.* **1993**, *103*, 649–660. [CrossRef] [PubMed]
15. Klughammer, C.; Schreiber, U. An improved method, using saturating light pulses, for the determination of photosystem I quantum yield via P700+-absorbance changes at 830 nm. *Planta* **1994**, *192*, 261–268. [CrossRef]
16. Golding, A.J.; Johnson, G.N. Down-regulation of linear and activation of cyclic electron transport during drought. *Planta* **2003**, *218*, 107–114. [CrossRef]
17. Miyake, C.; Miyata, M.; Shinzaki, Y.; Tomizawa, K.-I. CO2 Response of Cyclic Electron Flow around PSI (CEF-PSI) in Tobacco Leaves—Relative Electron fluxes through PSI and PSII Determine the Magnitude of Non-photochemical Quenching (NPQ) of Chl Fluorescence. *Plant Cell Physiol.* **2005**, *46*, 629–637. [CrossRef]
18. Eberhard, S.; Finazzi, G.; Wollman, F.-A. The Dynamics of Photosynthesis. *Annu. Rev. Genet.* **2008**, *42*, 463–515. [CrossRef]
19. Shaku, K.; Shimakawa, G.; Hashiguchi, M.; Miyake, C. Reduction-induced suppression of electron flow (RISE) in the photosynthetic electron transport system of Synechococcus elongates PCC 7942. *Plant Cell Physiol.* **2015**, *57*, 1443–1453.
20. Furutani, R.; Ifuku, K.; Suzuki, Y.; Noguchi, K.; Shimakawa, G.; Wada, S.; Makino, A.; Sohtome, T.; Miyake, C. P700 oxidation suppresses the production of reactive oxygen species in photosystem I. In *Advances in Botanical Research*; Elsevier: Amsterdam, The Netherlands, 2020; pp. 151–176.
21. Munekage, Y.; Hojo, M.; Meurer, J.; Endo, T.; Tasaka, M.; Shikanai, T. PGR5 Is Involved in Cyclic Electron Flow around Photosystem I and Is Essential for Photoprotection in Arabidopsis. *Cell* **2002**, *110*, 361–371. [CrossRef]
22. DalCorso, G.; Pesaresi, P.; Masiero, S.; Aseeva, E.; Schünemann, D.; Finazzi, G.; Joliot, P.; Barbato, R.; Leister, D. A Complex Containing PGRL1 and PGR5 Is Involved in the Switch between Linear and Cyclic Electron Flow in Arabidopsis. *Cell* **2008**, *132*, 273–285. [CrossRef] [PubMed]
23. Nuijs, A.M.; Shuvalov, V.A.; van Gorkom, H.J.; Plijter, J.J.; Duysens, L.N. Picosecond absorbance difference spectroscopy on the primary reactions and the antenna-excited states in Photosystem I particles. *Biochim. Biophys. Acta (BBA)-Bioenerg.* **1986**, *850*, 310–318. [CrossRef]
24. Trissl, H.-W. Determination of the quenching efficiency of the oxidized primary donor of Photosystem I, P700+: Implications for the trapping mechanism. *Photosynth. Res.* **1997**, *54*, 237–240. [CrossRef]

25. Bukhov, N.G.; Carpentier, R. Measurement of photochemical quenching of absorbed quanta in photosystem I of intact leaves using simultaneous measurements of absorbance changes at 830 nm and thermal dissipation. *Planta* **2003**, *216*, 630–638. [CrossRef]
26. Yamamoto, H.; Shikanai, T. PGR5-Dependent Cyclic Electron Flow Protects Photosystem I under Fluctuating Light at Donor and Acceptor Sides. *Plant Physiol.* **2019**, *179*, 588–600. [CrossRef]
27. Rantala, S.; Lempiäinen, T.; Gerotto, C.; Tiwari, A.; Aro, E.-M.; Tikkanen, M. PGR5 and NDH-1 systems do not function as protective electron acceptors but mitigate the consequences of PSI inhibition. *Biochim. Biophys. Acta (BBA)-Bioenerg.* **2020**, *1861*, 148154. [CrossRef]
28. Yamori, W.; Shikanai, T. Physiological Functions of Cyclic Electron Transport Around Photosystem I in Sustaining Photosynthesis and Plant Growth. *Annu. Rev. Plant Biol.* **2016**, *67*, 81–106. [CrossRef]
29. Kono, M.; Noguchi, K.; Terashima, I. Roles of the Cyclic Electron Flow Around PSI (CEF-PSI) and O2-Dependent Alternative Pathways in Regulation of the Photosynthetic Electron Flow in Short-Term Fluctuating Light in Arabidopsis thaliana. *Plant Cell Physiol.* **2014**, *55*, 990–1004. [CrossRef]
30. Kono, M.; Terashima, I. Elucidation of Photoprotective Mechanisms of PSI Against Fluctuating Light photoinhibition. *Plant Cell Physiol.* **2016**, *57*, 1405–1414. [CrossRef]
31. Tiwari, A.; Mamedov, F.; Grieco, M.; Suorsa, M.; Jajoo, A.; Styring, S.; Tikkanen, M.; Aro, E.-M. Photodamage of iron–sulphur clusters in photosystem I induces non-photochemical energy dissipation. *Nat. Plants* **2016**, *2*, 16035. [CrossRef]
32. Suorsa, M.; Järvi, S.; Grieco, M.; Nurmi, M.; Pietrzykowska, M.; Rantala, M.; Kangasjärvi, S.; Paakkarinen, V.; Tikkanen, M.; Jansson, S.; et al. Proton gradient regulation5 is essential for proper acclimation of Arabidopsis photosystem I to naturally and artificially fluctuating light conditions. *Plant Cell* **2012**, *24*, 2934–2948. [CrossRef]
33. Gollan, P.; Lima-Melo, Y.; Tiwari, A.; Tikkanen, M.; Aro, E.-M. Interaction between photosynthetic electron transport and chloroplast sinks triggers protection and signalling important for plant productivity. *Philos. Trans. R. Soc. B Biol. Sci.* **2017**, *372*, 20160390. [CrossRef]
34. Reiter, B.; Vamvaka, E.; Marino, G.; Kleine, T.; Jahns, P.; Bolle, C.; Leister, D.; Rühle, T. The Arabidopsis Protein CGL20 Is Required for Plastid 50S Ribosome Biogenesis. *Plant Physiol.* **2020**, *182*, 1222–1238. [CrossRef]
35. Takagi, D.; Amako, K.; Hashiguchi, M.; Fukaki, H.; Ishizaki, K.; Goh, T.; Fukao, Y.; Sano, R.; Kurata, T.; Demura, T.; et al. Chloroplastic ATP synthase builds up a proton motive force preventing production of reactive oxygen species in photosystem I. *Plant J.* **2017**, *91*, 306–324. [CrossRef]
36. Li, X.-P.; Björkman, O.; Shih, C.; Grossman, A.R.; Rosenquist, M.; Jansson, S.; Niyogi, K.K. A pigment-binding protein essential for regulation of photosynthetic light harvesting. *Nat. Cell Biol.* **2000**, *403*, 391–395. [CrossRef]
37. Clough, S.J.; Bent, A.F. Floral dip: A simplified method for Agrobacterium-mediated transformation of Arabidopsis thaliana. *Plant J.* **1998**, *16*, 735–743. [CrossRef]
38. Von Caemmerer, S.; Farquhar, G.D. Some relationships between the biochemistry of photosynthesis and the gas exchange of leaves. *Planta* **1981**, *153*, 376–387. [CrossRef]
39. Kramer, D.M.; Johnson, G.; Kiirats, O.; Edwards, G.E. New fluorescence parameters for the determination of Q A redox state and excitation energy fluxes. *Photosynt. Res.* **2004**, *79*, 209. [CrossRef]
40. Makino, A.; Sato, T.; Nakano, H.; Mae, T. Leaf photosynthesis, plant growth and nitrogen allocation in rice under different irradiances. *Planta* **1997**, *203*, 390–398. [CrossRef]
41. Arnon, D.I. Copper enzymes in isolated chloroplasts. Polyphenol-oxidase in B vulgaris. *Plant Physiol.* **1969**, *24*, 1–15. [CrossRef]
42. Munekage, Y.N.; Genty, B.; Peltier, G. Effect of PGR5 Impairment on Photosynthesis and Growth in Arabidopsis thaliana. *Plant Cell Physiol.* **2008**, *49*, 1688–1698. [CrossRef]
43. Teicher, H.B.; Møller, B.L.; Scheller, H.V. Photoinhibition of Photosystem I in field-grown barley (*Hordeum vulgare* L.): Induction, recovery and acclimation. *Photosynth. Res.* **2000**, *64*, 53–61. [CrossRef]
44. Lima-Melo, Y.; Gollan, P.J.; Tikkanen, M.; Silveira, J.A.G.; Aro, E.-M. Consequences of photosystem-I damage and repair on photosynthesis and carbon use in Arabidopsis thaliana. *Plant J.* **2018**, *97*, 1061–1072. [CrossRef] [PubMed]
45. Munekage, Y.; Hashimoto, M.; Miyake, C.; Tomizawa, K.-I.; Endo, T.; Tasaka, M.; Shikanai, T. Cyclic electron flow around photosystem I is essential for photosynthesis. *Nat. Cell Biol.* **2004**, *429*, 579–582. [CrossRef]
46. Kadota, K.; Furutani, R.; Makino, A.; Suzuki, Y.; Wada, S.; Miyake, C. Oxidation of P700 Induces Alternative Electron Flow in Photosystem I in Wheat Leaves. *Plants* **2019**, *8*, 152. [CrossRef]
47. Wada, S.; Yamamoto, H.; Suzuki, Y.; Yamori, W.; Shikanai, T.; Makino, A. Flavodiiron Protein Substitutes for Cyclic Electron Flow without Competing CO2 Assimilation in Rice. *Plant Physiol.* **2018**, *176*, 1509–1518. [CrossRef]
48. Shikanai, T. Chloroplast NDH: A different enzyme with a structure similar to that of respiratory NADH dehydrogenase. *Biochim. Biophys. Acta (BBA)-Bioenerg.* **2016**, *1857*, 1015–1022. [CrossRef]
49. Hashimoto, M.; Endo, T.; Peltier, G.; Tasaka, M.; Shikanai, T. A nucleus-encoded factor, CRR2, is essential for the expression of chloroplast ndhB in Arabidopsis. *Plant J.* **2003**, *36*, 541–549. [CrossRef] [PubMed]
50. Huang, W.; Zhang, S.-B.; Cao, K.-F. Stimulation of Cyclic Electron Flow During Recovery After Chilling-Induced Photoinhibition of PSII. *Plant Cell Physiol.* **2010**, *51*, 1922–1928. [CrossRef]
51. Li, L.; Nelson, C.; Trösch, J.; Castleden, I.; Huang, S.; Millar, A.H. Protein Degradation Rate in Arabidopsis thaliana Leaf Growth and Development. *Plant Cell* **2017**, *29*, 207–228. [CrossRef]

52. Sonoike, K. Photoinhibition of photosystemI: Its physiological significance in the chilling sensitivity of plants. *Plant Cell Physiol.* **1996**, *37*, 239–247. [CrossRef]
53. Sonoike, K.; Kamo, M.; Hihara, Y.; Hiyama, T.; Enami, I. The mechanism of the degradation of psaB gene product, one of the photosynthetic reaction center subunits of Photosystem I, upon photoinhibition. *Photosynth. Res.* **1997**, *53*, 55–63. [CrossRef]
54. Myouga, F.; Hosoda, C.; Umezawa, T.; Iizumi, H.; Kuromori, T.; Motohashi, R.; Shono, Y.; Nagata, N.; Ikeuchi, M.; Shinozaki, K. A Heterocomplex of Iron Superoxide Dismutases Defends Chloroplast Nucleoids against Oxidative Stress and Is Essential for Chloroplast Development in Arabidopsis. *Plant Cell* **2008**, *20*, 3148–3162. [CrossRef] [PubMed]
55. Gallie, D.R.; Chen, Z. Chloroplast-localized iron superoxide dismutases FSD2 and FSD3 are functionally distinct in Arabidopsis. *PLoS ONE* **2019**, *14*, e0220078. [CrossRef]
56. Thomas, D.J.; Avenson, T.J.; Thomas, J.B.; Herbert, S.K. A Cyanobacterium Lacking Iron Superoxide Dismutase Is Sensitized to Oxidative Stress Induced with Methyl Viologen but Is Not Sensitized to Oxidative Stress Induced with Norflurazon1. *Plant Physiol.* **1998**, *116*, 1593–1602. [CrossRef] [PubMed]

cells

MDPI

Article

Gradual Response of Cyanobacterial Thylakoids to Acute High-Light Stress—Importance of Carotenoid Accumulation

Myriam Canonico [1,2], Grzegorz Konert [1], Aurélie Crepin [1], Barbora Šedivá [1] and Radek Kaňa [1,2,*]

1 Centre Algatech, Institute of Microbiology of the Czech Academy of Sciences, Opatovický Mlýn, 379 81 Třeboň, Czech Republic; canonico@alga.cz (M.C.); konert@alga.cz (G.K.); crepin@alga.cz (A.C.); sediva@alga.cz (B.Š.)

2 Faculty of Science, University of South Bohemia in České Budějovice, Branišovská 31a, 370 05 České Budějovice, Czech Republic

* Correspondence: kana@alga.cz

Abstract: Light plays an essential role in photosynthesis; however, its excess can cause damage to cellular components. Photosynthetic organisms thus developed a set of photoprotective mechanisms (e.g., non-photochemical quenching, photoinhibition) that can be studied by a classic biochemical and biophysical methods in cell suspension. Here, we combined these bulk methods with single-cell identification of microdomains in thylakoid membrane during high-light (HL) stress. We used *Synechocystis* sp. PCC 6803 cells with YFP tagged photosystem I. The single-cell data pointed to a three-phase response of cells to acute HL stress. We defined: (1) fast response phase (0–30 min), (2) intermediate phase (30–120 min), and (3) slow acclimation phase (120–360 min). During the first phase, cyanobacterial cells activated photoprotective mechanisms such as photoinhibition and non-photochemical quenching. Later on (during the second phase), we temporarily observed functional decoupling of phycobilisomes and sustained monomerization of photosystem II dimer. Simultaneously, cells also initiated accumulation of carotenoids, especially γ-carotene, the main precursor of all carotenoids. In the last phase, in addition to γ-carotene, we also observed accumulation of myxoxanthophyll and more even spatial distribution of photosystems and phycobilisomes between microdomains. We suggest that the overall carotenoid increase during HL stress could be involved either in the direct photoprotection (e.g., in ROS scavenging) and/or could play an additional role in maintaining optimal distribution of photosystems in thylakoid membrane to attain efficient photoprotection.

Keywords: high light; thylakoid membrane; microdomains; carotenoids; photoprotection; *Synechocystis*; non-photochemical quenching; photoinhibition; photosystems

Citation: Canonico, M.; Konert, G.; Crepin, A.; Šedivá, B.; Kaňa, R. Gradual Response of Cyanobacterial Thylakoids to Acute High-Light Stress—Importance of Carotenoid Accumulation. *Cells* **2021**, *10*, 1916. https://doi.org/10.3390/cells10081916

Academic Editors: Suleyman Allakhverdiev, Alexander G. Ivanov and Marian Brestic

Received: 7 June 2021
Accepted: 25 July 2021
Published: 28 July 2021

1. Introduction

Photosynthesis is a key, light-driven bioenergetics process on Earth. The light-dependent photosynthetic reactions take place in the thylakoid membrane (TM). The membrane is located either in chloroplast of eukaryotic plants and algae or in the cytoplasm of prokaryotic cyanobacteria. Light-photosynthetic reactions in TM include various processes from light-absorption to electron/proton transport. For the purpose of light harvesting in cyanobacteria, several large and highly pigmented protein complexes localized in the TM are employed, including photosystem I (PSI), photosystem II (PSII), and phycobilisomes (PBS; the light-harvesting antennae bound on the TM surface). PSI, PSII, and PBS can then be for simplicity's sake called pigment–protein complexes (PPCs) [1].

The mechanisms behind the PPCs' activity (light absorption, charge separation, electron transport, etc.) have already been described in vitro for all these isolated complexes [2]. However, much less is known about overall cooperation and interactions between the PPCs in native TMs and how it is interlinked with their heterogeneous localization in vivo [1]. The meaning of the spatial PPCs' heterogeneity is an open question in photosynthesis [1,3–6]. It has been clearly shown almost 40 years ago [7–9] that, in higher plants, PSII is mainly present in TM regions called grana (stacked TM areas), while PSI is found in stroma lamellae (unstacked TM areas). A similar grana/stroma-like heterogeneity has been recently identified also in cyanobacteria [1], in line with several other works showing spatial variability in PPC localization in thylakoids [10–14]. In the work of Strašková and co-workers [1], a functional analogy between granal/stromal thylakoid of higher plants TM and so-called photosynthetic microdomains (MDs) in cyanobacteria has been proposed. Those MDs represent specific clusters of TM co-localization with a characteristic PSI/PSII/PBS ratio. Importantly, particular MDs do not fully segregate PSI/PSII and PBS from each other [1]. The importance of the MD mosaic or PPCs' heterogeneity in TM for cyanobacterial photosynthesis has been intensively studied [1,3–6]. However, its variability in PPCs' co-localization caused by fluctuation in light is still poorly understood (see some recent results [3,5,6,15–17]).

Generally, MDs' organization is rather stable in the range of minutes [1], but it can vary in the range of hours or days [6]. This slow process of TM acclimation of *Synechocystis* sp. PCC 6803 can be triggered at variable growth light conditions (continuous light vs. light/dark cycle) and it resembles higher plants grana/stroma reorganization defined as TM plasticity [6]. However, the stability/flexibility of TM organization remains an open question. Some authors suggested rather dynamic behavior of PPCs at extremely high-light (HL) irradiation [3,15] or static behavior of PPCs in a minute's time scale at more physiological HL intensities [1,5]. Therefore, the effect of HL stress on the overall TM and PPCs' organization is still not fully clear.

Light plays an essential role in photosynthesis but can be harmful in the case of excess. Therefore, all phototrophs including cyanobacterial cells can trigger several fast (in the range of minutes) photoprotective mechanisms that are well defined based on bulk measurements (i.e., on cell population level [18]). The three most important strategies are represented by photoinhibition [19], phycobilisome decoupling [20] and non-photochemical quenching (NPQ) [21]. Photoinhibition, in its broader and the original sense, represents a light-induced decrease in the photosynthetic rate or photosynthetic efficiency measured either as the oxygen evolution or CO_2 assimilation rates [22]. On the molecular level, photoinhibition is connected with PSII degradation, because PSII complexes are more sensitive to HL than PSI [23]. The decrease in PSII activity is caused by degradation of the 32-KDa D1 protein [24–30] that needs to be newly synthesized and assembled into PSII complex [31] to recover PSII activity. Therefore cyanobacteria, under photoinhibition, display a decrease in the levels of active PSII that can be measured indirectly as a change in variable chlorophyll (Chl) a fluorescence [32] or directly by the rate of D1 protein turnover [33,34].

The second photoprotective mechanism found in cyanobacteria is represented by PBS decoupling (reviewed in detail in [20]). In a situation of stress condition, such as HL or non-optimal (low or high) temperature, PBS can be partially decoupled [16,35], either functionally or physically detached from the reaction centers [16,17,35,36]. It has been proposed that the uncoupling process could be followed either by PBS redistribution between the two photosystems or disassembly of longer-term detached PBS from the TM [20]. Interestingly, this mechanism has been identified also during the long-term (in days) acclimation of thylakoid membrane to various cultivation light conditions [6]. The third photoprotective mechanism, NPQ, is also detectable based on changes in variable Chl a fluorescence [37]. Even though some previous works have discussed cyanobacterial NPQ as a PSII reaction center type quenching [38], the dominant part is represented by fluorescence quenching in PBS [39]. The mechanism is triggered by blue light and represents a unique cyanobacterial strategy to dissipate excessive light absorbed by PBS. The process requires protein called orange carotenoid protein (OCP) that acts as a blue light sensor [40]. The protein contains a carotenoid, echinenone that acts as a sensing molecule. Interestingly, carotenoids have an essential role in photoprotection in general, as shown for many other phototrophs [41–43]. Generally, carotenoids can be involved in the quenching of chlorophyll triplet state [44], in the scavenging of ROS [45] and changes in membrane properties [46], that can affect the efficiency of light harvesting [47,48]. However, the role of carotenoids in cyanobacterial photoprotection (except for the OCP-based mechanism) is less understood [21,49]. Cyanobacterial cells, indeed, contain a large amount of myxoxanthophyll, zeaxanthin and other carotenoids [50,51]. In contrast to higher plants, in which these pigments seem to act mainly as part of the PPCs, several studies pointed to their presence and importance in cyanobacteria in both TM and plasma membrane [38,40,41].

In the present paper, we studied the overall response of cyanobacterial thylakoids to acute HL stress. We identified a multiphase response of cyanobacteria to the HL stress (in line with [5]), with three main phases: (1) fast response phase (0–30 min), (2) intermediate phase (30–120 min), and (3) slow acclimation phase (120–360 min). During these three phases, cyanobacterial photosynthesis was affected on several levels including: (1) protein/pigment composition, (2) TM microdomain organization, and (3) activation of photoprotective mechanisms. In more details, we observed a more even distribution of PPCs on HL as it was visible in a smaller spatial variability in PSI/PSII/PBS distribution per cell. Simultaneously, we identified fast activation of photoinhibition and non-photochemical quenching (first phase) and temporal process of PBS decoupling (second phase). Finally, HL stress also caused the accumulation of carotenoids, myxoxanthophyll and γ-carotene. We suggest that the carotenoid accumulation during the acute HL stress (6 h) could help to maintain photosynthetic function either by direct photoprotection (e.g., ROS scavenging), or through indirect tuning in TM fluidity.

2. Materials and Methods

2.1. Growth Conditions and High-Light Treatment

We used the PSI-YFP tagged strain derived from the glucose-tolerant strain of *Synechocystis* sp. PCC 6803 [52] previously described in Tichý et al. [53]. Photosystem I was fluorescently tagged on its PsaF subunit as described previously [1,54] (hereafter *Synechocystis PSI-YFP)*. Cells were grown on an orbital shaker (T = 28 °C) in BG11 medium at continuous light (35 µmol photons m^{-2} s^{-1}, fluorescent tubes). Cells in the exponential growth phase and with low concentration (OD_{730} = 0.2–0.3) were used for the experiments. OD_{730} was measured with a WPA S800 Diode Array Spectrophotometer (Biochrom Ltd., Cambridge, England). High-light (HL) treatment was performed under white diodes (700 µmol photons m^{-2} s^{-1}) at 28 °C during continuous shaking. The control cells were kept at 35 µmol photons m^{-2} s^{-1} at 28 °C during continual shaking for the same time as for the HL treated cells. Cells were used for measurements at time intervals of 0, 10, 30, 60, 90, 120, 240, 360 min. The minimal effect of YFP tagging at PSI on the cell physiology and protein composition has been already proved before [1]; we additionally tested the effect of YFP tagging on light sensitivity of *Synechocystis* PCC 6803 strain. We did not observe significant changes between WT strain (without YFP tag at PSI) and *Synechocystis* PSI-YFP strains considering PSII photochemistry, pigment composition, single-cell fluorescence and others. Therefore, only the PSI-YFP tagged strain of *Synechocystis PSI-YFP* was used to study the HL effect.

2.2. Characterization of Cell Sizes, Types and Counts

Cell sizes, counts, and types during HL treatment were monitored by cell counter (Beckman, Multisizer 4, Indianapolis, IN, USA) and by confocal microscope (Zeiss LSM 880; Carl Zeiss Microscopy GmbH, Oberkochen, Germany). Cells were counted by a Coulter counter equipped with 50 µm aperture, where size threshold level was set up in the 1–4 µm range. A total of 50 µL of each sample was diluted in 10 mL of electrolyte solution (0.9% NaCl in deionized water) and measured 3 times. For comparison, the individual cell counts and types were also extracted from the confocal images according to Konert et al. [5] and Canonico et al. [6]. In line with Canonico and co-workers [6], we analyzed 4 characteristic cell shapes (regular, elongated, dividing, or string) separately. Other used parameters include fluorescence intensities and size.

2.3. Biochemical Analysis of Pigment–Protein Complexes

Proteins were separated from isolated thylakoid membranes. First, cells were pelleted, washed, and resuspended in buffer B (25 mM MES/NaOH, pH 6.5, 10 mM $CaCl_2$, 10 mM $MgCl_2$, and 25% (v/v) glycerol). Then, they were broken using balotina/zirconia beads and Mini-Beadbeater-24 (BioSpec Products, Bartlesville, Oklahoma, USA) for 5 cycles of 30 s of breaking and 1 min in ice. Thylakoid membrane fraction was separated by centrifugation at 20,000 *g* for 30 min at 4 °C. The fraction containing thylakoid membrane proteins (Chl content 5 μg) was loaded in the clear-native polyacrylamide gel electrophoresis (CN-PAGE). Protein complexes from the membrane were solubilized with 10% dodecyl-β-D-maltoside (DDM) in water to obtain the sample volume/DDM = 10 v/w). Native protein complexes were separated on 4%–14% gradient polyacrylamide gel (acrylamide to bis-acrylamide ratio was 60:1) according to Wittig et al. [55]. Native gels were color-scanned and chlorophyll a fluorescence was obtained by LAS 4000 camera (Fuji, Boston, MA, USA). The color gel pictures were analyzed by the ImageJ software (FIJI distribution). Each band corresponding either to PSI (trimer or monomer) or PSII (dimer or monomer) was taken per each time point. Areas of the peaks were analyzed and normalized to the time 0 min.

2.4. Confocal Microscopy

Cells for confocal imaging were prepared and images acquired by a method slightly modified from our previous papers [1,5]. We used a laser scanning confocal microscope (Zeiss LSM 880, Carl Zeiss Microscopy GmbH, Germany) equipped with a plan-apochromatic 63x/1.4 Oil DIC M27 objective. The three channel pictures were obtained by 2 sequential images with different parameters. PBS emission was excited by 633 nm laser (dichroic mirror: MBS 488/543/633) and detected at 642–677 nm (pixel dwell time between 8 and 33 μs). In the following sequence the chlorophyll a autofluorescence from PSII and YFP fluorescence from PSI were both excited with 488 nm Argon laser (dichroic mirror: MBS 488/543/633) and detected at 696–758 nm and 526–580 nm, respectively (pixel dwell time between 8 and 33 μs). Images were acquired with 8 bit, 512 × 512 pix).

Acquired images were separated into individual cell pictures by our ImageJ script (Fiji distribution, ver 1.53c) and their shapes were assigned according to equations described in detail by Canonico et al. [6]. Total cell fluorescence was calculated from whole cell area, and histograms for inner/outer area of cell were obtained by separation of cells into these two regions. Outer cell region was marked from end of cell towards inside of cell in thickness of 10 pixels (520 nm). The remaining core area was marked as inner region.

2.5. Measurements of Variable Chlorophyll a Fluorescence

The maximal quantum yield of PSII photochemistry (parameter F_v/F_m) was measured by AquaPen-C fluorometer (AP-C100, Photon Systems Instruments, Brno, Czech Republic) with 630 nm excitation and a 667–750 nm detection range. Maximal fluorescence (F_m) minimal (F_0) and variable ($F_v = F_m - F_0$) for Chl a fluorescence in dark adapted cells (for 20 min) were used to estimate maximal PSII efficiency F_v/F_m.

The extent of non-photochemical quenching (NPQ) was measured by a DUAL- PAM-100 fluorometer (Heinz Walz GmbH, Effeltrich, Germany) with a Dual-DR detector head (excitation 620 nm, detection above 700 nm). Cells were dark adapted for 10 min before the experiment. The values for estimation of NPQ were acquired from a standard protocol for cyanobacteria (see Figure S1). At first, the minimal fluorescence (F_0) was detected for dark adapted cells, in State II, by a red measuring light (20 Hz, 10 μmol photons m^{-2} s^{-1}, λ = 620 nm). Subsequently, cells were transited to State I by low intensity blue light (80 μmol photons μmol photons m^{-2} s^{-1}, λ = 460 nm, duration 180 s) and the maximal fluorescence (F_m) was obtained by a multiple turnover saturating flash (red light λ = 620 nm, 4000 μmol photons m^{-2} s^{-1}, duration 400 ms). The NPQ was then induced by high-intensity blue light (1374 μmol photons m^{-2} s^{-1}, duration 180 s) and subsequent recovery from NPQ state was then measured in low blue light (80 μmol photons μmol photons m^{-2} s^{-1}, λ = 460 nm, duration 360 s). The maximal value of fluorescence for the light adapted sample (F_m') was estimated by a multiple turnover flash (red light λ = 620 nm, 4000 μmol photons m^{-2} s^{-1}, duration 400 ms) every 10–30 s of the protocol. F_m' values were then used to estimate NPQ based on the Stern–Volmer relationship (NPQ = $(F_m - F_{m'})/F_{m'}$). The presented steady-state values of NPQ then represent F_m' after 180 s on high light and its recovery on low blue light for 360 s.

2.6. Fluorescence and Absorption Spectroscopy

Low temperature fluorescence emission spectra were recorded at 77K by an SM-9000 spectrofluorometer (Photon Systems Instruments, Brno, Czech Republic) (see, e.g., [56] for details). Cyanobacterial cells were dark adapted for 20 min and then concentrated on a GF–F filter (Whatman, Maidstone, UK). The filter with cells were then immersed into a Dewar flask with liquid nitrogen. Fluorescence was excited by monochromatic diodes (λ = 450 or 530 nm) and detected by diode array detector (spectral bandwidth 0.8 nm) in the spectral range of 200–980 nm. The dark current of the instrument was automatically subtracted before measurements, based on the method described previously for room temperature measurements [57]. For data processing, the spectra of cells were baseline-corrected using a blank sample (filter with BG-11 media).

Room temperature absorption spectra were measured by a Unicam UV/VIS 500 spectrometer (Thermo Spectronic, Cambridge, UK) equipped with an integrating sphere. Cells were collected on the membrane filters (pore size 0.8 mm; Pragochema, Prague, Czech Republic) and spectrum in the range 350–800 nm was detected.

2.7. Characterization of Pigment Composition

Pigment concentrations in cells were estimated spectroscopically (absolute chlorophyll concentrations by Ritchie's method [58]) and by HPLC analysis (relative pigment content). Pigments were isolated from 1 mL of cell suspension (OD range 0.3–0.4) that was centrifuged (10 min, 2000 g). The supernatant (990 μL) was removed, and the pellet was re-suspended in 990 μL of methanol (100%), vortexed and kept for 20 min on ice in the dark. The prepared sample was then centrifuged (10 min, 2000 g) and the supernatant with pigments in methanol solution was then used for absorption and for HPLC pigment analysis by an Agilent-1260 HPLC system (Perkin Elmer, Boston, MA, USA) equipped with a Radiomatic 150 TR scintillation detector and a diode-array detector.

The extracted pigments were injected into the HPLC system and pigments were separated on a Zorbax Eclipse Plus C18 column (4.6 μm particle size, 3.9 × 100 mm; Agilent, Santa Clara, CA, USA) on a linear gradient of two solvents: A (35% methanol, 15% acetonitrile in 0.25 M pyridine) and B (20% methanol, 20% acetone in acetonitrile). Pigments were eluted with solvent B (30%–95% in 25 min) followed by 95% solvent B at a flow rate of 0.8 mL min^{-1} at 40 °C, as described in [59]. Eluted pigments were identified based on their absorption spectra and retention times. The presence of γ-carotene in cells was further identified by comparison with the HPLC analysis of the commercially purchased standard of γ-carotene (analytical standard, purity ≥95%, Sigma Aldrich, Saint Louis, MO, USA). The pigments were identified based on their absorbance at 440 nm. Resulting values were normalized to total pigment absorption at 440 nm for all HPLC peaks. The absorption spectra of extracted pigments from cell suspension were collected by Unicam UV/VIS 500 spectrophotometer (Thermo Spectronic, Cambridge, UK). Absorptions at specific wavelengths (665 nm and 720 nm) were used to calculate chlorophyll a concentration (μg/mL): Chl a = 12.9447 ($A_{665} - A_{720}$) [58].

3. Results

3.1. High-Light Treatment and Cyanobacterial Cell Size

Changes in the size of *Synechocystis PSI-YFP* during 6 h of HL treatment were estimated by cell counting (Figure 1). Data were considered in light of the three-phase behavior published recently [5]. Total number of cells increased only slightly during HL, and there was only a slight difference between HL and the control samples (Figure 1a). The similar increase in the number of cells (for both HL and control) was also visible in the parameter of optical density (OD_{730}—Figure S2); however, that reflects both difference in cell sizes and/or cell counts [60]. Therefore, we checked the relative contribution of smaller (diameter ≤1.85 μm) and bigger (diameter >1.85 μm) cells during HL stress directly by cell counter (Figure 1b). Control cell values were stable over time, in contrast to the HL sample with the increased/decreased number of bigger/smaller cells (Figure 1b)—this was visible during the third phase (120–360 min) of HL response. We wanted to observe if we could see the same trend at the single-cell level. Our automatized approach in cell imaging (see Materials and Methods) allowed us to test whether the increase in the count of the bigger cells at HL was due to the higher amount of dividing cells [6]. However, we could not prove/disprove the hypothesis that the cell cycle was affected by HL treatment (Figure S3). Therefore, we did not obtain direct proof at the single-cell level to see if *Synechocystis PSI-YFP* cell division was delayed by HL treatments as it is indicated by changes in the cell sizes (Figure 1b).

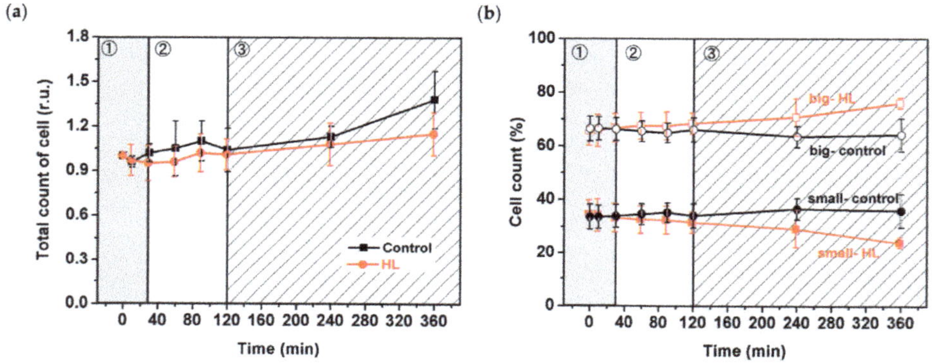

Figure 1. Changes in the number of *Synechocystis PSI-YFP* cells measured by cell counter during high-light (HL) treatment. (**a**) Relative changes in total cell count obtained independently of cell sizes. Data are normalized to time 0 min. (**b**) Relative changes in cell count of small (diameter ≤ 1.85 μm) and big (diameter > 1.85 μm) cells. Black lines = control; Red lines = HL treatment. The three phases of cyanobacterial cell response to HL (see Konert et al. 2019 [5]) are marked by different background colors and numbers (grey—fast response; white—intermediate phase; striped—acclimation phase). Data represent averages and SD from three biological replicates, the starting culture (t = 0 min) typically contains $n = 10^6$ cells per mL.

3.2. Changes in Organization of Pigment–Protein Complexes during HL Treatment

Because the HL treatment partially affected the proportion of bigger cells (Figure 1b), we checked whether it resulted in different protein heterogeneity in thylakoids. We estimated the effect of acute HL stress on the pattern of PPC co-localization into MDs (Figure 2), as defined by Strašková et al. (2019) [1]. We showed the existence of heterogeneous membrane areas in thylakoids called MDs based on co-localization of the three main PPCs: PSI (tagged by YFP), PSII, and PBS. Our data were in line with several previous results [1,5,6,61]. The three-channel imaging of these complexes gave rise to a mosaic of microdomains in TM that is visible as RGB (red–green–blue) pictures (see the color scheme in Figure 2a). It shows heterogeneous organization of PSI, PSII, and PBS inside thylakoids that is also confirmed by grey-scale pictures (Figure 2a). Particular MDs in the TM mosaic are then represented by membrane areas of specific PSI/PSII/PBS ratios and they represent specific cluster inside TM. MDs are visible as differently colored areas of TM (see RGB picture in the Figure 2a). In line with previous data [1,5,6,61], the two most dominant areas were represented by magenta MDs (with dominant PSII and PBS content, less PSI) and green MDs (with dominant PSI, less PSII and PBS content), both representing an analogue of granal/stromal areas of plant thylakoids [1].

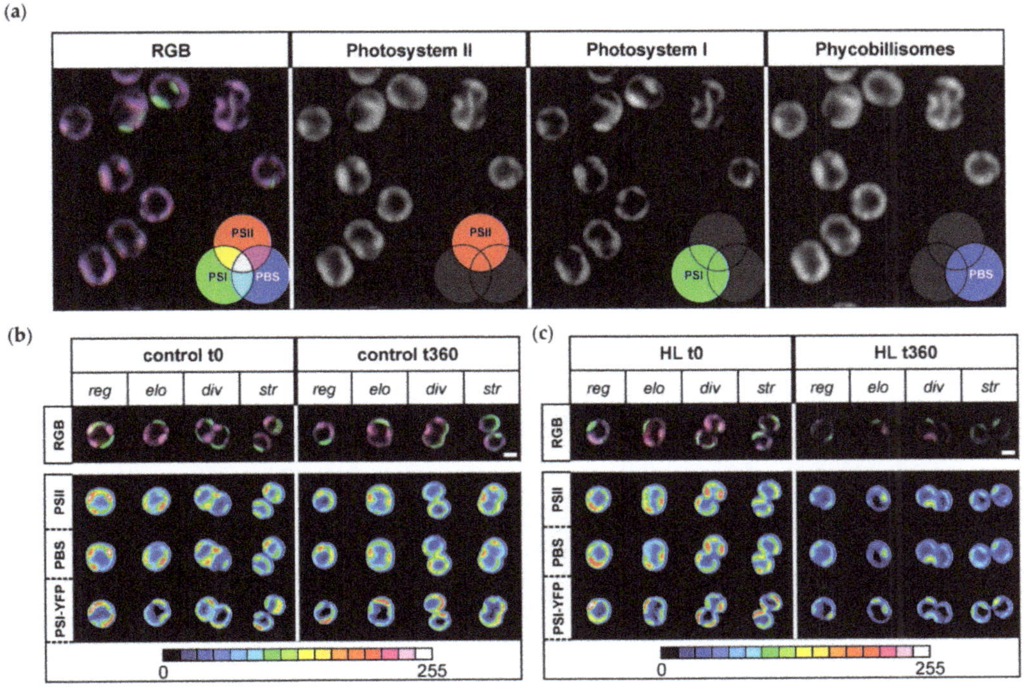

Figure 2. Microdomain (MD) organization of thylakoid membrane in *Synechocystis PSI-YFP* during HL treatment. (**a**) Typical RGB picture of *Synechocystis PSI-YFP* cells obtained by three channel confocal imaging (PSII channel—red; PSI–YFP channel—green; PBS channel—blue). Colors reflect PSI/PSII/PBS co-localization and the most dominant microdomains, namely magenta MD (dominant PBS and PSII, less PSI-YFP) and green (dominant PSI-YFP, less PSII and PBS) as described by Strašková et al. 2019. Panels (**b**) and (**c**) represent change in MD organization in control (**b**) and HL treated cells (**c**) at the beginning (t = 0 min) and the end of the experiment (t = 360 min). The four typical cell shapes are presented independently (reg = regular, elo = elongated, div = dividing, str = string). The first row shows three channels' pictures (RGB, 24-bit) with PSII, PBS, and PSI-YFP co-localization. The second, the third, and the fourth rows depict intensity of single-channel fluorescence of PSII, PBS, and PSI-YFP, respectively. Colors reflect intensity of fluorescence signal per channel (heatmap images) in the 8-bit scale 0–255 (see the color scale bar).

We studied changes in MDs during HL treatment. We have analyzed distribution of MDs inside single-cell thylakoids (see "RGB" line of Figure 2b,c) and changes in the intensity of their membrane areas for all three PPCs (see heatmap pictures of "PSI-YFP", "PSII", and "PBS"). Changes were analyzed separately for regular (reg), elongated (elo), dividing (div), and string (str) shapes of cells representing different stages in the cell cycle (see [6] for details). We observed a general decrease in fluorescence intensity of all three PPCs during HL (Figure 2c); the effect was absent in the control sample (Figure 2b). Further, we also observed a significant decrease in spatial PPCs' heterogeneity per single cell during HL (see disappearance of red spots in the heatmap of PSI, PSII, and PBS in Figure 2c). We noticed that all PPCs were more evenly distributed as a result of HL stress (compare redundancy of red spots in Figure 2b,c). This was further confirmed by the decrease in the fluorescence signal variability (visible as standard deviation value) per single cell for all three PPCs (data not shown). On the other hand, HL did not significantly affect presence of grana-like (magenta) and stroma-like (green) MDs (see RGB line in Figure 2c). In summary, our data suggest that the HL treatment induced only a loss in the intensity of high-emission spots (see red spots in the heatmap of Figure 2b) because PSI-YFP, PSII, and PBS became more evenly distributed in the thylakoids after acute HL stress.

3.3. Changes in the Single-Cell Fluorescence during HL Stress

The observed decrease in the fluorescence intensity visible in pictures (Figure 2) was further quantified and calculated for all three PPCs per single cell from pictures (Figure 3). Cells exposed to HL lost the fluorescence of PBS, PSI, and PSII over time (Figure 3b) in contrast to stable values for control cells (Figure 3a). The relative changes in the PSI/PSII/PBS fluorescence (relative to sum fluorescence values of all 3 PPCs) showed the three distinct phases in HL treated cells (Figure 3d), in line with our previous work [5]. The new set of data represented almost six thousand individual cells from six biological replicates (Figure 3). The first phase (fast response phase, up to 30 min) was characterized by an increase in relative PSI fluorescence ratio. The second phase (intermediate phase; 30–90 min) was characterized by stabilization of relative dynamics of all fluorescence ratios (Figure 3d). The following third phase (slow acclimation phase; 120–360 min) was then characterized by a fluorescence decrease in the relative PSI signal per cell and the concomitant increase in the relative PBS signal (Figure 3d). It is important to note that the relative PSII signal per cell was the most stable from all three PPCs during HL treatment (Figure 3d). In summary, the relative PSI-YFP fluorescence per cell initially increased and simultaneously PBS decreased during the first 30 min. These relative values seem to progressively come back to their initial points during the final slow acclimation phase (120–360 min; Figure 3d). It shows that, although we observed absolute fluorescence decay during HL due to quenching and degradation processes (Figure 3b), the relative ratios of PSI, PSII and PBS (all normalized to sum fluorescence of all three PPCs) per cell tend to recover to almost the same values as the initial non-stressed condition (Figure 3d).

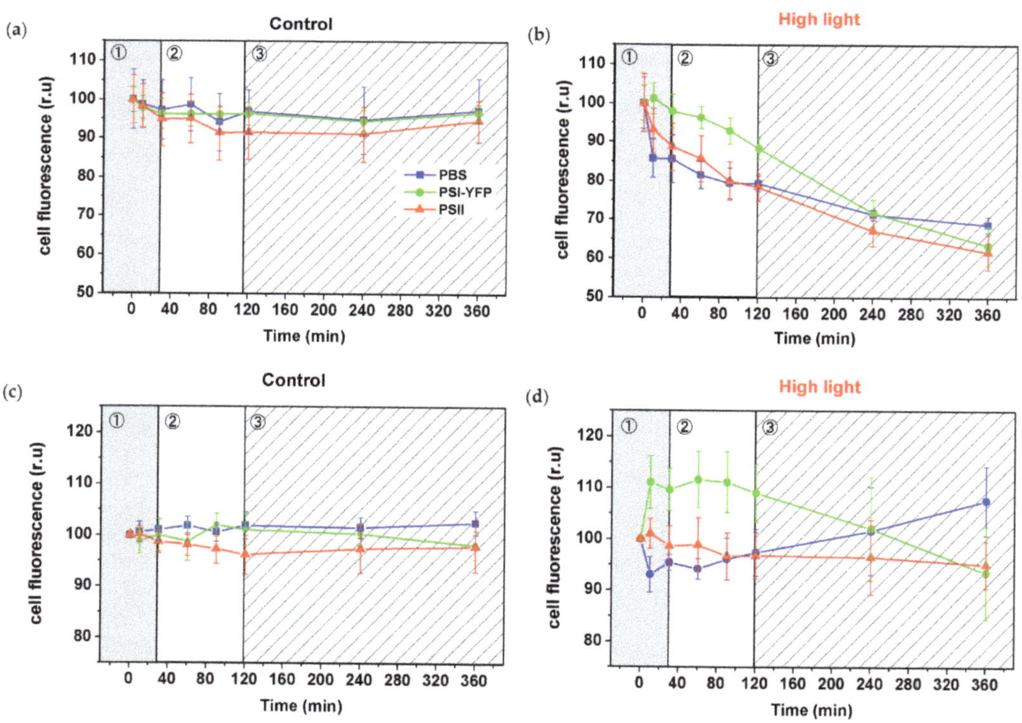

Figure 3. High-light induced changes in PSI-YFP, PSII, and PBS fluorescence emission detected from the single cells of *Synechocystis PSI-YFP*. Single-cell fluorescence intensities were calculated from three channels' confocal images (see, e.g., Figure 2) acquired at particular time spots during HL treatments. Panels (**a**) and (**b**) represent normalized fluorescence changes of single-cell PSII, PSI-YFP, and PBS emission for control (**a**) and HL treated sample (**b**). Data were normalized to fluorescence

intensity at t = 0 min for every channel independently. Panels (**c**) and (**d**) show relative fluorescence changes in PSI-YFP, PSII, and PBS when the presented values are normalized to the sum of all three PPCs' emission (PSI + PSII + PBS) and to the initial value at t = 0 min for each channel individually. Data represent control (**c**) and HL treated cells (**d**) of six independent biological replications. The total count of cells used for analysis was *n* = 5844. The three phases of cyanobacterial cell response to HL (see Konert et al. (2019) [5]) are marked by different background colors and numbers (grey—fast response; white—intermediate phase; striped—acclimation phase).

3.4. Effect of HL Treatment on Photosystem II Photochemistry

We identified a three-phase acclimation response of *Synechocystis PSI-YFP* at the single-cell level. We aimed to address whether the three-phase behavior on the single-cell level (Figure 3) correlates with some functional photoprotective processes known from bulk measurements of variable fluorescence. Therefore, we studied the progress of photoinhibition (measured as decrease in F_v/F_m—the maximum quantum yield of PSII, Figure 4a) and NPQ (Figure 4b). Indeed, F_v/F_m decreased significantly during the first phase (up to 30 min) due to HL treatment (Figure 4a). The kinetics of F_v/F_m decay slowed down in the second, intermediate phase (30–120 min), and remained constant in the slow acclimation phase (120–360 min), in line with our previous data [5]. These results point to the photoinhibition-induced increase in the D1-protein turnover [62] that is visible during the first 2 h on HL (Figure 4a).

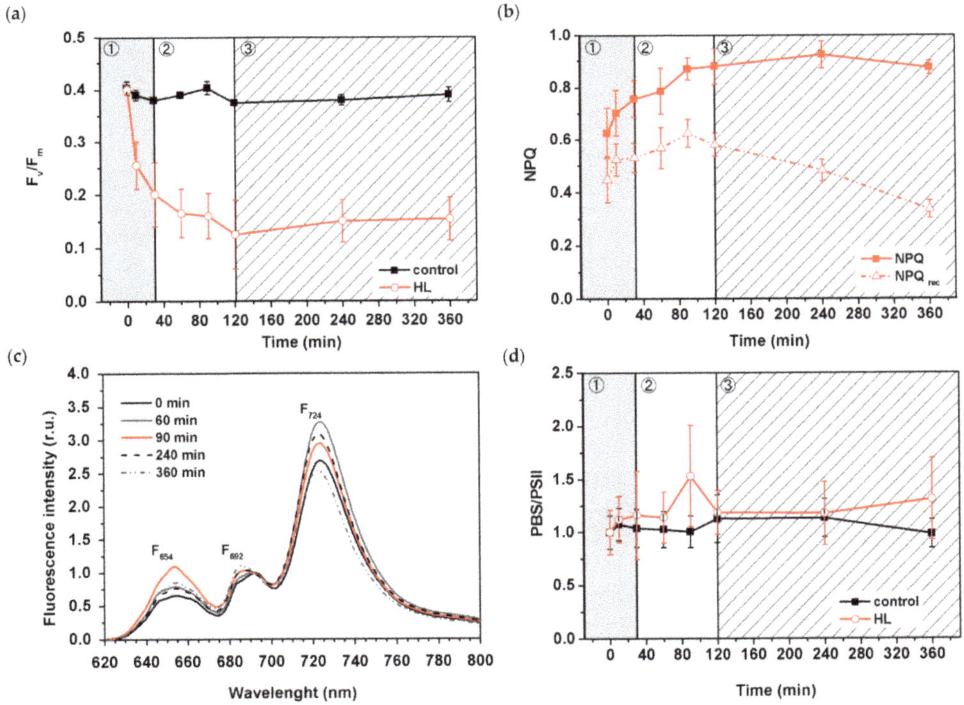

Figure 4. Effect of HL stress on photoprotective mechanism in *Synechocystis PSI-YFP* cells. (**a**) Change in maximal quantum yield of PSII (Fv/Fm) due to photoinhibition in cell suspension. Black/red lines represent control/HL treated sample; *n* = 3. (**b**) Extent of blue light-induced non-photochemical quenching on light (NPQ—squares) and its recovery in dark (NPQrec—triangles) during HL stress; *n* = 4. (**c**) Typical curve of low temperature (77K) fluorescence emission spectra (excitation 530 nm) for cells exposed to HL for different times (0 min—straight black line; 60 min—straight grey line; 90 min—red line; 240 min—dashed black line; 360 min—dashed grey line). F654, F724, F724 represent the positions of the

main emission peaks of PBS, PSII and PSI, respectively. (**d**) The PBS/PSII ratios estimated as F654/F692 from panel (**c**) for both HL and control cells; data were normalized to values at t = 0 min; $n = 4$. The three phases of cyanobacterial cell response to HL (see Konert et al. (2019) [5]) are marked by different background colors and numbers (grey—fast response; white—intermediate phase; striped—acclimation phase). All data in panels (**a**,**b**,**d**) represent averages and standard deviations.

We measured the NPQ parameter during HL treatment (Figure 4b) and found out that NPQ increased mostly during the first two phases (Figure 4b), when PBS emission per cell was relatively lowered (Figure 3d). Later, in the last slow acclimation phase (120–360 min), NPQ remained constant. Interestingly, when we calculated the extent of NPQ recovery in the dark (NPQ_{rec} calculated as remaining quenching after 802 s from the start of the protocol), we clearly saw that the recovery was even faster at the final (slow acclimation) phase of the HL treatment. It could indicate that the process of OCP reversibility from its active to inactive form was accelerated for long-term HL treatment (see discussion for details). We have recently identified a temporal importance of PBS decoupling during a long-term (weeks) TM reorganization [6]. We studied the same phenomenon considering the three-phase response of single cyanobacterial cells to HL (Figure 4c,d). The calculated F_{654}/F_{692} ratio of 77K fluorescence represented the ratio of PBS/PSII emission (Figure 4d). Its increase indicates the presence of an energetically decoupled PBS. This ratio only slightly increased during the first phase (0–30 min) but became significantly unstable and very high during the second phase (30–120 min) (Figure 4d), as it is visible as a change in the full fluorescence emission spectra (Figure 4c). It indicates that PBS decoupling becomes important mostly during the second phase of TM reorganization at HL (30–120 min, see the point at 90 min). We tried to address the process of PBS decoupling on the level of single *Synechocystis PSI-YFP* cell (Figure S4). We addressed the basic question of whether energetically decoupled PBS (visible in 77K fluorescence, Figure 4c) remains inside the TM area or if PBS physically move during the process of decoupling (see, e.g., [16,17]); in this case it should be visible as fluorescence increase in the central cell area. For that purpose, we calculated PBS fluorescence emission in these two cell areas (Figure S4d), but we found no significant differences between fluorescence in peripheral membrane area and central cell area (see Figure S4a,b). In both cases PBS fluorescence was quenched, as it is visible in the shifting in the histograms from higher to lower fluorescence. The recalculated relative changes in both areas (compared to time 0 min) finally proved that PBS fluorescence was rather quenched in both these areas both at 90 min and 360 min of HL treatment (Figure S4c). It indicates that PBS were not physically re-localized from the TM into the central part of cells during our HL treatment (see discussion).

3.5. Effect of HL Treatment on PPCs Levels

We analyzed pigment–protein composition during HL treatment by clear native gels (CN) and by 77K fluorescence spectra of intact *Synechocystis PSI-YFP* cells (Figure 5). CN gels were scanned as color pictures (Figure 5a) and as Chl autofluorescence (Figure 5b) to estimate the position of photosystems in the gel. The most pronounced change we observed was a clear vanishing of PSII dimer (PSII [2]) that remained less abundant until the end of experiment (Figure 5b). It was also supported by a detailed gel picture analysis by Image J (Figure S7b). This is in line with the maximal extent of photoinhibition visible in F_v/F_m (Figure 4a) that was noticeable just before the end of the second phase. Based on the similar quantification, we analyzed PSI bands' intensity. Here, we were also able to identify even smaller changes in PSI oligomerization state (trimer vs. monomer PSI—see PSI [3]/PSI [1] (Figure S7a)). The PSI oligomerization showed a three-phase pattern as PSI [3] initially increased (up to 30 min), and then more PSI [1] appeared later (time 90 min) in the second phase (Figure S7a). The oligomerization of PSI, then, did not change and remained stable in the final slow acclimation phase (120–360 min).

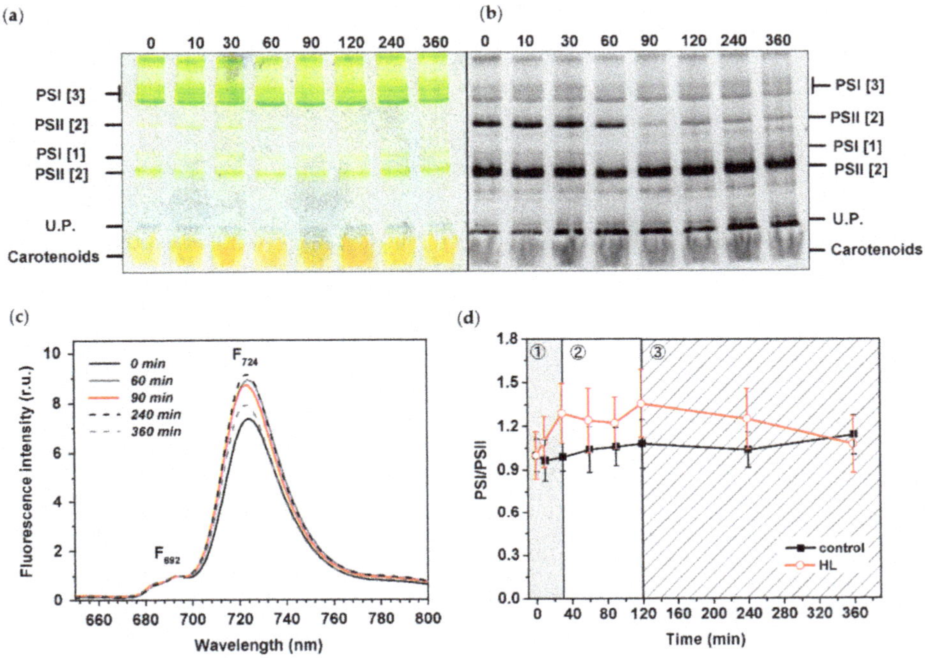

Figure 5. Biochemical and spectroscopic analysis of pigment–protein complexes of HL treated *Synechocystis PSI-YFP cells*. Thylakoid membranes were isolated from the cells and analyzed by CN-PAGE. Gels were imaged as a color-scanned picture (**a**) or as chlorophyll autofluorescence pictures (**b**). Numbers represent time points (in minutes) of HL treatment. Designation of complexes: PSI [3], PSI [1] trimeric and monomeric PSI complexes; PSII [2] and PSII [1] dimeric and monomeric PSII core complex; U.P. = unbound proteins. Each loaded sample contained 5 µg of Chl. (**c**) Normalized 77K fluorescence emission spectra at 692 nm (450 nm excitation) at different times of HL treatment. Straight black line = time 0 min; straight grey line = time 60 min; red line = time 90 min; dashed black line = time 240 min; dashed grey line = time 360 min. Positions of F_{692} and F_{724} representing PSII and PSI maximal emission, respectively, are marked. (**d**) Spectroscopic analyses of PSI/PSII ratio based on 77K fluorescence (excitation at 450 nm) for HL and control cells. PSI/PSII ratio deduced from the intensity of fluorescence of F_{692} (PSII maxim) and F_{724} (PSI maxim). Data represent averages and SD of four biological replicates and normalized to values at 0 min. The three phases of cyanobacterial cell response to HL (see Konert et al. (2019) [5]) are marked by different background colors and numbers (grey—fast response; white—intermediate phase; striped—acclimation phase).

We also investigated changes in PSI/PSII ratios based on the 77K fluorescence emission spectra (Figure 5c,d), as the F_{724}/F_{692} fluorescence ratio can be used as a proxy for PSI/PSII ratio (see, e.g., [56]). In line with the results observed on gel, the PSI/PSII initially increased during the first phase (0–30 min), then it stabilized during the second phase (30–120 min) and finally progressively decreased and returned to values similar to the initial levels (before starting the HL stress) during the third slow acclimation phase (120–360 min). It shows that the systems of protein complexes (in this case reflected as PSI/PSII ratio, Figure 5d) acclimated to the new conditions (HL stress); similar signs of successful acclimation are also visible on single cell measurements of PSI/PSII/PBS ratios that recover to their initial state at the end of the experiment (Figure 3d).

3.6. HL Effect on Pigment Composition

We surveyed the effect of HL treatment on the level of pigment concentration in *Synechocystis PSI-YFP*. The absorption spectra of intact cyanobacterial cells showed changes in the blue–green visible light region that indicated some HL-induced increase in carotenoid content (Figure 6a). A more detailed analysis of the chlorophyll content (calculated according to Ritchie (2006) [58]), showed an increasing trend in the control *Synechocystis PSI-YFP* cells while there was a stable trend in HL treated cells (Figure 6b). This is in line with the presence of slow growth of control sample that was rather absent in HL treated cells (Figure 1a). Interestingly, chlorophyll concentration per cell (Figure S5) slightly decreased during HL treatment, especially during the third (slow acclimation) phase. For that reason, we also analyzed the absorption spectra of the extracted pigment (Figure S6). These data were in line with the whole cell spectra (Figure 6a), showing an increase in carotenoids content, mostly during the third phase (120–360 min).

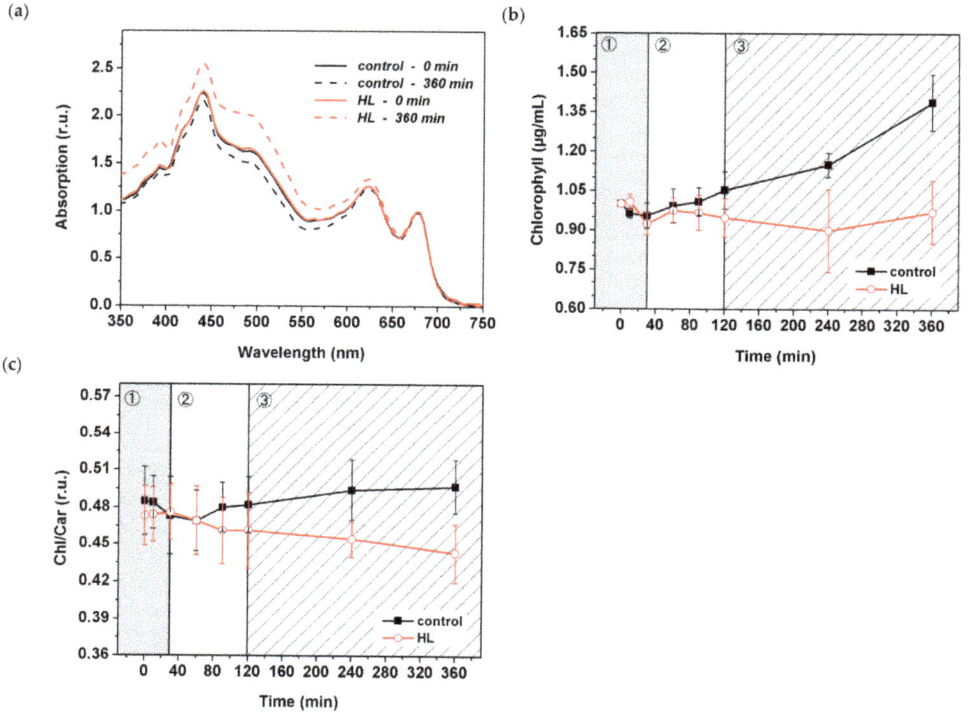

Figure 6. Spectroscopic and HPLC analysis of pigment composition during HL treatment of *Synechocystis PSI-YFP* cells. (**a**) Absorption spectra. Data were normalized to the peak corresponding to PSII (664–665 nm). Straight black line = control time 0 min; dashed black line = control 360 min. Straight red line = HL time 0 min; dashed red line = HL time 360 min. Average of four independent biological measurements. (**b**) Chlorophyll concentration was estimated spectroscopically according to Ritchie et al. (2006). Black line = control; red line = HL. Data represent average of four independent biological measurements that are normalized to time 0 min. (**c**) Chlorophyll/total carotenoids ratio was estimated based on HPLC analysis. All detected carotenoids were taken into consideration, $n = 4$. Black line = control; red line = HL. Presented HPLC data represent averages and SDs of four independent biological measurements. The three phases of cyanobacterial cell response to HL (see Konert et al. (2019) [5]) are marked by different background colors and numbers (grey—fast response; white—intermediate phase; striped—acclimation phase).

The HL-induced increase in carotenoids was analyzed by high-performance liquid chromatography (HPLC) (Figures 6c and 7). The method identified the type of carotenoids present in the *Synechocystis PSI-YFP* cells (Figure 7a) together with their relative concentration changes during HL treatment (Figure 7b). The Chl/carotenoids ratio significantly decreased during HL treatment (in this case pointing to an increase in carotenoid content) and remained stable for the control sample (Figure 6c). This was in line with previous absorbance measurements (Figure 6a and Figure S6). The HPLC analysis additionally identified individual types of carotenoids that were either accumulated or remained unaffected during HL treatment; it includes β-carotene, zeaxanthin, and echinenone (Figure 7a). Moreover, we also noticed an additional peak at 32 min retention time, present only in the HL treated cells (Figure 7a). The peak was assigned as γ-carotene based on its spectrum, retention time and by comparing it with a commercial analytical standard (Figure S8). It showed that γ-carotene, an intermediate in the carotenoid biosynthetic pathway, accumulates during HL.

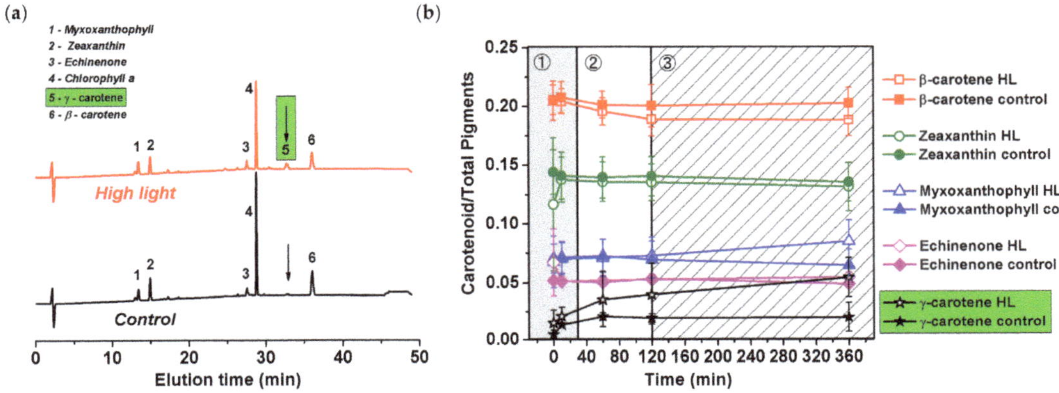

Figure 7. HPLC chromatogram and kinetic changes in carotenoid composition in *Synechocystis PSI-YFP* in HL and control culture. (**a**) HPLC chromatograms of the control and HL treated samples for 360 min. Position of the peaks was identified based on literature (see description). The peak presented only in HL cell (no. 5, see arrows, green bar), was identified by comparison of its retention time and absorbance with the analytical standard as γ-carotene (see Figure S8). (**b**) Kinetics of carotenoid accumulation during HL treatment: control sample—closed symbols; and HL—open symbols. Data represent averages and SD obtained from three biological replicates. The three phases of cyanobacterial cell response to HL (see Konert et al. (2019) [5]) are marked by different background colors and numbers (grey—fast response; white—intermediate phase; striped—acclimation phase).

We further estimated kinetics of concentration changes in all carotenoids including this newly identified γ-carotene (Figure 7b). We observed almost no HL-induced changes in zeaxanthin, echinenone and β-carotene content (Figure 7b). This contrasts with a small increase in myxoxanthophyll during the final slow acclimation phase (120–360 min). The most significant increase, however, was in γ-carotene (Figure 7b). γ-carotene increased significantly during the second phase (up to 30 min) and further increased during the third phase (Figure 7b). In total, the relative concentration of γ-carotene increased during HL from 0%–1% (typical for control samples) to 5% per total pigment (10% of total carotenoids). This suggests a new, yet unknown, role of γ-carotene accumulation during HL stress in cyanobacteria (for more details, see discussion).

4. Discussion

In this study, we observed how *Synechocystis PSI-YFP* copes during several hours of high-intensity light stress. We described kinetic changes of the cyanobacterial cells during HL treatment on several levels, starting from protein and pigment levels (e.g., carotenoid composition, see Figure 7), through sub-cellular and cellular level (changes in cell sizes, Figure 1b; in MDs, Figure 2), and to the whole cell suspension level (e.g., physiological parameters of photoprotection, i.e., NPQ and F_v/F_m in Figure 4). We thus extended our previously published data that focused solely on PSI/PSII/and PBS co-localization in thylakoids [5]. We raised the hypothesis that after the initial fast response of cyanobacterial cells to HL (until 120 min), cyanobacterial cells tend to acclimate to the new light regime by a progressive return to a similar state as the initial physiological one of the no-stress condition. This is visible in several parameters including single-cell data (PSI/PSII/PBS ratios in Figure 3d) and bulk data of PSI/PSII ratio (Figure 5c). Therefore, in line with our previous single-cell data [5], we could define a three-phase response of *Synechocystis PSI-YFP* to acute HL stress: (1) fast response (0–30 min); (2) intermediate (30–120 min); (3) slow acclimation (120–360 min).

The overall cell physiology, concerning cell volume, responded to the acute HL stress very slowly, and it was visible only in the third slow acclimation phase (see 120–360 min in Figure 1). Our data showed (Figure 1b) that during the HL treatment, the ratio between bigger and smaller cells was significantly affected by HL. It might be an indication that *Synechocystis PSI-YFP* cells exposed to HL slow down their cell cycle, as the fraction of bigger cells was more abundant. The cell size control in cyanobacteria is rather complex [63]. It seems that sizes of cyanobacterial cells are not fully synchronized on the single-cell level because the cyanobacterial clock produces distinctly sized and timed subpopulations of cyanobacteria in both constant and light–dark conditions [63]. It does not contradict the fact that the parameters of the whole culture show synchronized behavior (see, e.g., bioreactor experiments [6,64]). Looking at our data we could hypothesize that, due to the HL, cells were "waiting" for the precise time to divide whilst coping with the light stress. This regulation of the timing of cell divisions could give an advantage in avoiding division during the energetically unfavorable periods. We can only speculate that the *Synechocystis PSI-YFP* cell cycle might be delayed due to the energetically adverse time and the attempt to sustain the cell survival during the stress condition. Further studies will be necessary to confirm or disprove this.

We then stepped further from cell physiology to the sub-cellular level and explored TM organization of PPCs into MDs at the start and end of the experiment (Figure 2). We noticed that, although the fluorescence ratios of PPC emission of whole cells were similar between the initial and the final point of HL stress (Figure 3d), the spatial distributions inside the particular cells were altered (Figure 2c). These data are in line with the previous description of the TM mosaic in cyanobacteria [1,3–5]. The system of PSI/PSII/PBS organization into MDs changed significantly after 6 h of HL exposition; TM heterogeneity became less visible and all the three complexes were distributed more uniformly (compare Figure 2b,c, time 0 min and 360 min). The 6 h HL treatment decreased fluorescence in all three measured channels including PBS (Figure 3). This contrasts with the long-term acclimation (a shift during a week in light growth conditions, see [6]), when PBS fluorescence was the most stable, while PSI and PSII fluorescence were more equally distributed. Interestingly, the relative fluorescence emission of individual PPCs (normalized to sum of all three PPCs fluorescence) depicted the three-phase behavior that we noted before [5]. It shows that the three-phase response of the cyanobacterial thylakoids to 360 min of HL seems to be the typical strategy for *Synechocystis PSI-YFP* cells.

We compared the three-phase behavior detected at the single-cell level (Figure 3) with the analysis of the three photoprotective mechanisms: photoinhibition (F_v/F_m, Figure 4a); NPQ (Figure 4b) and PBS decoupling (Figure 4c,d—representing process of PBS detachment (energetical/physical) from PSII and/or PSI [18]). When PBS are energetically decoupled from photosystems, they cannot contribute to excessive excitation, protecting photosystems (see, e.g., [20]). The process can be measured as a relative increase in phycobilin fluorescence (Figure 4c). Indeed, we observed a significant increase in the PBS/PSII ratio at 90 min (during the intermediate phase) of the HL treatment (Figure 4d). These data indicate a temporal activation of PBS decoupling (around 90 min of HL) due to the acute light stress. However, our results did not show any sign of physical PBS decoupling from TM (Figure S4). Therefore, we suggest that the mechanism of PBS decoupling does not necessarily require a long-distance transport of PBS (e.g., from TM region into the center part of cell). Our data suggest that in some cases, only nanoscale re-arrangement changes near TM are necessary (Figure S4) to see signs of PBS decoupling in bulk measurements (Figure 4c,d). We proved that the process is triggered mostly during the second, intermediate phase.

In contrast to the PBS decoupling, the second photoprotective mechanism, blue light-induced NPQ, started to be activated already during the first phase of HL (0–30 min;

Figure 4b). This proves NPQ to be the fastest response of the photosynthetic membrane to excess light [65]. Our data showed a continuous increase in NPQ during the fast response and intermediate phases (up to 120 min; Figure 4b). During the third (slow acclimation) phase the absolute value of NPQ did not change; however, NPQ becomes more flexible as it is visible in the fast NPQ recovery (Figure 4b). In contrast to NPQ activation, which is triggered by OCP [66], the recovery process requires another protein—fluorescence recovery protein (FRP) [67]. The data suggest that the amount or activity of the proteins could increase during the third—slow acclimation—phase. This is just a hypothesis that will need to be confirmed/disproved by direct measurements.

The other photoprotective mechanism we studied during acute HL stress was photoinhibition. We showed a clear sign of photoinhibition during HL stress based on both fluorescence parameters (Figure 4a) and biochemistry (Figure 5b). A faster response was visible in F_V/F_m that decreased already at 10 min (fast phase in Figure 4a). However, the actual PSII degradation (visible as PSII monomerization, Figure S7b) was slower and appeared later during the intermediate phase (around 60 min of HL; Figure 5b and Figure S7b). These biochemical data were in line with previous observations (see, e.g., [68]); however, we have newly proved that the actual process of PSII monomerization is delayed after the first signs of photoinhibition, visible based on fluorescence.

The biochemical data showed that acute HL stress was linked with faster turnover of pigmented proteins of PSII (see, e.g., Figure 5). These protein complexes contain tens of pigments (chlorophyll, carotenoids) whose concentrations could be affected by protein degradation. Therefore, we have also estimated HL-induced changes in pigment composition by several methods (Figures 6 and 7). It is known that carotenoids in phototrophs have an essential role in photoprotection [31,32] and they are also involved in regulation of biological membranes (e.g., fluidity, see [69]) and pigment–proteins photochemistry (e.g., light harvesting efficiency), where they act as allosteric regulators (see, e.g., [47,48]). In cyanobacteria, the most typical carotenoid involved in photoprotection is echinenone, which is present in OCP protein involved in blue light-induced NPQ in PBS [67]. However, we did not detect any change in its concentration during 6 h of HL (Figure 7).

We found that the carotenoid content responds to acute HL stress in a surprisingly fast manner. For instance, γ-carotene started to accumulate significantly already during the second phase (60–120 min; Figure 7b). Subsequently, during the slowest acclimation phase (120–360 min) myxoxanthophyll content is also increased (Figure 7b). Interestingly, the carotenoid accumulation in the intermediate phase correlates with significant changes in PSII, specifically with the appearance of PSII monomers (Figure 5b and Figure S7b). We can tentatively suggest that higher myxoxanthophyll and γ-carotene contents can facilitate D1 protein synthesis and its assembly into PSII during accelerated photoinhibition (Figure 4a). It has been already suggested that proteins are mobilized during HL stress [15] or when they are disassembled from larger complexes [70]. Indeed, carotenoids (echinenone, zeaxanthin, myxoxanthophyll) were suggested to be involved in the PSII repair mechanism [71]. We thus hypothesize that the increased level of carotenoids (relative to chlorophylls, see Figure 6c) facilitates protein complexes' disassembly (especially PSII) and their movement into assembly centers, regarding which localization is still intensively discussed [13,72]. In fact, the structural importance of carotenoids for assembly of protein complexes, in this case PSI trimer, has been already proved [73].

Apart from the well-known photoprotective role of carotenoids (either direct photoprotection or in a tuning of protein functions [47,48]), they also have a clear structural effect in thylakoids [69]. Free carotenes and xanthophylls seem to be present in the hydrophobic region of the membrane bilayers. It has been hypothesized that they can affect membrane fluidizing effects [74]. For instance, myxoxanthophyll seems to be important for the TM structure [50,75]. Therefore, the increased myxoxanthophyll content we detected could affect membrane fluidity. The preferential synthesis of myxoxanthophylls in HL has already been observed [76], and they could possibly make the membrane more fluid, helping the photosynthetic apparatus during PSII disassembly/assembly in photoinhibition. Indeed, the "proteins mobilization" during HL stress has already been suggested for cyanobacteria [15]. For those reasons, we hypothesize that carotenoids, in addition to photoprotection, could play a role in modulation of membrane fluidity, important for protein mobility [77]. This hypothesis, however, needs to be tested by specific mutants without carotenoids.

The progressive γ-carotene accumulation on HL (Figure 7b) was the most surprising effect we observed. This carotenoid represents the last common precursor to the biosynthesis of all common carotenoids. Its HL-induced increase has not been reported in *Synechocystis* sp. PCC 6803 cells, yet γ-carotene was not usually included in HPLC pigment analyses, as its concentration (at non-stress condition) is often almost too insignificant to be detected (less than 1% of all carotenoids). However, we undoubtedly proved a small, but reproducible γ-carotene accumulation in HL, up to almost 10% of all carotenoids. Therefore, it does not seem to be caused by some malfunction of the carotenoids' biosynthesis pathway. Moreover, it is possible that γ-carotene increase affects also myxoxanthophyll accumulation. The monocyclic γ -carotene represents a branching point towards either β-carotene or myxoxanthophyll [78] that could possibly act as two competing routes during HL stress. The increased level of γ-carotene could thus represent a physiological response of cyanobacterial cell physiology during acute HL stress. One might suggest that its accumulation brings an advantage in that it provides a higher carotenoid pool during HL, a benefit for photoprotection (e.g., against ROS). Interestingly, the accumulation of carotenoid intermediates, (e.g., intermediates of the xanthophylls cycle) as an "economic" mechanism of photoprotection, was shown in eukaryotic phototrophs including the diatom *Phaeodactylum tricornutum* [79] and in the prasinophycean alga *Mantoniella squamata* [80]. In these organisms, the pathway intermediates were transformed into light harvesting and photoprotective pigments upon return to low light illumination, reducing the costs of a de novo synthesis of light-harvesting xanthophylls. Therefore, the increased pool of γ-carotene during HL (Figure 7b) in cyanobacteria could act in a similar way. The evaluation of the γ-carotene for photoprotection remains a task for future experiments.

5. Conclusions

We described a three-phase response of the *Synechocystis PSI-YFP* cell to acute HL stress, first noted by Konert et al. [5]. We surveyed the HL stress response of cyanobacterial cells at all possible levels, starting from the protein/pigment level, before analyzing the cellular level and finally the physiological parameters of photoprotection. The first phase, a "fast response phase" (0–30 min), showed a typical phenomenon of photoinhibition (Figures 4a and 5) and blue light-induced photoprotective non-photochemical quenching (NPQ) localized in PBS (Figure 4b). Later on, during the second phase (30–120 min)—"intermediate phase"—we detected accumulation of γ-carotene that correlated with depletion in PSII dimer and with functional decoupling of PBS from photosystems. The carotenoid accumulation was further stimulated in the last "acclimation phase" (120–360 min) when fluorescence properties of PPCs per cells progressively returned to their initial values (Figure 3d). It indicated transitions of cells into a new physiological equilibrium more suitable for the acute HL stress conditions. The new HL-acclimated state was identifiable by increased content of carotenoids (γ-carotene and myxoxanthophyll, Figure 7b) and by more even spatial distribution of PSI, PSII and PBS between microdomains (Figure 2b,c). We suggest that carotenoid accumulation might be involved in the maintenance of the new membrane and protein structure at HL condition. It is unclear whether the transition into the new steady state after 6 h of HL also requires a delay in the cyanobacterial cell cycle, as this was indicated by HL-induced changes in the cell sizes during HL treatment (Figure 1b). This hypothesis, however, requires additional experiments to be verified.

Supplementary Materials: The following are available online at https://www.mdpi.com/article/10.3390/cells10081916/s1, Figure S1: Example of Dual-PAM, Figure S2: Optical density of Synechocystis PSI-YFP culture measured at 730 nm, Figure S3: Cell type counts of Synechocystis PSI−YFP, Figure S4: Phycobilisomes fluorescence emission measured in the thylakoid membrane, Figure S5: Chlorophyll concentration per cell, Figure S6: Absorption spectra of extracted pigments, Figure S7: Protein composition analysis from CN gel of HL treated culture, Figure S8: Identification of γ-carotene.

Author Contributions: Conceptualization, R.K.; methodology, R.K., M.C. and G.K.; software, G.K.; formal analysis, G.K. and R.K.; investigation, M.C., G.K., A.C. and B.Š.; data curation, M.C., G.K., A.C., B.Š. and R.K.; writing—original draft preparation, M.C.; writing—review and editing, R.K., M.C., G.K., A.C. and B.Š.; visualization, M.C., G.K., A.C., B.Š. and R.K.; funding acquisition, R.K. All authors have read and agreed to the published version of the manuscript.

Funding: This research was funded by CZECH SCIENCE FOUNDATION (grant number 19-11494S), by the ERC project Photoredesign (No. 854126) and by Fondazione Cariplo (CYAO project, grant number 2016-0667). The instrumentation at center ALGATECH has been supported by the institutional project Algamic (CZ.1.05/2.1.00/19.0392).

Institutional Review Board Statement: Not applicable.

Informed Consent Statement: Not applicable.

Data Availability Statement: Not applicable.

Acknowledgments: We thank Jan Pilny for his help in HPLC analysis, Sara Romandini for helping with biochemical experiments and Filip Charvát for his assistance with variable chlorophyll a fluorescence measurements. Our special thanks is dedicated to Adéla Kaňová for here preliminary experiments in this project. We would like to thank Lenka (Ágnes) Bučinská for providing us EM picture for graphical abstract.

Conflicts of Interest: The authors declare no conflict of interest.

References

1. Strašková, A.; Steinbach, G.; Konert, G.; Kotabová, E.; Komenda, J.; Tichý, M.; Kaňa, R. Pigment-protein complexes are organized into stable microdomains in cyanobacterial thylakoids. *Biochim. Biophys. Acta (BBA)-Bioenerg.* **2019**, *1860*. [CrossRef] [PubMed]
2. Ruban, A.V.; Young, A.J.; Pascal, A.A.; Horton, P. The Effects of Illumination on the Xanthophyll Composition of the Photosystem II Light-Harvesting Complexes of Spinach Thylakoid Membranes. *Plant Physiol* **1994**, *104*, 227–234. [CrossRef]
3. Casella, S.; Huang, F.; Mason, D.; Zhao, G.Y.; Johnson, G.N.; Mullineaux, C.W.; Liu, L.N. Dissecting the Native Architecture and Dynamics of Cyanobacterial Photosynthetic Machinery. *Mol Plant* **2017**, *10*, 1434–1448. [CrossRef] [PubMed]
4. Grigoryeva, N.; Chistyakova, L. Fluorescence Microscopic Spectroscopy for Investigation and Monitoring of Biological Diversity and Physiological State of Cyanobacterial Cultures. *Cyanobacteria. Rij. IntechOpen* **2018**, 11–43. [CrossRef]
5. Konert, G.; Steinbach, G.; Canonico, M.; Kaňa, R. Protein arrangement factor: A new photosynthetic parameter characterizing the organization of thylakoid membrane proteins. *Physiol Plant.* **2019**, *166*, 264–277. [CrossRef]
6. Canonico, M.; Konert, G.; Kaňa, R. Plasticity of Cyanobacterial Thylakoid Microdomains Under Variable Light Conditions. *Front. Plant Sci.* **2020**, *11*. [CrossRef]
7. Andersson, B.; Anderson, J.M. Lateral heterogeneity in the distribution of chlorophyll-protein complexes of the thylakoid membranes of spinach chloroplasts. *Biochim. Biophys. Acta* **1980**, *593*, 427–440. [CrossRef]
8. Albertsson, P. A quantitative model of the domain structure of the photosynthetic membrane. *Trends Plant Sci.* **2001**, *6*, 349–358. [CrossRef]
9. Ruban, A.V.; Johnson, M.P. Visualizing the dynamic structure of the plant photosynthetic membrane. *Nat. Plants* **2015**, *1*, 15161. [CrossRef]
10. Vermaas, W.F.J.; Timlin, J.A.; Jones, H.D.T.; Sinclair, M.B.; Nieman, L.T.; Hamad, S.W.; Melgaard, D.K.; Haaland, D.M. In vivo hyperspectral confocal fluorescence imaging to determine pigment localization and distribution in cyanobacterial cells. *Proc. Natl. Acad. Sci. USA* **2008**, *105*, 4050–4055. [CrossRef]
11. Sherman, D.M.; Troyan, T.A.; Sherman, L.A. Localization of membrane-proteins in the cyanobacterium synechococcus sp pcc7942-radial asymmetry in the photosynthetic complexes. *Plant Physiol.* **1994**, *106*, 251–262. [CrossRef]
12. Huokko, T.; Ni, T.; Dykes, G.F.; Simpson, D.M.; Brownridge, P.; Conradi, F.D.; Beynon, R.J.; Nixon, P.J.; Mullineaux, C.W.; Zhang, P.; et al. Probing the biogenesis pathway and dynamics of thylakoid membranes. *Nat. Commun.* **2021**, *12*, 3475. [CrossRef]
13. Mahbub, M.; Hemm, L.; Yang, Y.; Kaur, R.; Carmen, H.; Engl, C.; Huokko, T.; Riediger, M.; Watanabe, S.; Liu, L.N.; et al. mRNA localization, reaction centre biogenesis and thylakoid membrane targeting in cyanobacteria. *Nat. Plants* **2020**, *6*, 1179–1191. [CrossRef] [PubMed]
14. Mullineaux, C.W.; Liu, L.-N. Membrane Dynamics in Phototrophic Bacteria. *Annu. Rev. Microbiol.* **2020**, *74*, 633–654. [CrossRef] [PubMed]
15. Sarcina, M.; Bouzovitis, N.; Mullineaux, C.W. Mobilization of photosystem II induced by intense red light in the cyanobacterium Synechococcus sp PCC7942. *Plant Cell* **2006**, *18*, 457–464. [CrossRef]

16. Tamary, E.; Kiss, V.; Nevo, R.; Adam, Z.; Bernat, G.; Rexroth, S.; Rogner, M.; Reich, Z. Structural and functional alterations of cyanobacterial phycobilisomes induced by high-light stress. *Biochim. Biophys. Acta-Bioenerg.* **2012**, *1817*, 319–327. [CrossRef]
17. Steinbach, G.; Kaňa, R. Automated microscopy: Macro language controlling a confocal microscope and its external illumination–adaptation for photosynthetic organisms. *Microsc. Microanal.* **2016**, *22*, 258–263. [CrossRef]
18. Kirilovsky, D. Modulating energy arriving at photochemical reaction centers: Orange carotenoid protein-related photoprotection and state transitions. *Photosynth. Res.* **2015**, *126*, 3–17. [CrossRef]
19. Adir, N.; Zer, H.; Shochat, S.; Ohad, I. Photoinhibition-a historical perspective. *Photosynth. Res.* **2003**, *76*, 343–370. [CrossRef]
20. Kirilovsky, D.; Kaňa, R.; Prášil, O. Mechanisms Modulating Energy Arriving at Reaction Centers in Cyanobacteria. In *Non-Photochemical Quenching and Energy Dissipation in Plants, Algae and Cyanobacteria*; Demmig-Adams, B., Garab, G., Adams, W., III, Govindjee, Eds.; Springer: Dordrecht, The Netherlands, 2014; pp. 471–501. [CrossRef]
21. Kirilovsky, D. Photoprotection in cyanobacteria: The orange carotenoid protein (OCP)-related non-photochemical-quenching mechanism. *Photosynth. Res.* **2007**, *93*, 7–16. [CrossRef]
22. Kok, B. On the Inhibition of Photosynthesis by Intense Light. *Biochim Biophys. Acta* **1956**, *21*, 234–244. [CrossRef]
23. Sonoike, K. Photoinhibition of Photosystem I: Its Physiological Significance in the Chilling Sensitivity of Plants. *Plant Cell Physiol.* **1996**, *37*, 239–247. [CrossRef]
24. Murata, N.; Takahashi, S.; Nishiyama, Y.; Allakhverdiev, S.I. Photoinhibition of photosystem II under environmental stress. *Biochim. Biophys. Acta* **2007**, *1767*, 414–421. [CrossRef]
25. Tyystjärvi, E. Photoinhibition of Photosystem II. *Int. Rev. Cell Mol. Biol.* **2013**, *300*, 243–303. [CrossRef]
26. Keren, N.; Krieger-Liszkay, A. Photoinhibition: Molecular mechanisms and physiological significance. *Physiol. Plant.* **2011**, *142*, 1–5. [CrossRef]
27. Oguchi, R.; Hikosaka, K.; Hirose, T. Does the photosynthetic light-acclimation need change in leaf anatomy? *Plant Cell Environ.* **2003**, *26*, 505–512. [CrossRef]
28. Takahashi, S.; Badger, M.R. Photoprotection in plants: A new light on photosystem II damage. *Trends Plant Sci.* **2011**, *16*, 53–60. [CrossRef]
29. Campbell, D.A.; Tyystjarvi, E. Parameterization of photosystem II photoinactivation and repair. *Biochim. Biophys. Acta* **2012**, *1817*, 258–265. [CrossRef]
30. Vass, I. Molecular mechanisms of photodamage in the Photosystem II complex. *Biochim. Biophys. Acta* **2012**, *1817*, 209–217. [CrossRef]
31. Komenda, J.; Sobotka, R.; Nixon, P.J. Assembling and maintaining the Photosystem II complex in chloroplasts and cyanobacteria. *Curr. Opin. Plant Biol.* **2012**, *15*, 245–251. [CrossRef] [PubMed]
32. Prasil, O.; Adir, N.; Ohad, I. Dynamics of photosystem II: Mechanism of photoinhibition and recovery processes. *Top. Photosynth.* **1992**, *11*, 295–348.
33. Komenda, J.; Tichy, M.; Prásil, O.; Knoppová, J.; Kuviková, S.; de Vries, R.; Nixon, P.J. The exposed N-terminal tail of the D1 subunit is required for rapid D1 degradation during photosystem II repair in Synechocystis sp PCC 6803. *Plant Cell* **2007**, *19*, 2839–2854. [CrossRef]
34. Komenda, J. Role of two forms of the D1 protein in the recovery from photoinhibition of photosystem II in the cyanobacterium Synechococcus PCC 7942. *Biochim. Biophys. Acta* **2000**, *1457*, 243–252. [CrossRef]
35. Stoitchkova, K.; Zsiros, O.; Jávorfi, T.; Páli, T.; Andreeva, A.; Gombos, Z.; Garab, G. Heat- and light-induced reorganizations in the phycobilisome antenna of Synechocystis sp. PCC 6803. Thermo-optic effect. *Biochim. Biophys. Acta (BBA)-Bioenerg.* **2007**, *1767*, 750–756. [CrossRef]
36. Kaňa, R.; Prášil, O.; Komárek, O.; Papageorgiou, G.; Govindjee. Spectral characteristic of fluorescence induction in a model cyanobacterium, *Synechococcus* sp (PCC 7942). *Biochim. Biophys. Acta* **2009**, *1787*, 1170–1178. [CrossRef] [PubMed]
37. Campbell, D.; Hurry, V.; Clarke, A.K.; Gustafsson, P.; Oquist, G. Chlorophyll fluorescence analysis of cyanobacterial photosynthesis and acclimation. *Microbiol. Mol. Biol. Rev. MMBR* **1998**, *62*, 667–683. [CrossRef]
38. Cser, K.; Vass, I. Radiative and non-radiative charge recombination pathways in Photosystem II studied by thermoluminescence and chlorophyll fluorescence in the cyanobacterium Synechocystis 6803. *Biochim. Biophys. Acta* **2007**, *1767*, 233–243. [CrossRef] [PubMed]
39. Stadnichuk, I.N.; Yanyushin, M.F.; Bernát, G.; Zlenko, D.V.; Krasilnikov, P.M.; Lukashev, E.P.; Maksimov, E.G.; Paschenko, V.Z. Fluorescence quenching of the phycobilisome terminal emitter LCM from the cyanobacterium Synechocystis sp. PCC 6803 detected in vivo and in vitro. *J. Photochem. Photobiol. B Biol.* **2013**, *125*, 137–145. [CrossRef]
40. Kirilovsky, D.; Kerfeld, C.A. The Orange Carotenoid Protein: A blue-green light photoactive protein. *Photochem. Photobiol. Sci.* **2013**, *12*, 1135–1143. [CrossRef]
41. Siefermann-Harms, D. The light-harvesting and protective functions of carotenoids in photosynthetic membranes. *Physiol. Plant.* **1987**, *69*, 561–568. [CrossRef]
42. Daddy, S.; Zhan, J.; Jantaro, S.; He, C.; He, Q.; Wang, Q. A novel high light-inducible carotenoid-binding protein complex in the thylakoid membranes of *Synechocystis* PCC 6803. *Sci. Rep.* **2015**, *5*, 9480. [CrossRef]
43. Kuthanová Trsková, E.; Belgio, E.; Yeates, A.M.; Sobotka, R.; Ruban, A.V.; Kana, R. Antenna proton sensitivity determines photosynthetic light harvesting strategy. *J. Exp. Bot.* **2018**, *69*, 4483–4493. [CrossRef]

44. Bilger, W.; Björkman, O. Role of the xanthophyll cycle in photoprotection elucidated by measurements of light-induced absorbance changes, fluorescence and photosynthesis in leaves of Hedera canariensis. *Photosynth. Res.* **1990**, *25*, 173–185. [CrossRef]
45. Cogdell, R. The function of pigments in chloroplasts. *Plant Pigment.* **1988**, 183–230.
46. Domonkos, I.; Kis, M.; Gombos, Z.; Ughy, B. Carotenoids, versatile components of oxygenic photosynthesis. *Prog. Lipid. Res.* **2013**, *52*, 539–561. [CrossRef]
47. Ruban, A.V.; Johnson, M.P. Xanthophylls as modulators of membrane protein function. *Arch. Biochem. Biophys.* **2010**, *504*, 78–85. [CrossRef]
48. Kaňa, R.; Kotabová, E.; Kopečná, J.; Trsková, E.; Belgio, E.; Sobotka, R.; Ruban, A.V. Violaxanthin inhibits nonphotochemical quenching in light-harvesting antennae of *Chromera velia*. *FEBS Lett.* **2016**, *590*, 1076–1085. [CrossRef]
49. Hirschberg, J.; Chamovitz, D. Carotenoids in Cyanobacteria. In *The Molecular Biology of Cyanobacteria*; Bryant, D.A., Ed.; Springer: Dordrecht, The Netherlands, 1994; pp. 559–579. [CrossRef]
50. Zakar, T.; Herman, E.; Vajravel, S.; Kovacs, L.; Knoppová, J.; Komenda, J.; Domonkos, I.; Kis, M.; Gombos, Z.; Laczko-Dobos, H. Lipid and carotenoid cooperation-driven adaptation to light and temperature stress in Synechocystis sp. PCC6803. *Biochim. Biophys. Acta (BBA)-Bioenerg.* **2017**, *1858*, 337–350. [CrossRef]
51. Tóth, T.N.; Chukhutsina, V.; Domonkos, I.; Knoppová, J.; Komenda, J.; Kis, M.; Lénárt, Z.; Garab, G.; Kovács, L.; Gombos, Z.; et al. Carotenoids are essential for the assembly of cyanobacterial photosynthetic complexes. *Biochim. Biophys. Acta (BBA)-Bioenerg.* **2015**, *1847*, 1153–1165. [CrossRef] [PubMed]
52. Williams, J.G.K. Construction of specific mutations in photosystem II photosynthetic reaction center by genetic engineering methods in Synechocystis 6803. *Methods Enzymol.* **1988**, *167*, 766–778.
53. Tichý, M.; Bečková, M.; Kopečná, J.; Noda, J.; Sobotka, R.; Komenda, J. Strain of Synechocystis PCC 6803 with Aberrant Assembly of Photosystem II Contains Tandem Duplication of a Large Chromosomal Region. *Front. Plant Sci.* **2016**, *7*. [CrossRef] [PubMed]
54. Strašková, A.; Knoppová, J.; Komenda, J. Isolation of the cyanobacterial YFP-tagged photosystem I using GFP-Trap®. *Photosynthetica* **2018**, *56*, 300–305. [CrossRef]
55. Wittig, I.; Karas, M.; Schägger, H. High Resolution Clear Native Electrophoresis for In-gel Functional Assays and Fluorescence Studies of Membrane Protein Complexes. *Mol. &Amp; Cell. Proteom.* **2007**, *6*, 1215–1225. [CrossRef]
56. Kaňa, R.; Kotabová, E.; Komárek, O.; Šedivá, B.; Papageorgiou, G.C.; Govindjee; Prášil, O. The slow S to M fluorescence rise in cyanobacteria is due to a state 2 to state 1 transition. *Biochim. Biophys. Acta* **2012**, *1817*, 1237–1247. [CrossRef]
57. Kaňa, R. Application of spectrally resolved fluorescence induction to study light-induced nonphotochemical quenching in algae. *Photosynthetica* **2018**, *56*, 132–138. [CrossRef]
58. Ritchie, R.J. Consistent sets of spectrophotometric chlorophyll equations for acetone, methanol and ethanol solvents. *Photosynth. Res.* **2006**, *89*, 27–41. [CrossRef]
59. Pazderník, M.; Mareš, J.; Pilný, J.; Sobotka, R. The antenna-like domain of the cyanobacterial ferrochelatase can bind chlorophyll and carotenoids in an energy-dissipative configuration. *J. Biol. Chem.* **2019**, *294*, 11131–11143. [CrossRef]
60. Sutton, S. Measurement of microbial cells by optical density. *J. Valid. Technol.* **2011**, *17*, 46–49.
61. Steinbach, G.; Schubert, F.; Kana, R. Cryo-imaging of photosystems and phycobilisomes in Anabaena sp PCC 7120 cells. *J. Photochem. Photobiol. B-Biol.* **2015**, *152*, 395–399. [CrossRef] [PubMed]
62. Aro, E.M.; McCaffery, S.; Anderson, J.M. Photoinhibition and D1 protein-degradation in peas acclimated to different growth irradiances. *Plant Physiol.* **1993**, *103*, 835–843. [CrossRef]
63. Martins, B.M.C.; Tooke, A.K.; Thomas, P.; Locke, J.C.W. Cell size control driven by the circadian clock and environment in cyanobacteria. *Proc. Natl. Acad. Sci. USA* **2018**, *115*, E11415–E11424. [CrossRef] [PubMed]
64. van Alphen, P.; Hellingwerf, K.J. Sustained Circadian Rhythms in Continuous Light in Synechocystis sp. PCC6803 Growing in a Well-Controlled Photobioreactor. *PLoS ONE* **2015**, *10*, e0127715. [CrossRef]
65. Demmig-Adams, B.; Garab, G.; Adams, W.W., III; Govindjee. *Non-Photochemical Quenching and Energy Dissipation in Plants, Algae and Cyanobacteria*; Demmig-Adams, B., Garab, G., Adams, W., III, Eds.; Springer: Dordrecht, The Netherlands, 2014; Volume XXXVIII, p. 649. [CrossRef]
66. Boulay, C.; Wilson, A.; D'Haene, S.; Kirilovsky, D. Identification of a protein required for recovery of full antenna capacity in OCP-related photoprotective mechanism in cyanobacteria. *Proc. Natl. Acad. Sci. USA* **2010**, *107*, 11620–11625. [CrossRef]
67. Gwizdala, M.; Wilson, A.; Kirilovsky, D. In Vitro Reconstitution of the Cyanobacterial Photoprotective Mechanism Mediated by the Orange Carotenoid Protein in *Synechocystis* PCC 6803. *Plant Cell* **2011**, *23*, 2631–2643. [CrossRef]
68. Kopecna, J.; Komenda, J.; Bucinska, L.; Sobotka, R. Long-term acclimation of the cyanobacterium Synechocystis sp. PCC 6803 to high light is accompanied by an enhanced production of chlorophyll that is preferentially channeled to trimeric photosystem I. *Plant Physiol.* **2012**, *160*, 2239–2250. [CrossRef]
69. Havaux, M. Carotenoids as membrane stabilizers in chloroplasts. *Trends Plant Sci.* **1998**, *3*, 147–151. [CrossRef]
70. Kaňa, R.; Steinbach, G.; Sobotka, R.; Vámosi, G.; Komenda, J. Fast Diffusion of the Unassembled PetC1-GFP Protein in the Cyanobacterial Thylakoid Membrane. *Life* **2021**, *11*, 15. [CrossRef]
71. Izuhara, T.; Kaihatsu, I.; Jimbo, H.; Takaichi, S.; Nishiyama, Y. Elevated Levels of Specific Carotenoids During Acclimation to Strong Light Protect the Repair of Photosystem II in Synechocystis sp. PCC 6803. *Front. Plant Sci.* **2020**, *11*. [CrossRef]
72. Rast, A.; Schaffer, M.; Albert, S.; Wan, W.; Pfeffer, S.; Beck, F.; Plitzko, J.M.; Nickelsen, J.; Engel, B.D. Biogenic regions of cyanobacterial thylakoids form contact sites with the plasma membrane. *Nat. Plants* **2019**, *5*, 436–446. [CrossRef] [PubMed]

73. Vajravel, S.; Kis, M.; Kłodawska, K.; Laczko-Dobos, H.; Malec, P.; Kovács, L.; Gombos, Z.; Toth, T.N. Zeaxanthin and echinenone modify the structure of photosystem I trimer in Synechocystis sp. PCC 6803. *Biochim. Biophys. Acta (BBA)-Bioenerg.* **2017**, *1858*, 510–518. [CrossRef] [PubMed]
74. Gruszecki, W.I.; Strzałka, K. Carotenoids as modulators of lipid membrane physical properties. *Biochim. Biophys. Acta* **2005**, *1740*, 108–115. [CrossRef]
75. Mohamed, H.E.; van de Meene, A.M.L.; Roberson, R.W.; Vermaas, W.F.J. Myxoxanthophyll Is Required for Normal Cell Wall Structure and Thylakoid Organization in the Cyanobacterium *Synechocystis* sp. Strain PCC 6803. *J. Bacteriol.* **2005**, *187*, 6883–6892. [CrossRef] [PubMed]
76. Steiger, S.; Schäfer, L.; Sandmann, G. High-light-dependent upregulation of carotenoids and their antioxidative properties in the cyanobacterium Synechocystis PCC 6803. *J. Photochem. Photobiol. B Biol.* **1999**, *52*, 14–18. [CrossRef]
77. Kaňa, R. Mobility of photosynthetic proteins. *Photosynth. Res.* **2013**, *116*, 465–479. [CrossRef]
78. Pagels, F.; Vasconcelos, V.; Guedes, A.C. Carotenoids from Cyanobacteria: Biotechnological Potential and Optimization Strategies. *Biomolecules* **2021**, *11*, 735. [CrossRef] [PubMed]
79. Lohr, M.; Wilhelm, C. Algae displaying the diadinoxanthin cycle also possess the violaxanthin cycle. *Proc. Natl. Acad. Sci. USA* **1999**, *96*, 8784–8789. [CrossRef]
80. Böhme, K.; Wilhelm, C.; Goss, R. Light regulation of carotenoid biosynthesis in the prasinophycean alga Mantoniella squamata. *Photochem. Photobiol. Sci.* **2002**, *1*, 619–628. [CrossRef]

cells

MDPI

Article

Use of Synchrotron Phase-Sensitive Imaging for the Investigation of Magnetopriming and Solar UV-Exclusion Impact on Soybean (*Glycine max*) Leaves

Anis Fatima [1,*,†], Sunita Kataria [2,*,†], Ashish Kumar Agrawal [1], Balwant Singh [1], Yogesh Kashyap [1], Meeta Jain [2], Marian Brestic [3,*], Suleyman I. Allakhverdiev [4] and Anshu Rastogi [5,6]

1 Technical Physics Division, Bhabha Atomic Research Centre, Trombay, Mumbai 400085, India; ashishka@rrcat.gov.in (A.K.A.); balwants@rrcat.gov.in (B.S.); yskashyap@barc.gov.in (Y.K.)
2 School of Biochemistry, Devi Ahilya Vishwavidyalaya, Khandwa Road, Indore 452001, India; mjjainmeeta@gmail.com
3 Department of Plant Physiology, Slovak University of Agriculture, A. Hlinku 2, 94976 Nitra, Slovakia
4 K.A. Timiryazev Institute of Plant Physiology, Russian Academy of Sciences, Botanicheskaya St. 35, 127276 Moscow, Russia; suleyman.allakhverdiev@gmail.com
5 Laboratory of Bioclimatology, Department of Ecology and Environmental Protection, Poznan University of Life Sciences, Piątkowska 94, 60-649 Poznan, Poland; anshu.rastogi@up.poznan.pl
6 Faculty of Geo-Information Science and Earth Observation (ITC), University of Twente, 7500 AE Enschede, The Netherlands
* Correspondence: anees349@gmail.com (A.F.); sunita_kataria@yahoo.com (S.K.); marian.brestic@gmail.com (M.B.)
† Anis Fatima and Sunita Kataria are co-first authors.

Citation: Fatima, A.; Kataria, S.; Agrawal, A.K.; Singh, B.; Kashyap, Y.; Jain, M.; Brestic, M.; Allakhverdiev, S.I.; Rastogi, A. Use of Synchrotron Phase-Sensitive Imaging for the Investigation of Magnetopriming and Solar UV-Exclusion Impact on Soybean (*Glycine max*) Leaves. *Cells* **2021**, *10*, 1725. https://doi.org/10.3390/cells10071725

Academic Editor: Sara Rinalducci

Received: 30 April 2021
Accepted: 3 July 2021
Published: 8 July 2021

Abstract: The combined response of exclusion of solar ultraviolet radiation (UV-A+B and UV-B) and static magnetic field (SMF) pre-treatment of 200 mT for 1h were studied on soybean (*Glycine max*) leaves using synchrotron imaging. The seeds of soybean with and without SMF pre-treatment were sown in nursery bags kept in iron meshes where UV-A+B (280–400 nm) and UV-B (280–315 nm) from solar radiation were filtered through a polyester filters. Two controls were planned, one with polythene filter controls (FC)- which allows all the UV (280–400 nm); the other control had no filter used (open control-OC). Midrib regions of the intact third trifoliate leaves were imaged using the phase-contrast imaging technique at BL-4, Indus-2 synchrotron radiation source. The solar UV exclusion results suggest that ambient UV caused a reduction in leaf growth which ultimately reduced the photosynthesis in soybean seedlings, while SMF treatment caused enhancement of leaf growth along with photosynthesis even under the presence of ambient UV-B stress. The width of midrib and second-order veins, length of the second-order veins, leaf vein density, and the density of third-order veins obtained from the quantitative image analysis showed an enhancement in the leaves of plants that emerged from SMF pre-treated seeds as compared to untreated ones grown in open control and filter control conditions (in the presence of ambient UV stress). SMF pre-treated seeds along with UV-A+B and UV-B exclusion also showed significant enhancements in leaf parameters as compared to the UV excluded untreated leaves. Our results suggested that SMF-pretreatment of seeds diminishes the ambient UV-induced adverse effects on soybean.

Keywords: phase-sensitive imaging; magnetopriming; UV exclusion; leaf venation; leaf hydraulics

1. Introduction

One of the non-ionizing parts of the electromagnetic spectrum of solar radiation is ultraviolet radiation. Ultraviolet (UV) radiations are further divided into three ranges: UV-A (315–400 nm), UV-B (280–315 nm), and UV-C (100–280 nm). The UV-C and major part of UV-B radiations are absorbed by the earth's ozone layer [1]. Even if around 20% of UV-B is able to pass through the ozone layer and reach the earth's surface, it may be harmful to biological systems due to its high energy content. Anthropogenic activities resulted in the reduction of the ozone layer, due to which the percentage of UV-B reaching the earth increased [2,3]. This further resulted in an increasing interest of scientists to understand how plants with a sessile nature react to this increased level of UV-B radiation [2–6]. The different responses of high UV-B radiation on plant structure, morphology, physiology, and genetics have been intensively studied previously [2,4,5,7] where UV-B radiations have been observed to adversely impact the cell membrane and caused changes in plant photosynthesis and enzyme activities [2,8].

Seed priming methods are the pre-treatment of seeds prior to sowing for the purpose of improving the physiological state of the seeds so that the seed germinates more efficiently [9,10]. There are several seed priming methods practiced in agronomy for increasing the seed germination, crop growth, and yield [10–13]. Static magnetic field (SMF) is a seed pre-treatment method based on the interactions of electromagnetic fields with seeds which act as bio-stimulators for the growth of seeds and plants [8,14–16]. The effect of SMF on plants has been extensively studied over the past few years as magnetic field pre-treatment may provide a non-chemical solution to the plants [16–18]. Some of the previous studies reported stimulatory effects of SMF treatment on crops including rice, maize, soybean, and sunflower [15,18–21], whereas the others reported slow development [22]. It is thus predicted that various plant species respond in different ways to varied frequencies and intensities of the magnetic field [23–25]. Plants showed reactions to magnetic fields based on the intensity, flux density, and exposure time [16,25,26]. The enhanced germination percentage improved plant growth, photosynthesis and yield were observed due to SMF pre-treatment of seeds as compared to the untreated seeds under non-stress as well as under abiotic stresses such as salt, water, UV-B, and arsenic toxicity [8,15–18,27–29]. The effect of magnetopriming on plants can be best understood in the framework of two mechanisms, namely the ion cyclotron-resonance (ICR) and the radical-pair models (RPM) [16,30]. The RPM is currently the only possible mechanism demonstrating the function of cryptochromes as a candidate for magneto-reception [16]. The experimental and theoretical studies provide evidence that the application of magnetic fields increases the average radical concentration, increases radical lifetime, and escalates the probability of radical reactions with cellular components [30]. The radical pair intermediates, triplet yields, and emission intensity that occur in Photosystem I and II of green plants can be modulated by an external magnetic field. The increased water uptake compared to untreated seeds is explained by the assumption that the magnetic field interacts with ionic currents in the cell membrane of the plant embryo [31]. In addition to these mechanisms, the interaction between environmental impacts such as ionizing radiation (ultraviolet-UV) and the magnetic field influence as a repair mechanism has also been reported previously in chick embryos [32].

Magnetic field treatment with low flux densities and the exclusion of solar UV radiation are the two parts of radiation biology that have positive stimulating effects on leaf growth, venation, and photosynthesis [8,18,29]. The network of leaf venation is composed of minor veins and a midrib (major conducting vein), which provides mechanical stability to the leaf structure. The venation network has the important function of transportation of water, nutrients, and carbon to different plant tissues [33–35]. The hydraulic system associated with plant leaf veins plays a key part in photosynthetic gas exchange and growth determination [36]. The width of midrib and minor veins, leaf vein density (LVD) (known as the vein length per leaf area), and the vein number density (which is the number of veins per leaf area) are all directly related to leaf hydraulic conductivity and photosynthesis [37–39]. Both magnetic treatment and exclusion of solar UV radiation change plant photosynthetic function which is related to the midrib of the leaf venation. The positive effects of solar UV exclusion and SMF on the leaf venation (midrib width) have been individually studied using synchrotron-based X-ray phase-contrast imaging [25,40]. However, there have been no reports on X-ray imaging of leaf venation to the combination of SMF pre-treatment of seeds and the exclusion of solar UV radiation.

The relationship of leaf venation and hence leaf hydraulics with photosynthesis is not yet explored completely. Advancements in non-destructive X-ray imaging techniques have overcome the limitations of manual sectioning and staining of leaves for imaging. So far, X-ray imaging studies for various parts of the plant have been reported [34]. X-ray radiography and micro-computed tomography (μCT) studies of intact plant parts with synchrotron radiation have contributed to the understanding of plant anatomical structures [37,41–46].

The phase-contrast imaging (PCI) technique relies on phase variations which occur when the X-ray wave front transmits through a sample [47–50]. The technique overcomes the limitations of conventional absorption-based techniques. It is well suited for imaging weakly absorbing samples like leaves in non-destructive ways [37]. In the present study, we have used the soybean (*Glycine max*) variety JS-335 an economically important crop to investigate the effects of exclusion of solar UV radiation in plants grown from the seeds pre-treated with SMF for 1 h. The aim of the present study was to determine the changes in the width of the midrib and minor veins, length of minor veins (2° and 3°) of leaves, and leaf vein density through high-resolution X-ray imaging and relate it to leaf growth, photosynthetic rate, and stomatal conductance.

2. Materials and Method

The soybean (*Glycine max* (L.) var. JS-335) seeds were procured from the Indian Institute of Soybean Research in Indore, India. The experiment was conducted under natural sunlight at the open terrace of the School of Biochemistry, in Devi Ahilya Vishwavidyalaya, Indore (22°44′ N, 75°50′ E), India. The experimental period was between October 2018 to December 2018. After moistening the SMF-pretreated (MT) and untreated (UT) soybean seeds were further mixed with recommended fungicides *viz*Bevistin and Diathane M at 2 gm kg^{-1} seeds and *Rhizobium* culture (provided by National Fertilizer limited, New-Delhi, India) at 3 g kg^{-1} seeds before sowing. The uniform shape and size of seeds were sown in plastic nursery bags of 34 × 34 cm. The nursery bags were filled with a mixture of soil, sand, and organic manure in a 2:2:1 ratio, and ten seeds of soybean were sown; three bags were prepared for each treatment. In each bag, six plants of uniform size were maintained after germination.

2.1. Magnetic Field Generation

An electromagnetic field generator ("AETec" Academy of Embedded Technology, Delhi, India) was used for the generation of magnetic field for seed pre-treatment, as previously described by Kataria et al. [51].

2.2. Magnetic Treatment

For the experiments, the seeds were exposed to SMF treatment of 200 mT for 1 h (MT) on the basis of our previous study on soybeans [25]. Through the Gauss meter, we can measure the magnetic field generated between the poles. The current in coils was regulated to obtain the exact magnetic field for the SMF pretreatment. At 50 mT, the variation in the applied field was observed to be 0.6% in the horizontal and 1.6% in the vertical direction, whereas, at 300 mT the variation decreased to 0.4% and 1.2% in both directions, respectively. A temperature of 25 ± 5 °C was maintained during seed exposure to SMF. The seeds from the same lot were kept under conditions without any influence of the magnetic field served as untreated (UT) seeds.

2.3. UV-A+B and UV-B Exclusion

The UV-A+B and UV-B radiations were cut-off from solar radiation by using band-pass polyester filters (Garware polyester Ltd., Mumbai) with cut-offs of <315 nm and <400 nm radiation. Two controls were designed for this study; one with a polythene filter transparent to all ambient light (filter control, FC) and the other grown on the terrace without any filter (open control, OC). Figure 1 shows the transparency of the filters used in the experiments. The transmission spectra of the filters were measured according to the method of Kataria et al. [8]. The filters were continuously used from seed germination to maturity, with a regular exchange of filters every two weeks due to the solar radiation effect on the filters. For proper ventilation, the lower sides of the cage (0.35 m above the surface) holding the filter were not covered. The experiments were placed in the corner where sunlight was available throughout the day without any shading. The temperature inside and outside the cage was monitored through thermometers. During the growing period, average temperature was raised from 25 °C to 32 °C. No significant difference in the inside and outside temperatures was observed due to proper ventilation.

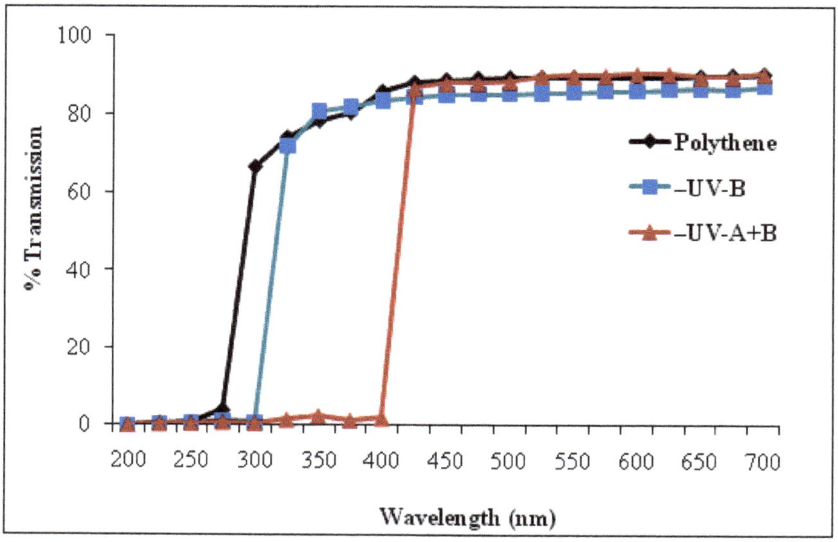

Figure 1. Transmission spectra of UV cut-off filters and polythene filter used for raising soybean plants under iron mesh cages [8] (Kataria et al. 2017a).

2.4. Radiation Measurement

At midday (around noon), a radiometer (Solar light Co. Inc. (PMA 2100), Glenside, PA, USA) was used to measure the intensity of solar spectra. The average photosynthetic active radiation (PAR) value at midday was observed to be 1450 μmol m^{-2} s^{-1} for the non-filter control, which decreased by 12.5% (1270 μmol m^{-2} s^{-1}) under the UV-B filter and 11.8% (1280 μmol m^{-2} s^{-1}) under the UV-A+B filter, whereas a decrease of 4.2% (1390 μmol m^{-2} s^{-1}) was observed for the filter control.

2.5. Growth Data Collection and Analysis

A random selection of plants was done after 45 days of seed germination (DAE). At least three plants in triplicates from each treatment were harvested and transferred to the laboratory for growth data analysis. The soil particles from roots were washed and different parts of the plant were measured through a portable laser leaf area meter CID-202 scanning leaf area meter (CID Inc., Camas, WA, USA).

2.6. Photosynthesis and Stomatal Conductance

The LI-COR photosynthetic system (Li-6200, LI-COR Inc., Lincoln, NE, Serial No. PPS 1332 USA) was used to measure net photosynthesis (*Pn*, μmol CO_2 m^{-2} s^{-1}) and stomatal conductance (*gs*, mol H_2O m^{-2} s^{-1}) for intact soybean plants from each experimental condition after 45 DAE. Photosynthetic measurements were performed on fully expanded third trifoliate leaves of soybean plants under ambient temperature and CO_2 concentration, on clear days. The photosynthetic photon flux density (PPFD) was observed to be in between 1300–1600 μmol m^{-2} s^{-1} with airflow of 500 μmol s^{-1} and CO_2 concentration of 350–380 ppm.

2.7. Phase Contrast Imaging Technique

The Imaging Beamline (BL-4), Indus-2 synchrotron radiation source [40,52] was used to generate the phase-contrast images. The experimental setup was previously described in [25].

The third trifoliate leaves of soybeans from all the groups were pressed flat and dried for two days at room temperature. The whole leaflets of the third trifoliate leaves were mounted in a rectangular metallic frame and phase-contrast images were acquired for middle regions in each leaf. The high-resolution X-ray microscope with 1.8 μm resolution (20 μm thick YAG-Ce scintillator, 4× objective, and PCO-2000 CCD camera) was used for image acquisition at 12keV energy, with a sample to detector distance (SDD) 50mm and an exposure time of 5 min.

2.8. Leaf Midrib Width Quantification

From the synchrotron images of the middle leaflet of third trifoliate leaves of soybeans, the midrib width was quantified at six places in the direction perpendicular to the length at fixed intervals with ImageJ [53]. The average width of the midrib vein and the adjoining minor vein (2°) was obtained for all the leaflets in the third trifoliate and an average value for the leaf was then calculated [25,40].

2.9. Leaf Minor Vein Length and Leaf Vein Density Quantification

The length of the minor vein (2°) was obtained using a freehand line in Image J. To obtain the total length and number of the (3°) minor vein in the entire phase contrast image of 2048 × 2048 pixel size, the objectJ plugin was used (plant-image-analysis.org/software/object (accessed on 12 April 2019). In the phase-contrast images, the vascular region above the midrib was selected with the freehand selection tool in Image J, and the area was measured. Similarly, the area of the vascular region below the midrib was acquired. To find the vascular area in the whole image, the area of the two regions measured were combined. Leaf vein density (LVD) was found by dividing the total length of all 3° veins (marked with red) in the image with the total area of the image. The total number of 3° veins in the images was divided with the total area to calculate the vein number density using ObjectJ.

2.10. Statistical Analysis

All data are presented in triplicate (n = 3); from each replica five plants were randomly taken for each treatment. The statistical analysis was performed on Microsoft Excel and Prism 4 (GrafPad Software, La Jolla, CA, USA) software where mean and standard errors were calculated, and the analysis of variance (ANOVA) followed by post hoc Newman–Keuls Multiple Comparison Test was performed. ### $p < 0.001$; ## $p < 0.01$; # $p < 0.05$ denotes statistically significant differences between seedlings that emerged from untreated (UT) seeds of OC with seedlings that emerged from untreated (UT) seeds of different treatment conditions-FC, UV-B and UV-A+B cutoff filters. *** $p < 0.001$; ** $p < 0.01$; * $p < 0.05$ denotes statistically significant differences between seedlings that emerged from SMF-pretreated (MT) and untreated (UT) seeds under each treatment.

3. Results and Discussion

In the present study, the individual effects of the exclusion of solar UV-A+B, UV-B radiation, and SMF pre-treatment as well as their combination were investigated on the growth, photosynthesis, and development of soybean leaves. Individual and joint exclusion of solar UV-A+B, UV-B radiation, and SMF pre-treatment significantly enhanced all leaf growth parameters studied in the present study, but the extent of enhancement was greater when the plants pre-treated with SMF were grown under ambient UV stress (OC and FC conditions).

A prominent increase was observed in the area and length of the middle leaflet of the third trifoliate leaves of soybean plants raised after SMF (200 mT for 1 h) priming with or without ambient UV radiations (Figure 2a,b). Similarly, solar UV exclusion also enhanced the area and length of middle leaflets of third trifoliate leaves of plants that emerged from untreated (UT) seeds (Figure 2a,b). The area of the middle leaflet increased by 44% and 50% through SMF-treatment respectively under OC and FC conditions as compared to their UT ones (Figure 2a).

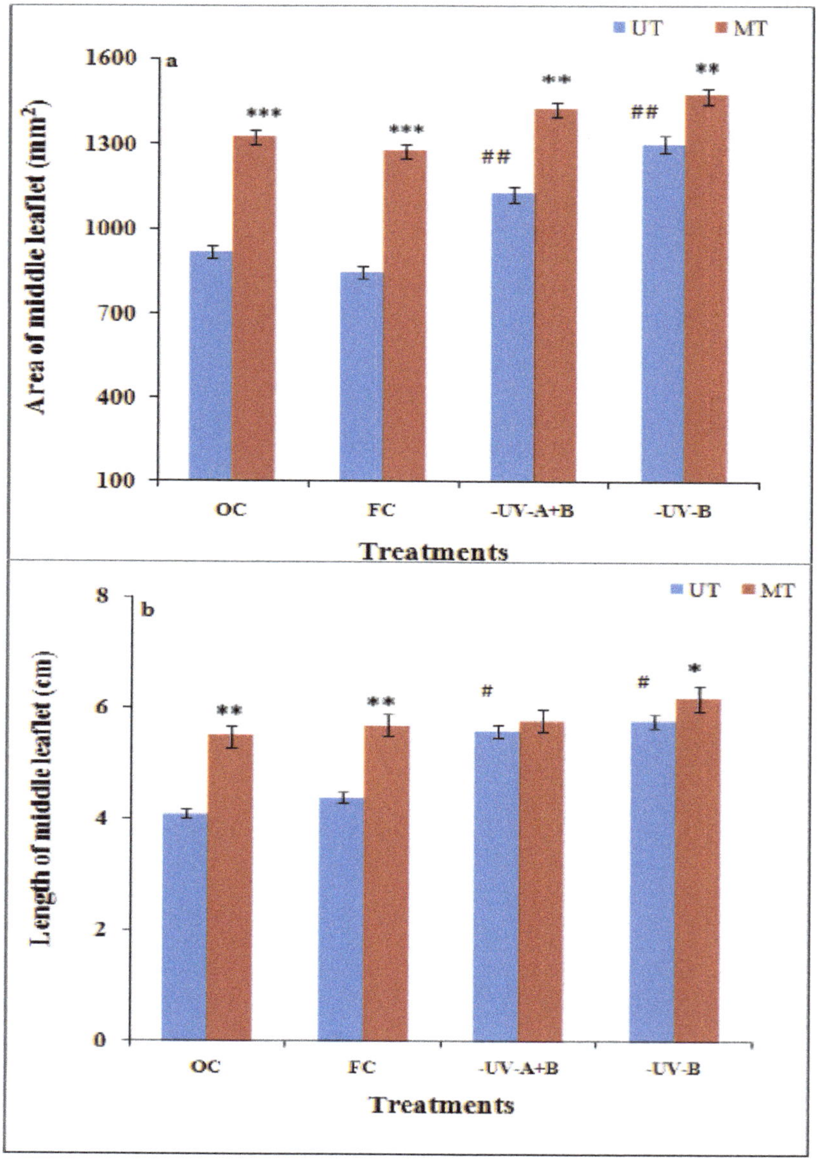

Figure 2. *Cont.*

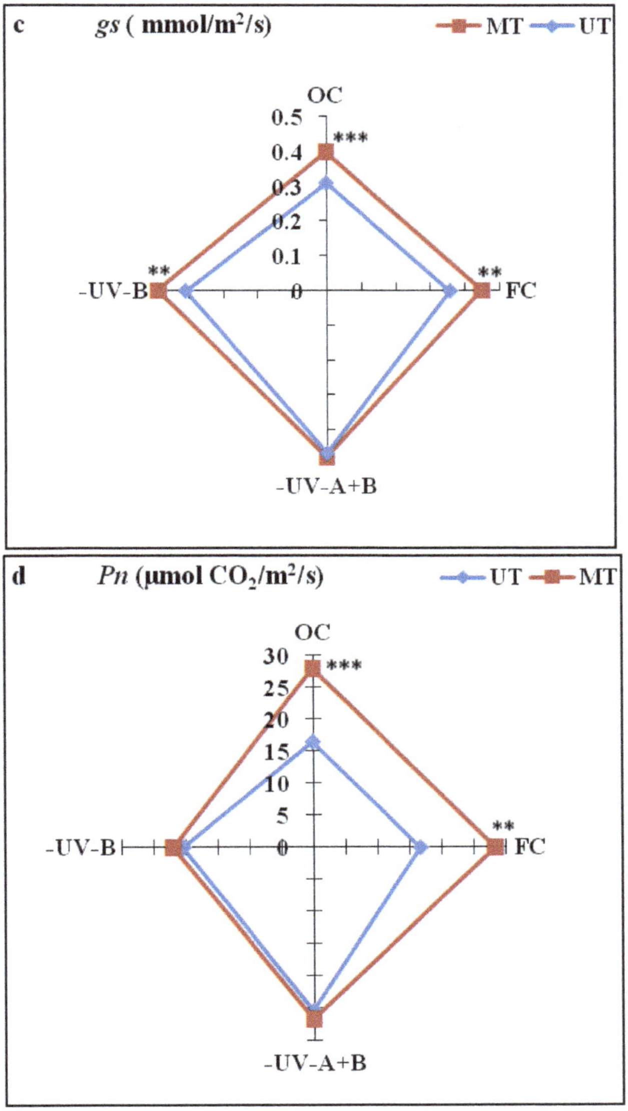

Figure 2. Leaf area (**a**), leaf length (**b**), stomatal conductance (**c**) and rate of photosynthesis (**d**) in middle leaflets of third trifoliate leaves of soybean after SMF pretreatment and solar UV exclusion in soybean. ## $p < 0.01$; # $p < 0.05$ denotes statistically significant differences between seedlings emerged from untreated (UT) seeds of OC with the seedlings emerged from untreated (UT) seeds of different treatments conditions-FC, UV-B and UV-A+B cutoff filters, *** $p < 0.001$; ** $p < 0.01$; * $p < 0.05$ denotes statistically significant differences between seedlings emerged from SMF-pretreated (MT) and untreated (UT) seeds under each treatment.

The enhancement in the length of middle leaflets of third trifoliate leaves of soybean after SMF treatment was 34% in OC and 30% in FC conditions as compared to their UT ones (Figure 2b). A significant increase in leaf length by 41% under solar UV-B exclusion and 37% under UV-A+B exclusion in UT was observed as compared to the plants from UT seeds under OC conditions (Figure 2b).

A significant enhancement in stomatal conductance and photosynthetic rate was observed for the plants pretreated with SMF of 200 mT for 1 h (Figure 2c,d). SMF caused a 28% and 26% increase in stomatal conductance and a 70% and 69% increase in the net photosynthetic rate as compared with untreated controls respectively in OC and FC (presence of ambient UV stress) conditions (Figure 2c,d). Enhancement of leaf area along with an increase in the rate of photosynthesis and stomatal conductance after the SMF pre-treatment (200 mT for 1 h) has been previously reported in soybean and maize [8,15,18,21].

A qualitative and quantitative comparison of phase-contrast images of untreated and SMF pre-treated leaves in OC, FC, UV-A+B, and UV-B showed enhancement in the midrib width, minor vein width, and leaf vascular region near the midrib (Figures 3–9). In the OC group which received all the ambient solar radiation (280–400 nm), the quantification of leaf veins in the phase-contrast images showed an enhancement of 44% in the width of the midrib in the plants grown from the SMF pre-treated seeds as compared to untreated seeds (Figures 3 and 4a). The visibility of vascular structures comprising of higher-order veins (3°) has also been improved in SMF pre-treated leaves (Figure 3b), which is due to a thinning effect [54].

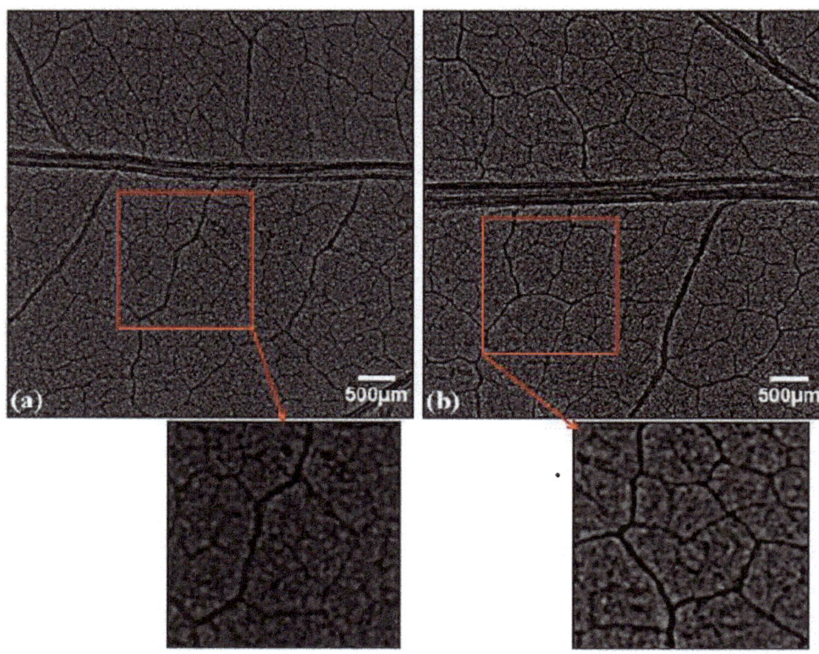

Figure 3. Phase contrast images of soybean leaves under open control (OC) receiving all ambient solar radiation: (**a**) emerging from untreated seeds, (**b**) emerging from seeds pre-treated with static magnetic field (SMF) of 200 mT strength for 1 h. The vascular region below the midrib region is highlighted in red and zoomed images are shown below the respective images.

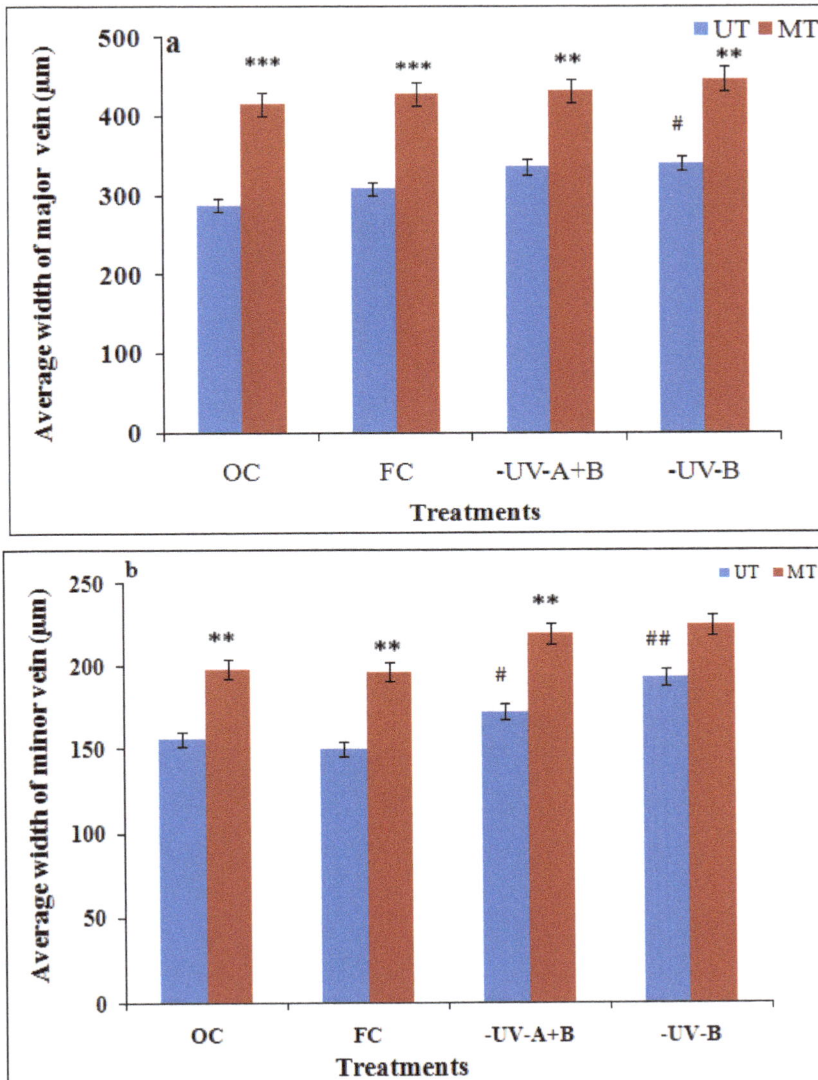

Figure 4. *Cont.*

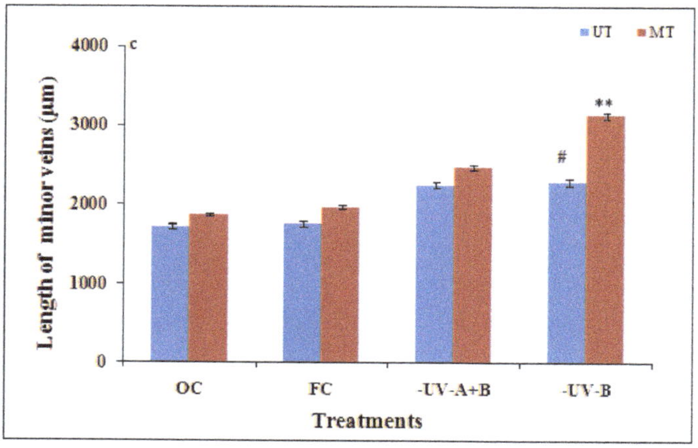

Figure 4. Width of midrib (**a**), width of minor veins (**b**) and length of minor veins (**c**) from X-ray images after SMF pretreatment and solar UV exclusion in middle leaflets of the third trifoliate leaves of soybean. ## $p < 0.01$; # $p < 0.05$ denotes statistically significant differences between seedlings emerged from untreated (UT) seeds of OC with the seedlings emerged from untreated (UT) seeds of different treatments conditions-FC, UV-B and UV-A+B cutoff filters. *** $p < 0.001$; ** $p < 0.01$ denotes statistically significant differences between seedlings that have emerged from SMF-pretreated (MT) and untreated (UT) seeds under each treatment.

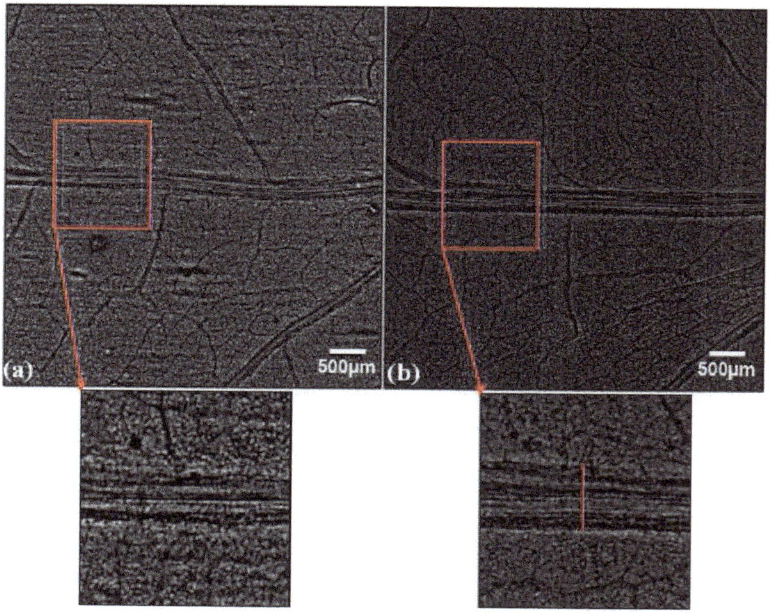

Figure 5. Phase contrast images of filter control (FC) soybean leaves grown with polythene filters which transmitted solar radiation: (**a**) emerging from untreated seeds, (**b**) emerging from seeds pretreated with static magnetic field (SMF) of 200 mT strength for 1 h. The midrib regions enclosed with the red square in the images are zoomed to show midrib enhancement. The midrib quantification was done as shown with the vertical line in the zoomed filter control of the magnetically treated leaf (FCMT) image.

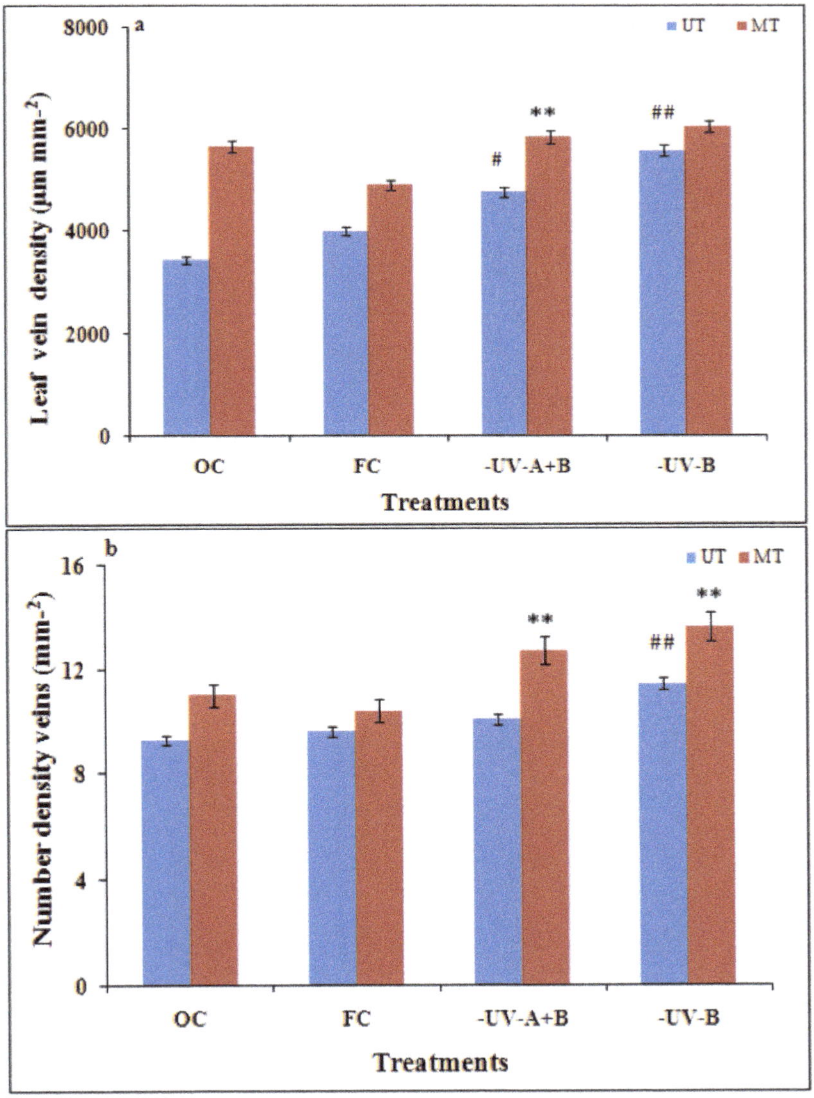

Figure 6. Leaf vein density of tertiary veins (**a**) and number density of veins (**b**) from X-ray images after SMF pretreatment and solar UV exclusion in soybeans. ## $p < 0.01$; # $p < 0.05$ denotes statistically significant differences between seedlings that emerged from untreated (UT) seeds of OC with the seedlings that emerged from untreated (UT) seeds of different treatment conditions; FC, UV-B, and UV-A+B cutoff filters. ** $p < 0.01$ denotes statistically significant differences between seedlings that emerged from SMF-pretreated (MT) and untreated (UT) seeds under each treatment.

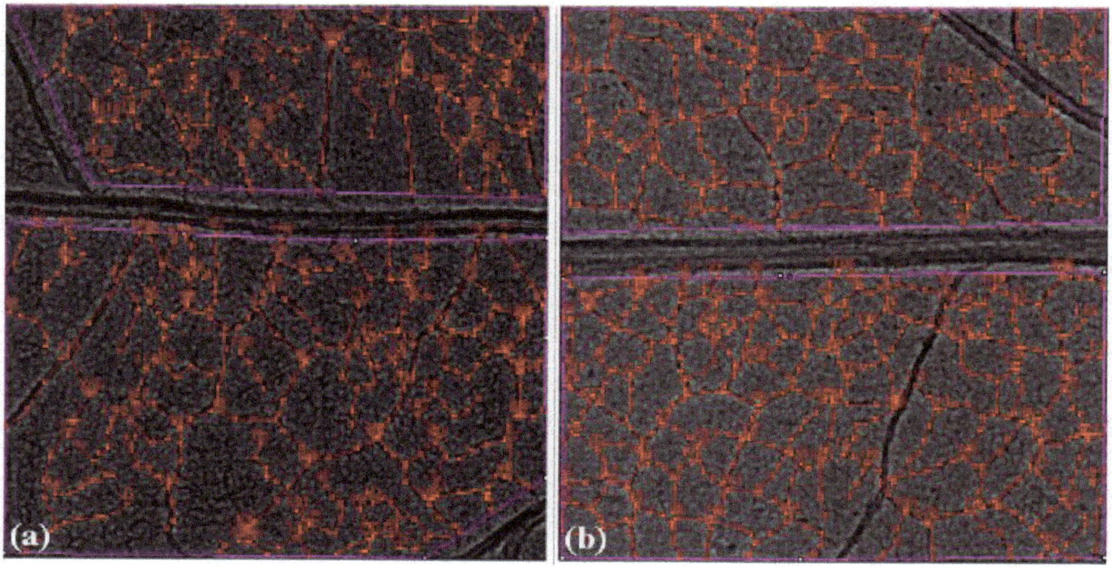

Figure 7. Phase-contrast images of soybean leaves from open control (OC) showing the 3° veins marked with red to obtain the total length of 3° veins and thus the leaf vein density (LVD) with the ObjectJ plugin. The number of minor veins in the images has been used to find the number density of the (3°) minor vein: (**a**) emerging from untreated seeds, (**b**) emerging from seeds pre-treated with a static magnetic field (SMF) of 200 mT strength for 1 h showing greater numbers of minor veins. Similar images for the quantification of 3° veins in other leaf groups have been obtained with ObjectJ.

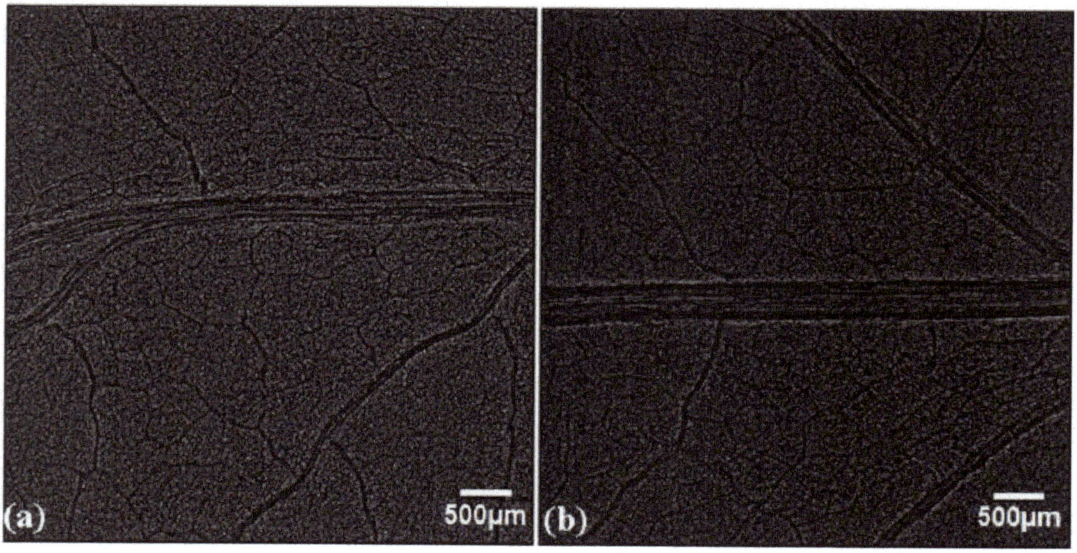

Figure 8. Phase contrast images of ultraviolet radiation excluded (UV-A+B) soybean leaves: (**a**) emerging from untreated seeds, (**b**) emerging from seeds pre-treated with static magnetic field (SMF) of 200 mT strength for 1 h.

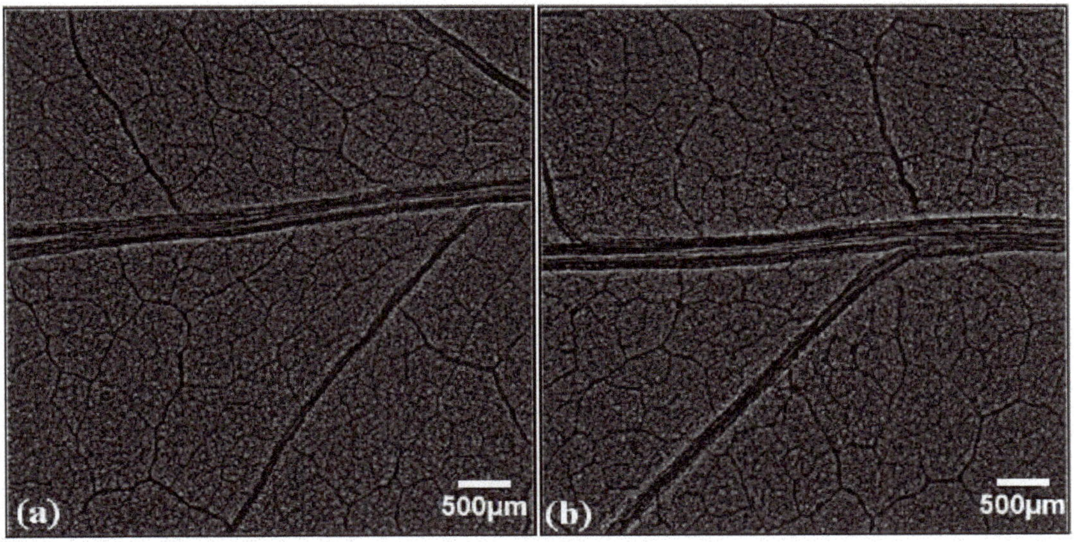

Figure 9. Phase contrast images of ultraviolet-B radiation excluded (UV-B) soybean leaves: (**a**) emerging from untreated seeds, (**b**) emerging from seeds pre-treated with static magnetic field (SMF) of 200 mT strength for 1 h.

The second-order (2°) minor veins also showed an increase of 27% in width and 8% in length by SMF treatment in the OC group (Figure 4b,c). Similar midrib enhancement in the SMF pre-treated group has been observed in the filter control leaves grown with polythene filters which received all the ambient solar radiation and also with UV cut-off filters (Figure 4a, Figure 5a,b, Figures 8 and 9a,b). A 28% increase by UV-A+B and 31% by UV-B filters in the average width of major veins was observed after SMF treatment as compared to their UT ones (Figure 4a).

The zoomed images of the midrib region enclosed with rectangles in red (Figure 5a,b) show enhancement of the midrib structure in the SMF pre-treated leaves. Apart from the first- and second-order leaf veins, quantification of the tertiary veins (3°) has also been done with the ObjectJ plugin to obtain leaf vein density (LVD) (μm mm^{-2}) and the number density of veins (mm^{-2}) in leaves of all groups (Figure 6a,b). The tertiary veins, which are visible in the untreated and SMF pre-treated open control leaf images, are shown in red color (Figure 7a,b).

Comparison showed a higher LVD and a higher number of 3° veins in the SMF pre-treated group compared to the untreated group (Figure 6a,b and Figure 7a,b). In the OC and FC groups receiving all the solar radiation, SMF pre-treatment led to better growth of the plants, as observed from the synchrotron imaging results and also supported by the area and length of leaves and along with rate of photosynthesis in the plants. Thus, it indicated that SMF pre-treatment alleviated the UV stress in plants grown under OC and FC conditions receiving ambient solar radiation.

In the UV-A+B and UV-B excluded group, the plants from untreated seeds (Figures 8a and 9a) showed enhancement as compared to plants receiving ambient solar radiation (OC and FC) in terms of the width of the midrib and 2° vein, length of the 2° vein, LVD and number density of 3°veins (Figures 4 and 6). The phase-contrast images for the combination of SMF pre-treatment and exclusion of solar UV-A+B and UV-B radiation (Figures 8b and 9b) have also shown significant enhancement in the width of the midrib by 28% and 31% respectively, as compared to leaves that emerged from untreated seeds under UV exclusion filters (Figures 4a, 8a and 9a). The enhancement in the width of the midrib observed in UV-excluded along with SMF pre-treated leaves is lesser than the enhancements of 44% and 38% which were obtained in leaves of SMF pre-treated plants receiving all solar radiation respectively in OC and FC conditions (Figures 3, 4a and 5).

An increase in the leaf vein density and number density of minor (3°) veins was seen in the SMF pre-treated control leaves receiving all UV and also in the UV-A+B, UV-B excluded leaves (Figure 6a,b). Leaf vein density, which is the total length of minor veins per unit area, accounts for >80% of the total vein length [34]. The increase in the LVD of a minor (3°) veins indicates increased hydraulic activity in the SMF pre-treated leaves as reported [34].

High LVD can enable higher stomatal conductance and also indicates higher rates of gas exchange per unit leaf area and photosynthesis [25,39]. The vein density, leaf mid rib and minor vein thickness, were strongly correlated with the hydraulic conductivity and higher photosynthetic rate of the leaves. Thus, the observation showed that SMF pretreatment and solar UV exclusion individually and together enhanced leaf hydraulic efficiency, which can be observed through the changes in leaf venation architecture. The leaves were observed to be expanded with thicker veins from SMF-treated and UV excluded plants which give good mechanical support, whereas transpiration cooling and improved photosynthesis were observed because of higher water transportation due to higher vein length per unit area of the leaves [39,55]. The mechanism by which plants perceive MFs and regulate the signal transduction pathway is not fully understood. It has been suggested that MF perception/signaling in plants is regulated by blue light photoreceptors-cryptochromes. It has also been found that reactive oxygen species (ROS) and nitric oxide (NO) are the signaling molecules for magnetopriming-induced seed germination, plant growth, and photosynthesis [29,56]. The participation of NO through nitric oxide synthase enzyme was confirmed in SMF-induced tolerance towards UV-B stress in soybean [56]. However, this aspect of magneto biology still deserves in-depth investigation during leaf growth and photosynthesis.

4. Conclusions

The exclusion of UV-A+B and UV-B radiation is advantageous, as it was suggested that plant growth, leaf area, and photosynthesis were inhibited by ambient UV-B stress. The exposure of seeds to SMF treatment prior to sowing is an eco-friendly method with the potential to alleviate the adverse effects of UV-B stress in the plants. Looking into the correlation between leaf venation and leaf hydraulic conductivity, we used X-ray imaging to study leaf venation (major and minor vein up to 3°) under UV-exclusion, SMF pre-treatment, and the combined effect of both. UV exclusion and SMF pre-treatment individually and jointly showed positive effects on plant growth, development, photosynthesis, and leaf venation parameters obtained from the X-ray images. To our knowledge, this is the first study on X-ray imaging of leaf venation under the combined effects of solar UV exclusion and SMF pre-treatment.

Author Contributions: Conceptualization, A.F., S.K., M.B. and S.I.A.; methodology, A.F. and S.K.; software, A.F.; validation, A.F., S.K., A.K.A., Y.K. and A.R.; formal analysis, S.K. and A.F.; investigation, S.K. and A.F.; resources, A.F., Y.K., M.J., A.K.A. and B.S.; data curation, A.F., A.R. and S.K.; writing—original draft preparation, A.F. and S.K.; writing—review and editing, A.F., S.K. and A.R.; visualization, A.F., S.K. and A.R.; supervision, Y.K., M.J., M.B. and S.I.A.; project administration, A.F., S.K., Y.K., M.J.; funding acquisition; A.F., S.K., A.R., M.B. and S.I.A. All authors have read and agreed to the published version of the manuscript.

Funding: Financial support by DST SERB National Post-Doctoral Fellowship-NPDF (PDF/2017/000643) to AF, SIA supported by the grant RFBR-NSFC (21-54-53015) and Department of Science Technology Women Scientists-A Scheme (SR/WOS-A/LS-17/2017) to SK are thankfully acknowledged. AR has performed this work while being on NAWA Bekker Programme under project No. PPN/BEK/2019/1/00090/U/00001.

Institutional Review Board Statement: Not applicable.

Informed Consent Statement: Not applicable.

Data Availability Statement: Users can obtain the datasets by being in touch with Anis Fatima (anees349@gmail.com), or Sunita Kataria (sunita_kataria@yahoo.com).

Conflicts of Interest: On behalf of all authors, the corresponding authors state that there is no conflict of interest.

References

1. McKenzie, R.L.; Bjorn, L.O.; Bais, A.; Ilyas, M. Changes in biologically active ultraviolet radiation reaching the Earth's surface. *Photochem. Photobiol. Sci.* **2003**, *2*, 5–15. [CrossRef]
2. Kataria, S.; Jajoo, A.; Guruprasad, K.N. Impact of increasing Ultraviolet-B (UV-B) radiation on photosynthetic processes. *J Photochem. Photobiol. B* **2014**, *137*, 55–66. [CrossRef] [PubMed]
3. Bornman, J.F.; Barnes, P.W.; Robson, T.M.; Robinson, S.A.; Jansen, M.A.K.; Ballare, C.L.; Flint, S.D. Linkages between stratospheric ozone, UV radiation and climate change and their implications for terrestrial ecosystems. *Photochem. Photobiol. Sci.* **2019**, *18*, 681–716. [CrossRef] [PubMed]
4. Kataria, S.; Guruprasad, K.N. Solar UV-B and UV-A/B exclusion effects on intraspecific variations in crop growth and yield of wheat varieties. *Field Crops Res.* **2012**, *125*, 8–13. [CrossRef]
5. Kataria, S.; Guruprasad, K.N. Intraspecific variations in growth, yield and photosynthesis of sorghum varieties to ambient UV (280–400 nm) radiation. *Plant Sci.* **2012**, *196*, 85–92. [CrossRef] [PubMed]
6. Rastogi, A.; Pospisil, P. Ultra-weak photon emission as a non-invasive tool for the measurement of oxidative stress induced by UVA radiation in *Arabidopsis thaliana*. *J. Photochem. Photobiol. B.* **2013**, *123*, 59–64. [CrossRef] [PubMed]
7. Huang, G.; Wang, L.; Zhou, Q. Lanthanum (III) regulates the nitrogen assimilation in soybean seedlings under ultraviolet-B radiation. *Biol. Trace Elem. Res.* **2013**, *151*, 105–112. [CrossRef]
8. Kataria, S.; Baghel, L.; Guruprasad, K.N. Alleviation ofAdverse effects ofambient UV stress ongrowth andsome potential physiological attributes insoybean (*Glycine max*) byseed pre-treatment with static magnetic field. *J. Plant Growth Regul.* **2017**, *36*, 550–565. [CrossRef]
9. Khan, M.N.; Zhang, J.; Luo, T.; Liu, J.; Rizwan, M.; Fahad, S.; Xu, Z.; Hu, L. Seed priming with melatonin coping drought stress in rapeseed by regulating reactive oxygen species detoxification: Antioxidant defense system, osmotic adjustment, stomatal traits and chloroplast ultrastructure perseveration. *Ind. Crop Prod.* **2019**, *140*, 111597. [CrossRef]
10. Sytar, O.; Kumari, P.; Yadav, S.; Rastogi, A. Phytohormone Priming: Regulator for Heavy Metal Stress in Plants. *J. Plant Growth Regul.* **2019**, *38*, 739–752. [CrossRef]
11. Alvarado-López, S.; Soriano, D.; Velázquez, N.; Orozco-Segovia, A.; Gamboa-de-Buen, A. Priming effects on seed germination in *Tecomastans* (Bignoniaceae) and *Cordiamegalantha* (Boraginaceae), two tropical deciduous tree species. *ActaOecologica* **2014**, *61*, 65–70.
12. Damalas, C.A.; Koutroubas, S.D.; Fotiadis, S. Hydro-priming effects on seed germination and field performance of Faba bean in spring sowing. *Agriculture* **2019**, *9*, 201. [CrossRef]
13. Blunk, S.; de Heer, M.I.; Malik, A.H.; Fredlund, K.; Ekblad, T.; Sturrock, C.J.; Mooney, S.J. Seed priming enhances early growth and improves area of soil exploration by roots. *Environ. Exp. Bot.* **2019**, *158*, 1–11. [CrossRef]
14. Shine, M.B.; Guruprasad, K.N. Impact of pre-sowing magnetic field exposure of seeds to stationary magnetic field on growth, reactive oxygen species and photosynthesis of maize under field conditions. *Acta Physiol. Plant* **2012**, *34*, 255–265. [CrossRef]
15. Kataria, S.; Baghel, L.; Guruprasad, K.N. Pre-treatment of seeds with static magnetic field improves germination and early growth characteristics under salt stress in maize and soybean. *Biocatal. Agric. Biotechnol.* **2017**, *10*, 83–90. [CrossRef]
16. Sarraf, M.; Kataria, S.; Taimourya, H.; Santos, L.O.; Menegatti, R.D.; Jain, M.; Ihtisham, M.; Liu, S. Magnetic field (MF) applications in plants: An overview. *Plants* **2020**, *9*, 1139. [CrossRef] [PubMed]
17. Kataria, S.; Jain, M. Magnetopriming Alleviates Adverse Effects of Abiotic Stresses on Plants. In *Plant Tolerance to Environmental Stress: Role of Phytoprotectants*, 1st ed.; Mirza, H., Masayuki, F., Hirosuke, O., Tofazzal-Islam, M., Eds.; CRC Press: Boca Raton, FL, USA, 2018; pp. 427–438.
18. Kataria, S.; Baghel, L.; Jain, M.; Guruprasad, K.N. Magnetopriming regulates antioxidant defense system in soybean against salt stress. *Biocatal. Agric. Biotechnol.* **2019**, *18*, 101090. [CrossRef]
19. Florez, M.; Carbonell, M.V.; Martínez, E. Early sprouting and first stages of growth of rice seeds exposed to a magnetic field. *Electro Magnetobiol.* **2004**, *23*, 167–176. [CrossRef]
20. Vashisth, A.; Nagarajan, S. Effect on germination and early growth characteristics in sunflower (*Helianthus annus*) seeds exposed to static magnetic field. *J. Plant Physiol.* **2010**, *167*, 149–156. [CrossRef]
21. Baghel, L.; Kataria, S.; Jain, M. Mitigation of adverse effects of salt stress on germination, growth, photosynthetic efficiency and yield in maize (*Zea mays* L.) through magnetopriming. *ActaAgrobot.* **2019**, *72*, 1757. [CrossRef]
22. Huang, H.H.; Wang, S.R. The Effects of 60Hz Magnetic Fields on Plant Growth. *Nat. Sci.* **2007**, *5*, 60–68.
23. Shine, M.B.; Guruprasad, K.N.; Anjali, A. Enhancement of germination, growth and photosynthesis in soybean by pre-treatment of seeds with magnetic field. *Bioelectromagnetics* **2011**, *32*, 474–484. [CrossRef]
24. Shine, M.B.; Guruprasad, K.N.; Anjali, A. Superoxide radical production and performance index of Photosystem II in leaves from magnetoprimed soybean seeds. *Plant Signal. Behav.* **2011**, *6*, 1635–1637.

25. Fatima, A.; Kataria, S.; Baghel, L.; Guruprasad, K.N.; Agrawal, A.K.; Singh, B.; Sarkar, P.S.; Shripathi, T.; Kashyap, Y. Synchrotron-based phase-sensitive imaging of leaves grown from magneto-primed seeds of soybean. *J. Synchrotron Rad.* **2017**, *24*, 232–239. [CrossRef] [PubMed]
26. Shine, M.B.; Guruprasad, K.N.; Anjali, A. Effect of stationary magnetic field strengths of 150 and 200 mT on reactive oxygen species production in soybean. *Bioelectromagnetics* **2012**, *33*, 428–437. [CrossRef]
27. Baghel, L.; Kataria, S.; Guruprasad, K.N. Static magnetic field treatment of seeds improves carbon and nitrogen metabolism under salinity stress in soybean. *Bioelectromagnetics* **2016**, *37*, 455–470. [CrossRef] [PubMed]
28. Fatima, A.; Kataria, S.; Prajapati, R.; Jain, M.; Agrawal, A.K.; Singh, B.; Kashyap, Y.; Tripathi, D.K.; Singh, V.P.; Gadre, R. Magnetopriming effects on arsenic stress-induced morphological and physiological variations in soybean involving synchrotron imaging. *Physiol. Plant* **2020**. [CrossRef]
29. Kataria, S.; Jain, M.; Rastogi, A.; Brestic, M. Static magnetic field treatment enhanced photosynthetic performance in soybean under supplemental ultraviolet-B (280-320 nm) radiation. *Photosynth. Res.* **2021**. [CrossRef]
30. Galland, P.; Pazur, A. Magnetoreception in plants. *J. Plant Res.* **2005**, *118*, 371–389. [CrossRef]
31. García-Reina, F.; Arza-Pascual, L. Influence of a stationary magnetic field on water relations in lettuce seeds. I: Theoretical considerations. *Bioelectromagnetics* **2001**, *22*, 589–595. [CrossRef]
32. Dicarlo, A.L.; Hargis, M.T.; Penafiel, L.M.; Litovitz, T.A. Short-term magnetic field exposure (60 Hz) induce protection against ultraviolet radiation damage. *Int. J. Rad. Biol.* **1999**, *75*, 1541–1549. [CrossRef] [PubMed]
33. Roth-Nebelsick, A.; Uhl, D.; Mosbrugger, V.; Kerp, H. Evolution and function of leaf venation architecture: A review. *Ann. Bot.* **2001**, *87*, 553–566. [CrossRef]
34. Brodribb, T.J.; Field, T.S.; Sack, L. Viewing leaf structure and evolution from hydraulic perspective. *Funct. Plant Biol.* **2010**, *37*, 488–498. [CrossRef]
35. Ellis, B.; Daly, D.C.; Hickey, L.J.; Mitchell, J.; Johnson, K.; Wilf, P.; Wing, S. *Manual of Leaf Architecture*; New York Botanical Garden: New York, NY, USA, 2009.
36. Brodribb, T.J.; Field, T.S.; Jordan, G.J. Leaf maximum photosynthetic rate and venation are linked by hydraulics. *Plant Physiol.* **2007**, *144*, 1890–1898. [CrossRef]
37. Blonder, B.; De Carlo, F.; Moore, J.; Rivers, M.; Enquist, B.J. X-ray imaging of leaf venation networks. *New Phytol.* **2012**, *196*, 1274–1282. [CrossRef]
38. Scoffoni, C.; Sack, L. Quantifying Leaf Vein Traits. Prometheus. 2011. Wiki. Available online: http://prometheuswiki.publish.csiro.au/tikiindex.php.Quantifying+leaf+vein+traits (accessed on 25 August 2014).
39. Sack, L.; Scoffoni, C. Leaf venation: Structure, function, development, evolution, ecology and applications in the past, present and future. *New Phytol.* **2013**, *198*, 983–1000. [CrossRef]
40. Fatima, A.; Kataria, S.; Guruprasad, K.N.; Agrawal, A.K.; Singh, B.; Sarkar, P.S.; Shripathi, T.; Kashyap, Y.; Sinha, A. Synchrotron X-ray phase contrast imaging of leaf venation in soybean (*Glycine max*) after exclusion of solar UV (280–400 nm) radiation. *J. Synchrotron Rad.* **2016**, *23*, 795–801. [CrossRef] [PubMed]
41. Verboven, P.; Herremans, E.; Helfen, L.; Ho, Q.T.; Abera, M.; Baumbach, T.; Wevers, M.; Nicolai, B.M. Synchrotron X-ray computed laminography of the three-dimensional anatomy of tomato leaves. *Plant J.* **2015**, *81*, 169–182. [CrossRef]
42. Lahlali, R.; Karunakaran, C.; Wang, L.; Willick, I.; Schmidt, M.; Liu, X.; Borondics, F.; Forseille, L.; Fobert, P.R.; Tanino, K.; et al. Synchrotron based phase contrast X-ray imaging combined with FTIR spectroscopy reveals structural and biomolecular differences in spikelets play a significant role in resistance to Fusarium in wheat. *BMC Plant Biol.* **2015**, *15*, 24. [CrossRef]
43. Lee, E.F.; Matthews, M.A.; McElrone, A.J.; Phillips, R.J.; Shackel, K.A.; Brodersen, C.R. Analysis of HRCT-derived xylem network reveals reverse flow in some vessels. *J. Theor. Biol.* **2013**, *333*, 146–155. [CrossRef] [PubMed]
44. Brodersen, C.R. Visualizing wood anatomy in three dimensions with high-resolution X-ray micro-tomography (μCT)—A review. *IAWAJ* **2013**, *34*, 408–424. [CrossRef]
45. Brodersen, C.R.; Lee, E.F.; Choat, B.; Jansen, S.; Phillips, R.J.; Shackel, K.A.; McElrone, A.J.; Matthews, M.A. Automated analysis of three dimensional xylem networks using high-resolution computed tomography. *New Phytol.* **2011**, *191*, 1168–1179. [CrossRef]
46. Brodersen, C.R.; McElrone, A.J.; Choat, B.; Lee, E.F.; Shackel, K.A.; Matthews, M.A. In vivo visualizations of dought-induced embolism spread in *Vitisvinifera*. *Plant Physiol.* **2013**, *161*, 1820–1829. [CrossRef]
47. Snigirev, A.; Snigireva, I.; Kohn, V.; Kuznetsov, S.; Schelokov, I. On the possibilities of X-ray phase contrast micro imaging by coherent high-energy synchrotron radiation. *Rev. Sci. Instrum.* **1995**, *66*, 5486–5492. [CrossRef]
48. Cloetens, P.; Barrett, R.; Baruchel, J.; Guigay, J.P.; Schlenker, M. Phase objects in synchrotron radiation hard X-ray imaging. *J. Phys. D Appl. Phys.* **1996**, *29*, 133–146. [CrossRef]
49. Mayo, S.C.; Stevenson, A.W.; Wilkins, S.W. In-Line Phase-Contrast X-ray Imaging and Tomography for Materials Science. *Materials* **2012**, *5*, 937–965.
50. Momose, A. Development toward high-resolution X-ray phase imaging. *Microscopy* **2017**, *66*, 155–166. [CrossRef] [PubMed]
51. Kataria, S.; Rastogi, A.; Bele, A.; Jain, M. Role of nitric oxide and reactive oxygen species in static magnetic field pre-treatment induced tolerance to ambient uv-b stress in soybean. *Physiol. Mol. Biol. Plants* **2020**, *26*, 931–945. [CrossRef] [PubMed]
52. Agrawal, A.K.; Singh, B.; Kashyap, Y.S.; Shukla, M.; Sarkar, P.S.; Sinha, A. Design, development and first experiments on the X-ray imaging beamline at Indus-2 synchrotron source RRCAT, India. *J. Synchrotron Rad.* **2015**, *221*, 531–1539. [CrossRef]

53. Rasband, W.S. *ImageJ*; U.S. National Institutes of Health: Bethesda, MD, USA, 2012. Available online: http://imagej.nih.gov/ij/ (accessed on 20 March 2019).
54. Schumaker, M.A.; Bassman, J.H.; Robberecht, R.; Radamaker, G.K. Growth, leaf anatomy, and physiology of Populus clones in response to solar ultraviolet-B radiation. *Tree Physiol.* **1997**, *17*, 617–626. [CrossRef]
55. Walls, R.L. Angiosperm leaf vein patterns are linked to leaf functions in a global scale data set. *Am. J. Bot.* **2011**, *98*, 244–253. [CrossRef] [PubMed]
56. Raipuria, R.; Kataria, S.; Watts, A.; Jain, M. Magneto-priming promotes nitric oxide via nitric oxide synthase to ameliorate the UV-B stress during germination of soybean seedlings. *J. Photochem. Photobiol.* **2021**, *220*, 112211. [CrossRef] [PubMed]

cells

MDPI

Article

Features of the Duckweed *Lemna* That Support Rapid Growth under Extremes of Light Intensity

Jared J. Stewart [1,*], William W. Adams III [1], Marina López-Pozo [1], Naiara Doherty Garcia [1], Maureen McNamara [1], Christine M. Escobar [2,3] and Barbara Demmig-Adams [1,*]

1 Department of Ecology and Evolutionary Biology, University of Colorado, Boulder, CO 80309, USA; william.adams@colorado.edu (W.W.A.III); marina.lopezpozo@colorado.edu (M.L.-P.); naiara.garcia@colorado.edu (N.D.G.); maureen.mcnamara@colorado.edu (M.M.)

2 Department of Aerospace Engineering Sciences, University of Colorado, Boulder, CO 80309, USA; chris@spacelabtech.com

3 Space Lab Technologies, LLC, Boulder, CO 80309, USA

* Correspondence: jared.stewart@colorado.edu (J.J.S.); barbara.demmig-adams@colorado.edu (B.D.-A.); Tel.: +1-303-492-5541 (B.D.-A.)

Abstract: This study addresses the unique functional features of duckweed via comparison of *Lemna gibba* grown under controlled conditions of 50 versus 1000 µmol photons m^{-2} s^{-1} and of a *L. minor* population in a local pond with a nearby population of the biennial weed *Malva neglecta*. Principal component analysis of foliar pigment composition revealed that *Malva* was similar to fast-growing annuals, while *Lemna* was similar to slow-growing evergreens. Overall, *Lemna* exhibited traits reminiscent of those of its close relatives in the family Araceae, with a remarkable ability to acclimate to both deep shade and full sunlight. Specific features contributing to duckweed's shade tolerance included a foliar pigment composition indicative of large peripheral light-harvesting complexes. Conversely, features contributing to duckweed's tolerance of high light included the ability to convert a large fraction of the xanthophyll cycle pool to zeaxanthin and dissipate a large fraction of absorbed light non-photochemically. Overall, duckweed exhibited a combination of traits of fast-growing annuals and slow-growing evergreens with foliar pigment features that represented an exaggerated version of that of terrestrial perennials combined with an unusually high growth rate. Duckweed's ability to thrive under a wide range of light intensities can support success in a dynamic light environment with periodic cycles of rapid expansion.

Keywords: antioxidants; carotenoids; chlorophyll fluorescence; photochemical efficiency; protein; tocopherol; xanthophyll cycle; zeaxanthin

check for updates

Citation: Stewart, J.J.; Adams, W.W., III; López-Pozo, M.; Doherty Garcia, N.; McNamara, M.; Escobar, C.M.; Demmig-Adams, B. Features of the Duckweed *Lemna* That Support Rapid Growth under Extremes of Light Intensity. *Cells* **2021**, *10*, 1481. https://doi.org/10.3390/cells10061481

Academic Editors: Suleyman Allakhverdiev and Marian Brestic

Received: 1 May 2021
Accepted: 9 June 2021
Published: 12 June 2021

1. Introduction

Small, floating plant species in the duckweed family (Lemnaceae) possess attractive nutritional features as they accumulate large quantities of high-quality protein (with all essential amino acids for humans) throughout the plant [1]. Furthermore, our group has highlighted the exceptional ability of *Lemna gibba* to accumulate high levels of the carotenoid zeaxanthin under conditions when the plant is growing rapidly [2,3]. Zeaxanthin (and its close isomer lutein) is an essential human micronutrient required to support brain function and fight systemic inflammation [4]. Duckweed also has potential uses in sustainable agricultural systems as food for humans, feed for animals (via conversion of wastewater to feed [5,6]), for other valuable products [7], or to improve nitrogen-use efficiency and yield of crops like rice [8]. Here, we present further insight into how *L. gibba* is able to combine fast growth across a range of environments with high nutritional content including pronounced zeaxanthin accumulation as well as other essential nutrients for humans or livestock.

We previously reported a notable ability of *L. gibba* to maintain uniformly high growth rates, paired with profound modulation of photoprotection, over a range of growth photon flux densities (PFDs) from 100 to 700 µmol m^{-2} s^{-1} of continuous light [2]. Plants grown under higher PFDs exhibited higher levels of the interconvertible xanthophyll cycle carotenoids (violaxanthin, antheraxanthin, and zeaxanthin) and pronounced conversion to zeaxanthin that dissipates potentially harmful excess absorbed light [3,9,10]. In the present study, we further broadened the range of growth PFDs to test whether duckweed's phenotypic plasticity with respect to photoprotective capacity and maintenance of a high growth rate may extend to even more extreme growth PFDs. We compared features of *L. gibba* grown under very low (50 μmol photons m^{-2} s^{-1}) or very high (1000 μmol photons m^{-2} s^{-1}) intensity of continuous light under otherwise common, controlled conditions. Continuous exposure (24 h per day) to the high growth PFD represented a greater total daily photon flux than that on the longest, brightest day on Earth. Beyond extending the range of PFDs versus the previous study [2], additional parameters were characterized in the present study including the light-use efficiency of biomass production as well as the production of protein as a key macronutrient and α-tocopherol (vitamin E) as an additional micronutrient. Moreover, CO_2-saturated photosynthetic capacity was characterized under both saturating light and the respective contrasting growth PFDs, and photosynthesis as well as protein and all micronutrients were expressed on multiple reference bases (per frond area, biomass, and chlorophyll [Chl] content) for a fuller evaluation of both plant function and nutritional quality for the consumer.

Furthermore, the present study tested the hypothesis that the combination of exceptionally rapid growth with a remarkable ability to grow under a wide range of light intensities in duckweed may be associated with pigment patterns not seen in other fast-growing species. In particular, prior studies of leaf pigment composition in slow-growing evergreens or perennials versus fast-growing annual species often reported an inverse relationship between growth rate and accumulation of photoprotective pigments (for a review, see [3]). The present study compared a population of *Lemna minor* in an open outdoor pond with a nearby population of the fast-growing terrestrial biennial weed *Malva neglecta* that was previously shown to exhibit a pigment composition and photoprotective capacity similar to that of fast-growing annual crop species [11]. Foliar pigment composition of *M. neglecta* and *L. minor* growing in full sun outdoors as well as that of *L. gibba* grown in low versus high PFD under controlled conditions were compared via principal component analysis to foliar pigment data for other species groups (including annuals as well as evergreens and other perennials).

2. Materials and Methods

2.1. Plant Material and Growth Conditions

2.1.1. Controlled Conditions

Cultures of *Lemna gibba* L. 7741 (G3) obtained from Rutgers Duckweed Stock Cooperative (http://ruduckweed.org; accessed on 10 June 2021) were grown under controlled conditions in Conviron PGR15 and E15 growth chambers (Controlled Environments Ltd., Winnipeg, MB, Canada). Plants were grown in 150 × 75 mm PYREX Crystallizing Dishes (Corning Inc., Corning, NY, USA) that contained 1000 mL of freshly prepared Schenk and Hildebrandt medium (bioWORLD, Dublin, OH, USA [12]) at a concentration of 1.6 g L^{-1} (pH adjusted to 5.5 via 1% [*w*/*v*] KOH). Plants were cultivated continuously under 50 μmol photons m^{-2} s^{-1} (provided by F72T12/CW/HO fluorescent bulbs [Philips, Somerset, NJ, USA] and 100 W, 130 V incandescent bulbs [EiKO, Shawnee, KS, USA]) at 25 °C, and a subset of plants from each dish were transferred to clean dishes containing freshly prepared media at least once per week.

Plants from the cultures under 50 μmol photons m^{-2} s^{-1} were acclimated to 1000 μmol photons m^{-2} s^{-1} (provided by M47/E 1000 W metal halide bulbs; Philips, Somerset, NJ, USA) following the procedure described by Stewart et al. [2]. A subset of plants from the stock cultures (approximately 20 fronds per dish) were transferred to 200 μmol photons m^{-2} s^{-1} (supplied by C503C-WAN white LEDs; CREE Inc., Durham, NC, USA) for three days and then transferred to 1000 μmol photons m^{-2} s^{-1} for three days. After each of these three-day acclimation phases, a subset of plants (approximately 20 fronds per dish) that had developed under the prevailing light intensities were transferred to clean dishes with freshly prepared and filtered media. This process ensured that the characterized plant material developed under 1000 μmol photons m^{-2} s^{-1} and had not developed under a lower PFD and then been transferred to 1000 μmol photons m^{-2} s^{-1}. Following this six-day acclimation process, plants were characterized over the course of four days while growing under 1000 μmol photons m^{-2} s^{-1} and a media temperature of 25 °C.

2.1.2. Field Conditions

Populations of *Lemna minor* L. and *Malva neglecta* Wallr. plants (identification based on morphology and geographic distribution) growing naturally in Superior, CO, USA (39°56′28″ N, 105°09′02″ W) were characterized. *Malva neglecta* is a relatively fast-growing herbaceous biennial species that remains photosynthetically active throughout the year in this area (see [13]). The *L. minor* plants were growing in a slow-moving section of a small stream, and the *M. neglecta* plants were growing on a south-facing slope immediately north of the stream. Both locations received direct, midday sunlight. Samples for pigment analysis were collected during exposure to full sun (maximal PFD of 1600 μmol m^{-2} s^{-1}) prior to solar noon on 17 May 2019. Samples were imaged (for quantification of frond/leaf area via ImageJ [14]) and then submerged and stored in liquid nitrogen at the field site. The four characterized samples of *L. minor* each consisted of multiple fronds from multiple plants (i.e., multiple biological replicates per sample), whereas the four characterized samples from *M. neglecta* each consisted of one leaf segment from four separate plants (i.e., four biological replicates).

2.2. *Growth Metrics*

Under controlled conditions, the dishes containing *L. gibba* plants were imaged from directly overhead once per day during the four-day period of characterization, and the frond area was quantified from these images using MATLAB Image Processing Toolbox (MathWorks, Natick, MA, USA) as previously described [2]. Dry mass of whole plants (i.e., fronds with intact roots) and only fronds (i.e., fronds with excised roots) was measured from samples that had been dried at 70 °C for seven days. Prior to drying, each sample was imaged from directly overhead, and the frond area was determined from these images using ImageJ [14].

Relative growth rate was calculated by dividing the difference in natural logarithm–transformed frond areas at the end and the beginning of the four-day experimental period by the time elapsed between the two measurements. Doubling time was calculated as the natural logarithm of 2 divided by relative growth rate. Light-use efficiency of frond area production was calculated as the accumulated frond area (i.e., the difference between the final and initial frond areas) divided by the number of incident photons on the frond surface (calculated as described in detail by Stewart et al. [2]). Light-use efficiency of biomass production was estimated as the accumulated biomass (i.e., product of accumulated frond area [m^2] and whole-plant dry mass per unit frond area [g m^{-2}]) divided by the number of available photons during this time period.

2.3. *Photosynthesis and Respiration*

Rates of photosynthetic oxygen evolution were determined as described by Stewart et al. [2] with saturating CO_2 (5% CO_2, 21% O_2, balance N_2) using leaf disc oxygen electrodes (Hansatech Instruments Ltd., Norfolk, United Kingdom; see [15]) and a circulating water bath set to 25 °C. Fronds from *L. gibba* plants grown under 50 μmol photons m^{-2} s^{-1} and 1000 μmol photons m^{-2} s^{-1} were assayed under their respective growth PFDs as well as a saturating PFD of 1500 μmol photons m^{-2} s^{-1}. Respiration rates were determined following measurements of photosynthesis as the rate of oxygen consumption in darkness.

Photochemical and photoprotective processes were assessed via measurements of chlorophyll fluorescence with a PAM-101 chlorophyll fluorometer (Walz, Effeltrich, Germany) following the procedures described in detail by Stewart et al. [2] and using calculations described by Demmig-Adams et al. [16]. To ascertain the maximal level of fluorescence (F_m or F_m') in high-light flashes, two or three flashes were given in rapid succession [17], which revealed that the maximal attainable level was reached during the first flash under all conditions used here.

2.4. Protein and Starch

Total protein content was quantified spectrophotometrically with the Total Protein Kit, Micro Lowry, Peterson's Modification (Sigma-Aldrich, Saint Louis, MO, USA), which follows a modified version [18] of the procedure described by Lowry et al. [19]. Whole plants with approximately three fronds per dish, which had been imaged and immediately frozen in liquid nitrogen, were homogenized via mortar and pestle, combined with 1 mL of water, vortexed, and centrifuged for 10 min at 10,000 rpm. The resulting supernatant was decanted, combined with 0.1 mL of deoxycholate, vortexed, and maintained at room temperature for 10 min. Subsequently, 0.1 mL of trichloroacetic acid was added, and this solution was vortexed and centrifuged for 10 min at 10,000 rpm. The resulting pellet was re-suspended in 1 mL of Lowry reagent, transferred to cuvettes, and then mixed with an additional 1 mL of water that was used to rinse the microcentrifuge tube. After 20 min of incubation at room temperature, 0.5 mL of the Folin–Ciocalteu phenol reagent was added, and this solution was mixed via pipette and incubated at room temperature for 30 min. Absorbance at 660 nm was determined with a Beckman DU 640 Spectrophotometer (Beckman Instruments, Inc., Fullterton, CA, USA) and these values were converted to protein levels (in μg mL^{-1}) using a standard calibration curve based on a gradient of bovine serum albumin.

The abundance of starch in *L. gibba* plants was detected qualitatively with a diluted iodine-potassium iodide solution (Lugol's solution; Sigma-Aldrich, St. Louis, MO, USA). Plants were cleared in 70% (*v*/*v*) ethanol, stained for 5 min, and then immediately mounted and imaged with a high-resolution scanner (Perfection 3200 Photo; Epson America, Inc., Long Beach, CA, USA).

2.5. Chlorophyll, Carotenoid, and α-Tocopherol Levels

Chlorophylls *a* & *b*, lutein, zeaxanthin (Z), antheraxanthin (A), violaxanthin (V), neoxanthin, β-carotene, and α-tocopherol levels were quantified via high-performance liquid chromatography as previously described in detail [2,20]. Under controlled conditions, samples of approximately 10 fronds per dish were collected under the respective growth PFDs, imaged, and then frozen and stored in liquid nitrogen.

Pigment data from a previously conducted survey of multiple species in the same area by Demmig-Adams and Adams [21], which included *M. neglecta*, were used in the present study and divided into three groups: (i) shade-grown perennials; (ii) sun-grown perennials; and (iii) sun-grown annuals (for detail, see Table S1). To ensure the datasets were comparable, pigments were characterized from samples of *Vinca minor* plants growing in complete shade (shade-grown perennials) as well as plants growing exposed to full sun (sun-grown perennials) on the University of Colorado campus (Boulder, CO, USA) collected during the afternoon of 16 April 2019.

2.6. Statistical Analyses

Comparisons of means were preceded by Levene's tests to assess the equality of variances. Comparison of two means were made with Student's (equal variances) or Welch's (unequal variances) *t*-tests, and comparisons of more than two means were made with one-way ANOVAs and post-hoc Tukey–Kramer HSD. Linear relationships between two variables were evaluated with Pearson correlations. Comparisons with multiple parameters were conducted with principal component analysis on correlations. All statistical analyses were made using JMP Pro 15.0.0 (SAS Institute Inc., Cary, NC, USA).

3. Results

3.1. Growth and Photosynthesis of Lemna gibba under Extremes in Growth PFD

Despite vastly different light availability (a 20-fold difference between 50 versus 1000 μmol photons m^{-2} s^{-1} of continuous light) during growth, a similar amount of duckweed frond area accumulated over time in the two controlled conditions (Figure 1A). This represented a similar frond area doubling time of just under 1.5 days for either of the two growth PFDs (Figure 1B), which also corresponded to similar average relative growth rates of 0.48 ± 0.04 and 0.50 ± 0.03 day^{-1} (average daily increase in natural logarithm–adjusted frond area over the four-day experimental phase) for plants grown under 50 and 1000 μmol photons m^{-2} s^{-1}, respectively. As a consequence of the similar frond area accumulation under the two vastly different growth PFDs, the ratio of frond area produced per incident PFD, which can be considered the light-use efficiency of frond area production, was dramatically greater (1733%) in fronds grown under 50 versus 1000 μmol photons m^{-2} s^{-1} (Figure 1C), in other words, almost proportional to the 20-fold (or 2000%) difference in incident PFD.

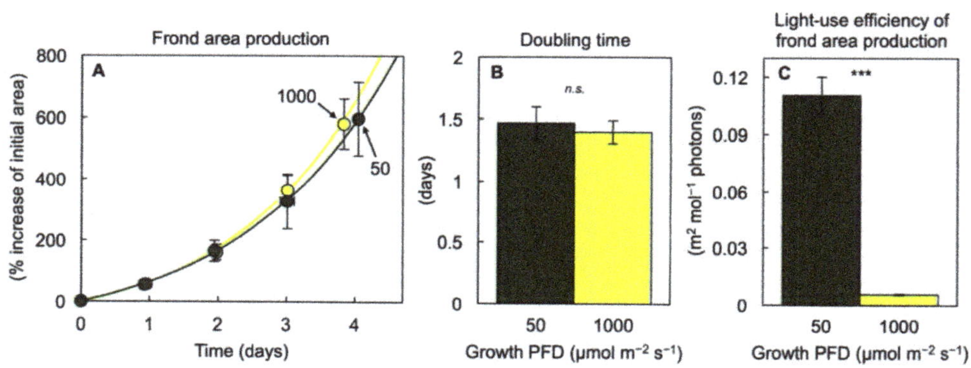

Figure 1. (**A**) Accumulation of frond area (% relative to frond area at the beginning of the experiment) over four days of growth, (**B**) average doubling time in frond area over this four-day period, and (**C**) light-use efficiency of frond area production (total frond area produced relative to incident PFD during the four-day growth period) in *Lemna gibba* plants under growth PFDs of 50 (green) or 1000 (yellow) μmol m^{-2} s^{-1}. Mean values ± standard deviations; $n = 7$ for 50 μmol photons m^{-2} s^{-1}; $n = 3$ for 1000 μmol photons m^{-2} s^{-1}. A significant difference between the growth PFDs is denoted by asterisks in (**C**); *** = $p < 0.001$; *n.s.* = not significantly different.

A greater amount of dry biomass (Figure 2A) and protein (Figure 2B) was accumulated on a frond area basis in plants growing under the higher PFD, but the fraction of dry biomass (% biomass in g g^{-1}) that consisted of protein (Figure 2C) was greater under the lower PFD, which resulted in a remarkable 46% of dry biomass consisting of protein under the low growth PFD. There was also evidence for greater starch accumulation (Figure 2D,E) under the higher PFD. Another contributing factor to the difference in total dry biomass in plants grown under 1000 versus 50 μmol photons m^{-2} s^{-1} was a difference in root dry biomass, which accounted for 16% versus 6%, respectively. As was the case for area production, light-use efficiency of total biomass production on an incident light basis was much greater in fronds grown under 50 versus 1000 μmol photons m^{-2} s^{-1}, albeit at a less pronounced percentage (672%) than seen for frond area (Figure 1C) due to the fact that, unlike frond area accumulation, dry mass accumulation was almost 3× greater at the higher growth PFD (Figure 2A).

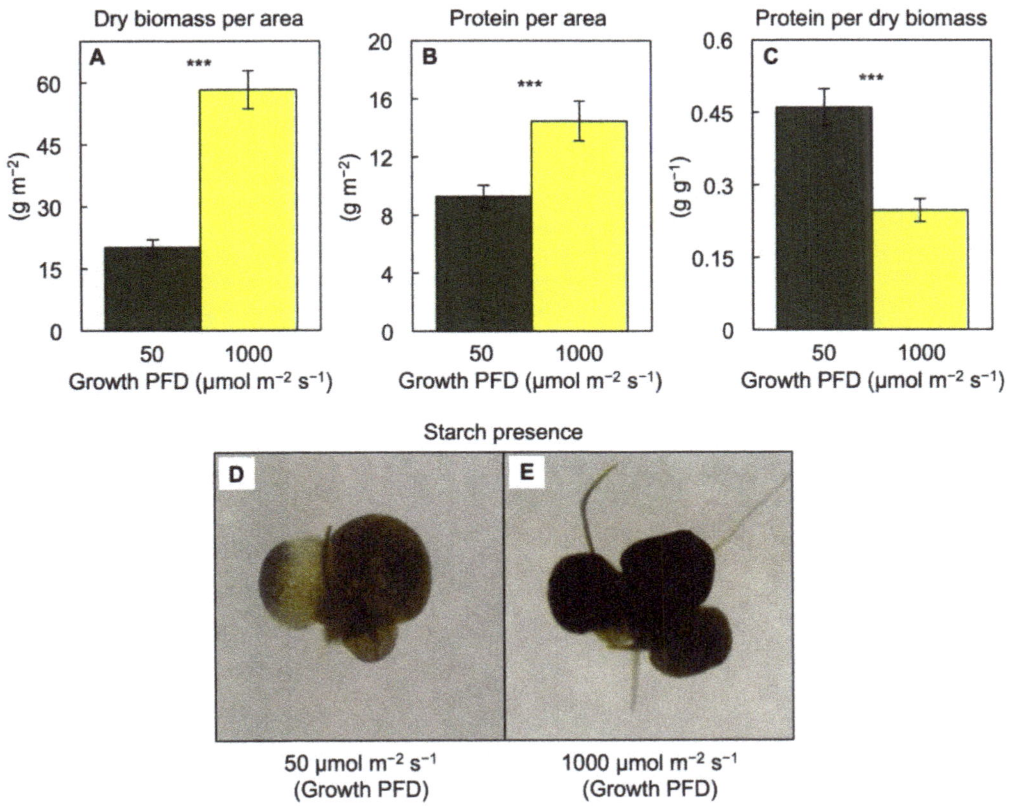

Figure 2. (**A**) Dry biomass per frond area, (**B**) protein per frond area, (**C**) protein per dry biomass, and (**D**,**E**) presence of starch (detected via diluted iodine-potassium iodide solution) in *Lemna gibba* plants grown under 50 (green) or 1000 (yellow) µmol photons m^{-2} s^{-1}. Mean values ± standard deviations; $n = 4$. Significant differences between the growth PFDs are denoted by asterisks; *** = $p < 0.001$. Dimensions of each image (**D**,**E**) are 1 cm × 1 cm.

CO_2-saturated photosynthesis rate was determined for plants grown under their respective growth PFDs of 50 and 1000 µmol m^{-2} s^{-1} as well as under a common saturating PFD of 1500 µmol m^{-2} s^{-1} (Figure 3). Furthermore, the resulting photosynthesis rates were expressed on three different reference bases (i.e., per frond area (Figure 3A), per frond dry biomass (Figure 3B), and per frond Chl content (Figure 3C)), revealing different trends. Photosynthesis on a frond area basis was higher in plants grown under the high PFD (Figure 3A) when measured under the respective PFDs (1000 versus 50 µmol m^{-2} s^{-1}). Since fronds grown under high PFD accumulated a greater amount of biomass than fronds grown under the low PFD, the photosynthesis rate on a dry mass basis was similar in the fronds when measured at their respective growth PFDs (Figure 3B). Likewise, respiration rates were higher on an area basis (3.36 ± 0.39 versus 1.42 ± 0.82 µmol O_2 m^{-2} s^{-1}, $p < 0.05$) and similar on a dry mass basis (0.08 ± 0.01 versus 0.07 ± 0.04 µmol O_2 g^{-1} s^{-1}, $p > 0.05$) in fronds grown under 1000 versus 50 µmol photons m^{-2} s^{-1}, respectively.

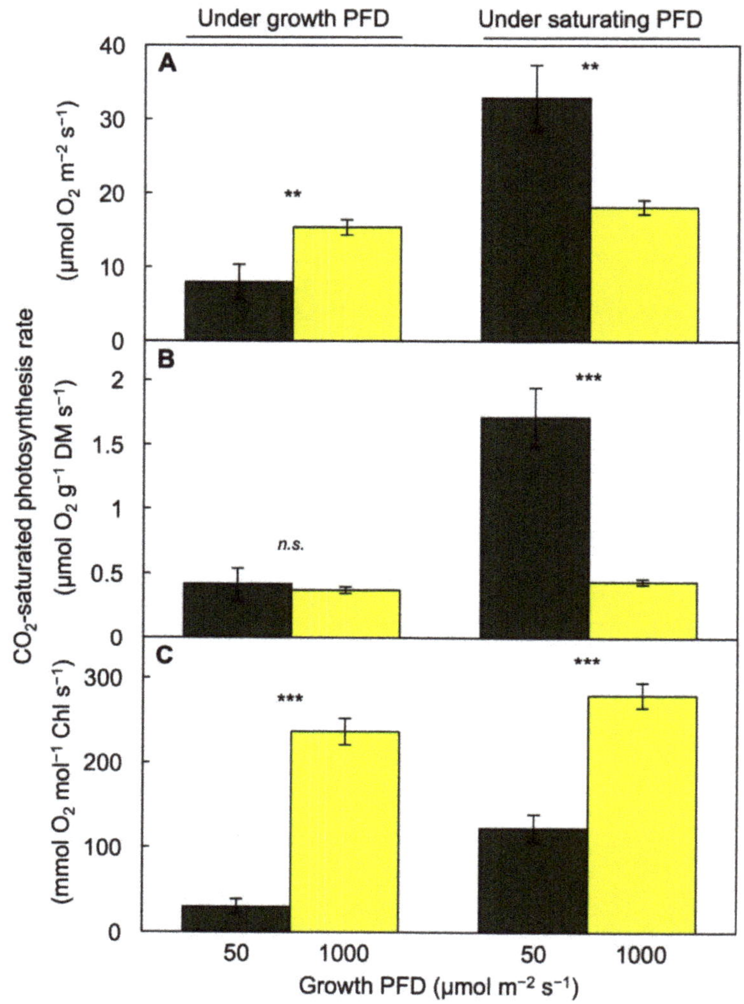

Figure 3. CO_2-saturated rates of photosynthetic oxygen evolution of fronds on the bases of (**A**) area, (**B**) dry mass (DM), and (**C**) chlorophyll (Chl) *a* + *b* levels from *Lemna gibba* plants grown under PFDs of 50 (green) or 1000 (yellow) µmol m^{-2} s^{-1} and measured under the respective growth PFDs (left columns) as well as a common saturating PFD of 1500 µmol m^{-2} s^{-1} (right columns). Mean values ± standard deviations; $n = 3$. Significant differences between growth PFDs are denoted by asterisks; ** = $p < 0.01$, *** = $p < 0.001$, *n.s.* = not significantly different.

In contrast, light- and CO_2-saturated photosynthetic capacity on an area basis (and even more so on a dry mass basis) measured under a saturating PFD of 1500 µmol m^{-2} s^{-1} was higher in the fronds grown under the low versus the high PFD (Figure 3A,B). Finally, photosynthesis rate on a Chl basis was much higher in the fronds grown under the high versus the low PFD when measured either under the respective growth PFDs or under saturating PFD (Figure 3C).

3.2. Pigment Composition, Light-Use Efficiency, and Photoprotection of Lemna gibba under Extremes in Growth PFD

The fronds grown under the low PFD were green (Figure 4A) whereas those grown under the high PFD were bright yellow (Figure 4B). This difference in visual appearance was associated with a much lower Chl content, but similar total carotenoid content, on a frond area basis in the fronds grown under 1000 versus 50 µmol photons m^{-2} s^{-1} (Figure 4C). This difference in Chl content is, furthermore, consistent with the much higher ratios of O_2 evolution relative to Chl (Figure 3C) in the fronds grown under the high versus the low PFD (which contrasted with the similar or lower rates of O_2 evolution on the frond area or dry mass bases, respectively). Despite the lower Chl content, the ratio of O_2 evolution per absorbed photons would likely be considerably lower at the high versus the low PFD, but this ratio cannot be computed since the fraction of absorbed light available to photosynthesis (by way of chlorophyll as opposed to carotenoids, at least some of which may not harvest light for photosynthesis) in a yellow frond is unknown.

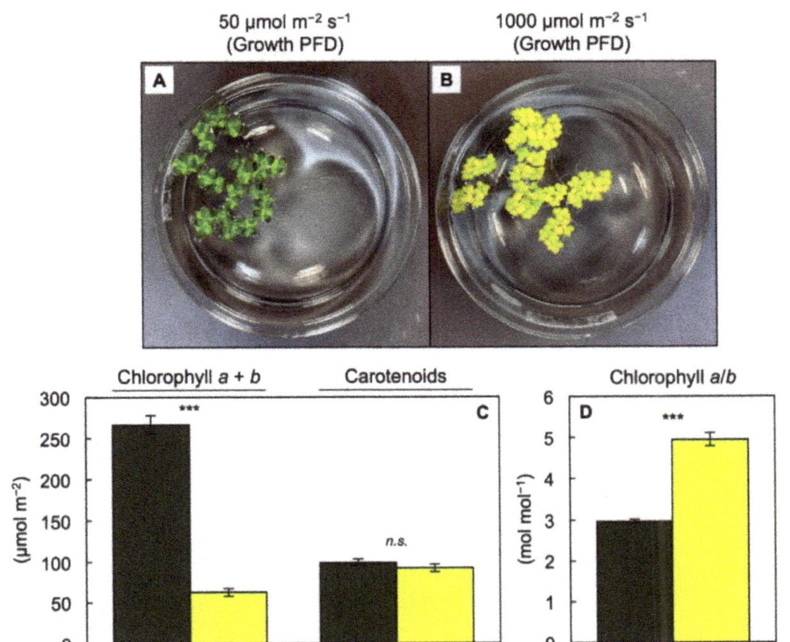

Figure 4. (**A**,**B**) Images of crystallizing dishes (1L volume) with representative cultures (starting from 20 fronds per dish) at the end of the four-day growth period, (**C**) chlorophyll $a + b$ and carotenoid levels per frond area, and (**D**) the molar ratio of chlorophyll a to chlorophyll b in fronds of *Lemna gibba* grown under PFDs of 50 (green) or 1000 (yellow) µmol m^{-2} s^{-1}. Mean values ± standard deviations; n = 3 or 4. Significant differences between growth PFDs are denoted by asterisks; *** = $p < 0.001$; *n.s.* = not significantly different.

Figure 5A shows an estimation (from chlorophyll fluorescence) of the allocation of absorbed light to PSII photochemistry (Photochemistry) and non-photochemical routes (Dissipation) as well as the fraction of excitation energy dissipated neither via photochemical or regulated non-photochemical routes (Remainder). The combination of photochemical and non-photochemical routes of excitation energy utilization or dissipation apparently was sufficient to prevent any major build-up of excitation energy.

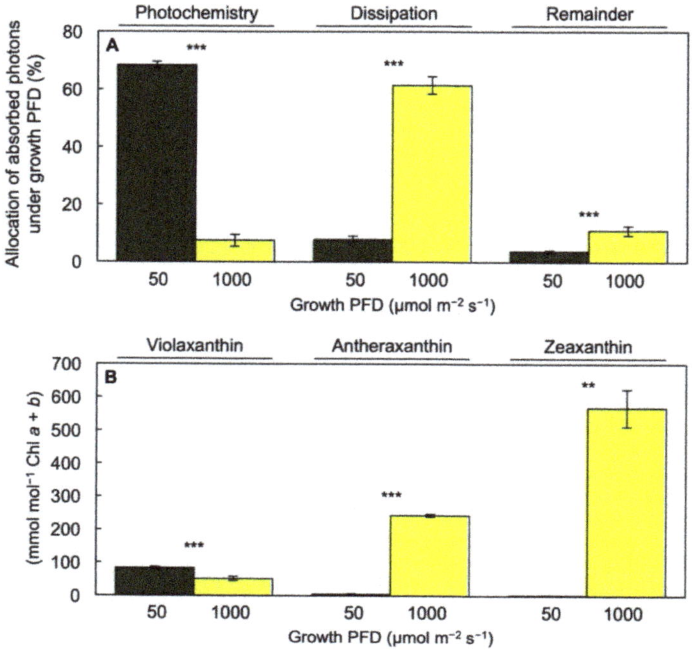

Figure 5. (**A**) Estimated percentages of absorbed light (under the respective growth PFDs) allocated to PSII photochemistry ($F_v'/F_m' \times q_P$), dissipation ($0.8 - F_v'/F_m'$), and the fraction of excitation energy dissipated neither via photochemical or regulated non-photochemical routes (remainder; $F_v'/F_m' \times [1 - q_P]$) and (**B**) levels of xanthophyll cycle pool constituents violaxanthin, antheraxanthin, and zeaxanthin relative to chlorophyll (Chl) $a + b$ in fronds of *Lemna gibba* plants grown under PFDs of 50 (green) or 1000 (yellow) μmol m^{-2} s^{-1}. The conversion state of the xanthophyll cycle to zeaxanthin and antheraxanthin in (**B**) should be compared to the level of dissipation in (**A**). Mean values ± standard deviations; $n = 3$ or 4. Significant differences between growth PFDs are denoted by asterisks; ** = $p < 0.01$, *** = $p < 0.001$.

The much greater estimated dissipation of absorbed light via regulated non-photochemical routes (Figure 5A, Dissipation) in fronds grown under 1000 versus 50 μmol photons m^{-2} s^{-1} was mirrored by strong accumulation of the photoprotective xanthophylls zeaxanthin and antheraxanthin relative to Chl $a + b$ (Figure 5B). Furthermore, Figure 6 shows that either the estimated fraction of absorbed light allocated to PSII photochemistry (Figure 6A) or the fraction of the interconvertible xanthophyll cycle pool (violaxanthin + antheraxanthin + zeaxanthin, V + A + Z) converted to the de-epoxidized forms zeaxanthin and antheraxanthin (Figure 6B) exhibited a significant positive relationship with the energy-use efficiency of total frond dry biomass production over a range of six growth PFDs from 50 to 1000 μmol m^{-2} s^{-1} [2]. For the energy-use efficiency of frond area production, significant correlations were likewise seen (not shown) with the estimated fraction of absorbed light allocated to PSII photochemistry ($p = 0.008$; $r^2 = 0.862$) and xanthophyll cycle pool conversion state ($p = 0.039$; $r^2 = 0.694$), respectively.

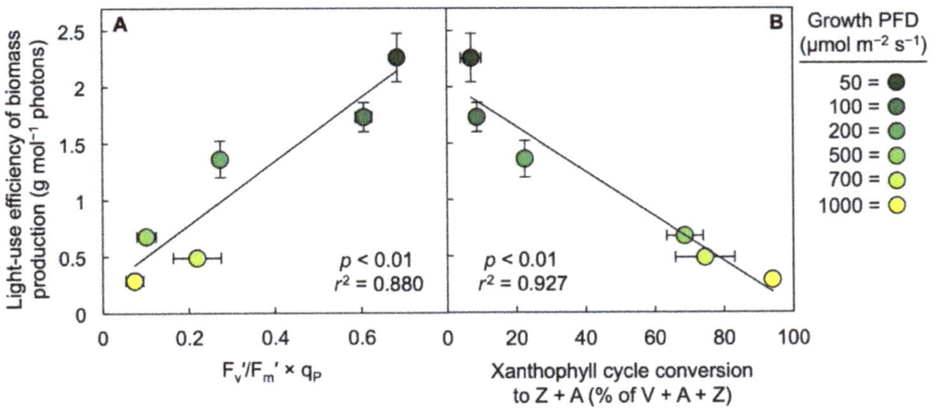

Figure 6. Correlations between the light-use efficiency of biomass production (total dry biomass produced per incident PFD during the four-day growth period) and (**A**) $(F_m' - F)/F_m' = F_v'/F_m' \times q_P$ (as an estimate of the light-use efficiency of PSII photochemistry) or (**B**) the percent of the xanthophyll cycle pool converted to zeaxanthin (Z) and antheraxanthin (A) in *Lemna gibba* fronds grown under six PFDs ranging from 50 to 1000 µmol m^{-2} s^{-1}. Mean values ± standard deviations, $n = 3$ or 4. Data for PFDs of 100 to 700 µmol m^{-2} s^{-1} were calculated from [2]. V = violaxanthin.

Figure 7 shows that the estimated fraction of absorbed light allocated to photosystem II photochemistry in fronds grown under 1000 µmol photons m^{-2} s^{-1} was lowered to less than 20% ($F_v'/F_m' < 0.2$) under this growth PFD and rose rather quickly upon transfer of fronds to 10 µmol photons m^{-2} s^{-1} to over 60% ($F_v/F_m > 0.6$) within 30 min. In contrast, there was only a small (albeit significant; $p < 0.001$) difference between the estimated fraction of absorbed light allocated to photosystem II photochemistry (not shown) under the growth PFD of 50 µmol photons m^{-2} s^{-1} (at 72.2 ± 1.2%) versus 30 min of recovery (78.1 ± 0.8%) in 10 µmol photons m^{-2} s^{-1}.

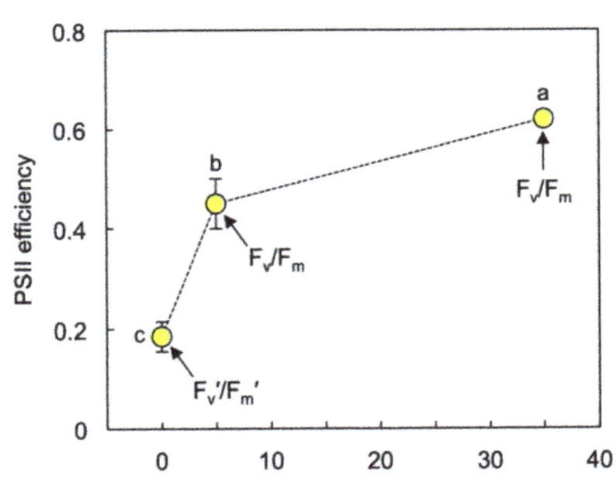

Figure 7. F_v'/F_m' (as an estimate of intrinsic PSII efficiency during exposure to growth PFD) and F_v/F_m (as an estimate of intrinsic PSII efficiency after 5 min darkness) for *L. gibba* fronds grown under 1000 µmol photons m^{-2} s^{-1}. F_v/F_m was determined at two time points (i.e., after 5 min of darkness upon removal from growth PFD and following a recovery period of 30 min under 10 µmol photons m^{-2} s^{-1} followed by 5 min darkness). Mean values ± standard deviations; $n = 3$. Significant differences between time points are denoted by different lower-case letters.

A full characterization of pigment and α-tocopherol composition of *L. gibba* fronds grown under the extremes of 50 and 1000 μmol photons m^{-2} s^{-1} is presented in Tables 1 and 2. Table 1 shows the levels of carotenoids and α-tocopherol on both frond area and dry biomass (DM) bases. Table 2 shows the ratios of carotenoids and α-tocopherol to Chl *a* + *b* or Chl *b* only as well as other ratios. While total carotenoid concentration under the high versus the low growth PFD was the same on a frond area basis (Figure 4), it was only about half on a DM basis (Table 1), but about 4× higher on a Chl *a* + *b* basis (Table 2). All individual carotenoids except zeaxanthin and antheraxanthin were present at lower concentrations on an area basis, and even lower on a DM basis (Table 1), while being enhanced relative to total Chl *a* + *b*, and even more so relative to Chl *b* alone, at the high versus the low growth PFD (Table 2). Zeaxanthin, antheraxanthin, and the total xanthophyll cycle pool (V + A + Z) were all greater on an area basis (Table 1) and relative to Chl *a* + *b* or Chl *b* alone (Figure 5B, Table 2). Zeaxanthin and antheraxanthin (Z + A), but not the total xanthophyll cycle pool, were also greater on a DM basis (Table 1). Moreover, higher ratios were seen at the high growth PFD for zeaxanthin or the total xanthophyll cycle pool to lutein, all xanthophylls to β-carotene, and the proportion of the total xanthophyll cycle pool that was converted to either zeaxanthin alone or the sum of zeaxanthin and antheraxanthin (Table 2). The only carotenoid present in the same ratio to Chl *a* + *b*, thus exhibiting a proportional decline as Chl *a* + *b* at the high growth PFD, was neoxanthin (Table 2). However, the neoxanthin level was not lowered as much as the Chl *b* level at the high growth PFD (Table 2). The only compound among those considered in Table 1 that was not significantly different in concentration on an area basis between the two growth PFDs was α-tocopherol, which resulted in a strong increase in the ratio of α-tocopherol to Chl *a* + *b* (Table 2). α-Tocopherol concentration was about half per DM at the high versus the low growth PFD (Table 1).

Table 1. Levels of carotenoids and α-tocopherol per frond area and dry mass (DM) in fronds of *Lemna gibba* grown under PFDs of 50 or 1000 μmol m^{-2} s^{-1}.

Units	Compound(s)	Growth PFD (μmol m^{-2} s^{-1})		*p*
		50	1000	
(μmol m^{-2})	Zeaxanthin	1 ± 0	35 ± 2	***
	Antheraxanthin	1 ± 0	15 ± 1	***
	Violaxanthin	22 ± 1	3 ± 0	***
	Lutein	43 ± 2	26 ± 2	***
	Neoxanthin	13 ± 0	3 ± 0	***
	β-Carotene	19 ± 3	9 ± 1	**
	Z + A	2 ± 1	50 ± 3	***
	V + A + Z	24 ± 1	54 ± 3	***
	α-Tocopherol	5 ± 1	5 ± 1	*n.s.*
(mmol g^{-1} DM)	Zeaxanthin	30 ± 12	843 ± 60	***
	Antheraxanthin	57 ± 26	362 ± 23	***
	Violaxanthin	1156 ± 56	77 ± 11	***
	Lutein	2214 ± 124	627 ± 37	***
	Neoxanthin	692 ± 20	78 ± 10	***
	β-Carotene	995 ± 157	212 ± 13	***
	Z + A	88 ± 38	1205 ± 64	***
	V + A + Z	1255 ± 55	1282 ± 62	*n.s.*
	Carotenoids	5192 ± 220	2198 ± 113	***
	α-Tocopherol	241 ± 38	113 ± 18	**

Mean values ± standard deviations; *n* = 3 or 4. Significant differences between growth PFDs are denoted by asterisks; ** = $p < 0.01$; *** = $p < 0.001$; *n.s.* = not significantly different. A = antheraxanthin, V = violaxanthin, Z = zeaxanthin.

Table 2. Levels of carotenoids and α-tocopherol relative to chlorophylls (Chl) or other carotenoids in fronds of *Lemna gibba* grown under PFDs of 50 or 1000 μmol m^{-2} s^{-1}.

Units	Compound(s)	Growth PFD (μmol m^{-2} s^{-1})		p
		50	1000	
(mmol mol^{-1} Chl $a + b$)	Lutein	160 ± 8	421 ± 25	***
	Neoxanthin	50 ± 2	52 ± 3	*n.s.*
	β-Carotene	72 ± 11	142 ± 5	***
	Z + A	6 ± 3	811 ± 57	**
	V + A + Z	90 ± 5	862 ± 52	***
	Carotenoids	371 ± 14	1478 ± 75	**
	α-Tocopherol	17 ± 3	73 ± 12	**
(mmol mol^{-1} Chl b)	Zeaxanthin	9 ± 4	3372 ± 301	**
	Antheraxanthin	17 ± 8	1445 ± 61	***
	Violaxanthin	330 ± 14	307 ± 44	*n.s.*
	Lutein	632 ± 24	2504 ± 125	***
	Neoxanthin	198 ± 10	308 ± 12	***
	β-Carotene	284 ± 47	847 ± 52	***
	Z + A	25 ± 12	4818 ± 306	**
	V + A + Z	356 ± 20	5125 ± 289	***
	Carotenoids	1470 ± 61	8784 ± 416	***
	α-Tocopherol	69 ± 11	446 ± 72	***
(mol mol^{-1})	Xanthophylls/β-Carotene	4.25 ± 0.71	9.39 ± 0.67	***
	Zeaxanthin/Lutein	0.01 ± 0.01	1.35 ± 0.05	***
	(V + A + Z)/Lutein	0.56 ± 0.03	2.05 ± 0.02	***
(% of V + A + Z)	Zeaxanthin	2 ± 1	66 ± 3	***
	Z + A	7 ± 3	94 ± 1	***

Mean values ± standard deviations; $n = 3$ or 4. Significant differences between growth PFDs are denoted by asterisks; ** = $p < 0.01$; *** = $p < 0.001$; *n.s.* = not significantly different. A = antheraxanthin, V = violaxanthin, Z = zeaxanthin.

3.3. Unique Pigment Patterns in Lemna Compared to Other Species

Under exposure to full sunlight, fronds of *L. minor* floating on a local pond exhibited conversion of a greater percentage of the total pool of interconvertible xanthophylls (V + A + Z) to the photoprotective forms zeaxanthin or zeaxanthin + antheraxanthin compared to leaves of a nearby population of the weed *M. neglecta* (Figure 8). Whereas the conversion state to the photoprotective, de-epoxidized components was higher in *L. minor* (Figure 8), the total pool of the xanthophyll cycle relative to Chl $a + b$ was smaller in *L. minor* compared to *M. neglecta* (Figure 9A).

Lutein was present at a greater level on a Chl basis in *L. minor* compared to *M. neglecta* (Figure 9B). In contrast, β-carotene per Chl (Figure 9C), α-tocopherol per Chl (Figure 9D), Chl $a + b$ per area (Figure 9E), and the Chl a/b molar ratio (Figure 9F) were all lower in *L. minor* than in *M. neglecta*.

The herbaceous weed *M. neglecta* (a biennial) fell into a cluster comprised of fast-growing annual species that accumulated rather modest amounts of zeaxanthin in full sun, while *L. minor* fell into a cluster of slow-growing perennial species (Figure 10A) using principal component analysis based on foliar pigment composition (Figure 10B). *Lemna minor*'s pigment composition was thus similar to that of slow-growing perennials, which accumulated large amounts of zeaxanthin and lutein in full sun. Remarkably, even *L. gibba* grown under 50 μmol photons m^{-2} s^{-1} fell into this cluster (Figure 10A), despite the fact that it did not accumulate zeaxanthin under this low growth PFD (Figure 5B). The yellow fronds of *L. gibba* grown under 1000 μmol photons m^{-2} s^{-1} that maintained carotenoids while strongly lowering Chl content (Figure 4C) fell far to the right of all three clusters (Figure 10A).

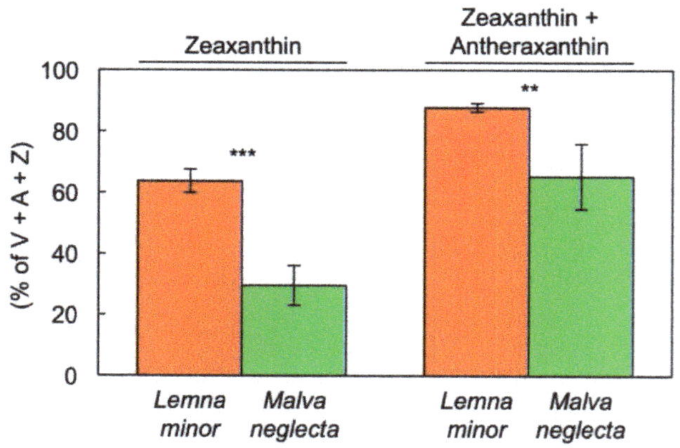

Figure 8. Conversion state of the xanthophyll cycle pool to zeaxanthin (Z) or zeaxanthin + antheraxanthin (A) under full sun (1600 µmol photons m^{-2} s^{-1}) in fronds of *Lemna minor* (orange) and leaves of *Malva neglecta* (green) growing naturally in Superior, CO, USA. Mean values ± standard deviations; $n = 4$. Significant differences are denoted by asterisks; ** = $p < 0.01$, *** = $p < 0.001$. V = violaxanthin.

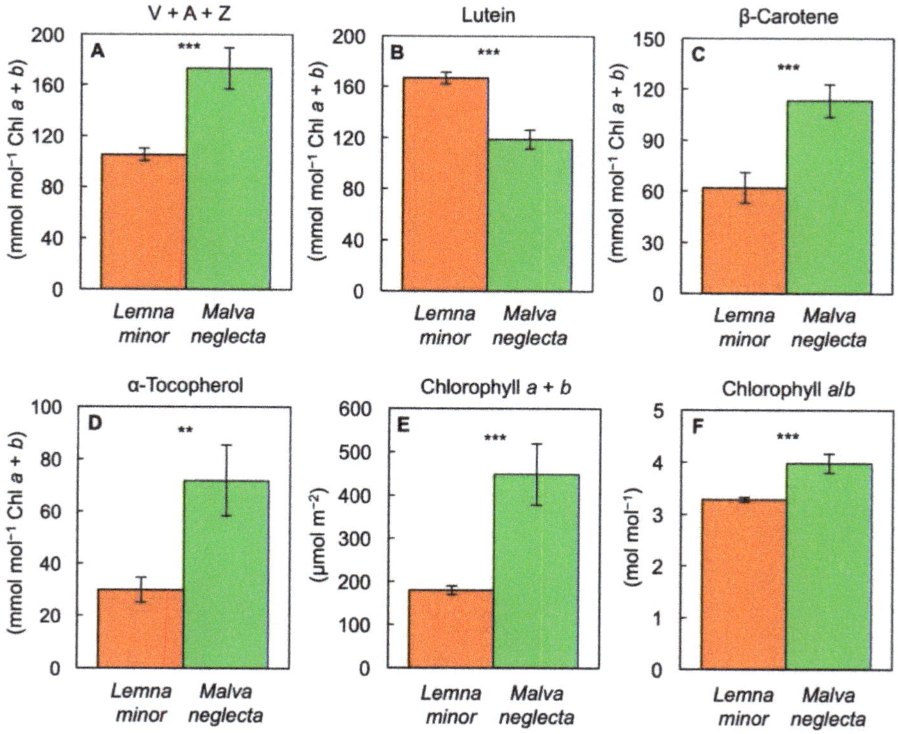

Figure 9. Levels of (**A**) the xanthophyll cycle pool constituents, violaxanthin, antheraxanthin, and zeaxanthin (V + A + Z), (**B**) lutein, (**C**) β-carotene, and (**D**) α-tocopherol per chlorophyll (Chl) $a + b$, (**E**) chlorophyll $a + b$ levels per frond/leaf area, and (**F**) the molar ratio of chlorophylls a to b in fronds of *Lemna minor* (orange) and leaves of *Malva neglecta* (green) growing naturally in Superior, Colorado, USA. Mean values ± standard deviations; $n = 4$. Significant differences are denoted by asterisks; ** = $p < 0.01$, *** = $p < 0.001$.

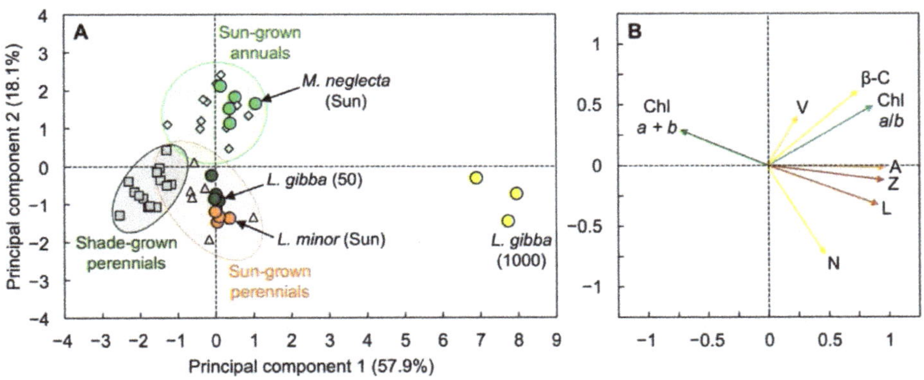

Figure 10. (**A**) Score and (**B**) loading plots of the first two principal components from principal component analysis on correlations of chlorophyll (Chl) $a + b$ per leaf/frond area, Chl a/b molar ratio, and levels of violaxanthin (V), β-carotene (β-C), antheraxanthin (A), zeaxanthin (Z), lutein (L), and neoxanthin (N) per Chl $a + b$ in fronds of *Lemna gibba* grown under PFDs of 50 (dark green circles) and 1000 (bright yellow circles) μmol m^{-2} s^{-1} in growth chambers, and fronds of sun-grown *Lemna minor* (bright orange circles), leaves of sun-grown *Malva neglecta* (bright green circles), and a variety of sun-grown annuals (light green diamonds), shade-grown perennials (gray squares), and sun-grown perennials (light orange triangles). For details, see Tables S1–S3 and [21].

4. Discussion

The results of this study extend the conclusions reported in [2] of a notable ability of the duckweed *L. gibba* to grow rapidly across a range of growth PFDs from 100 to 700 μmol m^{-2} s^{-1}. The present study extended this growth PFD range to include an even lower intensity of 50 μmol photons m^{-2} s^{-1} and an even higher intensity of 1000 μmol photons m^{-2} s^{-1} and documented the same high growth rate under the latter two extremes. This ability of duckweed to thrive under a wide range of light intensities makes sense in the context of duckweed ecology. Duckweed thrives in ponds where light environments can range from deep shade (at the pond's edge where emergent macrophytes and/or overhanging willows, etc., may provide considerable shade) to full sun in the middle of an open pond, with rapid cycles of vegetative expansion (e.g., during spring upon activation of overwintering turions, after a pond is disrupted by inclement weather involving wind and/or recharge with a major precipitation event, etc.) experienced periodically.

4.1. Interspecies Comparison

Foliar pigment composition of a closely related species, *L. minor*, growing on an open pond in full sunlight was similar to those of slow-growing evergreens, in particular with respect to the high maximal conversion of the xanthophyll cycle pool to zeaxanthin and antheraxanthin at midday in sunny habitats (see, e.g., [21,22]). Duckweed thus exhibited a combination of the traits of fast-growing annuals and slow-growing evergreens with foliar pigment features reminiscent of evergreens but coupled with a growth rate that exceeds that of the fastest-growing terrestrial plants [23].

Duckweeds are fast-growing aquatic plants that are members of the monocotyledonous order Arales; duckweeds were previously considered a subfamily (Lemnoideae) of the Araceae, but are now grouped as their own family (Lemnaceae), which is closely related to the Araceae (for recent reviews of its taxonomy, see [24,25]). Terrestrial Araceae are common in shaded rainforest environments, possessing adaptations for high shade tolerance that make them suitable as house plants (e.g., genera such as *Alocasia, Dieffenbachia, Monstera, Philodendron* [26–29]). The foliar pigment composition, and its response to growth PFD, seen here in duckweed, was reminiscent of a member of the Araceae, *Monstera deliciosa* (subfamily Monsteroideae), which possesses a remarkable ability to acclimate to both deep shade and full sunlight [30]. This makes sense in the context of the ecology of some aroids as hemi-epiphytic vines that germinate in soil in deep shade on the rainforest floor, grow toward the darkest area (a tree trunk), climb into the forest canopy, lose connection to soil/ground, and thrive in dappled to full sunlight within the canopy [31–33].

On the other hand, their exceptionally high growth rates set duckweeds apart from terrestrial Araceae. The high growth rates of duckweed may be associated with genome reductions in these diminutive, floating plants that are associated with loss of controls placed on growth and stomatal conductance in terrestrial species [34,35]. Terrestrial species typically curb growth and stomatal opening under limiting water as a defense strategy [36]. This would appear less necessary for species that float on water, and duckweeds indeed impose much less control on either growth or stomata [35]. In addition, terrestrial plants curb growth under limiting nitrogen supply in the soil [37,38], whereas duckweeds have a particular propensity for effective nitrogen uptake from the growth medium [39] and an expanded set of genes for nitrogen uptake and amino acid synthesis [34].

4.2. Specific Features That May Contribute to Duckweed's High Shade Tolerance

As noted by Stewart et al. [2], *L. gibba* cultivated under favorable controlled conditions exhibited thin leaves with apparent minimal self-shading.

Concerning foliar pigment composition, principal component analysis revealed that neither of the two *Lemna* species examined here clustered with other fast-growing species, and instead clustered with slow-growing, highly shade-tolerant evergreens, and perennials growing in full sun (pond *L. minor* and *L. gibba* grown under the low PFD) or fell outside either cluster (*L. gibba* grown under the extremely high PFD). Foliar pigment patterns of evergreen and perennial species can be differentiated from those of fast-growing, herbaceous annuals, and biennials by comparatively high total Chl contents with lower levels of VAZ pool carotenoids and β-carotene relative to total Chl, and lower Chl a/b ratios but comparatively greater levels of lutein relative to Chl in the evergreens and perennials [21]. These features are all consistent with a high light-harvesting capacity associated with comparatively large Chl *b*- and lutein-containing light-harvesting antennae and smaller β-carotene-binding inner antennae [40]. One can thus describe *Lemna* as being unusual in combining fast growth with high shade tolerance. The shade tolerance of duckweeds and their ability to maintain a high growth rate under very low growth PFD may also be due to the fact that floating plants with much-reduced root structures do not need to support a significant proportion of non-photosynthetic tissue, which is challenging in very low light.

Another aspect of duckweed physiology that would support high growth rates in low light, where photosynthetic light-use efficiency must be high, is a rapid return to high photochemical efficiency upon transfer from high to low PFD. This was seen in the present study in the form of a rapid lowering of the rate of thermal energy dissipation upon transfer of high-light-grown *L. gibba* to low light with little to no sustained depression of photosystem II photochemical efficiency or photoinhibition. Notably, such rapid return to a high PSII photochemical efficiency is also seen in shade-tolerant species subsequent to exposure to rapid sunflecks in natural understory settings [3,41] as well as in sun-grown plants of terrestrial Araceae upon return to low light after extended exposure to high light ([42]; see also [43]).

Yet another feature *L. gibba* shares with terrestrial Araceae is maintenance of a similar photosynthetic capacity on an area basis across a wide range of growth PFDs, as reported by Stewart et al. [2] for *L. gibba* grown under PFDs from 100 to 700 μmol photons m^{-2} s^{-1}. This trend is also reminiscent of what was reported for *Monstera deliciosa*. When grown under high versus low PFDs, *M. deliciosa* maintained a similar photosynthetic capacity and adjusted its capacity for regulated, photoprotective dissipation of excess absorbed light (not utilized in photochemistry), whereas fast-growing annuals strongly adjusted their photosynthetic capacities with little to no difference in the capacity for photoprotective energy dissipation [30]. The finding of the present study that photosynthetic capacity as well as relative growth rate in fronds grown under 50 μmol photons m^{-2} s^{-1} was similarly high as those of fronds grown over a range of 100 to 700 μmol photons m^{-2} s^{-1} [2] further supports *L. gibba*'s tendency to maintain a similar photosynthetic capacity across a wide range of growth PFDs (see more below on what that means for its high-light tolerance). Concerning *L. gibba*'s shade tolerance, one could speculate that its ability to support a high photosynthetic capacity under low growth PFD may contribute to its high growth rate under very low PFD. It is possible that a high level of the CO_2-fixing enzyme Rubisco may be associated with duckweed's propensity to accumulate vegetative storage protein throughout the plant rather than storing protein only in the seed like soybean (duckweed can produce 20× more protein per hectare than soybean [6]). Martindale and Bowes [44] described an unusual propensity of duckweed to accumulate high levels of Rubisco across a range of growth PFDs. While *L. gibba* plants contained more protein on an area basis under 1000 versus 50 μmol photons m^{-2} s^{-1}, protein level relative to dry biomass was actually higher under the low growth PFD (biomass was 46% protein) compared to the high growth PFD (biomass was 25% protein on a gram per gram basis). High-quality plant-based protein from duckweeds could thus be produced highly efficiently under low growth PFD.

4.3. Features That Likely Contribute to Duckweed's Tolerance of High Light

Evergreens and perennials often exhibit relatively lower maximal electron transport rates associated with very high fractions (around 90%) of absorbed light allocated to non-photochemical routes as well as very high fractions of the xanthophyll cycle pool converted to zeaxanthin at midday in sunny, but otherwise favorable, habitats [3,16,22,45]. In contrast, annuals and biennials often exhibit relatively higher maximal electron transport rates and relatively lower fractions (around 50%) of absorbed light allocated to non-photochemical routes and fractions of the xanthophyll cycle pool converted to zeaxanthin at midday in the same habitats [3,16,22,45]. In a comparison of the response of terrestrial annual species with the evergreen *M. deliciosa* to a range of growth PFDs, the annuals exhibited pronounced differences in photosynthetic capacity on a leaf area basis with no or only modest differences in photoprotective dissipation of excess excitation energy over a wide range of growth PFDs (with midday peaks of 300 versus 1500 μmol m^{-2} s^{-1}), whereas *M. deliciosa* showed no difference in photosynthetic capacity on a leaf area basis but a higher level of both thermal energy dissipation and zeaxanthin content at the higher growth PFD [30]. Duckweed exhibited similar features as *M. deliciosa*, with a pronounced increase in the fraction of absorbed light allocated to energy dissipation via regulated non-photochemical routes and of zeaxanthin accumulation, but no change in photosynthetic capacity, with increased growth PFD. This profoundly greater non-photochemical dissipation of absorbed light in the fronds grown under 1000 μmol photons m^{-2} s^{-1} was apparently highly effective in limiting the build-up of excitation energy (absorbed light not utilized via either photochemistry or non-photochemical routes), as demonstrated in the present study.

At the same time, foliar pigment composition of *Lemna* was distinctive; even shade-grown fronds not containing zeaxanthin exhibited an overall pigment pattern similar to that of sun-grown terrestrial perennials. Furthermore, yellow fronds grown under an extremely high light supply exhibited a much-exaggerated version of this pattern. These features further illustrate that *Lemna* is unusual in combining fast growth with a distinct pigment composition. The low Chl $a + b$ content, and concomitant high Chl a/b ratio, in *L gibba* grown under continuous light of 1000 μmol photons m^{-2} s^{-1} is consistent with a strong downregulation of antenna size, which is different from the response of evergreens that, as their name indicates, exhibit limited variation of chlorophyll content. Further support for a small antenna size in *L. gibba* grown under the high PFD comes from the much lower neoxanthin concentration (lowered in proportion of Chl $a + b$) and the lower levels of β-carotene and lutein on a frond area basis. The fact that xanthophyll cycle pool, the concentrations of antheraxanthin and zeaxanthin, and the ratio of total xanthophylls to β-carotene were all greater on a frond area basis at the high PFD is presumably due to strong upregulation of zeaxanthin-based photoprotection that may take place not only in pigment-binding protein complexes, but also in the membrane phospholipid bilayer [46].

Since light supply was 20× greater at 1000 versus 50 μmol photons m^{-2} s^{-1} and total dry biomass produced was only just under 3× greater, the biomass produced per mol photons (i.e., the light-use efficiency of biomass production) was dramatically lower at the high PFD. The significant linear relationships between the light-use efficiency of total biomass production and either the fluorescence-derived parameter $F_v'/F_m' \times q_P$ or the fraction of the xanthophyll cycle pool converted to its de-epoxidized components showed that features associated with primary photosynthetic processes can serve as indicators of duckweed productivity across a range of growth PFDs, irrespective of possible variations of biomass composition with respect to the proportion of, for example, protein, starch, or pigments. Duckweed biomass is particularly valuable with high levels of protein and starch, as previously noted [2]. These relationships may also be as straight-forward in duckweed because this species consists of a one-layer canopy of fronds with minimal non-photosynthetic tissue.

Both F_v/F_m and $F_v'/F_m' \times q_P$ were also correlated with measures of productivity in rice (see [47]). Prediction of productivity of other systems including whole ecosystems, from parameters associated with primary photosynthetic events is possible but requires consideration of additional features [48,49]. Whereas dark F_v/F_m was shown to be closely correlated with light-use efficiency of photosynthetic electron transport (from O_2 evolution [50,51]), and $F_v'/F_m' \times q_P$ is frequently used to estimate photochemical efficiency under illumination [52], these relationships can be tenuous [17] as was also recently discussed by Sipka et al. [53]. Nevertheless, our result that either $F_v'/F_m' \times q_P$ or xanthophyll cycle pool conversion correlated closely with the light-use efficiency of plant productivity in duckweed is consistent with the assumption that the activity of any additional dissipative processes varies in proportion with the regulated non-photochemical dissipation of excitation energy associated with de-epoxidized xanthophyll cycle components in this species. Future research should examine a possible involvement of alternative photochemical sinks for excitation energy (other than carbon fixation [54]; see also [55,56]) such as oxygen reduction by electron transport, photorespiration, and nitrogen reduction (especially by plastid nitrite reductase). In *Lemna*, nitrogen reduction could be of interest because of the demonstrated enrichment in the duckweed genome of core enzymes in amino acid synthesis [34] and the propensity of duckweed to produce vegetative storage protein.

In the present study, light- and CO_2-saturated maximal photosynthetic capacity on a frond area or dry mass basis was lower in fronds grown under 1000 μmol photons m^{-2} s^{-1} compared to fronds grown under either 50 μmol photons m^{-2} s^{-1} (this report) or under 100 to 700 μmol photons m^{-2} s^{-1} [2]. This lower maximal photosynthetic capacity did not, however, lead to a lower growth rate in the fronds grown under 1000 versus 50 μmol photons m^{-2} s^{-1}, which indicates that at the very high growth PFD of 1000 μmol m^{-2} s^{-1} of continuous light, the somewhat lower photosynthetic capacity was sufficient to support the same high growth rate as at 50 μmol photons m^{-2} s^{-1}. The lower photosynthetic capacity on an area or dry mass basis in fronds grown under 1000 μmol photons m^{-2} s^{-1} thus represents an adjustment that allows the plants to maintain a similar growth rate with a lesser photosynthetic capacity and a much lower chlorophyll content in a growth environment with a very high light supply. This feature could also be associated with duckweed accumulating Rubisco levels in excess of what is needed for CO_2 fixation at a given time. Total Rubisco level may be associated with light- and CO_2-saturated photosynthetic capacity, but varying proportions of this capacity may be sufficient to support growth under different growth PFDs [44]. The observed extremely high capacity of photosynthetic O_2 evolution on a Chl basis in fronds grown under 1000 μmol photons m^{-2} s^{-1} of continuous light is consistent with a strong preferential downregulation of antenna size relative to the components of photosynthetic electron transport under this high growth PFD.

Furthermore, *L. gibba* accumulated considerable starch at higher growth PFDs including under 700 μmol photons m^{-2} s^{-1} where no decline in photosynthetic capacity was seen [2], which suggests that the lower photosynthetic capacity in plants grown under

1000 μmol photons m^{-2} s^{-1} may not be due to downregulation associated with starch accumulation. *Lemna*'s propensity for unabated growth and photosynthetic activity over a wide range of growth environments is consistent with the reported reduction of control by water and nutrient level in duckweeds coupled with their genome reduction and associated permanently open stomates and highly effective nutrient acquisition [35].

In a nutshell, *L. gibba* plants grown under continuous high light of 1000 μmol photons m^{-2} s^{-1} effectively counteracted build-up of potentially dangerous excess excitation through a combination of strong downregulation of antenna size (i.e., how much light is absorbed) with strong non-photochemical dissipation of excess absorbed light under the high growth PFD. These photoprotective processes did not interfere with the ability of photosynthetic electron transport to support similar area production and higher dry biomass production in the plants grown under 1000 versus 50 μmol photons m^{-2} s^{-1}. This high light tolerance of duckweed is reminiscent to that of the alga *Chlorella ohadii*, which can thrive under a light intensity equivalent to twice that of sunlight [57]. Under such high light conditions, this alga also exhibited unabated fast growth, a high photosynthesis rate, a small antenna size (a constitutive small antenna and an extremely high Chl *a*/*b* ratio of 13:1), and photoprotective energy dissipation that, however, relied on mechanisms other than zeaxanthin-associated non-photochemical energy dissipation [57]. The latter results and those reported here illustrate the existence of photosynthetic systems that grow at high rates under extremely high light and use unique combinations of photoprotective mechanisms. Duckweed is presumably well adapted to a range of natural environments that include predominantly either continuously shaded or high-light-exposed sites (where antenna size modulation should be particularly beneficial) as well as sites with rapidly fluctuating light (where the rapid reversibility of non-photochemical dissipation should be particularly beneficial).

In conclusion, the duckweed *L. gibba* was evidently able to acclimate to very high growth light intensity through a combination of a high growth rate with pronounced starch and protein accumulation, decreased light absorption (presumably by downregulation of antenna size), pronounced non-photochemical dissipation of excess light associated with zeaxanthin as well as other forms of photoprotection provided by the xanthophyll lutein (that can remove excitation energy from Chl in the triplet state [58]), β-carotene (that can contribute to the photoprotection of photosystem I [59]), and α-tocopherol (that can scavenge singlet oxygen and lipid peroxy radicals [60,61]). The greater levels of especially zeaxanthin, α-tocopherol, lutein, and to some extent β-carotene relative to Chl under high growth PFD are consistent with an enhanced need for photoprotection on part of the plant. From the standpoint of human/animal nutrition, however, production of micronutrients per area or as a percent of biomass matters most. Whereas zeaxanthin production required high light irrespective of reference basis, α-tocopherol (vitamin E), lutein, β-carotene (provitamin A), and protein levels as a percent of biomass were all lower under the high growth PFD. Therefore, a mixed lighting protocol with mainly low background PFD, supplemented with brief exposures to high light (see, e.g., [62]) might be attractive to produce high-quality nutrition for the consumer.

Supplementary Materials: The following are available online at https://www.mdpi.com/article/10.3390/cells10061481/s1, Table S1: Source, species, and grouping of data used for principal component analysis of pigment composition (see Figure 10), as well as scores for all eight of the resulting principal components; Table S2: Loading values of the eight pigment parameters used for principal component analysis (see Figure 10) for the eight resulting principal components; Table S3: Correlation matrix for the eight variables used in the principal component analysis of pigment composition (see Figure 10).

Author Contributions: B.D.-A., C.M.E. and W.W.A.III wrote the grant on which this study is based. B.D.-A., W.W.A.III and J.J.S. planned the experiments with contributions from C.M.E., J.J.S., W.W.A.III, M.L.-P., N.D.G. and M.M. carried out the experiments and biochemical assays with contributions from B.D.-A., C.M.E., B.D.-A., W.W.A.III, J.J.S. and M.L.-P. analyzed and interpreted the data. B.D.-A., W.W.A.III and J.J.S. wrote the manuscript. All authors have read and agreed to the published version of the manuscript.

Funding: This work was funded by the Translational Research Institute for Space Health through Cooperative Agreement NNX16AO69A, the National Science Foundation award number IOS-1907338, and the University of Colorado.

Institutional Review Board Statement: Not applicable for this work using only plants.

Informed Consent Statement: Not applicable for this work using only plants.

Data Availability Statement: The data presented in this study are available from the corresponding author upon reasonable request.

Acknowledgments: We thank Adam Escobar and Paul Fourounjian for the valuable discussion and feedback, and Gabrielle Glime for assistance with the data collection.

Conflicts of Interest: C.M.E. has a financial interest in Space Lab Technologies, LLC, a company that may be affected by this research. A Memorandum of Understanding approved by the University of Colorado manages potential conflicts arising from this relationship. The remaining authors declare no conflicts of interest. The sponsors had no role in the design, execution, interpretation, or writing of the study.

References

1. Appenroth, K.J.; Sree, K.S.; Böhm, V.; Hammann, S.; Vetter, W.; Leiterer, M.; Jahreis, G. Nutritional value of duckweeds (Lemnaceae) as human food. *Food Chem.* **2017**, *15*, 266–273. [CrossRef] [PubMed]
2. Stewart, J.J.; Adams, W.W., III; Escobar, C.M.; López-Pozo, M.; Demmig-Adams, B. Growth and essential carotenoid micronutrients in *Lemna gibba* as a function of growth light intensity. *Front. Plant Sci.* **2020**, *11*, 480. [CrossRef]
3. Demmig-Adams, B.; Stewart, J.J.; López-Pozo, M.; Polutchko, S.K.; Adams, W.W., III. Zeaxanthin, a molecule for photoprotection in many different environments. *Molecules* **2020**, *25*, 5825. [CrossRef]
4. Demmig-Adams, B.; López-Pozo, M.; Stewart, J.J.; Adams, W.W., III. Zeaxanthin and lutein: Photoprotectors, anti-inflammatories, and brain food. *Molecules* **2020**, *25*, 3607. [CrossRef]
5. Xu, J.; Shen, G. Growing duckweed in swine wastewater for nutrient recovery and biomass production. *Bioresour. Technol.* **2011**, *102*, 848–853. [CrossRef]
6. Mohedano, R.A.; Costa, R.H.R.; Tavares, F.A.; Filho, P.B. High nutrient removal rate from swine wastes and protein biomass production by full-scale duckweed ponds. *Bioresour. Technol.* **2012**, *112*, 98–104. [CrossRef]
7. Liu, Y.; Xu, H.; Yu, C.; Zhou, G. Multifaceted roles of duckweed in aquatic phytoremediation and bioproducts synthesis. *GCB Bioenergy* **2021**, *13*, 70–82. [CrossRef]
8. Sun, H.; A, D.; Feng, Y.; Vithanage, M.; Mandal, S.; Shaheen, S.; Rinklebe, J.; Shi, W.; Wang, H. Floating duckweed mitigated ammonia volatilization and increased grain yield and nitrogen use efficiency of rice in biochar amended paddy soil. *Chemosphere* **2019**, *237*, 124532. [CrossRef]
9. Demmig, B.; Winter, K.; Krüger, A.; Czygan, F.-C. Photoinhibition and zeaxanthin formation in intact leaves. A possible role of the xanthophyll cycle in the dissipation of excess light energy. *Plant Physiol.* **1987**, *84*, 218–224. [CrossRef]
10. Park, S.; Steen, C.J.; Lyska, D.; Fischer, A.L.; Endelman, B.; Iwai, M.; Niyogi, K.K.; Fleming, G.R. Chlorophyll–carotenoid excitation energy transfer and charge transfer in *Nannochloropsis oceanica* for the regulation of photosynthesis. *Proc. Natl. Acad. Sci. USA* **2019**, *116*, 3385–3390. [CrossRef]
11. Adams, W.W., III; Demmig-Adams, B. Operation of the xanthophyll cycle in higher plants in response to diurnal changes in incident sunlight. *Planta* **1992**, *186*, 390–398. [CrossRef]
12. Schenk, R.U.; Hildebrandt, A.C. Medium and techniques for induction and growth of monocotyledonous and dicotyledonous plant cell cultures. *Can. J. Bot.* **1972**, *50*, 199–204. [CrossRef]
13. Muller, O.; Stewart, J.J.; Cohu, C.M.; Polutchko, S.K.; Demmig-Adams, B.; Adams, W.W., III. Leaf architectural, vascular and photosynthetic acclimation to temperature in two biennials. *Physiol. Plant.* **2014**, *152*, 763–772. [CrossRef] [PubMed]
14. Schneider, C.A.; Rasband, W.S.; Eliceiri, K.W. NIH image to ImageJ: 25 years image analysis. *Nat. Methods* **2012**, *9*, 671–675. [CrossRef]
15. Delieu, T.; Walker, D.A. Polarographic measurement of photosynthetic oxygen evolution by leaf discs. *New Phytol.* **1981**, *89*, 165–178. [CrossRef]
16. Demmig-Adams, B.; Adams, W.W., III; Barker, D.H.; Logan, B.A.; Bowling, D.R.; Verhoeven, A.S. Using chlorophyll fluorescence to assess the fraction of absorbed light to thermal dissipation of excess excitation. *Physiol. Plant.* **1996**, *98*, 253–264. [CrossRef]

17. Logan, B.A.; Adams, W.W., III; Demmig-Adams, B. Avoiding common pitfalls of chlorophyll fluorescence analysis in the field. *Funct. Plant Biol.* **2007**, *34*, 853–859. [CrossRef]
18. Peterson, G.L. A simplification of the protein assay method of Lowry et al. which is more generally applicable. *Anal. Biochem.* **1977**, *83*, 346–356. [CrossRef]
19. Lowry, O.H.; Rosebrough, N.J.; Farr, A.L.; Randall, R.J. Protein measurement with the Folin phenol reagent. *J. Biol. Chem.* **1951**, *193*, 265–275. [CrossRef]
20. Stewart, J.J.; Adams, W.W., III; Cohu, C.M.; Polutchko, S.K.; Lombardi, E.M.; Demmig-Adams, B. Differences in light-harvesting, acclimation to growth-light environment, and leaf structural development between Swedish and Italian ecotypes of *Arabidopsis thaliana*. *Planta* **2015**, *242*, 1277–1290. [CrossRef]
21. Demmig-Adams, B.; Adams, W.W., III. Carotenoid composition in sun and shade leaves of plants with different life forms. *Plant Cell Environ.* **1992**, *15*, 411–419. [CrossRef]
22. Demmig-Adams, B.; Cohu, C.M.; Muller, O.; Adams, W.W., III. Modulation of photosynthetic energy conversion efficiency in nature: From seconds to seasons. *Photosynth. Res.* **2012**, *113*, 75–88. [CrossRef] [PubMed]
23. Ziegler, P.; Adelmann, K.; Zimmer, S.; Schmidt, C.; Appenroth, K.J. Relative in vitro growth rates of duckweeds (*Lemnaceae*)—The most rapidly growing higher plants. *Plant Biol.* **2015**, *17*, 33–41. [CrossRef]
24. Bog, M.; Appenroth, K.-J.; Sree, K.S. Duckweed (*Lemnaceae*): Its molecular taxonomy. *Front. Sustain. Food Syst.* **2019**, *3*, 117. [CrossRef]
25. Tippery, N.P.; Les, D.H. Tiny plants with enormous potential: Phylogeny and evolution of duckweeds. In *The Duckweed Genomes. Compendium of Plant Genomes*; Cao, X., Fourounjian, P., Wang, W., Eds.; Springer: Cham, Switzerland, 2020; pp. 19–38. [CrossRef]
26. Demmig-Adams, B.; Adams, W.W., III. Capacity for energy dissipation in the pigment bed in leaves with different xanthophyll cycle pools. *Aust. J. Plant Physiol.* **1994**, *21*, 575–588. [CrossRef]
27. Logan, B.A.; Barker, D.H.; Adams, W.W., III; Demmig-Adams, B. The response of xanthophyll cycle-dependent energy dissipation in *Alocasia brisbanensis* to sunflecks in a subtropical rainforest. *Aust. J. Plant Physiol.* **1997**, *24*, 27–33. [CrossRef]
28. Skillman, J.B.; Garcia, M.; Virgo, A.; Winter, K. Growth irradiance effects on photosynthesis and growth in two co-occurring shade-tolerant neotropical perennials of contrasting photosynthetic pathways. *Am. J. Bot.* **2005**, *92*, 1811–1819. [CrossRef]
29. Tan, C.L.; Wong, N.H.; Tan, P.Y.; Wee, L.Y. Growth light provision for indoor greenery: A case study. *Energy Build.* **2017**, *144*, 207–217. [CrossRef]
30. Demmig-Adams, B.; Ebbert, V.; Mellman, D.L.; Mueh, K.E.; Schaffer, L.; Funk, C.; Zarter, C.R.; Adamska, I.; Jansson, S.; Adams, W.W., III. Modulation of PsbS and flexible vs sustained energy dissipation by light environment in different species. *Physiol. Plant.* **2006**, *127*, 670–680. [CrossRef]
31. Ray, T.S. Foraging behaviour in tropical herbaceous climbers (*Araceae*). *J. Ecol.* **1992**, *80*, 189–203. [CrossRef]
32. Strong, D.R., Jr.; Ray, T.S., Jr. Host tree location behavior of a tropical vine (*Monstera gigantea*) by skototropism. *Science* **1975**, *190*, 804–806. [CrossRef]
33. Mantovani, A.; Brito, C.; Mantuano, D. Does the same morphology mean the same physiology? Morphophysiological adjustments of *Philodendron hederaceum* (Jacq.) Schott, an isomorphic aroid, to groudcanopy transition. *Theor. Exp. Plant Physiol.* **2018**, *30*, 89–101. [CrossRef]
34. Wang, W.; Haberer, G.; Gundlach, H.; Gläβer, C.; Nussbaumer, T.; Luo, M.C.; Lomsadze, A.; Borodovsky, M.; Kerstetter, R.A.; Shanklin, J.; et al. The *Spirodela polyrhiza* genome reveals insights into its neotenous reduction fast growth and aquatic lifestyle. *Nat. Commun.* **2013**, *5*, 3311. [CrossRef] [PubMed]
35. Michael, T.P.; Ernst, E.; Hartwick, N.; Chu, P.; Bryant, D.; Gilbert, S.; Ortleb, S.; Baggs, E.L.; Sree, K.S.; Appenroth, K.J.; et al. Genome and time-of-day transcriptome of *Wolffia australiana* link morphological minimization with gene loss and less growth control. *Genome Res.* **2021**, *31*, 225–238. [CrossRef] [PubMed]
36. Hsiao, T.C. Plant responses to water stress. *Annu. Rev. Plant Physiol.* **1973**, *24*, 519–570. [CrossRef]
37. Paul, M.J.; Driscoll, S.P. Sugar repression of photosynthesis: The role of carbohydrates in signalling nitrogen deficiency through source: Sink imbalance. *Plant Cell Environ.* **1997**, *20*, 110–116. [CrossRef]
38. Logan, B.A.; Demmig-Adams, B.; Rosenstiel, T.N.; Adams, W.W., III. Effect of nitrogen limitation on foliar antioxidants in relationship to other metabolic characteristics. *Planta* **1999**, *209*, 213–220. [CrossRef]
39. Cedergreen, N.; Madsen, T.V. Nitrogen uptake by the floating macrophyte *Lemna minor*. *New Phytol.* **2002**, *155*, 285–292. [CrossRef]
40. Yamamoto, H.Y.; Bassi, R. Carotenoids: Localization and function. In *Oxygenic Photosynthesis: The Light Reactions. Advances in Photosynthesis and Respiration*; Ort, D.R., Yocum, C.F., Heichel, I.F., Eds.; Springer: Dordrecht, The Netherlands, 1996; Volume 4, pp. 539–563. [CrossRef]
41. Adams, W.W., III; Demmig-Adams, B.; Logan, B.A.; Barker, D.H.; Osmond, C.B. Rapid changes in xanthophyll cycle-dependent energy dissipation and photosystem II efficiency in two vines, *Stephania japonica* and *Smilax australis*, growing in the understory of an open *Eucalyptus* forest. *Plant Cell Environ.* **1999**, *22*, 125–136. [CrossRef]
42. Demmig-Adams, B.; Muller, O.; Stewart, J.J.; Cohu, C.M.; Adams, W.W., III. Chloroplast thylakoid structure in evergreen leaves employing strong thermal energy dissipation. *J. Photochem. Photobiol. B* **2015**, *152*, 357–366. [CrossRef] [PubMed]
43. Demmig-Adams, B.; Moeller, D.L.; Logan, B.A.; Adams, W.W., III. Positive correlation between levels of retained zeaxanthin plus antheraxanthin and degree of photoinhibition in shade leaves of *Schefflera arboricola* (Hayata) Merrill. *Planta* **1998**, *205*, 367–374. [CrossRef]

44. Martindale, W.; Bowes, G. The effects of irradiance and CO_2 on the activity and activation of ribulose-1,5-bisphosphate carboxylase/oxygenase in the aquatic plant *Spirodela polyrhiza*. *J. Exp. Bot.* **1996**, *47*, 781–784. [CrossRef]
45. Demmig-Adams, B.; Adams, W.W., III; Ebbert, V.; Logan, B.A. Ecophysiology of the xanthophyll cycle. In *The Photochemistry of Carotenoids. Advances in Photosynthesis and Respiration*; Frank, H.A., Young, A.J., Britton, G., Cogdell, R.J., Eds.; Kluwer Academic Publishers: Dordrecht, The Netherlands, 1999; Volume 8, pp. 245–269. [CrossRef]
46. Havaux, M.; García-Plazaola, J.I. Beyond non-photochemical fluorescence quenching: The overlapping antioxidant functions of zeaxanthin and tocopherols. In *Non-Photochemical Quenching and Energy Dissipation in Plants, Algae and Cyanobacteria, Advances in Photosynthesis and Respiration*; Demmig-Adams, B., Garab, G., Adams, W.W., III, Govindjee, G.H., Eds.; Springer: Dordrecht, The Netherlands, 2014; Volume 40, pp. 583–603. [CrossRef]
47. Huang, M.; Shan, S.; Zhou, X.; Chen, J.; Cao, F.; Jiang, L.; Zou, Y. Leaf photosynthetic performance related to higher radiation use efficiency and grain yield in hybrid rice. *Field Crops Res.* **2016**, *193*, 87–93. [CrossRef]
48. Garbulsky, M.F.; Peñuelas, J.; Gamon, J.; Inoue, Y.; Filella, I. The photochemical reflectance index (PRI) and the remote sensing of leaf, canopy and ecosystem radiation use efficiencies a review and meta-analysis. *Remote Sens. Environ.* **2011**, *115*, 281–297. [CrossRef]
49. Porcar-Castell, A.; Tyystjärvi, E.; Atherton, J.; van der Tol, C.; Flexas, J.; Pfündel, E.E.; Moreno, J.; Frankenberg, C.; Berry, J.A. Linking chlorophyll *a* fluorescence to photosynthesis for remote sensing applications: Mechanisms and challenges. *J. Exp. Bot.* **2014**, *65*, 4065–4095. [CrossRef] [PubMed]
50. Björkman, O.; Demmig, B. Photon yield of O_2 evolution and chlorophyll fluorescence characteristics at 77 K among vascular plants of diverse origins. *Planta* **1987**, *170*, 489–504. [CrossRef] [PubMed]
51. Adams, W.W., III; Demmig-Adams, B.; Winter, K.; Schreiber, U. The ratio of variable to maximum chlorophyll fluorescence from photosystem II, measured in leaves at ambient temperature and at 77 K, as an indicator of the photon yield of photosynthesis. *Planta* **1990**, *180*, 166–174. [CrossRef]
52. Maxwell, K.; Johnson, G.N. Chlorophyll fluorescence—A practical guide. *J. Exp. Bot.* **2000**, *51*, 659–666. [CrossRef] [PubMed]
53. Sipka, G.; Magyar, M.; Mezzetti, A.; Akhtar, P.; Zhu, Q.; Xiao, Y.; Han, G.; Santabarbara, S.; Shen, J.-R.; Lambrev, P.H.; et al. Light-adapted charge-separated state of photosystem II: Structural and functional dynamics of the closed reaction center. *Plant Cell* **2021**, *33*, 1286–1302. [CrossRef]
54. Walker, B.J.; Kramer, D.M.; Fisher, N.; Fu, X. Flexibility in the energy balancing network of photosynthesis enables safe operation under changing environmental conditions. *Plants* **2020**, *9*, 301. [CrossRef] [PubMed]
55. Logan, B.A.; Demmig-Adams, B.; Adams, W.W., III. Acclimation of photosynthesis to the environment. In *Concepts in Plant Photobiology: Photosynthesis and Photomorphogenesis*; Singhal, G.S., Renger, G., Sopory, S.K., Irrgang, K.-D., Govindjee, G.H., Eds.; Narosa Publishing House: New Dehli, India, 1999; pp. 477–512. [CrossRef]
56. Logan, B.A.; Demmig-Adams, B.; Adams, W.W., III; Bilger, W. Context, quantification, and measurement guide for non-photochemical quenching of chlorophyll fluorescence. In *Non-Photochemical Quenching and Energy Dissipation in Plants, Algae and Cyanobacteria, Advances in Photosynthesis and Respiration*; Demmig-Adams, B., Garab, G., Adams, W.W., III, Govindjee, G.H., Eds.; Springer: Dordrecht, The Netherlands, 2014; Volume 40, pp. 187–201. [CrossRef]
57. Treves, H.; Raanan, H.; Kedem, I.; Murik, O.; Keren, N.; Zer, H.; Berkowicz, S.M.; Giordano, M.; Norici, A.; Shotland, Y.; et al. The mechanisms whereby the green alga *Chlorella ohadii*, isolated from desert soil crust, exhibits unparalleled photodamage resistance. *New Phytol.* **2016**, *210*, 1229–1243. [CrossRef] [PubMed]
58. Dall'Osto, L.; Lico, C.; Alric, J.; Giuliano, G.; Havaux, M.; Bassi, R. Lutein is needed for efficient chlorophyll triplet quenching in the major LHCII antenna complex of higher plants and effective photoprotection in vivo under strong light. *BMC Plant Biol.* **2006**, *6*, 32. [CrossRef]
59. Cazzaniga, S.; Li, Z.; Niyogi, K.K.; Bassi, R.; Dall'Osto, L. The Arabidopsis *szl1* mutant reveals a critical role of β-carotene in photosystem I photoprotection. *Plant Physiol.* **2012**, *159*, 1745–1758. [CrossRef]
60. Munné-Bosch, S.; Alegre, L. The function of tocopherols and tocotrienols in plants. *Crit. Rev. Plant Sci.* **2002**, *21*, 31–57. [CrossRef]
61. Krieger-Liszkay, A.; Fufezan, C.; Trebst, A. Singlet oxygen production in photosystem II and related protection mechanism. *Photosynth. Res.* **2008**, *98*, 551–564. [CrossRef] [PubMed]
62. Cohu, C.M.; Lombardi, E.; Adams, W.W., III; Demmig-Adams, B. Increased nutritional quality of plants for long-duration space missions through choice of plant variety and manipulation of growth conditions. *Acta Astronaut.* **2014**, *94*, 799–806. [CrossRef]

cells

MDPI

Article

Comparative Study between Exogenously Applied Plant Growth Hormones versus Metabolites of Microbial Endophytes as Plant Growth-Promoting for *Phaseolus vulgaris* L.

Mohamed A. Ismail [1], Mohamed A. Amin [1], Ahmed M. Eid [1], Saad El-Din Hassan [1,*], Hany A. M. Mahgoub [1], Islam Lashin [1,2], Abdelrhman T. Abdelwahab [1,3], Ehab Azab [4], Adil A. Gobouri [5], Amr Elkelish [6,7] and Amr Fouda [1,*]

1 Department of Botany and Microbiology, Faculty of Science, Al-Azhar University, Nasr City, Cairo 11884, Egypt; alazharbotany5617@gmail.com (M.A.I.); mamin7780@azhar.edu.eg (M.A.A.); aeidmicrobiology@azhar.edu.eg (A.M.E.); Saad.el-din.hassan@umontreal.ca or h.mahgoub@azhar.edu.eg (H.A.M.M.); islam79@azhar.edu.eg (I.L.); omarabdot@yahoo.com (A.T.A.)
2 Department of Biology, Faculty of Science and Arts, Al Mandaq, Albaha University, Al-Baha 1988, Saudi Arabia
3 Department of Botany Science, Faculty of Science, Northern Border University, Arar 73211, Saudi Arabia
4 Department of Nutrition and Food Science, College of Science, Taif University, P.O. Box 11099, Taif 21944, Saudi Arabia; e.azab@tu.edu.sa
5 Department of Chemistry, College of Science, Taif University, P.O. Box 11099, Taif 21944, Saudi Arabia; a.gobouri@tu.edu.sa
6 Botany and Microbiology Department, Faculty of Science, Suez Canal University, Ismailia 41511, Egypt; amr.elkelish@science.suez.edu.eg or amr.elkelish@uni-jena.de
7 Department of Plant Physiology, Matthias Schleiden Institute of Genetics, Bioinformatics and Molecular Botany, Friedrich-Schiller-University Jena, 07743 Jena, Germany
* Correspondence: saad_hassan@azhar.edu.eg (S.E.-D.H.); amr_fh83@azhar.edu.eg (A.F.); Tel.: +20-102-3884804 (S.E.-D.H.); +20-111-3351244 (A.F.)

Citation: Ismail, M.A.; Amin, M.A.; Eid, A.M.; Hassan, S.E.-D.; Mahgoub, H.A.M.; Lashin, I.; Abdelwahab, A.T.; Azab, E.; Gobouri, A.A.; Elkelish, A.; et al. Comparative Study between Exogenously Applied Plant Growth Hormones versus Metabolites of Microbial Endophytes as Plant Growth-Promoting for *Phaseolus vulgaris* L. *Cells* **2021**, *10*, 1059. https://doi.org/10.3390/cells10051059

Academic Editor: Suleyman Allakhverdiev

Received: 18 March 2021
Accepted: 26 April 2021
Published: 29 April 2021

Abstract: Microbial endophytes organize symbiotic relationships with the host plant, and their excretions contain diverse plant beneficial matter such as phytohormones and bioactive compounds. In the present investigation, six bacterial and four fungal strains were isolated from the common bean (*Phaseolus vulgaris* L.) root plant, identified using molecular techniques, and their growth-promoting properties were reviewed. All microbial isolates showed varying activities to produce indole-3-acetic acid (IAA) and different hydrolytic enzymes such as amylase, cellulase, protease, pectinase, and xylanase. Six bacterial endophytic isolates displayed phosphate-solubilizing capacity and ammonia production. We conducted a field experiment to evaluate the promotion activity of the metabolites of the most potent endophytic bacterial (*Bacillus thuringiensis* PB2 and *Brevibacillus agri* PB5) and fungal (*Alternaria sorghi* PF2 and, *Penicillium commune* PF3) strains in comparison to two exogenously applied hormone, IAA, and benzyl adenine (BA), on the growth and biochemical characteristics of the *P. vulgaris* L. Interestingly, our investigations showed that bacterial and fungal endophytic metabolites surpassed the exogenously applied hormones in increasing the plant biomass, photosynthetic pigments, carbohydrate and protein contents, antioxidant enzyme activity, endogenous hormones and yield traits. Our findings illustrate that the endophyte *Brevibacillus agri* (PB5) provides high potential as a stimulator for the growth and productivity of common bean plants.

Keywords: endophytes; culture filtrate; exogenously hormone; Phaseolus vulgaris; antioxidant enzymes

1. Introduction

Common beans (*Phaseolus vulgaris* L.) are important legumes as a source of food and money for small farm owners worldwide [1]. Common beans are an essential source of micronutrients and protein, in addition to their caloric content. They are of interest in developing countries as an inexpensive protein source (compared to animal protein sources) and because of their long-term storage life [2]. The popularity of this crop stems from its relative ease of production, variety of uses, good source of nutrition, and deliciousness. Beans are widely adapted to moderate temperature growing environments, with an average precipitation of approximately 400 mL and a growing season of 60–120 days [3]. By the year 2050, the world's population is expected to increase by 34% (approximately 9.1 billion people), and thus, we need approximately 70% more food production [4].

Plant hormones, or phyto-regulators, have the leading role in a plant's life cycle [5]. Plant hormones are involved in most of development stages such as seed germination, fruit maturation, seedling, and leaf senescence [6]. Auxin is the most common class of plant hormones, the most prominent of which is indole-3-acetic acid (IAA), a key regulator of plant growth and development [7], involved in the phototropic response, fruit development, cell differentiation, division, and elongation, and has a major role in plant physiology [7]. IAA increases stomatal conductance, the photosynthetic average, plant pigmentation, and the piling up of total soluble sugars such as fructose and glucose [8].

Other frequently used plant hormones are cytokinin's, which affect cell division and many other aspects of the plant development and growth processes involving shoot growth and initiation, seed germination, abscission, senescence, and apical dominance [9,10]. Benzyl adenine (BA) is one of the commonly used cytokinin's [11], which has promotional effects that have been demonstrated by several studies. IAA is the most common endogenous hormone and has an effective role in root elongation and growth. It results from the metabolism of L-tryptophan by various microorganisms, including plant growth-promoting bacteria (PGPB) [12].

Fungi and plant-related bacteria produce many plant hormones, including IAA, as one of the mechanisms to manage plant development and growth. Microbial indole modifies the structure of roots to boost the availability of nutrients to these microbes that can produce plant hormones [13].

PGPB, which have many useful effects on their host plant, involve the growth activation, metabolism regulation, and modulation of phytohormone signaling that contribute to the adaptation to adverse environmental conditions [14]. Endophytes are being studied for their ability to create novel biologically active compounds, as well as their ability to preserve plant health through their biofertilizer and biocontrol properties [15].

Endophytic fungi as symbiosis are well recognized, and approximately 1.5 million different fungal species have been identified in various host plants, with one million of them being endophytic in nature [16]. Both endophytic fungi and the host plants profit from their mutualistic relationship. Fungal endophytes have been documented to produce a variety of bioactive compounds, which improve plant tolerance to various biotic stresses [7].

Moreover, endophytes exhibit a disease-suppressive capacity, which might be induced by the production of various secondary metabolites, including salicylic acid, siderophores, enzymes, proteins, elicitors and/or antibiotics [17]. Recently, these diverse metabolites can be utilized for various biotechnological applications [18–20]. Beneficial endophytic bioinoculants must possess plant growth-promoting properties such as the production of ammonia, phytohormones, hydrogen cyanide, and siderophore, as well as phosphate solubilization [21]. Utilizing endophytes as biofertilizers and/or biocontrol agents in the organic farming of diverse crops has become interesting due to their richness in naturally bioactive products [22].

The efficacy of microbial endophytes in promoting plant-growth and their comparison with exogenously applied hormones have not been investigated until now. Therefore, the aim of the current investigation was the isolation, identification, and characterization of bacterial and fungal endophytes from the roots of the *Phaseolus vulgaris* plant, and the exploration of their potentiality compared to two common exogenously applied hormones on the growth and biochemical properties of *P. vulgaris* plants to explore the possibility of applying these microbial isolates as biofertilizers for the improvement of the growth performance and metabolites of crops.

2. Materials and Methods

2.1. Plant Sampling and Microbial Endophytes Isolation

Unearthed roots of the common bean (*Phaseolus vulgaris* L.) plant were collected from El-Menofia governorate (30°38′33.5″ N 30°56′54.4″ E), Egypt. Plant roots were preserved in plastic pages at 4°C for 48 h before being processed. The roots are thoroughly rinsed with tap water and then sterilized distilled water. Roots were surface sterilized by immersing them in sterile distilled water for 1 min, 70% ethanol for 1 min, 2.5 percent chlorine bleach for 4 min, 70% ethanol for 30 s, and then washing them three times in sterile distilled water. To ensure that the sterilization process is working properly; the water of last washing was streaked onto nutrient (readymade, Oxoid) and Czapek Dox (CD) (readymade, Oxoid) agar media, the absence of microbial growth in the cultured media verified the sterilization process' success [23].

The sterilized roots were cut into small segments (5 mm/segment) aseptically, and fifteen root segments were put on three nutrient agar plates (5 segments/plate). The plates were daily observed to detect bacterial growth. The bacterial isolates that appeared around the root segment are picked up and streaked on another nutrient plate for complete purification. The same previous procedure was repeated by putting root segment on Czapek Dox (CD) agar media for fungal isolation. The fungal colony that appears around the root segment are picked up and undergo purification steps on CD agar media [24]. The purified fungal and bacterial isolates were stored on slants at 4 °C for future studies.

2.2. Molecular Identification of Microbial Endophytes

The endophytic bacterial and fungal isolates obtained from the surface-sterilized roots were subjected to molecular identification based on amplification and sequencing of bacterial 16S rRNA and fungal ribosomal DNA internal transcribed spacer (ITS) region.

For bacterial identification, the genomic DNA was extracted according to the modified method [25], and the PCR protocol was achieved according to Fouda et al. [26]. Briefly, individual bacterial colonies were picked up using a sterile toothpick and resuspended in 50 μL of sterile deionized H_2O. The cell suspension was heated for 10 min in a water bath at 97 °C, then the cell lysate was forced (15,000× *g*, 10 min) and the upper layer containing the DNA was retrieved. The content of extracted DNA was calculated by detecting its absorbance at 260 nm using a UV spectrophotometer. A fragment of 16S rDNA was PCR-amplified using the bacterial universal primers 27f (50-AGAGTTTGATCCTGGCTCAG-30) and 1492r (50-GGTTACCTTGTTACGACTT-30). The PCR tube contained 1 × PCR buffer, 0.5 mM $MgCl_2$, 2.5 U Taq DNA polymerase (QIAGEN Inc., Germantown, MD, USA), 0.25 mM dNTP (Deoxynucleoside triphosphate), 0.5 μM primers, and approximately 5 ng of bacterial genomic DNA.

In the case of fungi, genomic DNA was extracted using the Gene Jet Plant genomic DNA purification Kit (Thermo procedure). Using genomic DNA as a template and primers for ITS1 (5′-TCCGTAGGTGAACCTGCGG-3′) and ITS4 (5′-TCCTCCGCTTATTGATATGC-3′), the ITS region was amplified in PCR reaction. Maxima Hot Start PCR Master Mix (Thermo), 0.5 μM of each primer, and 1 μL of extracted fungal gDNA were included in the PCR mixture (50 μL) [27].

The PCR was carried out in a DNA Engine Thermal Cycler by Sigma Scientific Services Company (Cairo, Egypt) with a hot starting performed at 94 °C for 3 min, followed by 30 cycles of 94 °C for 30 s, 55 °C for 30 s, and 72 °C for 60 s, followed by a final extension performed at 72 °C for 10 min. GATC Company (Germany) used an ABI 3730x1 DNA sequencer to perform the sequencing. Using the NCBI BLAST software, the obtained sequences were compared to the Gene Bank database. BLASTN was used to compare sequences to 16S rRNA and ITS sequences in the Gene Bank database, and bootstrap analysis was used to build phylogenetic trees. The sequences obtained in this analysis were deposited in Gene Bank under accession numbers MW485939 to MW485944 for bacteria and MW485935 to MW485938 for fungi.

2.3. Screening of Plant Growth Promotion Potential of Microbial Endophytes

2.3.1. Colorimetric Determination of Indole-3-Acetic Acid (IAA)

Bacterial endophytes were inoculated in 20 mL of LB broth (contain g L^{-1}: tryptone, 10; yeast extract, 5; NaCl, 10; 1.0 L dis. H_2O, adjusted to pH 7) amended with 5 g L^{-1} of L-tryptophan (chosen as the best concentration from our previous study) and incubated in orbital shaker (180 rpm) at 30 ± 2 °C for 7 days. Also, fungal discs (10 mm in diameter) covered with hyphae were cultured in 20 mL CD broth containing 5 g L^{-1} of L-tryptophan and incubated in the same manner at 28 °C for 10 days. At the end of incubation period, five mL of each endophytic culture were collected and centrifuged at 6000 rpm at 4 °C for 30 min. Two mL of Salkowski's reagent was added to 1.0 mL of each culture supernatant and measured at 530 nm (Jenway 6305 UV spectrophotometer). IAA was quantified with pure IAA standard [28].

2.3.2. Ammonia Production

The potency of isolated microbial endophytes to produce ammonia in peptone broth (containing g L^{-1}: peptone, 10; NaCl, 5; 1.0 L dis. H_2O) was evaluated and detected by Nessler's reagent. The degree of color change indicates the potency of the endophytic isolates to produce ammonia.

2.3.3. Phosphate Solubilization

To determine the phosphate solubilizing strains, bacterial and fungal endophytes were cultured on Pikovskaya's agar medium (containing g L^{-1}: dextrose, 10; yeast extract, 0.5; $MgSO_4{\cdot}7H_2O$, 0.1; $Ca_3(PO_4)_2$, 5; $(NH_4)_2SO_4$, 0.5; $FeSO_4{\cdot}7H_2O$, 0.0001; $MnSO_4{\cdot}H_2O$, 0.0001; KCl, 0.2; and agar 15 g/1000 mL dis. H_2O). The development of clear zones around bacterial colonies and fungal discs considered as positive [29].

2.3.4. Extracellular Enzymatic Activities

Extracellular enzymes produced by bacterial endophytes were assessed based on minimal salt (MS) media (Containing g L^{-1}: KH_2PO_4, 1.5; $NaNO_3$, 6; KCl, 5; $MgSO_4{\cdot}7H_2O$, 0.5; $FeSO_4$, 0.01; $ZnSO_4$, 0.01; and agar 15 g/1000 mL dis. H_2O) containing the specific dissolved indicative substrate. The inhibition zones induced around microbial endophytes growth after adding the specific reagent was measured as indicative for the enzymatic activity. Estimation of the endophytic fungal enzymes was achieved under the same conditions with addition of yeast extract (5 g L^{-1}) to the MS media [30]. Endophytic enzyme activities of pectinase, amylase, xylanase, cellulase and protease were estimated by culturing the bacterial and fungal isolates on MS media containing 1% of pectin, soluble starch, xylan, carboxy-methylcellulose (CMC) and gelatin, respectively. The specific reagents hexadecyl-trimethyl ammonium bromide, iodine, absolute ethyl alcohol, and acidic mercuric chloride were used to visualize pectinolytic, amylolytic, and cellulolytic, xylanolytic, and proteolytic enzymatic activities, respectively [31].

2.3.5. In Vitro Evaluation of Antagonistic Activity of Endophytic Isolates against Selected Phytopathogenic Fungi

All the endophytic isolates were screened for their antagonistic properties against the fungal phytopathogens based on dual-culture assay method [32]. The widely prevalent fungal pathogens *Fusarium oxysporum*, *Verticillium dahlia*, *Alternaria alternata* and *Pythium ultimum* were obtained from the Plant Pathology Department, Faculty of Agriculture, Zagazig University, Egypt. Each bacterial isolate was spotted at 3 equidistant points along the perimeter of the potato dextrose agar (PDA, readymade, Oxoid) plates and incubated in dark for 48 h at 28 °C. The fungal pathogens were previously grown on PDA at 28 °C for 7 days. Every fungal pathogen had one plug (6 mm) cut from the leading edge and inserted in the center of the incubated bacterial plates; plates without bacterial inoculation served as controls. The dual culture plates were incubated for 5 days at 28 °C, after incubation. The percentage of mycelial growth inhibition was calculated by the formula:

$$\text{Inhibition percentages (I\%)} = \frac{\text{R1} - \text{R2}}{\text{R1}} \times 100 \quad (1)$$

where R1 is the radial growth of the phytopathogen without bacterial endophyte (control), while in endophytic inoculated plates, R2 represents pathogen radial development (dual culture plates).

Likewise, the four days-old fungal plugs (6 mm) of the endophytic isolates were inoculated in the dual culture PDA plates as the bacterial isolates.

2.4. *Field Experiment Using Phaseolus vulgaris L. Plant*

2.4.1. Experimental Design

A field experiment was conducted under field conditions during the winter season of 2019 at the garden of the Faculty of Science, Al-Azhar University, Cairo, Egypt, to evaluate the impact of foliar spray of two exogenously plant hormone [i.e. indole-3-acetic acid (IAA) and benzyl adenine (BA)] compared to the metabolites of most potent endophytic bacterial strains (PB2 and PB5, separately or in a consortium (BM)) and fungal strains (PF2 and PF3 separately or in a consortium (FM)) on growth, biochemical and yield parameters of bean (*Phaseolus vulgaris* L.) plant.

The *Phaseolus vulgaris* seeds were surface disinfested by immersing them in ethyl alcohol (70%) for one minute, then in 4% sodium hypochlorite for 15 min, followed by sterilized distilled water washing.

The soil of the experimental site was classified as sandy soil containing sand, silt, and clay of 95.16%, 3.45%, 1.39%, respectively, in the 0–30 cm soil layer. The chemical compositions of soil used were Na, Ca, K, P, and Cl with values of 185.25, 25.0, 16.20, 24.30, and 132.50 mg kg^{-1}, respectively. The plant was grown under field conditions and irrigated when needed. The field trial was set up in a completely randomized design (CRD). A randomized complete block design with 5 blocks was set up. Within each block were 9 plots: treated with C (control: distilled water), H1 (IAA, 100 ppm), H2 (BA, 100 ppm), endophytic metabolites (100 ppm) of PB2, PB5, BM, PF2, PF3, and FM. Each planted plot contained one row with 5 plants per treatment and 30 cm between each plant, in total 225 were grown with 45 plants for each block. For analysis, five plant samples were randomly collected per treatment.

2.4.2. Culture Conditions and Extraction of Endophytic Secondary Metabolites

Endophytic bacterial isolates (PB2, PB5, and BM) were grown in LB medium for 24 h, at the end of incubation period, adjusted the bacterial concentrations at 10^8 cell/mL. About 200 µL of the previous bacterial concentration were inoculated into 500 mL of LB broth amended with 5 g L^{-1} tryptophan. In a shaking incubator, the cultures were incubated for 6 days at 28 °C. The bacterial broth media was centrifuged for 30 min at 6000 rpm, and the supernatant was harvested after incubation period. On the other hand, the fungal endophytes (PF2, PF3, and FM) were grown in CD agar medium at 28 °C, two fungal discs (10 mm) of a 5-day old cultures of each fungus was inoculated in 500 mL CD broth and incubated at 28 °C in shaking incubator. After incubation for ten days, fungal cultures were filtered through gauze cloth, forced for 30 min at 6000 rpm, and the supernatant was harvested.

The endophytic metabolites were extracted using the method of Kim et al. [33], with minor modifications: the obtained supernatant was mixed with an equivalent volume of ethyl acetate (1:1 *v/v*) and kept for 10 h at 4 °C, the solvent layer (containing metabolites) was separated by a separation funnel and evaporated in a rotary evaporator (40 °C/90 rpm) to get the crude metabolites. Ten mg of the concentrated residue was re-dissolved in 100 mL distilled water to get a final concentration of 100 ppm.

2.4.3. Foliar Spraying of Phaseolus Vulgaris Plants with Endophytic Metabolites and Exogenously Applied Hormone

Suspensions of the exogenously hormones (IAA and BA), the endophytic metabolites (PB2, PB5, PF2, and PF3), and their consortium (BM and FM) were prepared and adjusted to a concentration of 100 ppm. During foliar spray, the soil was coated with aluminum foil to ensure that the solution only reached the tops of the plants, not the ground. Spraying was done up to down, a spray atomizer was used for foliar application, the final volume of liquid to spray the leaves was 2 mL per plant. The first spraying was done when the seedlings were 15 days old, while the second and third foliar spraying was performed when the seedlings were 30 and 50 days old, respectively. The harvest was achieved for three times described as follows: after 35 days of planting (first harvest) as a vegetative stage; after 55 days (second harvest) as a flowering stage; and the third harvest was achieved after 90 days of planting (yield stage).

2.4.4. Growth and Vegetative Parameters Measurement

The shoots of 35 and 55-day-old bean plants (n = 5) were separated from the plants and were putted in an oven at 70 °C until a constant weight, then, the heights and dry weights of the shoots were measured. Fresh apical of shoot subsamples were stored in refrigerator for assessment the plant enzymatic activities.

2.4.5. Physiological Parameters Measurements

Determination of Photosynthetic Pigments

The quantitative determination of chlorophylls was done using the method of Vernon and Seely [34]. One gram aliquot of fresh leaves was cut into small pieces. The pigments were extracted by grinding the cut tissue with suitable amount of glass powder in mortar using 100 mL of 80% aqueous acetone (*v/v*). The homogenate was transferred quantitatively to a Buchner filter with Whatman No. 1 filter paper. The filtrate was transferred quantitatively to 100 mL volumetric flask and made up to a total volume of 100 mL using 80% acetone.

The optical density of the plant extract was measured using spectrophotometer of two wavelengths (649 and 665 nm). These are positions in the spectrum where maximum absorption by chlorophyll (a) and (b) occurs. The concentrations of chlorophyll (a), (b) and total chlorophyll in leaf plant tissue were calculated using the following equations:

$$\text{Mg. chlorophyll (a)/g tissue} = 11.63\ (\text{A } 665) - 2.39\ (\text{A } 649) \tag{2}$$

$$\text{Mg. chlorophyll (b)/g tissue} = 20.11\ (\text{A } 649) - 5.18\ (\text{A } 665) \tag{3}$$

$$\text{Mg. chlorophyll (a + b)/g tissue} = 6.45\ (\text{A } 665) + 17.72\ (\text{A } 649) \tag{4}$$

For carotenoids, the concentration was carried out according to equation [35]:

$$\text{Car.} = 1000 \times (\text{A } 470) - 1.82\ \text{Chl (a)} - 85.02\ \text{Chl (b)}/198 = \text{mg/g fresh wt.} \tag{5}$$

where (A) denotes the optical density, Chl (a) denotes the chlorophyll a; Chl (b) denotes the chlorophyll b.

Determination of Total Soluble Carbohydrates and Proteins

The content of total soluble carbohydrates in the dry shoot were determined using the Umbriet et al. method [36]. Contents of soluble proteins in the dry shoot were estimated according to the methods of lowery et al. [37].

Determination of Amylase and Protease Enzymes Activity

Amylase activities were estimated using a method modified from that described by Afifi et al. [38]. In this method, the enzyme-substrate reaction mixture consisted of 1.0 mL of the supernatant to be assayed for amylase activity and 1.0 mL of soluble starch solution

(10 mg mL^{-1}) into 0.5 M sodium acetate buffer pH 5.3, containing 20 mM $CaCl_2$. The reaction mixture was allowed to react for 15 min at 40 °C. The reaction was stopped by adding 5.0 mL of 0.5 M acetic acid. To 1.0 mL of the incubated mixture, 3.0 mL of iodine reagent (0.01% Potassium iodide and 0.001% iodine) were added to obtain starch iodine complex, which showed blue color. The amylase activity was estimated by measuring O.D. at 660 nm. Proteases were determined using the method of Ong & Gaucher [39]. In this method, 0.25 mL of the supernatant to be assayed for proteases activities was added to 1.75 mL of 0.25% casein in 0.2 M phosphate buffer pH 7 and incubated at 37 °C for 60 min. At the end of the incubation period, 3.0 mL of 5% trichloroacetic acid was then added, and after standing at room temperature for 30 min, the solution was filtered. To 1.0 mL of the filtrate, 5.0 mL of 0.4 N NaOH and 1.0 mL diluted (1:2 *v/v*) BDH were added consecutively and after color development for 30 min. The O.D. at 660 nm was determined using UV-visible spectrophotometer.

Determination of Antioxidant Enzymes Activity

In 50 mL sodium phosphate buffer (pH 7.0) containing 1% soluble polyvinyl pyrrolidine (PVP), five grams of fresh leaf tissue including the terminus buds in addition to the first and second young leaves were macerated. The homogeneous agent has been centrifugated at a rate of 15,000 rpm for 20 min at 4 °C and used for enzyme activity testing. The activity of catalase (EC 1.11.1.6) enzyme was determined using the method defined by Chen et al. [40]. The mixture of final reaction (10 mL) containing 40 μL enzyme extract was added to 9.96 mL H_2O_2 phosphate buffer pH 7.0 (0.16 mL of 30% H_2O_2 to 100 mL of 50 mM phosphate buffer). The rate of change in H_2O_2 absorbance over 60 s was calculated using a spectrophotometer set to 250 nm and expressed in EU mg^{-1} protein. The amount of enzyme that reduced 50% of H_2O_2 in 60 s at 25 °C was described as one unit of enzyme activity. Peroxidase (POD, EC 1.11.1.7) activity was calculated using the method defined by Kar and Mishra [41]. In a final volume of 5 mL, the sample mixture included 125 μM phosphate buffer (pH 6.8), 50 μM pyrogallol, 50 μM H_2O_2, and 1.0 mL of the 20 diluted enzyme extract. The enzyme activity was expressed as EU mg^{-1} protein, and the volume of purpurogallin produced was calculated by calculating the absorbance at 420 nm.

Extraction, Purification, and Determination of Endogenous Hormones

The methods for extraction, purification, and determination of the hormones were carried out according to [42] with some modifications.

Hormone extraction: Around 1.0 g apical buds have been melted with liquid nitrogen; 10.0 mL acetonitrile extraction medium has been applied with an antioxidant of 50 mg L^{-1} diethyldithiocarbamate. The homogenous product was incubated at 4 °C for 12 h in the dark.The extracts were forced for 10 min at 10,000× *g*. The residue was removed with the same solvent twice.

Hormone purification: The supernatant was mixed and condensed to residual through rotatory evaporation under low pressure at 37 °C, then re-dissolved in 10.0 mL 0.5 mol L^{-1} phosphate buffer (pH 8.0), and then applied 8.0 mL chloroform and oscillated to eliminate pigment. The chloroform phase was eliminated. To eliminate hydroxybenzene, 0.15 g insoluble polyvinylpyrrolidone (PVP) was applied to the aqueous phase and centrifuged at 10,000× *g* for 10 min. Pipette 5.0 mL supernatant and switch pH to 3.0 with pure formic acid. The aqueous layer was removed twice with 3.0 mL ethyl acetate. By rotatory evaporation under low pressure, the ethyl acetate phase was condensed and re-dissolved in a 1 mL combined solution of acetonitrile/methanol/0.6 percent acetic acid solution (5:50:45, V:V:V). Finally, 0.22 mm hydrophobic membranes were used to filter the hormone extraction.

Hormone determination: A high-performance liquid chromatography (HPLC) method was used to analyze the hormones. For loading into a Waters symmetry C18 column (4.6 mm × 150 mm, 5 mm), samples (20 μL) were loaded into a specific 20 μL loop. Acetonitrile: methanol: 0.6% acetic acid solution (5:50:45, V:V:V) have been used as a mobile phase. A Waters series 515 pump was used to elute samples from the column at a flow rate of 0.6 mL min^{-1} at 25 °C. Hormone peaks were observed using a photodiode array detector (Waters 2998 Sparations Module, USA) with absorbance set at 218 nm for IAA, 270 nm for BA, 200 nm for GAs, and 262 nm for ABA.

Hormone quantification: By comparing the retention time and characteristic absorption wavelength of various standard hormones (Sigma Chemical, Co., St. Louis, MO, USA), the sample hormones were identified. The hormone concentrations were determined by comparing the peak areas of the samples to those of standard samples. Both of the results were adjusted to account for the known recovery rates of various hormone concentrations. BA, IAA, GAs, and ABA had recovery rates of 86.35 ± 3.17, 84.79 ± 3.48, 83.65 ± 4.15, and 88.01 ± 2.98%, respectively (calculating process of the recovery rate: two equivalent samples were taken, one of which was added with a quantitative reference material of the component to be checked, and then the two samples were purified and measured simultaneously using the same procedure). The recovery rate was calculated as the ratio of the difference between the two samples' measured quantities and the added standard material quantity.

Yield Parameters Measurements

The number and weight of pods per plant were measured after 90 days of sowing from five single plants of the row in each plot. The number and weight of seeds per pod were assessed from five single plants of the row in each plot. One hundred-seed weight (g) was determined using a sensitive digital balance. Contents of the total soluble carbohydrate and the total soluble protein were determined in seed yield that are calculated form the five single plants of the row of each plot using sensitive digital balance and corrected based on seed moisture content.

2.5. Statistical Analysis

Data were statistically analyzed using SPSS v17 (SPSS Inc., Chicago, IL, USA). One-way analysis of variance (ANOVA) was used to evaluate the In-vitro antagonistic activity, extracellular enzymes production, IAA, and ammonia production potential, P-solubilization ability of endophytic isolates and comparing the effect of spraying the metabolites of these endophytes and spraying exogenously IAA, BA on *Phaseolus vulgaris* growth performance. Tukey's range tests were used to do an a posteriori multiple comparison at $p < 0.05$. As mentioned above, all analyses are the averages of three to six independent replicates.

3. Results and Discussion

3.1. Isolation and Identification of Bacterial and Fungal Endophytes Isolated from Root of Common Bean Plant

In this study, a total of twenty-five bacterial isolates and twelve fungal isolates were isolated from the surface-sterilized roots of common bean (*Phaseolus vulgaris* L.) as culturable endophytes (as mentioned in the material and method section). Out of bacterial and fungal isolates, six bacterial species (PB1, PB2, PB3, PB4, PB5, and PB6) and four fungal species (PF1, PF2, PF3, and PF4) were selected as different strains based on morphological and cultural characteristics, as well as microscopic examination. These isolates were subjected to molecular identifications based on gene sequence analysis of 16S rRNA for bacterial isolates and internal transcribed spacers (ITS) for fungal isolates. The gene sequence analysis revealed that the isolated bacterial endophytes were related to two phyla. The Proteobacteria (class: γ-Proteobacteria) and Firmicutes (class: Bacilli) had, percentages of 33.3% and 66.7%, respectively. The γ-Proteobacteria strains were identified as *Enterobacter asburiae* PB1 and *Acinetobacter radioresistens* PB3, while the Firmicutes strains were identified as *Bacillus thuringiensis* PB2, *Brevibacillus brevis* PB4, *Brevibacillus agri* PB5, and *Bacillus subtilis* PB6 with similarity percentages ranging between 98.88% and 99.84% (Table 1, Figure 1).

Table 1. The 16S rRNA and ITS sequence identifications of the bacterial and fungal endophytes isolated from sterilized root of common bean plants.

Microbial Strain Code	GenBank Accession Number	Homolog Sequences	Sequence Identity (%)	Closest Accession Number
PB1	MW485939	*Enterobacter asburiae*	99.16	BBED01000197
PB2	MW485940	*Bacillus thuringiensis*	99.69	ACNF01000156
PB3	MW485941	*Acinetobacter radioresistens*	99.70	BAGY01000082
PB4	MW485942	*Brevibacillus brevis*	98.88	AB271756
PB5	MW485943	*Brevibacillus agri*	99.84	D78454
PB6	MW485944	*Bacillus subtilis*	98.99	ABQL01000001
PF1	MW485935	*Penicillium crustosum*	99.62	NR077153
PF2	MW485936	*Alternaria sorghi*	98.63	NR160246
PF3	MW485937	*Penicillium commune*	98.69	NR111143
PF4	MW485938	*Aspergillus flavus*	98.84	NR111041

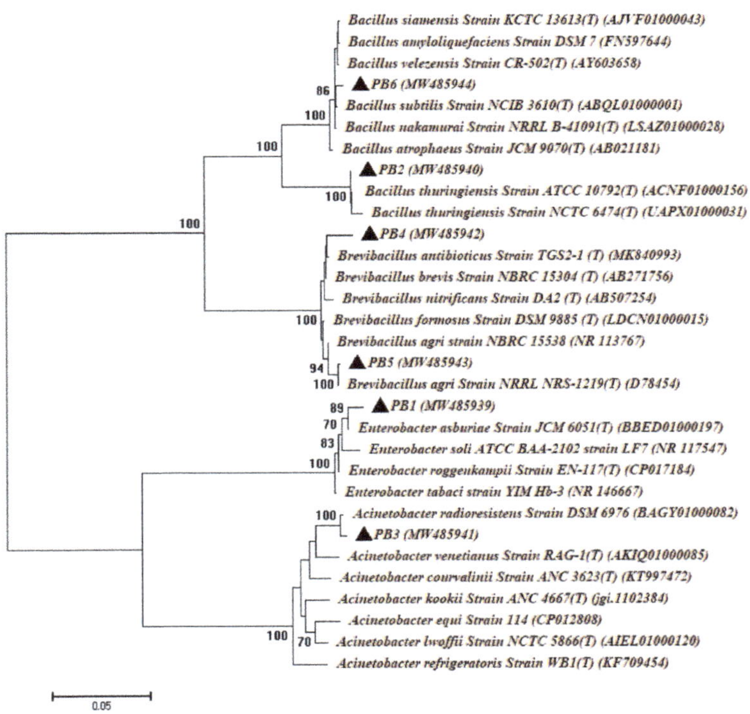

Figure 1. Phylogenetic tree of the 16S rRNA sequences of bacteria isolated from the root of the common bean plant. PB1–PB6 refers to the 16S rRNA sequences of bacterial isolates. The bacterial isolates and their identities were listed in Table 1. MEGA 6 was used to carry out the analysis, which used the neighbor-joining approach with a bootstrap value (1000 replicates).

Consistent with our data, López-López et al. [43] recorded *Bacillus* spp. and *Acinetobacter radioresistens* as bacterial endophytes isolated from the surface-sterilized root of *Phaseolus vulgaris*. Recently, Sendi et al. [44] showed that most bacterial endophytes associated with the roots of the common bean belong to *Bacillus* spp. with a percentage of 58.33%. Moreover, the bacterial endophytes isolated from sterilized common bean leaves are related to the Proteobacteria (36.7%), Firmicutes (32.9%), and Actinobacteria (29.7%) of the total bacterial isolates [45]. The *Bacillus* species are considered the most predominant endophytic strains isolated from different plant species, having been isolated from *Glycyrrhiza* spp., *Lycium chinense, Belamcanda chinensis, Pinellia ternata, Achillea fragrantissima, Taxus yunnanensis, Bletilla striata, Pulicaria incisa, Fagonia mollis, Leonurus heterophyllus, Digitalis purpurae* and *Pinellia pedatisecta* as described previously [29,46].

In the current study, four fungal endophytic strains were obtained from the sterilized root of the *Phaseolus vulgaris* plant belonging to Ascomycota, defined as *Penicillium crustosum* PF1, *Alternaria sorghi* PF2, *Penicillium commune* PF3 and *Aspergillus flavus* PF4, with identification percentages ranging between 98.63% and 99.62% (Table 1, Figure 2). To the best of our knowledge, this is the first record of *Alternaria sorghi* being isolated as an endophytic strain. In our recent study, *Penicillium* spp., *Alternaria* spp., and *Aspergillus* spp. were identified as endophytic fungi isolated from the medicinal plant *Ephedra pachyclada* [47]. These fungi have a variety of plant growth-promoting properties, including IAA production, antimicrobial activity, ammonia production, and phosphate solubilization. To date, there have been few published studies concerning fungal endophytes isolated from the root of *Phaseolus vulgaris* as plant growth-promoting.

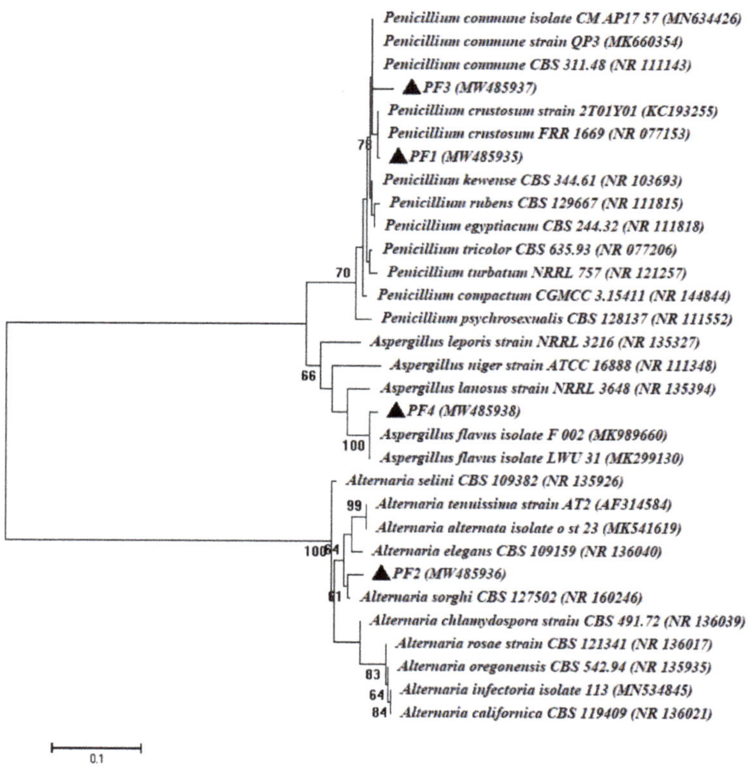

Figure 2. Phylogenetic tree of fungal strains' ITS sequences against NCBI reference sequences (National Center for Biotechnology Information). PF1–PF4 refers to the ITS sequences of fungal isolated from common bean plants. The fungal isolates and their identities were listed in Table 1. The analysis was carried out in MEGA 6 using the neighbor-joining approach.

3.2. *In Vitro* Assessment of the Plant Growth-Promoting Activities of Bacterial and Fungal Endophytes

3.2.1. Colorimetric Determination of IAA

Our results showed the ability of all the endophytic isolates to produce IAA to varying degrees in media containing tryptophan (5 g L^{-1}). The fungal endophytes exhibited higher bio-fabrication of IAA than those made by endophytic bacteria. The fungal endophytes of *Alternaria sorghi* PF2 and *Penicillium commune* PF3 displayed the maximum IAA production, recording 235.33 ± 7.51 and 183.33 ± 7.53 μg mL^{-1}, respectively, while elevated bacterial IAA synthesis was carried out by isolating *Bacillus thuringiensis* PB2 and *Brevibacillus agri* PB5, with values of 181.66 ± 7.62 and 164.66 ± 6.38 μg mL^{-1}, respectively ($p \leq 0.001$; Figure 3). In the same regard, ALKahtani et al. [29] reported the potency of the endophytic *Brevibacillus* sp. for production of IAA (59.7 μg mL^{-1}) in media containing 5 mg mL^{-1} of tryptophan. Moreover, HPLC analysis confirmed IAA production by the endophytic fungi *Colletotrichum fructicola* associated with *Arabica coffee*, recording 1205.58 μg mL^{-1} after 26 days of incubation [48]. Previous studies have indicated the ability of microbial endophytes to produce IAA to varying degrees, whereas in the current study, isolated endophytes produced IAA in the range of 81.6 ± 2.6–235.3 ± 7.5 μg mL^{-1}. IAA has a vital role in plant microbial interactions; it has been suggested as a communication method between the endophytic microbiome and the roots of the host plant [49]. Therefore, our isolates may affect plant growth and are candidates for use as surrogates to improve crop quality and productivity.

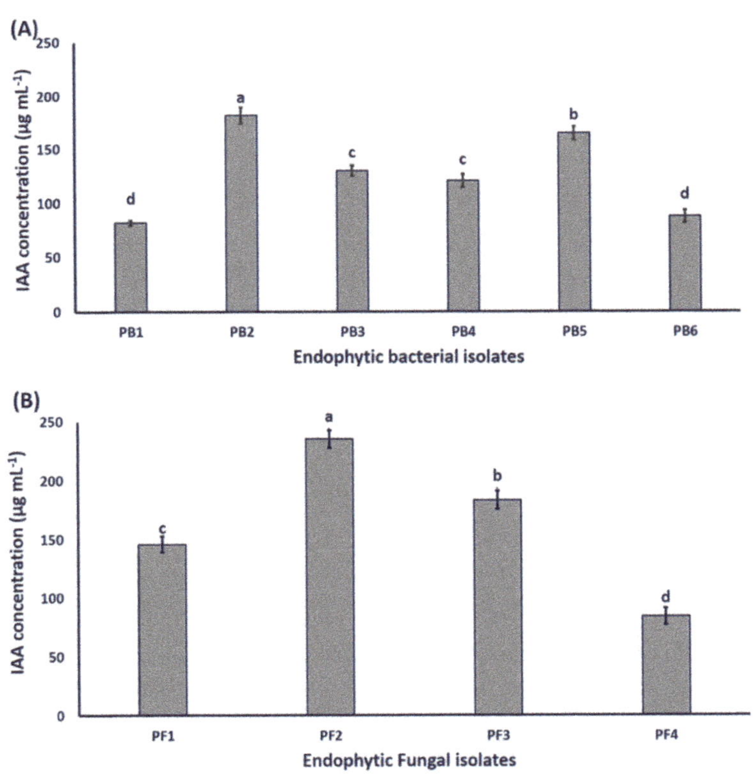

Figure 3. IAA production by (**A**) endophytic bacterial isolates and (**B**) endophytic fungal isolates; with 5 g L^{-1} tryptophan. Bars with the same letter for each endophytic isolate did not differ significantly at a significant level of ($p \leq 0.05$); Error bars indicate means ± SE by LSD test (n = 6). PB1 is *E. asburiae;* PB2 is *B. thuringiensis;* PB3 is *A. radioresistens;* PB4 is *B. brevis;* PB5 is *B. agri;* PB6 is *B. subtilis;* PF1 is *P. crustosum;* PF2 is *A. sorghi;* PF3 is *P. commune;* PF4 is *A. flavus.*

3.2.2. Ammonia Production and Phosphate Solubilization

Adding nitrogen fertilizers to the soil is costly, so ammonia-producing endophytes that are likely to supply plants with nitrogen constitute a desirable feature of soil fertility and improved plant growth. Additionally, ammonia production helps plants resist colonization by pathogens [50]. Herein, a varying degree of ammonia production capacity was observed in 50% of the bacterial isolates *Bacillus thuringiensis* PB2, *Brevibacillus brevis* PB4, and *Brevibacillus agri* PB5, and in 75% of the fungal isolates *Penicillium crustosum* PF1, *Alternaria sorghi* PF2, and *Penicillium commune* PF3 (Supplementary Table S1). The high proportion of isolates producing ammonia in this study is consistent with the study conducted by Szilagyi-Zecchini et al. [51]. Most of the endophytic bacteria of the genera *Pseudomonas*, *Enterobacter*, and *Bacillus* isolated from chickpea plants have been identified as ammonia producers [52]. Recently, Ripa et al. [53] isolated and identified the endophytic fungal inhabitance of wheat plants and declared that 34% of the isolated fungi displayed ammonia production activities; these isolates belong to the genera *Aspergillus*, *Fusarium*, *Alternaria* and *Trichoderma*. Endophytic bioinoculants must possess plant growth-enhancing properties such as hydrogen cyanide, phytohormones, siderophores, ammonia production, and phosphate solubilization [21].

With the exception of nitrogen, phosphate is the most important macronutrient needed by plant, and despite the abundance of phosphates in agricultural soil, plants do not have access to them, because they are incorporated into organic compounds or in the form of mineral salts of scarce solubility [54]. Therefore, we explored the ability of our isolates to dissolve phosphate and the possibility of using them as alternatives for sustainable agriculture. The phosphate-solubilizing activity of isolated endophytes was indicated by the clearance zones developed on Pikovskaya's media containing tricalcium phosphate. Three bacterial isolates and three fungal isolates were potent phosphate solubilizers. However, the fungal endophytic isolates manifested higher phosphate-solubilizing capacity than the bacterial endophytes, with the highest phosphate-solubilizing activity listed for fungal isolates *Alternaria sorghi* PF2 and *Penicillium commune* PF3 (Supplementary Table S1).

In the same regard, 11 fungal isolates associated with wheat plants (*Triticum aestivum* L.), showed phosphate-solubilizing potential to varying degrees and were defined as phosphate solubilizers based on the halo zones that were developed around the fungal growth on Pikovskaya's medium [53]. Interestingly, one of the most essential properties of endophytes is their ability to dissolve insoluble phosphate in the soil. Haidar et al. [55] isolated 27 endophytes of bacteria from the endosphere of *Corchorus olitorius* and specified the isolates, *Bacillus* sp., *Bacillus firmus*, *Staphylococcus pasteuri* and *Kocuria* sp. as robust phosphate solubilizers, and they estimated the solubilized phosphate in broth media in the range of 113.52 - 218.47 µg P/mL. Microorganisms dissolve inorganic phosphate by acidification due to the production of inorganic and organic acids and the excretion of H^+, in addition to siderophore and exopolysaccharide production, while organic phosphate solubilization is mediated by phytase, C–P lyase, and phosphatase enzymes [56].

3.2.3. Extracellular Enzymatic Activities

Practically, fungal and bacterial endophytes produce various key enzymatic groups such as tyrosinase, xylanases, amylases, pectinases, hemicellulases, gelatinase, phytases, cellulases, proteases, chitinase and asparaginase. The majority of these enzymes are produced by the endophytes residing in crops or medicinal plants, and the enzymatic activities of these endophytes have been detected based on agar-based procedures [57].

The bacterial and fungal endophytic isolates showed different activities in the production of one or more of the tested enzymes. In general, the fungal isolates produced more enzymes than the bacterial isolates; this was evidenced by recording larger clear degrading sectors than those recorded for the bacterial isolates on the indicative substrate plates. Bacterial isolates *Bacillus thuringiensis* PB2 and *Brevibacillus agri* PB5 produced the highest amounts of amylase, cellulase, and protease, while the isolate *Acinetobacter radioresistens* PB3 was the largest producer of bacterial pectinase and xylanase ($p \leq 0.001$). The maximum fungal enzymatic production was assigned to the isolate *Bacillus thuringiensis* FB2 with degrading clear zones of 37 ± 1.7, 42 ± 1, 38.3 ± 1.2, 42.3 ± 1.3, and 31.3 ± 0.8 mm for amylase, cellulase, protease, pectinase, and xylanase respectively (Supplementary Table S1).

Compatible with our results, ALKahtani. et al. [29] isolated 13 culturable bacterial endophytes from plants that have therapeutic properties, namely, *Fagonia mollis* Delile and *Achillea fragrantissima* (Forssk) Sch. Bip. These isolates belong to genera *Bacillus, Paenibacillus* and *Brevibacillus* and showed varying activity of pectinase, xylanase, CMCase, gelatinase, cellulase and amylase. Furthermore, 52 endophytic fungal isolates were obtained from *Thai orchids;* at least 25% of these isolates showed enzymatic activity by the production of one or more of the exo-enzymes: Lipase, amylase, protease, cellulase and pectinase [58]. Endophytic enzymes play a major role in the hydrolysis processes that aid their colonization of the selected host plant and then gain their food from the plant tissues and their interactions with plant pathogens [59]. On the contrary, during long periods of co-evaluation, plants may try to limit the growth of endophytes by producing many toxic substances, but endophytes try to gradually form some tolerance mechanisms by producing mycotoxins and exoenzymes. In addition, these enzymes are valuable in many industrial, agricultural, and medical applications [60].

3.2.4. In Vitro Antagonistic Activities

Previous studies have given increasing attention to the endophytic microbes as a distinct choice for the biological control of some plant diseases for being environmentally safe. Likewise, our results showed the potency of all the isolated endophytes to reduce the radial growth of the phytopathogenic fungi cultured on dual culture plates. The bacterial and fungal isolates inhibited more than 20% of the fungal growth of *Fusarium oxysporum, Alternaria alternata, Verticillium dahlia,* while *Pythium ultimum* growth was reduced by more than 9%. The maximum growth inhibition of *A. alternata* and *V. dahlia* was assigned to the bacterial isolate *Brevibacillus agri* PB5, recording antagonistic indices of 51.6 ± 0.5% and 62.6 ± 2%, respectively. The fungal isolate *Penicillium commune* PF3 showed the highest antagonistic activity against *F. oxysporum* with an inhibition percentage of 61.3 ± 2.3%. In addition, the extreme suppression of *P. ultimum* (55.3 ± 2.6%) was attributed to the fungal isolate *Alternaria sorghi* PF2 (Table 2).

Table 2. The antagonistic activity of isolated endophytes against selected phytopathogenic fungi.

Endophytic Isolates	**Percentage of Growth Inhibition (%)**			
	F. oxysporum	***A. alternata***	***V. dahlia***	***P. ultimum***
Control	0^{i}	0^{h}	0^{h}	0^{h}
PB1	31.6 ± 1.4^{g}	45.6 ± 0.8^{c}	28.6 ± 0.3^{f}	12.3 ± 0.8^{f}
PB2	54 ± 0.57^{c}	49 ± 1.1^{b}	57.6 ± 2^{b}	24.3 ± 1.4^{d}
PB3	24.6 ± 0.33^{h}	34.6 ± 0.8^{f}	22.6 ± 1.4^{g}	13.3 ± 1.4^{ef}
PB4	41.6 ± 1.4^{e}	44.6 ± 0.8^{c}	33.6 ± 0.8^{e}	15 ± 1.1^{e}
PB5	58.6 ± 0.6^{ab}	51.6 ± 0.5^{a}	62.6 ± 2^{a}	22.6 ± 2.6^{d}
PB6	26.6 ± 2^{h}	21 ± 0.57^{g}	36.3 ± 0.3^{d}	9.6 ± 0.3^{g}
FP1	46.6 ± 1.45^{d}	37.3 ± 1.2^{e}	21.6 ± 1.3^{g}	41 ± 1.7^{c}
FB2	57.3 ± 0.6^{b}	42.3 ± 0.8^{d}	38.6 ± 0.8^{c}	55.3 ± 2.6^{a}
FB3	61.3 ± 2.3^{a}	45.6 ± 1.4^{c}	31.6 ± 1.4^{e}	48.3 ± 3.1^{b}
FB4	35.6 ± 1.4^{f}	33.3 ± 0.8^{f}	26.6 ± 0.8^{f}	24.6 ± 0.8^{d}

Values in the same column with different letters are significantly different ($p \leq 0.05$), values refer to the means ± SE (n = 3).

In agreement with our findings, Egamberdieva et al. [61] isolated endophytic bacteria from plants with therapeutic properties, that is, *Hypericum perforatum* and *Ziziphora capitata*, the isolates of which belonged to the genera *Stenotrophomonas, Pantoea, Achromobacter, Serratia, Enterobacter, Bacillus, Erwinia, Arthrobacter,* and *Pseudomonas* and showed diverse antagonistic activity against a group of pathogenic soil-borne fungi, including the genera *Fusarium, Botrytis, Pythium, Gaeumannomyces* and *Alternaria*.

Bacterial endophytes should have various properties that manage the control of phytopathogens such as siderophores, antibiotics and antifungal production, the secretion of extracellular enzymes that degrade fungal cell wall and fusaric acid, and the synthesis of volatile organic components, in addition to the induction of host plant-induced systemic resistance [62].

In addition, various investigations have reported the contrasted antagonistic activity of different fungal endophytes against the phytopathogenic fungi; *Fusarium solani, Ganoderma boninense, Sclerotinia sclerotiorum* and *Colletotrichum acutatum* [63,64].

3.3. *Evaluation of the Promotion Effect of the Most Potent Bacterial and Fungal Endophytic Strains on the Growth and Metabolic Activities of the Phaseolus vulgaris Plant under Field Conditions*

A field study was performed in the Botany and Microbiology Department, Faculty of Science, Al-Azhar University, Cairo, Egypt, during the winter season in 2019 to investigate the effect of two exogenously plant growth-promoting hormones (IAA and BA) compared to the metabolites of the most potent bacterial (*Bacillus thuringiensis* PB2 and *Brevibacillus agri* PB5) and fungal (*Alternaria sorghi* PF2, *Penicillium commune* PF3) endophytic strains isolated from the roots of common bean plants, or their combination (BM and FM) as a foliar spraying on the growth and metabolic activities of *Phaseolus vulgaris* L. plants at three harvested stages. The first harvest (35 days of planting) as a vegetative stage; the second harvest (55 days) as a flowering stage; the third harvest as a yield stage (90 days of planting). The vegetative growth parameters, as well as all physiological parameters, were analyzed from the first and second harvest, whilst the yield parameters were analyzed only from the third harvest.

The foliar spraying approach is preferring over other treatment method such as soaking because of symmetrical spread of different treatments over the whole crop, in addition, the spraying method increase the response of plant to treatments. Recently, the spraying method was more effective than the soaking method during the study of CuO-NPs on the morphological and physiological growth traits of *Triticum aestivum* L. [65]. Moreover, Sarkar et al. [66] showed that the foliar spraying of micronutrients was more effective due to it is overcomes the losses of nutrient in soil amendment other treatment methods.

3.3.1. Effect of the Foliar Spraying of Exogenously Applied Hormone and Microbial Endophyte Metabolites on the Vegetative Growth Traits of Common Bean Plants

Foliar spraying of common bean plants with 100 ppm of IAA or BA led to a non-significant increase in shoot and root lengths compared to control plants. However, plants treated with IAA showed a non-significant decrease in root length at the first stage, while the shoots of plants treated with BA significantly increased in the first stage of growth compared to non-treated plants. On the contrary, plants treated with the metabolites of the endophytic isolates displayed a significant increase in shoot length during the first stage of growth, and the root length significantly increased in the second stage of growth compared to control plants. Neither exogenously hormones nor bacterial metabolites had a significant effect on the leaf number of plants with respect to control plants (Table 3). In the same context, Turbat and colleagues performed a plant bioassay to investigate the effect of fungal endophytic extracts on some of the growth parameters of *Arabidopsis thaliana* and recorded a boost in the length of the primary root, and the significant modifications in the photosynthetic pigments were due to endophytic fungal extract treatments. However, they reported a decrease in the plant biomass [67]. Moreover, the fermentation filtrate from the endophytic bacterial complex (*Pestalotiopsis* aff. *neglecta/Luteibacter* sp.) significantly increased the length of tomato seedlings with considerably longer roots compared to controls in a seedling assay conducted in vitro in petri dishes [68].

Table 3. Effect of foliar spraying of exogenously applied hormone and microbial endophyte metabolites on root, shoot length and leave numbers of common bean plants.

Treatments		Shoot Length (cm)		Root Length (cm)		Leaf Numbers	
		1st Stage	2nd Stage	1st Stage	2nd Stage	1st Stage	2nd Stage
Control		22.4 ± 0.5 b	27.4 ± 0.3 ab	9.3 ± 0.6 ab	8.98 ± 0.3 b	5.0 ± 0.0 a	5.67 ± 0.2 a
Exogenously Applied Hormones	IAA	23.3 ± 0.5 b	29.3 ± 0.9 ab	8.0 ± 0.4 b	9.8 ± 0.7 ab	4.6 ± 0.3 a	6.0 ± 0.0 a
	BA	25.3 ± 0.4 a	28.2 ± 0.7 ab	9.7 ± 0.4 ab	8.9 ± 0.4 b	4.4 ± 0.2 a	6.0 ± 0.0 a
Bacterial Endophytes	PB2	26.9 ± 0.8 a	34.0 ± 0.9 a	10.03 ± 0.3 ab	11.5 ± 0.6 a	4.6 ± 0.2 a	6.0 ± 0.0 a
	PB5	28.7 ± 0.9 a	33.8 ± 1.6 a	9.4 ± 0.4 ab	10.5 ± 0.9 ab	4.6 ± 0.3 a	6.0 ± 0.0 a
	BM	25.1 ± 0.8 a	25.5 ± 0.7 b	9.5 ± 0.2 ab	11.5 ± 0.4 a	4.1 ± 0.1 a	6.0 ± 0.0 a
Fungal Endophytes	PF2	26.1 ± 0.4 a	29.2 ± 0.7 ab	8.4 ± 0.3 b	12.7 ± 0.7 a	4.0 ± 0.0 a	6.0 ± 0.0 a
	PF3	26.1 ± 0.7 a	25.5 ± 0.3 b	10.2 ± 0.5 ab	12.5 ± 0.5 a	4.1 ± 0.1 a	6.0 ± 0.0 a
	FM	25.3 ± 0.7 a	27.2 ± 0.8 ab	10.4 ± 0.3 ab	11.0 ± 0.5 a	4.0 ± 0.0 a	6.0 ± 0.0 a

Different letters between columns denote that mean values are significantly different ($p \leq 0.05$) by Tukey's test.

Furthermore, foliar spraying with exogenously applied hormones or endophytic extracts resulted in a comparable increase in the biomass (fresh and dry root and shoot weight) of treated plants at all stages of development, and this increase was significant in most treatments. However, the root fresh weight of the *Penicillium commune* (PF3) and fungal consortium (FM) treatments recorded insignificant weight loss, while the root dry weight of *Alternaria sorghi* (PF2) plants significantly decreased in the first stage of growth compared to controls (Supplementary Table S2).

Analysis of the pure culture filtrates of the fungal endophytes *Phoma glomerata* LWL2 and *Penicillium* sp. LWL3 proved the existence of various concentrations of gibberellins (GAs) and IAA. Moreover, the application of these culture filtrates significantly improved the shoot length, chlorophyll content, and fresh and dry weight of shoots of both the dwarf mutant Waito-C (GAs-deficient) and the common rice cultivar (Dongjin-beyo) [69]. The endophytic fungi associated with mangroves were also evaluated for their ability to promote *Oryza sativa* L. growth. All of the isolated endophytic fungi were found to promote the growth of *O. sativa* L. "Cempo Ireng" [70]. Concerning the effect of exogenously applied hormones on plant growth, spraying exogenous auxins and cytokinin such as IAA and 6-benzylaminopurine resulted in an asymmetrical response of the treated alfalfa plants during their vegetative growth, while their application as a mixture was most effective and increased the biomass of plant shoots; however, the clearest effect was for the rate of total chlorophyll to carotenoid content [71]. Some endophytic fungi isolated from maize and rice plants, such as *Fusarium*, *Sarocladium*, *Aspergillus* and *Penicillium*, have been confirmed to be determining factors in plant growth improvement [72]. Fouda et al. [31] found that *Penicillium chrysogenum* and *Alternaria alternata* endophytic fungi, isolated from the *Asclepias sinaica* plant, experience enhanced root growth and root elongation attributed to the production of ammonia and IAA. Abdallah et al. [73] investigated the ability of endophytic bacteria isolated from *Withania somnifera* fruits to promote plant growth. According to the findings, the most active isolate, *Alcaligenes faecalis*, was found to produce indole-3-acetic acid and improve phosphate solubilization.

Generally, microbial endophytes can be used as biofertilizers, which reduce the reliance on chemical fertilizers while also increasing nutrient availability, thus enhancing plant growth.

3.3.2. Effect of the Foliar Spraying of Exogenously Applied Hormones and Microbial Endophyte Metabolites on the Chlorophyll and Carotene Contents of Common Bean Plants

The photosynthetic is a key process for life, it is meaning consumption of solar energy and transformed to energy for chemical compounds synthesis, also, it is strongly affected by various environmental stressors such as drought, high or low temperature, salinity, and intensity of light [74]. Plant photosynthetic potential is correlated with chlorophyll concentration, which provides some insight into the physiological state of the plant system. Regarding chlorophyll a (Chl. a), no treatments had a significant effect except for BA treatment, which led to a significant increase in the Chl. a content during the two harvest stages. In agreement with our results, Hanaa and Safaa [75] concluded that foliar application of IAA (100 ppm) can enhance the chlorophyll content of *Triticum aestivum* L. (bread wheat). Ma and co-authors used chlorophyll and carotenoids as indicator for investigate the effect of exogenous application of abscisic acid and 1-naphthaleneacetic acid on the coloration of fruits [76].

In addition, the metabolites of *P. commune* (PF3) were the only treatment that significantly increased chlorophyll b (Chl. b) in the first harvest stage, whereas in the second harvest stage, plants treated with IAA, PB5, BM, and PF3 showed a significant increase in their Chl. b content. The total chlorophyll content was increased for plants treated with *B. thuringiensis* (PB2) and *P. commune* (PF3) in the first harvest stage, while the following treatments of PB5, BM, PF3, and FM positively affected the total chlorophyll content of plants in the second harvest stage. These findings are consistent with those of Heidari and Golpayegani [77], who discovered that rhizobacteria (*Pseudomonades* sp., *Bacillus lentus*, *Azospirillum brasilens* and their combinations) improved photosynthetic pigments in basil (*Ocimum basilicum* L.) plants, where a combined inoculation of *Pseudomonas* sp. and *Bacillus lentus* in plant stimulated chlorophyll synthesis as well as photosynthetic electron transport. Timmusk et al. [78] found that inoculating wheat seedlings with *Bacillus thuringiensis* AZP2 under drought stress resulted in increased plant biomass and five-fold higher survival under extreme drought due to lower volatile emissions and increased photosynthesis, implying that bacterial inoculation enhanced plant stress resistance. Naveed et al. [79] reported that *Burkholderia phytofirmans* strain PsJN and *Enterobacter* sp. FD17 increased plant biomass and photosynthetic rate of the Mazurka maize cultivar up to 48% and 45%, respectively, as compared with the control under normal irrigation. Kumar et al. [80] observed a decline in chlorophyll content in chickpea leaves due to drought stress; however, inoculation with PGPR mitigated the negative effects of drought on chlorophyll content.

Based on our results, none of the treatments appeared to have effect on the plant carotene contents in the first harvest stage. However, common bean plants treated by BA, PB2, PF2, PF3, and FM experienced slight increases in plant carotene in the second stage, while IAA, PB5, and BM negatively affected plant carotene content (Supplementary Table S3).

The growth stimulatory effect of IAA and BA could improve plant nutrient content, so the augmented pigmentation in our treated plants could be attributed to a potential increase in Mg and the other essential elements required for chlorophyll synthesis, in addition to down-regulation of the chlorophyll degradation process [81]. Moreover, the foliar spraying of IAA (10^{-8} M) elevated the chlorophyll level of *Brassica juncea* by 13.27%, 18.90%, and 17.87% after 30, 45, and 60 days, respectively, from sowing [82]. In addition, enhanced growth involves increased leaf area and a consequent improvement in photosynthetic properties [82], keeping in mind the role of cytokinin's in boosting the activity of the ribulose-1, 5-bisphosphate carboxylase/oxygenase (Rubisco) enzyme in the photosynthesis process, raising the photosynthetic capability [83]. Sosnowski et al. [71] observed that cytokinin increased the content of plastid pigments in alfalfa leaves. However, a combination of auxin and cytokinin increased nitrate reductase production in alfalfa roots and increased the ratio of overall chlorophyll content to carotenoids.

3.3.3. Effect of the Foliar Spraying of Exogenously Applied Hormones and Microbial Endophyte Metabolites on the Total Soluble Carbohydrate and Protein Contents of Common Bean Plants

Our results indicate that all treatments, whether of exogenously applied hormones or endophytic metabolites, led to a significant increase in the carbohydrate and protein contents of the roots and shoots in all treated plants during the two growth stages. However, the plants treated with IAA, BA, or the *Bacillus thuringiensis* PB2 and *Alternaria sorghi* PF2 endophytes led to an insignificant increase in root proteins in both growth stages. Moreover, the plants sprayed with the endophytic culture filtrates manifested maximum increases in carbohydrate and protein contents in all growth stages, while plants treated with BA showed a maximum increase in the root proteins for all growth stages (Table 4).

Table 4. Effect foliar spraying of exogenously applied hormones and microbial endophyte metabolites on carbohydrates and protein contents of common bean plants.

Treatments		**Root Carbohydrate (mg g^{-1} DW)**		**Root Protein (mg g^{-1} DW)**		**Shoot Carbohydrate (mg g^{-1} DW)**		**Shoot Protein (mg g^{-1} DW)**	
		1st Stage	**2nd Stage**	**1st Stage**	**2nd Stage**	**1st Stage**	**2nd Stage**	**1st Stage**	**2nd Stage**
Control		90.7 ± 0.4^{c}	70.4 ± 0.7^{e}	58.8 ± 0.6^{b}	60.9 ± 0.9^{b}	75.7 ± 1.2^{d}	77.2 ± 1.5^{d}	49.8 ± 0.3^{d}	56.4 ± 0.7^{b}
Exogenously Applied Hormones	IAA	108.4 ± 1.7^{b}	86.6 ± 1.1^{d}	59.8 ± 0.7^{b}	64.2 ± 0.9^{b}	98.7 ± 0.6^{c}	101.9 ± 2.9^{c}	53.7 ± 0.4^{c}	59.3 ± 0.8^{a}
	BA	113.2 ± 3.2^{ab}	90.6 ± 4.2^{cd}	61.2 ± 0.8^{b}	67.4 ± 1.5^{ab}	115.6 ± 3.5^{ab}	112.4 ± 1.4^{ab}	65.5 ± 0.3^{a}	59.3 ± 0.4^{a}
Bacterial Endophytes	PB2	108.4 ± 1.7^{bb}	99.5 ± 0.7^{bc}	61.1 ± 0.7^{b}	64.6 ± 0.9^{b}	113.2 ± 4.5^{ab}	107.6 ± 4.9^{bc}	58.6 ± 0.7^{b}	60.7 ± 1.1^{a}
	PB5	118.2 ± 1.7^{a}	95.5 ± 1.4^{c}	64.8 ± 0.1^{a}	69.3 ± 1.3^{a}	110.0 ± 5.5^{b}	117.3 ± 2.4^{ab}	58.9 ± 0.6^{b}	59.8 ± 0.2^{a}
	BM	107.6 ± 3.7^{b}	105.9 ± 3.5^{b}	64.9 ± 0.6^{a}	70.8 ± 0.9^{a}	122.9 ± 2.02^{a}	116.5 ± 1.6^{ab}	59.0 ± 0.5^{b}	59.3 ± 0.6^{a}
Fungal Endophytes	PF2	103.5 ± 4.5^{b}	121.4 ± 4.2^{a}	61.5 ± 0.8^{b}	66.2 ± 0.8^{b}	119.8 ± 6.9^{a}	111.2 ± 1.2^{ab}	58.3 ± 0.9^{b}	59.8 ± 0.3^{a}
	PF3	104.4 ± 2.1^{b}	106.1 ± 2.9^{b}	64.01 ± 0.8^{a}	69.85 ± 1.21^{a}	103.6 ± 2.9^{b}	107.6 ± 4.9^{bc}	59.3 ± 0.4^{b}	60.3 ± 0.2^{a}
	FM	109.9 ± 2.8^{b}	110.03 ± 2.8^{ab}	63.9 ± 0.8^{a}	70.7 ± 1.5^{a}	113.2 ± 4.5^{ab}	120.5 ± 0.7^{a}	58.9 ± 0.6^{b}	60.3 ± 0.4^{a}

Different letters between columns denote that mean values are significantly different ($p \leq 0.05$) by Tukey's test.

Frequent studies have reported the favorable effect of applying exogenously applied hormones on the contents of soluble carbohydrates and proteins in plants. In this context, treating the in vitro cultures of *Aechmea blanchetiana* with the auxins 1-naphthaleneacetic acid (NAA) and indole-3-butylic acid (IBA) enhanced the dry and fresh mass of the roots. Moreover, the roots of the 70-day-old plantlets treated with IBA showed increased concentrations of its soluble protein content. After 90 days of culture, the roots of the plantlets treated with NAA displayed high concentrations of soluble carbohydrates [84]. In general, sucrose is the main photosynthetic product; it is the essential substrate of sink metabolism and the major form of transported carbon. Thus, IAA, BA, and endophytic extracts have a fundamental role in increasing the carbohydrate content of sprayed plants [85].

Our results confirm the findings of Ismail et al. [86], who reported that the association between the fungal endophyte *Aspergillus japonicus* and the roots of soybeans causes a considerable increase in growth traits and significantly increased the plant content of total sugars, lipids, and proteins under normal and heat-stressed growth conditions. Existing evidence suggests the role of cytokinin's and auxin in managing endosperm cell division through the grain-filling phase in rice. Thus, these leading hormones might control the sink size (grain capacity) for carbohydrate accumulation [87]. Moreover, Bhatia and Singh [88] used ^{14}C and proved the favorable role of IAA in transforming sugars into starch in grain.

3.3.4. Effect of the Foliar Spraying of Exogenously Applied Hormones and Microbial Endophyte Metabolites on the Amylase, Protease, Catalase, and Peroxidase Activities of Common Bean Plants

The common bean plants treated with 100 ppm IAA manifested a significant increase in catalase and peroxidase activities in both growth stages. Meanwhile, protease activity showed a non-significant increase, and although amylase activity showed a non-significant increase in the first stage of growth, it experienced a significant increase in the second stage of growth. On the contrary, *P. vulgaris* plants sprayed with 100 ppm BA showed a significant increase in peroxidase activity in all growth stages, as well as a significant decrease in catalase activity in the second growth stage.

Interestingly, the common bean plants treated with endophytic metabolites displayed varying degrees of enzymatic activities. Plants sprayed with 100 ppm PB5 showed a maximum increase in amylase, protease, catalase, and peroxidase activities in all growth stages. The obtained results in Table 5 clearly show the insignificant responses of protease activity to microbial endophytes individually or mixed at both stages of growth. In harmony with our results, Sadar et al. [89] determined that the crown leaves and fruit pulp of pineapples (*Ananasa comosus*) harbor 28 fungal endophytes belonging to the genera *Aspergillus, Penicillium, Cladosporium, Fusarium, Colletotrichum* and *Alternaria*. These fungal endophytes produce one or more of the enzymes amylase, protease, and lipase in varying degrees. Extracellular enzymes are the byproducts of microbial cell formation, and they play a role in a variety of biological and environmental processes outside of cells. Plants use antioxidant enzymes such as catalase and peroxidase to ameliorate the toxic substances such as reactive oxygen species (ROS), biotic, and abiotic stresses [90].

Table 5. Effect of foliar spraying of exogenously applied hormones and microbial endophyte metabolites on amylase, protease, catalase, and peroxidase of common bean plants.

Treatments		Amylase (Unit/μg FW)		Protease (Unit/μg FW)		Catalase (Unit/μg FW)		Peroxidase (Unit/μg FW)	
		1st Stage	2nd Stage	1st Stage	2nd Stage	1st Stage	2nd Stage	1st Stage	2nd Stage
Control		1.2 ± 0.1^{b}	1.3 ± 0.2^{c}	0.14 ± 0.01^{a}	0.15 ± 0.0^{b}	66.7 ± 1.4^{a}	95.0 ± 2.9^{c}	58.33 ± 1.7^{d}	58.33 ± 1.7^{c}
Exogenously Applied Hormones	IAA	1.7 ± 0.1^{b}	1.9 ± 0.02^{b}	0.15 ± 0.01^{a}	0.18 ± 0.0^{ab}	82.0 ± 1.7^{b}	106.6 ± 1.7^{b}	95.00 ± 5.0^{ab}	96.67 ± 3.3^{b}
	BA	1.8 ± 0.1^{b}	1.9 ± 0.1^{b}	0.14 ± 0.0^{a}	0.16 ± 0.01^{b}	63.3 ± 1.3^{a}	63.3 ± 3.3^{d}	101.6 ± 1.7^{a}	105.0 ± 2.9^{a}
Bacterial Endo-phytes	PB2	2.7 ± 0.2^{a}	2.9 ± 0.1^{a}	0.16 ± 0.01^{a}	0.15 ± 0.01^{b}	83.3 ± 0.4^{b}	123.3 ± 3.3^{b}	86.7 ± 3.3^{b}	103.3 ± 1.7^{a}
	PB5	2.7 ± 0.1^{a}	3.1 ± 0.1^{a}	0.17 ± 0.0^{a}	0.19 ± 0.0^{a}	88.3 ± 1.7^{b}	155.0 ± 5.0^{a}	101.7 ± 1.7^{a}	111.6 ± 4.4^{a}
	BM	1.7 ± 0.1^{b}	2.8 ± 0.3^{a}	0.15 ± 0.0^{a}	0.19 ± 0.0^{a}	83.3 ± 1.4^{b}	113.3 ± 3.3^{b}	103.3 ± 1.7^{a}	96.67 ± 3.3^{b}
Fungal En-dophytes	PF2	2.13 ± 0.2^{a}	1.9 ± 0.1^{b}	0.18 ± 0.0^{a}	0.18 ± 0.01^{ab}	86.7 ± 0.3^{b}	111.6 ± 4.4^{b}	76.67 ± 1.7^{c}	103.3 ± 3.3^{ab}
	PF3	1.7 ± 0.1^{b}	2.2 ± 0.1^{b}	0.18 ± 0.0^{a}	0.18 ± 0.01^{ab}	75.0 ± 0.4^{c}	83.33 ± 4.4^{c}	98.33 ± 4.4^{a}	91.67 ± 1.7^{b}
	FM	1.76 ± 0.1^{b}	2.3 ± 0.1^{b}	0.15 ± 0.0^{a}	0.18 ± 0.0^{ab}	85.0 ± 0.9^{b}	91.67 ± 1.7^{c}	38.33 ± 4.4^{e}	66.67 ± 4.4^{c}

Different letters between columns denote that mean values are significantly different ($p \leq 0.05$) by Tukey's test.

In reality, the xylanase, hemicellulase, phytase, protease, asparaginase, cellulase, pectinase, tyrosinase, gelatinase, chitinase, amylase, and other enzymes formed by the endophytic bacteria and fungi that inhabit medicinal or crop plants, are among the most important enzymes [57].

3.3.5. Effect of the Foliar Spraying of Exogenously Applied Hormones and Microbial Endophyte Metabolites on the Endogenous Hormones of Common Bean Plants

Exogenous hormones play an important role in stimulating plant growth under various biotic and abiotic stresses, but little is known about the relationship of exogenous hormones to their endogenous concentrations [91]. Our tests showed that the application of exogenously hormones as well as endophytic extracts had varying effects on the hormonal content of the common bean plants, and some of these treatments caused a significant increase in the hormonal content of the plant. The maximum increase in the plant contents of IAA, GA3, and BA (0.8 ± 0.5, 18.4 ± 1.3, and 3.4 ± 0.1 mg/100 g fresh weight (FW), respectively) was achieved due to treatment by the metabolites of *B. thuringiensis* (PB2), while the foliar spray using metabolites of *A. sorghi* (PF2) led to the maximum increase in ABA (6.2 ± 0.5 µg/100 g) (Table 6).

Table 6. Effect of exogenous application of exogenously applied hormones and microbial endophyte metabolites on endogenous levels of indole-3-acetic acid (IAA), benzyl Adenine (BA), gibberellic acid (GA3), and abscisic acid (ABA) of common bean plants.

Treatments		IAA (mg/100 g FW)	BA (mg/100 g FW)	GA3 (mg/100 g FW)	ABA (µg/100 g FW)
Control		0.3 ± 0.02 ^{b}	1.3 ± 0.7 ^{cd}	2.6 ± 0.9 ^{c}	1.02 ± 0.7 ^{c}
Exogenously Applied Hormones	IAA	0.3 ± 0.05 ^{b}	0.85 ± 0.6 ^{d}	14.8 ± 2.4 ^{b}	2.1 ± 0.3 ^{c}
	BA	0.4 ± 0.05 ^{ab}	1.9 ± 0.7 ^{bc}	3.2 ± 1.7 ^{c}	1.6 ± 0.6 ^{c}
Bacterial Endophytes	PB2	0.8 ± 0.5 ^{a}	3.4 ± 0.1 ^{a}	18.4 ± 1.3 ^{a}	2.2 ± 0.5 ^{c}
	PB5	0.4 ± 0.1 ^{b}	2.0 ± 0.8 ^{b}	8.3 ± 1.9 ^{d}	2.7 ± 0.2 ^{c}
	BM	0.31 ± 0.06 ^{b}	1.4 ± 0.5 ^{c}	3.8 ± 1.9 ^{c}	4.1 ± 0.3 ^{b}
Fungal Endophytes	PF2	0.4 ± 0.04 ^{b}	2.6 ± 0.6 ^{ab}	16.9 ± 1.02 ^{ab}	6.2 ± 0.5 ^{a}
	PF3	0.34 ± 0.04 ^{b}	1.8 ± 0.1 ^{bc}	17.1 ± 1.7 ^{a}	5.7 ± 0.7 ^{ab}
	FM	0.6 ± 0.05 ^{a}	3.0 ± 0.9 ^{a}	2.9 ± 1.5 ^{c}	4.3 ± 0.8 ^{b}

Different letters between columns denote that mean values are significantly different ($p \leq 0.05$) by Tukey's test.

In a comparative study, El-Mergawi and El-Wahed [92] conducted a greenhouse experiment to determine how foliar sprays of IAA or salicylic acid (SA) affect maize plant growth and the concentrations of endogenous IAA and SA. They reported that spraying plants with IAA and SA (0.25–2 mM) significantly increases the endogenous content of IAA and SA in a concentration-dependent manner, and the maximum increase (10–34%) was recorded after two days of treatment. Hormones applied exogenously alter plant growth and development by causing changes in their intrinsic contents. However, it is unclear whether exogenous hormones have direct effects on growth or are linked to changes in endogenous hormones [93].

Regarding endophytic extracts, the culture filtrates of the endophytic fungi *Fusarium proliferatum* BRL1 and *Asprgillus fumigatus* TS1 were examined for their plant growth-promoting traits on the mutant rice cultivar Waito-C. It was demonstrated that these culture filtrates improve biomass, chlorophyll content, and shoot-root length. In addition, a considerable increase in endogenous gibberellin (GA_1) has been reported in treated rice plants [94]. Our results might be attributed to the master role of IAA in promoting plant development and growth by inducing various processes involving apical dominance, tissue growth, cell division, differentiation of vascular tissues, senescence and ripening, gravitropism and phototropism, embryogenesis, and lateral root initiation [82]. In addition, BA stimulates plant cell division, promotes plant growth, sets blossoms, and improves fruit quality [95].

The reason the IAA concentration was sometimes higher/lower or the same in the control and IAA-treated plants (Table 6) could be attributed to PGPR influencing plant auxin homeostasis by various mechanisms depending on the strains of bacteria. In addition, PGPR can affect the expression of the plant genes that are part of auxin synthesis, transport, or signaling machinery. For instance, PGPR *Pseudomonas putida* 1290 can utilize IAA as a nutritive substrate, thus eliminating the inhibiting effect of high levels of exogenous IAA on plant. The extracellular applications of both IAA and IAA-producing strains was used as a source of exogenous IAA [96].

Furthermore, PGPR can synthesize substances with auxin-like activity such as cyclopeptides and indole. Cyclopeptides stimulate the expression of auxin-induced marker constructs DR5/uidA and BA3/uidA and cause the degradation of Aux/IAA, the repressor of auxin-dependent genes, enhancing auxin signal transduction [97]. A small fraction of the indole that is produced by the PGPR can be converted into IAA. Interestingly, indole at higher concentrations can act as an auxin antagonist [98].

In addition, some biologically active substances produced by PGPR do not have auxin-like activity but can still affect plant auxin homeostasis such as the volatile compounds acetoin, 3,4-butanediol [99], albuterol and 1,3- propanediol, which can affect the expression of IAA transporter genes and can enhance the expression of auxin synthesis genes [100].

In a greenhouse, a soya bean seedling was inoculated with *Porostereum spadiceum* AGH786 endophytic fungus under NaCl stress. Phytohormones such as GAs, JA, and ABA, as well as isoflavones, tended to be secreted, but GAs was secreted in greater quantities than in the control plants [101]. Different phytohormones produced by endophytes reduce the dependence on synthetic fertilizers.

3.3.6. Effect of the Foliar Spraying of Exogenously Applied Hormones and Microbial Endophyte Metabolites on the Yield Parameters of Common Bean Plants

In the current study, yield analysis demonstrated that all treatments (IAA, BA, PB2, PB5, BM, PF2, PF3, and FM) caused a significant increase in pod count, seed count, and protein content; the maximum increase in these parameters was recorded for the FM treatment. In addition, IAA and all endophytic metabolites significantly increased pods and seeds weight, with a maximum increase recorded for *Brevibacillus agri* PB5 treatment.

The exogenous application of IAA (100 ppm) caused significant increases in the weight of 100 seeds, while the outcomes of the other treatments seemed similar to that of the control. By contrast, all treatments significantly increased the carbohydrate content of the treated common beans, and the maximum increase was assigned for the *Brevibacillus agri* PB5 treatment (Table 7).

Table 7. Effect of foliar spraying of exogenously applied hormones and microbial endophyte metabolites on yield parameters and seed yield components of common bean plants.

Treatments		Pods/Plant		Seeds/Pod		100 Seeds Weight	Seed Yield Components	
		Number	Weight (g)	Number	Weight (g)	Weight (g)	Carbohydrate (mg/g DW)	Protein (mg/g DW)
Control		2.0 ± 0.02^{c}	7.3 ± 0.1^{b}	6.3 ± 0.3^{d}	6.0 ± 0.0^{b}	45.2 ± 4.8^{b}	81.5 ± 0.9^{c}	54.6 ± 0.2^{b}
Exogenously applied Hormones	IAA	3.0 ± 0.6^{b}	8.8 ± 0.4^{ab}	11.3 ± 2.03^{c}	7.7 ± 0.3^{ab}	71.2 ± 9.9^{a}	110.8 ± 3.5^{b}	60.1 ± 1.0^{a}
	BA	3.7 ± 0.3^{ab}	7.7 ± 0.3^{b}	13.7 ± 0.9^{b}	6.7 ± 0.3^{b}	49.3 ± 4.7^{b}	123.1 ± 4.3^{ab}	63.5 ± 1.9^{a}
Bacterial Endophytes	PB2	4.0 ± 0.1^{a}	8.5 ± 0.3^{ab}	15.0 ± 0.6^{b}	7.1 ± 0.1^{ab}	47.6 ± 1.2^{b}	134.6 ± 2.1^{a}	64.9 ± 1.2^{a}
	PB5	4.7 ± 0.3^{a}	10.4 ± 0.2^{a}	18.0 ± 0.6^{a}	9.0 ± 0.1^{a}	50.1 ± 1.3^{b}	138.9 ± 1.9^{a}	62.8 ± 0.1^{a}
	BM	4.7 ± 0.3^{a}	8.7 ± 0.3^{ab}	16.3 ± 2.2^{ab}	7.7 ± 0.3^{ab}	48.3 ± 5.1^{b}	120.7 ± 0.7^{b}	64.9 ± 1.2^{a}
Fungal Endophytes	PF2	5.0 ± 1.0^{a}	8.9 ± 0.2^{ab}	18.7 ± 0.9^{a}	7.7 ± 0.3^{ab}	41.3 ± 3.02^{b}	118.4 ± 1.2^{b}	61.2 ± 0.8^{a}
	PF3	5.3 ± 0.3^{a}	9.5 ± 0.4^{a}	17.0 ± 2.7^{ab}	8.3 ± 0.3^{a}	51.6 ± 8.3^{b}	127.5 ± 4.5^{ab}	65.9 ± 0.5^{a}
	FM	5.4 ± 0.3^{a}	9.9 ± 0.3^{a}	20.3 ± 0.3^{a}	8.7 ± 0.3^{a}	42.7 ± 2.3^{b}	107.6 ± 3.7^{b}	67.1 ± 1.3^{a}

Different letters between columns denote that mean values are significantly different ($p \leq 0.05$) by Tukey's test.

Another experiment was carried out to see how different growth regulators affected the growth and yield of lentil plants. Giannakoula and colleagues reported that plants treated with gibberellic acid showed a 26% reduction in the 1000-seed weight. Spraying plants with prohexadione-calcium and topflor, on the contrary, increased the weight of 1000 seeds by 16% percent and 30%, respectively [102]. It has been observed that the application of culture filtrate of the endophytic fungus *Piriformospora indica* to pots containing *Helianthus annus* Sun gold plants and growing plants for 90 days in a greenhouse increased the seed number, seed dry weight, and seed oil content by 9.12%, 45.89% and 51.13% respectively, in addition to increasing the total biomass (roots, stems, leaves, flowers, and seeds) by 36.7% compared to the control [103]. Several studies have reported that endophytes enhance legume crop yield by producing phytohormones such as IAA and GA3 [104], and ethylene [105].

4. Conclusions

Endophytic microbes have been shown to play an important role in plant growth, development, and fitness by producing growth-promoting substances and secondary metabolites. Therefore, six bacterial and four fungal endophytes were isolated from the sterilized roots of the *Phaseolus vulgaris* plant, and they were subjected to molecular identification based on the amplification and sequencing of bacterial 16S rRNA and the fungal ribosomal DNA internal transcribed spacer (ITS) region. The obtained microbial species exhibited plant growth-promoting activities, including phosphate solubilizing, ammonia production, nitrogen fixation, biocontrol of phytopathogen, extracellular enzymatic activities, and IAA production. A field trial was conducted to investigate the effect of two exogenously plant growth-promoting hormones (IAA and BA) compared to the metabolites of the most potent bacterial (*Bacillus thuringiensis* PB2 and *Brevibacillus agri* PB5) and fungal (*Alternaria sorghi* PF2, *Penicillium commune* PF3) endophytic species individually or in combination (BM and FM) as a mode of foliar spraying on the growth and biochemical properties of common bean plants in order to evaluate the possibility of applying these microbial isolates as enhancers for legume plants. This study represents the first investigation of the efficacy of microbial endophytes as plant growth-promoters in comparison to exogenously applied hormones. Our investigations demonstrated that bacterial and fungal endophytic metabolites surpassed the exogenously applied hormones in increasing the plant biomass, photosynthetic pigments, carbohydrate and protein contents, antioxidant enzyme activity, endogenous hormones, and yield traits. Our findings illustrate that the endophyte *Brevibacillus agri* (PB5) had the greatest effect on both the vegetative growth and metabolism of the common bean plants. In the future, the use of endophytic microbes for plant growth promotion will eliminate the need for organic fertilizers and synthetic growth regulators in various crops. More research on different plant species is required to see whether these are plant-specific characteristics and to learn more about the processes at work in the interactions between microbial metabolites and the host that enable plants to optimize their responses under aggressive conditions.

Supplementary Materials: The following are available online at https://www.mdpi.com/article/10.3390/cells10051059/s1, Table S1: Extracellular enzymatic activities of bacterial and fungal endophytic strains isolated from roots of common bean plants. Table S2: Effect of foliar spraying of exogenously applied hormones and microbial endophyte metabolites on fresh and dry weight of shoot and root of common bean plants. Table S3: Effect of foliar spraying of exogenously applied hormones and microbial endophyte metabolites on chlorophyll and carotene content of common bean plants. Figure S1. Example of plant growth-promoting activities of some endophytic bacteria and fungi. (A) ammonia production; (B) phosphate solubilization; (C and D) amylase activity for some bacterial and fungal strains; (E) Protease activity for some bacterial strain; (F) cellulase activity for some fungal strains; (G) Xylanase activity for some bacterial strains. PB1 is *E. asburiae;* PB2 is *B. thuringiensis;* PB3 is *A. radioresistens;* PB4 is *B. brevis;* PB5 is *B. agri;* PB6 is *B. subtilis;* PF1 is *P. crustosum;* PF2 is *A. sorghi;* PF3 is *P. commune;* PF4 is *A. flavus.* Figure S2: In vitro antagonistic activity of some endophytic bacterial species obtained in the current study against against *Fusarium oxysporum* (A), *Pythium* *ultimum* (B), and *Alternaria alternata* (C) as plant pathogenic fungi. The selected endophytic bacterial species denotes as (A) *Brevibacillus agri PB5;* (B) *Bacillus thuringiensis* PB2; (C) *Brevibacillus brevis* PB4; (D) *Brevibacillus agri PB5;* (E) *Acinetobacter radioresistens* PB3; (F) *Brevibacillus agri PB5.* Figure S3. (A) photo of greenhouse experiemnt; (B) photo showed gradually shhot and root length for different treatment.

Author Contributions: Conceptualization, M.A.I., M.A.A., A.M.E., S.E.-D.H. and A.F.; methodology, M.A.I., M.A.A., A.M.E., S.E.-D.H., H.A.M.M., I.L., A.T.A. and A.F.; software, M.A.I., M.A.A., A.M.E, S.E.-D.H., H.A.M.M., I.L., A.T.A., A.E. and A.F.; validation M.A.I., M.A.A., A.M.E., S.E.-D.H., H.A.M.M., I.L., A.T.A. and A.F.; formal analysis, M.A.I., M.A.A., A.M.E., S.E.-D.H., H.A.M.M, I.L., A.T.A., A.E. and A.F.; investigation, M.A.I., M.A.A., A.M.E., S.E.-D.H., H.A.M.M., I.L., A.T.A. and A.F.; resources, M.A.I., M.A.A., A.E., S.E.-D.H. and A.F.; data curation, M.A.I., M.A.A., S.E.-D.H. and A.F.; writing—original draft preparation, M.A.I., M.A.A., A.E., H.A.M.M. and A.F.; writing—review and editing, M.A.I., M.A.A., A.M.E, S.E.-D.H., H.A.M.M., I.L., A.T.A. and A.F. funding acquisition, E.A. and A.A.G. All authors have read and agreed to the published version of the manuscript.

Funding: This research received no external funding.

Institutional Review Board Statement: Not applicable.

Informed Consent Statement: Not applicable.

Data Availability Statement: The data presented in this study are available on request from the corresponding author.

Acknowledgments: We thank Taif University Researchers Supporting Project number (TURSP-2020/13), Taif University, Taif, Saudi Arabia.

Conflicts of Interest: The authors declare no conflict of interest.

References

1. Nassary, E.K.; Baijukya, F.; Ndakidemi, P.A. Assessing the Productivity of Common Bean in Intercrop with Maize across Agro-Ecological Zones of Smallholder Farms in the Northern Highlands of Tanzania. *Agriculture* **2020**, *10*, 117. [CrossRef]
2. Castro-Guerrero, N.A.; Isidra-Arellano, M.C.; Mendoza-Cozatl, D.G.; Valdés-López, O. Common Bean: A Legume Model on the Rise for Unraveling Responses and Adaptations to Iron, Zinc, and Phosphate Deficiencies. *Front. Plant Sci.* **2016**, *7*, 600. [CrossRef] [PubMed]
3. Myers, J.R.; Kmiecik, K. Common Bean: Economic Importance and Relevance to Biological Science Research. In *The Common Bean Genome;* Pérez de la Vega, M., Santalla, M., Marsolais, F., Eds.; Springer International Publishing: Cham, Switzerland, 2017; pp. 1–20.
4. Loboguerrero, A.M.; Campbell, B.M.; Cooper, P.J.M.; Hansen, J.W.; Rosenstock, T.; Wollenberg, E. Food and Earth Systems: Priorities for Climate Change Adaptation and Mitigation for Agriculture and Food Systems. *Sustainability* **2019**, *11*, 1372. [CrossRef]
5. Trewavas, A. How do plant growth substances work? *Plant Cell Environ.* **1981**, *4*, 203–228. [CrossRef]
6. Gan, L.; Wei, Z.; Yang, Z.; Li, F.; Wang, Z. Updated Mechanisms of GCN5—The Monkey King of the Plant Kingdom in Plant Development and Resistance to Abiotic Stresses. *Cells* **2021**, *10*, 979. [CrossRef]
7. Samak, D.H.; El-Sayed, Y.S.; Shaheen, H.M.; El-Far, A.H.; Abd El-Hack, M.E.; Noreldin, A.E.; El-Naggar, K.; Abdelnour, S.A.; Saied, E.M.; El-Seedi, H.R.; et al. Developmental Toxicity of Carbon Nanoparticles during Embryogenesis in Chicken. *Environ Sci. Pollut. Res.* **2020**, *27*, 19058–19072. [CrossRef] [PubMed]
8. Li, H.; Wang, J.; Lin, L.; Liao, M.; Lv, X.; Tang, Y.; Wang, X.; Xia, H.; Liang, D.; Ren, W.; et al. Effects of mutual grafting on cadmium accumulation characteristics of first post-generations of *Bidens pilosa* L. and Galinsoga parviflora Cav. *Environ. Sci. Pollut. Res. Int.* **2019**, *26*, 33228–33235. [CrossRef]
9. Dewitte, W.; Chiappetta, A.; Azmi, A.; Witters, E.; Strnad, M.; Rembur, J.; Noin, M.; Chriqui, D.; Van Onckelen, H. Dynamics of Cytokinins in Apical Shoot Meristems of a Day-Neutral Tobacco during Floral Transition and Flower Formation. *Plant Physiol.* **1999**, *119*, 111. [CrossRef]
10. Werner, T.; Motyka, V.; Strnad, M.; Schmülling, T. Regulation of plant growth by cytokinin. *Proc. Natl. Acad. Sci. USA* **2001**, *98*, 10487–10492. [CrossRef]
11. Elkelish, A.A.; Alnusaire, T.S.; Soliman, M.H.; Gowayed, S.; Senousy, H.H.; Fahad, S. Calcium Availability Regulates Antioxidant System, Physio-Biochemical Activities and Alleviates Salinity Stress Mediated Oxidative Damage in Soybean Seedlings. *J. Appl. Bot. Food Qual* **2019**, *92*, 258–266.
12. Çakmakçı, R.; Mosber, G.; Milton, A.H.; Alatürk, F.; Ali, B. The Effect of Auxin and Auxin-Producing Bacteria on the Growth, Essential Oil Yield, and Composition in Medicinal and Aromatic Plants. *Curr. Microbiol.* **2020**, *77*, 564–577. [CrossRef]

13. Soliman, M.; Elkelish, A.; Souad, T.; Alhaithloul, H.; Farooq, M. Brassinosteroid Seed Priming with Nitrogen Supplementation Improves Salt Tolerance in Soybean. *Physiol. Mol. Biol. Plants* **2020**, *26*, 501–511. [CrossRef]
14. Collenburg, L.; Beyersdorf, N.; Wiese, T.; Arenz, C.; Saied, E.M.; Becker-Flegler, K.A.; Schneider-Schaulies, S.; Avota, E. The Activity of the Neutral Sphingomyelinase Is Important in T Cell Recruitment and Directional Migration. *Front. Immunol.* **2017**, *8*, 1007. [CrossRef]
15. Hassan, S.E.L.D.; Salem, S.S.; Fouda, A.; Awad, M.A.; El-Gamal, M.S.; Abdo, A.M. New approach for antimicrobial activity and bio-control of various pathogens by biosynthesized copper nanoparticles using endophytic actinomycetes. *J. Radiat. Res. Appl. Sci.* **2018**, *11*, 262–270. [CrossRef]
16. Refat, M.S.; Ibrahim, H.K.; Sowellim, S.Z.A.; Soliman, M.H.; Saeed, E.M. Spectroscopic and Thermal Studies of Mn(II), Fe(III), Cr(III) and Zn(II) Complexes Derived from the Ligand Resulted by the Reaction between 4-Acetyl Pyridine and Thiosemicarbazide. *J. Inorg. Organomet. Polym. Mater.* **2009**, *19*, 521–531. [CrossRef]
17. Saied, E.M.; Banhart, S.; Bürkle, S.E.; Heuer, D.; Arenz, C. A Series of Ceramide Analogs Modified at the 1-Position with Potent Activity against the Intracellular Growth of Chlamydia Trachomatis. *Future Med. Chem.* **2015**, *7*, 1971–1980. [CrossRef]
18. Hassan, S.E.; Fouda, A.; Radwan, A.A.; Salem, S.S.; Barghoth, M.G.; Awad, M.A.; Abdo, A.M.; El-Gamal, M.S. Endophytic actinomycetes *Streptomyces* spp mediated biosynthesis of copper oxide nanoparticles as a promising tool for biotechnological applications. *J. Biol. Inorg. Chem.* **2019**, *24*, 377–393. [CrossRef]
19. Salem, S.S.; El-Belely, E.F.; Niedbała, G.; Alnoman, M.M.; Hassan, S.E.; Eid, A.M.; Shaheen, T.I.; Elkelish, A.; Fouda, A. Bactericidal and In-Vitro Cytotoxic Efficacy of Silver Nanoparticles (Ag-NPs) Fabricated by Endophytic Actinomycetes and Their Use as Coating for the Textile Fabrics. *Nanomaterials* **2020**, *10*, 2082. [CrossRef]
20. Eid, A.M.; Fouda, A.; Niedbała, G.; Hassan, S.E.; Salem, S.S.; Abdo, A.M.; Hetta, H.F.; Shaheen, T.I. Endophytic *Streptomyces laurentii* Mediated Green Synthesis of Ag-NPs with Antibacterial and Anticancer Properties for Developing Functional Textile Fabric Properties. *Antibiotics* **2020**, *9*, 641. [CrossRef]
21. Passari, A.K.; Mishra, V.K.; Leo, V.V.; Gupta, V.K.; Singh, B.P. Phytohormone production endowed with antagonistic potential and plant growth promoting abilities of culturable endophytic bacteria isolated from *Clerodendrum colebrookianum* Walp. *Microbiol. Res.* **2016**, *193*, 57–73. [CrossRef]
22. Molina, G.; Pimentel, M.R.; Bertucci, T.C.; Pastore, G.M. Application of fungal endophytes in biotechnological processes. *Chem. Eng. Trans.* **2012**, *27*. [CrossRef]
23. Fouda, A.; Hassan, S.E.; Abdo, A.M.; El-Gamal, M.S. Antimicrobial, Antioxidant and Larvicidal Activities of Spherical Silver Nanoparticles Synthesized by Endophytic *Streptomyces* spp. *Biol. Trace Elements Res.* **2020**, *195*, 707–724. [CrossRef]
24. Hassan, S.E.-D. Plant growth-promoting activities for bacterial and fungal endophytes isolated from medicinal plant of *Teucrium polium* L. *J. Adv. Res.* **2017**, *8*, 687–695. [CrossRef]
25. Miller, D.N.; Bryant, J.E.; Madsen, E.L.; Ghiorse, W.C. Evaluation and optimization of DNA extraction and purification procedures for soil and sediment samples. *Appl. Environ. Microbiol.* **1999**, *65*, 4715–4724. [CrossRef]
26. Fouda, A.; Abdel-Maksoud, G.; Abdel-Rahman, M.A.; Salem, S.S.; Hassan, S.E.-D.; El-Sadany, M.A.-H. Eco-friendly approach utilizing green synthesized nanoparticles for paper conservation against microbes involved in biodeterioration of archaeological manuscript. *Int. Biodeterior. Biodegrad.* **2019**, *142*, 160–169. [CrossRef]
27. Fouda, A.; Abdel-Maksoud, G.; Abdel-Rahman, M.A.; Eid, A.M.; Barghoth, M.G.; El-Sadany, M.A.-H. Monitoring the effect of biosynthesized nanoparticles against biodeterioration of cellulose-based materials by *Aspergillus niger*. *Cellulose* **2019**, *26*, 6583–6597. [CrossRef]
28. Ahmad, F.; Ahmad, I.; Khan, M. Indole Acetic Acid Production by the Indigenous Isolates of *Azotobacter* and Fluorescent *Pseudomonas* in the Presence and Absence of Tryptophan. *Turk. J. Biol.* **2005**, *29*, 29–34.
29. ALKahtani, M.D.F.; Fouda, A.; Attia, K.A.; Al-Otaibi, F.; Eid, A.M.; Ewais, E.E.-D.; Hijri, M.; St-Arnaud, M.; Hassan, S.E.-D.; Khan, N.; et al. Isolation and Characterization of Plant Growth Promoting Endophytic Bacteria from Desert Plants and Their Application as Bioinoculants for Sustainable Agriculture. *Agronomy* **2020**, *10*, 1325. [CrossRef]
30. Da Silva Ribeiro, A.; Polonio, J.C.; Costa, A.T.; Dos Santos, C.M.; Rhoden, S.A.; Azevedo, J.L.; Pamphile, J.A. Bioprospection of Culturable Endophytic Fungi Associated with the Ornamental Plant *Pachystachys lutea*. *Curr. Microbiol.* **2018**, *75*, 588–596. [CrossRef]
31. Fouda, A.H.; Hassan, S.E.-D.; Eid, A.M.; Ewais, E.E.-D. Biotechnological applications of fungal endophytes associated with medicinal plant *Asclepias sinaica* (Bioss.). *Ann. Agric. Sci.* **2015**, *60*, 95–104. [CrossRef]
32. Ghildiyal, A.; Pandey, A. Isolation of cold tolerant antifungal strains of *Trichoderma* sp. from glacial sites of Indian Himalayan region. *Res. J. Microbiol.* **2008**, *3*, 559–564.
33. Kim, H.Y.; Choi, G.J.; Lee, H.B.; Lee, S.W.; Lim, H.K.; Jang, K.S.; Son, S.W.; Lee, S.O.; Cho, K.Y.; Sung, N.D.; et al. Some fungal endophytes from vegetable crops and their anti-oomycete activities against tomato late blight. *Lett. Appl. Microbiol.* **2007**, *44*, 332–337. [CrossRef] [PubMed]
34. Vernon, L.P.; Seely, G.R. *The Chlorophylls*; Academic Press: Cambridge, MA, USA, 2014.
35. Lichtenthaler, H.K. Chlorophylls and carotenoids: Pigments of photosynthetic biomembranes. In *Methods in Enzymology*; Academic Press: Cambridge, MA, USA, 1987; Volume 148, pp. 350–382.
36. Umbreit, W.W.; Burris, R.H.; Stauffer, J.F. *Manometric Techniques: A Manual Describing Methods Applicable to the Study of Tissue Metabolism*; Burgess: Minneapolis, MN, USA, 1964.

37. Lowry, O.H.; Rosebrough, N.J.; Farr, A.L.; Randall, R.J. Protein measurement with the Folin phenol reagent. *J. Biol. Chem.* **1951**, *193*, 265–275. [CrossRef]
38. Afifi, W.; Ahmed, M.; Moussa, Z.; Abd El-Hamid, M. Effect of gamma irradiation and GA3 on amylase activity of pea seedlings. *Ann. Agric. Sci. Moshtohor.* **1986**, *24*, 2047–2057.
39. Ong, P.S.; Gaucher, G.M. Protease production by thermophilic fungi. *Can. J. Microbiol.* **1973**, *19*, 129–133. [CrossRef] [PubMed]
40. Chen, C.; Yu, R.; Owuor, E.D.; Kong, A.N. Activation of antioxidant-response element (ARE), mitogen-activated protein kinases (MAPKs) and caspases by major green tea polyphenol components during cell survival and death. *Arch. Pharm. Res.* **2000**, *23*, 605–612. [CrossRef]
41. Kar, M.; Mishra, D. Catalase, Peroxidase, and Polyphenoloxidase Activities during Rice Leaf Senescence. *Plant Physiol.* **1976**, *57*, 315. [CrossRef]
42. Yang, R.; Yang, T.; Zhang, H.; Qi, Y.; Xing, Y.; Zhang, N.; Li, R.; Weeda, S.; Ren, S.; Ouyang, B.; et al. Hormone profiling and transcription analysis reveal a major role of ABA in tomato salt tolerance. *Plant Physiol. Biochem. PPB* **2014**, *77*, 23–34. [CrossRef]
43. López-López, A.; Rogel, M.A.; Ormeño-Orrillo, E.; Martínez-Romero, J.; Martínez-Romero, E. Phaseolus vulgaris seed-borne endophytic community with novel bacterial species such as *Rhizobium endophyticum* sp. nov. *Syst. Appl. Microbiol.* **2010**, *33*, 322–327. [CrossRef]
44. Sendi, Y.; Pfeiffer, T.; Koch, E.; Mhadhbi, H.; Mrabet, M. Potential of common bean (*Phaseolus vulgaris* L.) root microbiome in the biocontrol of root rot disease and traits of performance. *J. Plant Dis. Prot.* **2020**, *127*, 453–462. [CrossRef]
45. De Oliveira Costa, L.E.; de Queiroz, M.V.; Borges, A.C.; de Moraes, C.A.; de Araújo, E.F. Isolation and characterization of endophytic bacteria isolated from the leaves of the common bean (*Phaseolus vulgaris*). *Braz. J. Microbiol.* **2012**, *43*, 1562–1575. [CrossRef]
46. Fouda, A.; Eid, A.M.; Elsaied, A.; El-Belely, E.F.; Barghoth, M.G.; Azab, E.; Gobouri, A.A.; Hassan, S.E.-D. Plant Growth-Promoting Endophytic Bacterial Community Inhabiting the Leaves of *Pulicaria incisa* (Lam.) DC Inherent to Arid Regions. *Plants* **2021**, *10*, 76. [CrossRef]
47. Khalil, A.M.A.; Hassan, S.E.-D.; Alsharif, S.M.; Eid, A.M.; Ewais, E.E.-D.; Azab, E.; Gobouri, A.A.; Elkelish, A.; Fouda, A. Isolation and Characterization of Fungal Endophytes Isolated from Medicinal Plant *Ephedra pachyclada* as Plant Growth-Promoting. *Biomolecules* **2021**, *11*, 140. [CrossRef]
48. Numponsak, T.; Kumla, J.; Suwannarach, N.; Matsui, K.; Lumyong, S. Biosynthetic pathway and optimal conditions for the production of indole-3-acetic acid by an endophytic fungus, *Colletotrichum fructicola* CMU-A109. *PLoS ONE* **2018**, *13*, e0205070. [CrossRef]
49. Mehmood, A.; Hussain, A.; Irshad, M.; Hamayun, M.; Iqbal, A.; Khan, N. In vitro production of IAA by endophytic fungus *Aspergillus awamori* and its growth promoting activities in Zea mays. *Symbiosis* **2019**, *77*, 225–235. [CrossRef]
50. Wang, F.; Li, H.; Liu, Q.; Li, Z.; Li, R.; Zhang, H.; Liu, L.; Emelchenko, G.A.; Wang, J. A graphene oxide/amidoxime hydrogel for enhanced uranium capture. *Sci. Rep.* **2016**, *6*, 1–8. [CrossRef]
51. Szilagyi-Zecchin, V.J.; Ikeda, A.C.; Hungria, M.; Adamoski, D.; Kava-Cordeiro, V.; Glienke, C.; Galli-Terasawa, L.V. Identification and characterization of endophytic bacteria from corn (*Zea mays* L.) roots with biotechnological potential in agriculture. *Amb Express* **2014**, *4*, 26. [CrossRef]
52. Brígido, C.; Singh, S.; Menéndez, E.; Tavares, M.J.; Glick, B.R.; Félix, M.D.R.; Oliveira, S.; Carvalho, M. Diversity and Functionality of Culturable Endophytic Bacterial Communities in Chickpea Plants. *Plants* **2019**, *8*, 42. [CrossRef]
53. Ripa, F.A.; Cao, W.-D.; Tong, S.; Sun, J.-G. Assessment of Plant Growth Promoting and Abiotic Stress Tolerance Properties of Wheat Endophytic Fungi. *BioMed Res. Int.* **2019**, *2019*, 6105865. [CrossRef]
54. Miller, S.H.; Browne, P.; Prigent-Combaret, C.; Combes-Meynet, E.; Morrissey, J.P.; O'Gara, F. Biochemical and genomic comparison of inorganic phosphate solubilization in *Pseudomonas* species. *Environ. Microbiol. Rep.* **2010**, *2*, 403–411. [CrossRef]
55. Haidar, B.; Ferdous, M.; Fatema, B.; Ferdous, A.S.; Islam, M.R.; Khan, H. Population diversity of bacterial endophytes from jute (*Corchorus olitorius*) and evaluation of their potential role as bioinoculants. *Microbiol Res.* **2018**, *208*, 43–53. [CrossRef]
56. Prabhu, N.; Borkar, S.; Garg, S. Chapter 11—Phosphate solubilization by microorganisms: Overview, mechanisms, applications and advances. In *Advances in Biological Science Research*; Meena, S.N., Naik, M.M., Eds.; Academic Press: Cambridge, MA, USA, 2019; pp. 161–176.
57. Khan, A.L.; Shahzad, R.; Al-Harrasi, A.; Lee, I.-J. Endophytic Microbes: A Resource for Producing Extracellular Enzymes. In *Endophytes: Crop Productivity and Protection: Volume 2*; Maheshwari, D.K., Annapurna, K., Eds.; Springer International Publishing: Cham, Switzerland, 2017; pp. 95–110.
58. Sopalun, K.; Iamtham, S. Isolation and screening of extracellular enzymatic activity of endophytic fungi isolated from *Thai orchids*. *S. Afr. J. Bot.* **2020**, *134*, 273–279. [CrossRef]
59. Sunitha, V.; Nirmala Devi, D.; Srinivas, C. Extracellular enzymatic activity of endophytic fungal strains isolated from medicinal plants. *World J. Agric. Sci.* **2013**, *9*, 1–9.
60. Naik, S.; Abrar, D.S.; Krishnappa, M. Industrially important enzymes from fungal endophytes. *Recent Adv. White Biotechnol. Fungi* **2019**, 263–280. [CrossRef]
61. Egamberdieva, D.; Wirth, S.; Behrendt, U.; Ahmad, P.; Berg, G. Antimicrobial Activity of Medicinal Plants Correlates with the Proportion of Antagonistic Endophytes. *Front. Microbiol.* **2017**, *8*, 199. [CrossRef]

62. Zhao, L.; Xu, Y.; Lai, X. Antagonistic endophytic bacteria associated with nodules of soybean (*Glycine max* L.) and plant growth-promoting properties. *Braz. J. Microbiol.* **2018**, *49*, 269–278. [CrossRef]
63. Chowdhary, K.; Sharma, S. Plant Growth Promotion and Biocontrol Potential of Fungal Endophytes in the Inflorescence of *Aloe vera* L. *Proc. Natl. Acad. Sci. USA* **2020**, *90*, 1045–1055. [CrossRef]
64. Hamzah, T.N.T.; Lee, S.Y.; Hidayat, A.; Terhem, R.; Faridah-Hanum, I.; Mohamed, R. Diversity and Characterization of Endophytic Fungi Isolated From the Tropical Mangrove Species, *Rhizophora mucronata*, and Identification of Potential Antagonists Against the Soil-Borne Fungus, *Fusarium solani*. *Front. Microbiol.* **2018**, *9*, 1707. [CrossRef]
65. Badawy, A.A.; Abdelfattah, N.A.H.; Salem, S.S.; Awad, M.F.; Fouda, A. Efficacy Assessment of Biosynthesized Copper Oxide Nanoparticles (CuO-NPs) on Stored Grain Insects and Their Impacts on Morphological and Physiological Traits of Wheat (*Triticum aestivum* L.) Plant. *Biology* **2021**, *10*, 233. [CrossRef]
66. Sarkar, D.; Mandal, B.; Kundu, M.C. Increasing use efficiency of boron fertilisers by rescheduling the time and methods of application for crops in India. *Plant Soil* **2007**, *301*, 77–85. [CrossRef]
67. Turbat, A.; Rakk, D.; Vigneshwari, A.; Kocsubé, S.; Thu, H.; Szepesi, Á.; Bakacsy, L.; Škrbić, B.D.; Jigjiddorj, E.-A.; Vágvölgyi, C.; et al. Characterization of the Plant Growth-Promoting Activities of Endophytic Fungi Isolated from *Sophora flavescens*. *Microorganisms* **2020**, *8*, 683. [CrossRef] [PubMed]
68. Hoffman, M.T.; Gunatilaka, M.K.; Wijeratne, K.; Gunatilaka, L.; Arnold, A.E. Endohyphal bacterium enhances production of indole-3-acetic acid by a foliar fungal endophyte. *PLoS ONE* **2013**, *8*, e73132. [CrossRef] [PubMed]
69. Waqas, M.; Khan, A.L.; Kamran, M.; Hamayun, M.; Kang, S.M.; Kim, Y.H.; Lee, I.J. Endophytic fungi produce gibberellins and indoleacetic acid and promotes host-plant growth during stress. *Molecules* **2012**, *17*, 10754–10773. [CrossRef] [PubMed]
70. Tumangger, B.; Nadilla, F.; Baiduri, N.; Fitriani; Mardina, V. In vitro screening of endophytic fungi associated with mangroveas biofertilizer on the growth of black rice (*Oryza sativa* L. "Cempo Ireng"). *IOP Conf. Ser. Mater. Sci. Eng.* **2018**, *420*, 012080. [CrossRef]
71. Sosnowski, J.; Malinowska, E.; Jankowski, K.; Król, J.; Redzik, P. An estimation of the effects of synthetic auxin and cytokinin and the time of their application on some morphological and physiological characteristics of Medicago x varia T. Martyn. *Saudi J. Biol. Sci.* **2019**, *26*, 66–73. [CrossRef]
72. Potshangbam, M.; Devi, S.I.; Sahoo, D.; Strobel, G.A. Functional Characterization of Endophytic Fungal Community Associated with *Oryza sativa* L. and *Zea mays* L. *Front. Microbiol.* **2017**, *8*, 325. [CrossRef]
73. Abdallah, R.A.B.; Mejdoub-Trabelsi, B.; Nefzi, A.; Jabnoun-Khiareddine, H.; Daami-Remadi, M. Isolation of endophytic bacteria from *Withania somnifera* and assessment of their ability to suppress *Fusarium wilt* disease in tomato and to promote plant growth. *J. Plant Pathol. Microbiol.* **2016**, *7*, 5. [CrossRef]
74. Sukhov, V.; Sukhova, E.; Sinitsyna, Y.; Gromova, E.; Mshenskaya, N.; Ryabkova, A.; Ilin, N.; Vodeneev, V.; Mareev, E.; Price, C. Influence of Magnetic Field with Schumann Resonance Frequencies on Photosynthetic Light Reactions in Wheat and Pea. *Cells* **2021**, *10*, 149. [CrossRef]
75. Hanaa, H.; Safaa, A. In Foliar application Foliar application of IAA at different growth stages and their influenced on growth and productivity of bread Wheat *(triticum aestivum* L.). *J. Phys. Conf. Ser.* **2019**, *1294*, 092029. [CrossRef]
76. Ma, G.; Zhang, L.; Kudaka, R.; Inaba, H.; Furuya, T.; Kitamura, M.; Kitaya, Y.; Yamamoto, R.; Yahata, M.; Matsumoto, H.; et al. Exogenous Application of ABA and NAA Alleviates the Delayed Coloring Caused by Puffing Inhibitor in Citrus Fruit. *Cells* **2021**, *10*, 308. [CrossRef]
77. Heidari, M.; Golpayegani, A. Effects of water stress and inoculation with plant growth promoting rhizobacteria (PGPR) on antioxidant status and photosynthetic pigments in basil (*Ocimum basilicum* L.). *J. Saudi Soc. Agric. Sci.* **2012**, *11*, 57–61. [CrossRef]
78. Timmusk, S.; Abd El-Daim, I.A.; Copolovici, L.; Tanilas, T.; Kännaste, A.; Behers, L.; Nevo, E.; Seisenbaeva, G.; Stenström, E.; Niinemets, Ü. Drought-tolerance of wheat improved by rhizosphere bacteria from harsh environments: Enhanced biomass production and reduced emissions of stress volatiles. *PLoS ONE* **2014**, *9*, e96086. [CrossRef]
79. Naveed, M.; Mitter, B.; Reichenauer, T.G.; Wieczorek, K.; Sessitsch, A. Increased drought stress resilience of maize through endophytic colonization by *Burkholderia phytofirmans* PsJN and Enterobacter sp. FD17. *Environ. Exp. Bot.* **2014**, *97*, 30–39. [CrossRef]
80. Kumar, A.; Bahadur, D.; Maurya, B.; Raghuwanshi, R.; Meena, V.; Singh, S.; Dixit, J. Does a plant growth promoting rhizobacteria enhance agricultural sustainability? *J. Pure Appl. Microbiol.* **2015**, *9*, 715–724.
81. Leopold, A.C.; Noodén, L.D. Hormonal Regulatory Systems in Plants. In *Hormonal Regulation of Development II: The Functions of Hormones from the Level of the Cell to the Whole Plant*; Scott, T.K., Ed.; Springer: Berlin/Heidelberg, Germany, 1984; pp. 4–22.
82. Naeem, M.; Bhatti, I.R.A.M.; Ahmad, R.H.; Ashraf, M.Y. Effect of some growth hormones (GA3, IAA and Kintin) on the Morphology and early or delayed iniation of bud of lintil (Lens culinaris Medik). *Pak. J. Bot.* **2004**, *36*, 801–809.
83. Di Benedetto, A.; Galmarini, C.; Tognetti, J. Effects of combined or single exogenous auxin and/or cytokinin applications on growth and leaf area development in *Epipremnum aureum*. *J. Hortic. Sci. Biotechnol.* **2015**, *90*, 643–654. [CrossRef]
84. Chu, E.; Tavares, A.; Kanashiro, S.; Giampaoli, P.; Yokota, E. Effects of auxins on soluble carbohydrates, starch and soluble protein content in *Aechmea blanchetiana* (Bromeliaceae) cultured in vitro. *Sci. Hortic.* **2010**, *125*, 451–455. [CrossRef]
85. Farrar, J.; Pollock, C.; Gallagher, J. Sucrose and the integration of metabolism in vascular plants. *Plant Sci.* **2000**, *154*, 1–11. [CrossRef]

86. Ismail; Hamayun, M.; Hussain, A.; Iqbal, A.; Khan, S.A.; Lee, I.-J. Endophytic Fungus *Aspergillus japonicus* Mediates Host Plant Growth under Normal and Heat Stress Conditions. *BioMed Res. Int.* **2018**, *2018*, 7696831. [CrossRef]
87. Dietrich, J.; Kaminek, M.; Blevins, D.; Reinbott, T.; Morris, R. Changes in cytokinins and cytokinin oxidase activity in developing-maize kernels and the effects of exogenous cytokinin on kernel development. *Plant Physiol. Biochem.* **1995**, *33*, 327–336.
88. Bhatia, S.; Singh, R. Phytohormone-mediated transformation of sugars to starch in relation to the activities of amylases, sucrose-metabolising enzymes in sorghum grain. *Plant Growth Regul.* **2002**, *36*, 97–104. [CrossRef]
89. Sadar, P.; Jadhav, N.; Kulkarni, N.; Pachori, R. Diversity and enzymatic activity of endophytic fungi isolated from pine apple (*Ananasa comosus*). *Eur. J. Biomed. Pharm. Sci.* **2017**, *4*, 647–652.
90. Naliwajski, M.; Skłodowska, M. The Relationship between the Antioxidant System and Proline Metabolism in the Leaves of Cucumber Plants Acclimated to Salt Stress. *Cells* **2021**, *10*, 609. [CrossRef]
91. López-Ruiz, B.A.; Zluhan-Martínez, E.; Sánchez, M.D.; Álvarez-Buylla, E.R.; Garay-Arroyo, A. Interplay between Hormones and Several Abiotic Stress Conditions on Arabidopsis thaliana Primary Root Development. *Cells* **2020**, *9*, 2576. [CrossRef]
92. El-Mergawi, R.A.; Abd El-Wahed, M.S.A. Effect of exogenous salicylic acid or indole acetic acid on their endogenous levels, germination, and growth in maize. *Bull. Natl. Res. Cent.* **2020**, *44*, 167. [CrossRef]
93. Szalai, G.; Horgosi, S.; Soós, V.; Majláth, I.; Balázs, E.; Janda, T. Salicylic acid treatment of pea seeds induces its de novo synthesis. *J. Plant Physiol.* **2011**, *168*, 213–219. [CrossRef]
94. Bilal, L.; Asaf, S.; Hamayun, M.; Gul, H.; Iqbal, A.; Ullah, I.; Lee, I.-J.; Hussain, A. Plant growth promoting endophytic fungi *Asprgillus fumigatus* TS1 and *Fusarium proliferatum* BRL1 produce gibberellins and regulates plant endogenous hormones. *Symbiosis* **2018**, *76*, 117–127. [CrossRef]
95. Polisetty, R.; Paul, V.; Deveshwar, J.J.; Khetarpal, S.; Suresh, K.; Chandra, R. Multiple shoot induction by benzyladenine and complete plant regeneration from seed explants of chickpea (*Cicer arietinum* L.). *Plant Cell Rep.* **1997**, *16*, 565–571. [CrossRef] [PubMed]
96. Leveau, J.H.; Lindow, S.E. Utilization of the plant hormone indole-3-acetic acid for growth by *Pseudomonas putida* strain 1290. *Appl. Environ. Microbiol.* **2005**, *71*, 2365–2371. [CrossRef] [PubMed]
97. Ortiz-Castro, R.; Díaz-Pérez, C.; Martínez-Trujillo, M.; del Río, R.E.; Campos-García, J.; López-Bucio, J. Transkingdom signaling based on bacterial cyclodipeptides with auxin activity in plants. *Proc. Natl. Acad. Sci. USA* **2011**, *108*, 7253–7258. [CrossRef] [PubMed]
98. Bailly, A.; Groenhagen, U.; Schulz, S.; Geisler, M.; Eberl, L.; Weisskopf, L. The inter-kingdom volatile signal indole promotes root development by interfering with auxin signalling. *Plant J. Cell Mol. Biol.* **2014**, *80*, 758–771. [CrossRef]
99. Ryu, C.M.; Farag, M.A.; Hu, C.H.; Reddy, M.S.; Wei, H.X.; Paré, P.W.; Kloepper, J.W. Bacterial volatiles promote growth in *Arabidopsis*. *Proc. Natl. Acad. Sci. USA* **2003**, *100*, 4927–4932. [CrossRef]
100. Tahir, H.A.; Gu, Q.; Wu, H.; Raza, W.; Hanif, A.; Wu, L.; Colman, M.V.; Gao, X. Plant Growth Promotion by Volatile Organic Compounds Produced by *Bacillus subtilis* SYST2. *Front. Microbiol.* **2017**, *8*, 171. [CrossRef]
101. Hamayun, M.; Hussain, A.; Khan, S.A.; Kim, H.Y.; Khan, A.L.; Waqas, M.; Irshad, M.; Iqbal, A.; Rehman, G.; Jan, S.; et al. Gibberellins Producing Endophytic Fungus *Porostereum spadiceum* AGH786 Rescues Growth of Salt Affected Soybean. *Front. Microbiol.* **2017**, *8*, 686. [CrossRef]
102. Giannakoula, A.E.; Ilias, I.F.; Dragišić Maksimović, J.J.; Maksimović, V.M.; Živanović, B.D. The effects of plant growth regulators on growth, yield, and phenolic profile of lentil plants. *J. Food Compos. Anal.* **2012**, *28*, 46–53. [CrossRef]
103. Bagde, U.S.; Prasad, R.; Varma, A. Influence of culture filtrate of Piriformospora indica on growth and yield of seed oil in Helianthus annus. *Symbiosis* **2011**, *53*, 83–88. [CrossRef]
104. Khan, A.L.; Waqas, M.; Kang, S.M.; Al-Harrasi, A.; Hussain, J.; Al-Rawahi, A.; Al-Khiziri, S.; Ullah, I.; Ali, L.; Jung, H.Y.; et al. Bacterial endophyte *Sphingomonas* sp. LK11 produces gibberellins and IAA and promotes tomato plant growth. *J. Microbiol.* **2014**, *52*, 689–695. [CrossRef]
105. Straub, D.; Yang, H.; Liu, Y.; Tsap, T.; Ludewig, U. Root ethylene signalling is involved in *Miscanthus sinensis* growth promotion by the bacterial endophyte *Herbaspirillum frisingense* GSF30(T). *J. Exp. Bot.* **2013**, *64*, 4603–4615. [CrossRef]

cells

MDPI

Article

Influence of Magnetic Field with Schumann Resonance Frequencies on Photosynthetic Light Reactions in Wheat and Pea

Vladimir Sukhov [1,2,*], Ekaterina Sukhova [1], Yulia Sinitsyna [2,3], Ekaterina Gromova [1], Natalia Mshenskaya [2,3], Anastasiia Ryabkova [1,2], Nikolay Ilin [2], Vladimir Vodeneev [1,2], Evgeny Mareev [2] and Colin Price [2,4]

1 Department of Biophysics, N.I. Lobachevsky State University of Nizhny Novgorod, 603950 Nizhny Novgorod, Russia; n.catherine@inbox.ru (E.S.); kater333@inbox.ru (E.G.); nastay2903@bk.ru (A.R.); v.vodeneev@mail.ru (V.V.)
2 Earth's Electromagnetic Environment Laboratory, Institute of Applied Physics of Russian Academy of Sciences, 603600 Nizhny Novgorod, Russia; jsin@inbox.ru (Y.S.); tasya.mshanka@yandex.ru (N.M.); ilyin@appl.sci-nnov.ru (N.I.); evgeny.mareev@gmail.com (E.M.); cprice@flash.tau.ac.il (C.P.)
3 Department of Biochemistry and Biotechnology, N.I. Lobachevsky State University of Nizhny Novgorod, 603950 Nizhny Novgorod, Russia
4 Department of Geophysics, Porter School of the Environment and Earth Sciences, Tel Aviv University, Tel Aviv-Yafo 6997801, Israel
* Correspondence: vssuh@mail.ru; Tel.: +7-909-292-8653

Abstract: Photosynthesis is an important target of action of numerous environmental factors; in particular, stressors can strongly affect photosynthetic light reactions. Considering relations of photosynthetic light reactions to electron and proton transport, it can be supposed that extremely low frequency magnetic field (ELFMF) may influence these reactions; however, this problem has been weakly investigated. In this paper, we experimentally tested a hypothesis about the potential influence of ELFMF of 18 μT intensity with Schumann resonance frequencies (7.8, 14.3, and 20.8 Hz) on photosynthetic light reactions in wheat and pea seedlings. It was shown that ELFMF decreased non-photochemical quenching in wheat and weakly influenced quantum yield of photosystem II at short-term treatment; in contrast, the changes in potential and effective quantum yields of photosystem II were observed mainly under chronic action of ELFMF. It is interesting that both short-term and chronic treatment decreased the time periods for 50% activation of quantum yield and non-photochemical quenching under illumination. Influence of ELFMF on pea was not observed at both short-term and chronic treatment. Thus, we showed that ELFMF with Schumann resonance frequencies could influence photosynthetic light processes; however, this effect depends on plant species (wheat or pea) and type of treatment (short-term or chronic).

Keywords: extremely low frequency magnetic field; Schumann resonance frequencies; photosynthetic light reactions; non-photochemical quenching; quantum yield of photosystem II; wheat; pea

Citation: Sukhov, V.; Sukhova, E.; Sinitsyna, Y.; Gromova, E.; Mshenskaya, N.; Ryabkova, A.; Ilin, N.; Vodeneev, V.; Mareev, E.; Price, C. Influence of Magnetic Field with Schumann Resonance Frequencies on Photosynthetic Light Reactions in Wheat and Pea. *Cells* **2021**, *10*, 149. https://doi.org/10.3390/cells10010149

Received: 13 December 2020
Accepted: 11 January 2021
Published: 13 January 2021

Publisher's Note: MDPI stays neutral with regard to jurisdictional claims in published maps and institutional affiliations.

1. Introduction

Photosynthesis, participating in solar energy consumption and its transformation into energy of chemical compounds, is a key process in plant life; however, it is also an important target of action of environmental factors and, thereby, can participate in the development of stress changes in plants. There are numerous investigations, which show that photosynthetic processes can be affected by action of environmental factors on plant organism (light with excess intensity [1–3], low or high temperatures [4–6], drought [7–9], salinity [10–12], etc.) or by propagation of long-distance stress signals through plant body (e.g., electrical signals [13–15]). These photosynthetic changes have intricate character including photosynthetic damage and decrease of photosynthetic intensity (e.g., decrease of CO_2 assimilation and quantum yield of photosystem II (Φ_{PSII})) [3,9,12] and induction of numerous adaptive responses, which participate in photosynthetic protection and can also modify photosynthetic activity (e.g., increase of non-photochemical quenching (NPQ) [12,16,17], activation of cyclic electron flow [8,17], changes in localization of ferredoxin-NADP-reductase [18,19], etc.).

Photosynthetic light reactions are strongly related to transport of electrons through chloroplast's electron transport chain (ETC) [16,17] and ion fluxes through thylakoid membranes [20,21]; considering that both processes are charge transfer, they can probably be influenced by magnetic fields (MFs). MFs are important physical factors of environment; they include stationary geomagnetic fields, stationary artificial magnetic fields, and non-stationary MFs. In particular, extremely low frequency magnetic fields (ELFMFs) are widely produced as a result of direct human activity (mainly, MFs with industrial frequencies, equaling to 50 and 60 Hz) and as result of natural environmental events, mainly magnetospheric substorms and lightning [22]. From the beginning of its development, the Earth's biosphere has been influenced by various electromagnetic fields, among which a special place is occupied by fields at frequencies of Schumann resonances formed by the Earth–ionosphere resonator [22]. The Schumann resonance frequencies equal 7.8, 14.3, 20.8, 27.3, and 33.8 Hz [23,24].

It is important that ELFMFs can strongly influence physiological processes in living organisms (e.g., see works [22,25,26]). There are studies [27–30] which show the influence of these MFs on growth, seed germination, ions transport, tolerance to environmental stressors, and other processes in plants; however, changes in these processes can be controversial. In particular, numerous investigations (see reviews [27–30]) focus on the influence of ELFMFs on plant growth processes because positive effects can be important for plant agricultural cultivation, in contrast, negative effects should be considered in plant protection. It is shown that ELFMFs [27–30] mostly stimulate plant growth and production of biomass; however, their suppression is also observed. Size and (or) direction of the effect is dependent on plant species (e.g., 60 Hz frequency MF strongly stimulates growth of radish, weakly influences the growth in barley and does not influence this process in mustard [31]), duration of long-term ELFMF treatment (e.g., 50 Hz frequency MF stimulates wheat growth at 17–24-h treatment and suppresses it at 2-day treatment [32]), duration of short-term exposition and magnitude of the magnetic fields (e.g., 50 Hz frequency MF induces different growth stimulation at 5, 10, and 15 min of exposition [33]), etc. It is interesting that growth changes can be also observed under low frequencies and magnitudes of MFs; e.g., sinusoidal ELFMF with about 16 Hz frequency and 20 μT magnitude increases root dry weights in wheat and does not influence this parameter in sunflower seedlings [34]. Considering the influence of ELFMFs on processes of plant growth and production of biomass, it can be expected that these magnetic fields influences photosynthetic processes.

There are a number of works (see reviews [27–30]), which show the influence of ELFMFs on photosynthesis and related processes. In particular, treatment by ELFMF can change the content of chlorophylls and carotenoids [35–38], stimulate gene expression of ribulose 1,5-bisphosphate carboxylase/oxygenase small subunit [36], influence photosynthetic CO_2 assimilation and transpiration [36,38,39]. However, there are only few studies, which are devoted to investigation of the influence of ELFMFs on plant seedlings [36,39]; other works are devoted to the influence of these magnetic fields on seeds [35,37,38]. Results of these works are rather contradictory: Yano et al. [39] show that chronic action of ELFMFs can decrease photosynthetic CO_2 assimilation while the short-term action of these magnetic fields (2 h) does not influence the assimilation; in contrast, Alemán et al. [36] show that short-term action of ELFMFs (3 min) both suppresses and stimulates net photosynthesis (the type of effect is dependent on the stage of development). There are many potential reasons of these differences, including the study of different plants (radish [39] and *Coffea arabica* [36]) and use of very different magnetic field strengths (50 μT [35] and 2 mT [36]). It should be also noted that both papers [36,39] are devoted to the investigation of only MFs with 50 Hz industrial frequency and do not analyze photosynthetic light reactions.

Thus, the problem of the influence of ELFMFs on photosynthetic processes remains poorly investigated; moreover, investigations of the influence of MFs with Schumann resonance frequencies (7.8, 14.3, 20.8, 27.3, and 33.8 Hz [23,24]) on photosynthetic processes are absent. However, ELFMFs with Schumann resonance frequencies are ubiquitously generated as a result of global lightning activity [22]; amplitude of these natural environmental frequencies can be dependent on global climate changes (i.e., it can be indirectly related to human's activity). Considering these points, problem of the influence of ELFMFs with Schumann resonance frequencies on photosynthetic processes in plants can be important for plant biology. The present work is devoted to the analysis of the influence of ELFMFs with Schumann resonance frequencies (the first, second, and third harmonics that equal 7.83, 14.3, and 20.8 Hz) on photosynthetic light reactions in seedling of wheat and pea.

2. Materials and Methods

2.1. Materials

Wheat (*Triticum aestivum* L., cultivar "Zlata") and pea (*Pisum sativum* L., cultivar "Albumen") seedlings were used in experiments with treatment by magnetic fields. Seeds were soaked for 3 days before planting. Plants were cultivated (up to 9–13-days age) in vegetation room in pots with soil at 24 °C and 16/8 h (light/dark) photoperiod; luminescent lamps FSL YZ18RR (Foshan Electrical And Lighting Co., Ltd., Foshan, China) were used for illumination.

2.2. Short-Term and Chronic Treatments by Magnetic Fields with Schumann Resonance Frequencies

Two systems for plant treatment by artificial ELFMFs were manufactured: the first system was used for photosynthetic measurements in leaves of plants under simultaneous treatment by MFs (volume of homogenous magnetic field was about 20 × 20 × 20 cm^3) (Figure 1a); the second system was used for plant cultivation under chronic action of ELFMFs (volume of homogenous magnetic field was about 30 × 30 × 30 cm^3) (Figure 1b). Both systems were based on Helmholtz coils (100 loops for first system and 130 loops for second system for each coil with corrected input impedance of 50 ohm) with different radii (0.3 m for the first system and 0.5 m for the second system). Positions of Helmholtz coils supported the direction of ELFMFs, which was perpendicular to the direction of a geomagnetic field (about 50 μT). RIGOL DG1032 Waveform Generator (RIGOL Technology Co., Ltd., Suzhou, China) was used for the generation of sinusoidal electrical signals with frequencies equaling 7.8, 14.3, and 20.8 Hz (for experiments with short-term treatment) or 14.3 Hz (for experiments with chronic treatment). Magnitude of ELFMFs was 18 μT in all variants of the experiments.

Short-term treatment of wheat seedlings by ELFMFs (in 9–13 days after pea planting and in 9–11 days after wheat planting) was initiated after fixation of plants in the first system; treatment was continued during the measurement of parameters of photosynthetic light reactions (30 min) (Figure 1a). In the control variant, seedlings were fixed in the system and their photosynthetic parameters were measured; however, the ELFMFs did not act on plants.

Both systems were used for chronic treatment of seedlings by the investigated ELFMF: the plants were cultivated in the second system with MF treatment (Figure 1b) from initiation of soaking of seeds to plant planting (3 days) and from plant planting to transfer of seedlings into the first system for photosynthetic measurements (on the 9th day of cultivation) (Figure 1a). The time taken for plant translocation from the second system to the first one was less than 30 s. Parameters of ELFMF in the first system were identical to the ones in the second system in this experimental variant (chronic treatment by MF). Photosynthetic parameters were measured by the action of this field. In control, seedlings were cultivated and measured under similar conditions (analogical system for plant growth was used); however, ELFMF treatment was absent.

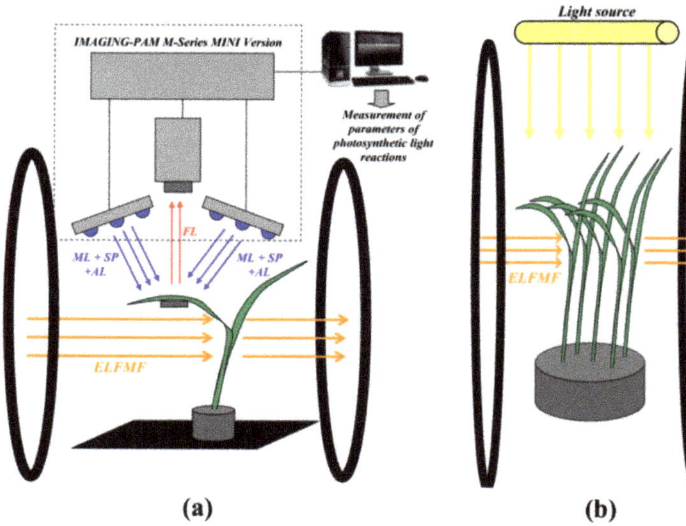

Figure 1. (**a**) Schema of plant localization in experiments with simultaneous action of artificial extremely low frequency magnetic field (ELFMF) and measurements of parameters of photosynthetic light reactions using IMAGING-PAM M-Series MINI Version. ML measuring light, SP saturation pulse, and AL actinic light. Blue light (450 nm) was used for ML, SP, and AL. FL is chlorophyll fluorescence. (**b**) Schema of localization of plants in experiments with chronic action of ELFMF. Luminescent lamps FSL YZ18RR were used as a light source for growth. Wheat plants were used as examples in both figures. Localization of plants in control experiments was identical for both variants; however, they were not treated by ELFMF. The direction of ELFMF was perpendicular to the direction of the geomagnetic field (about 50 μT).

2.3. Measurements of Parameters of Photosynthetic Light Reactions

All measurements of photosynthetic parameters (excluding control variants) were performed simultaneously with the plant treatment by magnetic fields; duration of the photosynthetic measurements was about 30 min. Only one photosynthetic measurement was performed for each seedling. In experiments with chronic action of ELFMF, seedlings were measured on the 9th day after plant planting; 9–11 days age wheat and 9–13 days age pea seedlings were used for photosynthetic measurements in experiments with short-term action of ELFMFs.

A system of pulse-amplitude-modulation (PAM) fluorescence imaging (IMAGING-PAM M-Series MINI Version, Heinz Walz GmbH, Effeltrich, Germany) was used for measurements of parameters of photosynthetic light reactions in the second mature leaves of wheat and pea seedlings (Figure 1a). The pulses of measuring light (ML) with low average intensity (less than 1 μmol $m^{-2}s^{-1}$), saturation pulses (SP) with 800 ms duration and 6000 μmol $m^{-2}s^{-1}$ intensity, and actinic light (AL) with 625 μmol $m^{-2}s^{-1}$ intensity were used for analysis. Only, blue light (450 nm) was used for the photosynthetic investigations. Five wheat leaves were simultaneously investigated in each experiment; photosynthetic parameters were calculated as averaging of ones in three areas in each leaf (Figure 2a). Single pea leaf was investigated in each experiment; photosynthetic parameters were calculated as averaging of ones in six areas in the leaf (Figure 2b). Standard round areas (which were formed by software of IMAGING-PAM) were used; their localizations were approximately similar for all investigated leaves in the control and ELFMF-treated plants. The procedure minimized errors, which could be related by different shapes and areas of the investigates leaves.

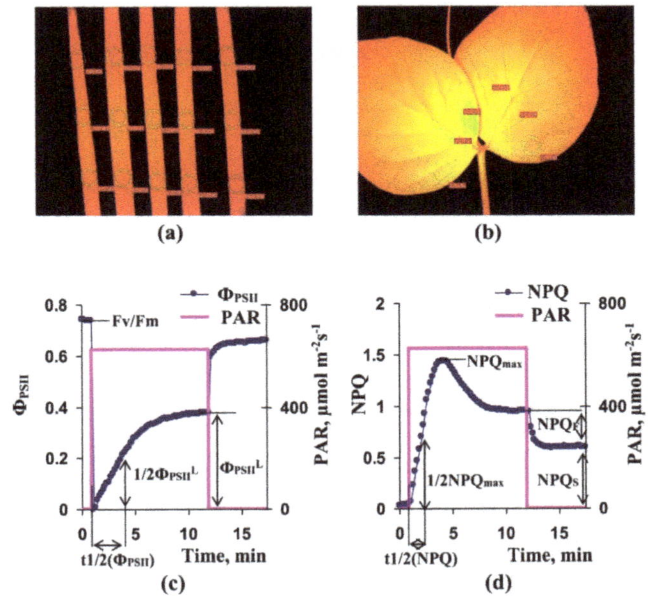

Figure 2. (**a**) Localizations of investigated areas (ROIs) in wheat leaves at PAM-imaging. Photosynthetic parameters were averaged for each wheat leaf (3 ROIs were averaged). (**b**) Localizations of investigated areas (ROIs) in pea leaves at PAM-imaging. Photosynthetic parameters were averaged for pea leaf (6 ROIs were averaged). (**c**) Record of changes in quantum yield of photosystem II (Φ_{PSII}) under action of actinic light (its intensity is marked as PAR). Fv/Fm is the potential quantum yield of photosystem II, $\Phi_{PSII}{}^{L}$ is the effective quantum yield of photosystem II after 10 min of illumination by actinic light, and $t_{1/2}(\Phi_{PSII})$ is the time taken for 50% increase of Φ_{PSII} under illumination. Wheat leaf is used for this record. (**d**) Record of changes in non-photochemical quenching (NPQ) under action of actinic light. NPQ_F is fast-relaxing component of NPQ after 10 min of illumination, NPQ_S is slow-relaxing component of NPQ after this illumination, NPQ_{max} is maximal value of NPQ, and $t_{1/2}$(NPQ) is time taken for 50% increase of NPQ under illumination. Wheat leaf is used for this record.

Seedlings were adapted to dark conditions for 15 min after fixation in the measuring system. After that, ML was turned on and SPs were periodically generated every 10 s. The first SP was used for estimation of initial and maximum rates of photosystem II fluorescence (F_0 and F_m, respectively); following SPs were used for estimation of the current rates of fluorescence (F) and maximum fluorescence rates under light conditions ($F_m{}'$). AL was turned on for 80 s after the first SP; duration of illumination by actinic light was about 10 min. Periodical SPs were generated for 5 min after termination of illuminations. Parameters of photosystem II, including Fv/Fm, the potential quantum yield of photosystem II, Φ_{PSII}, the effective quantum yield of photosystem II, and NPQ, the non-photochemical quenching of chlorophyll fluorescence, were calculated on the basis of F_0, F_m, F, and $F_m{}'$ in accordance to standard equations [16,40,41].

For further analysis, we used Fv/Fm, $\Phi_{PSII}{}^{L}$ (effective quantum yield of photosystem II after 10 min of illumination by actinic light), and $t_{1/2}(\Phi_{PSII})$ (time taken for 50% increase of Φ_{PSII} under illumination); estimation of these parameters are shown in Figure 2c. In addition, we used NPQ_F (fast-relaxing component of NPQ after 10 min of illumination), NPQ_S (slow-relaxing component of NPQ after this illumination), NPQ_{max} (maximal value of NPQ during illumination), and $t_{1/2}$(NPQ) (time taken for 50% increase of NPQ under illumination); estimation of these parameters were shown in Figure 2d.

2.4. Statistics

We used 30 wheat seedlings and 9 pea seedlings for each variant of the experiment (control, 7.8, 14.3, and 20.8 Hz) with short-term treatments by magnetic fields. 30 Wheat seedlings and 6 pea seedlings were exposed to chronic treatment by ELFMF (14.3 Hz); same quantities of plants were used in the control. Mean values, standard errors, and Pearson's correlation coefficients were presented. The Student's t-test was used for estimation of significance of differences between control plants and plants treated by ELFMFs.

3. Results

3.1. Investigation of the Influence of Short-Term Treatment by Magnetic Fields with Schumann Resonance Frequencies on Parameters of Photosynthetic Light Reactions

Investigation of the influence of short-term treatment by magnetic fields with Schumann resonance frequencies (the first, second, and third harmonics) on parameters of photosynthetic light reactions in wheat and pea seedlings was the first task of our work.

Figure S1 shows light-induced changes in Φ_{PSII} and NPQ, which were different in the control plants and those treated by ELFMF. Figure 3a shows that short-term treatment by ELFMFs (with frequencies equaling to 7.8, 14.3, and 20.8 Hz) did not influence the potential quantum yield of photosystem II, which was related to the maximal photochemical efficiency of photosystem II [16,40,41], in leaves of wheat seedlings. Effective quantum yield under illumination by AL, which was related to the photochemical efficiency of photosystem II under used light conditions [16,40,41], was also not affected by ELFMFs (Figure 3b). In contrast, significant decrease in time taken for 50% increase of Φ_{PSII} under illumination, which can likely be related to light-induced activation of ETC [42,43], was observed under action of MFs with 14.3 and 20.8 Hz frequencies (Figure 3c). This decrease was not significant under the action of ELFMF with 7.8 Hz frequency; however, the tendency for this decrease was also observed at this variant of MF treatment.

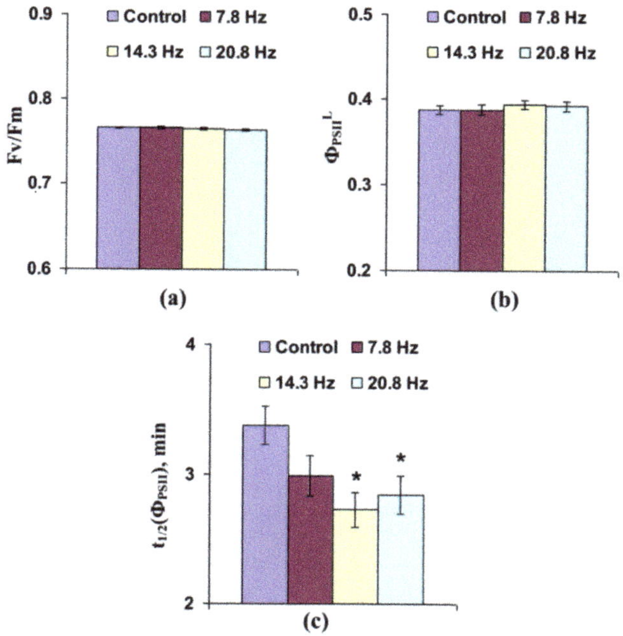

Figure 3. Influence of short-term treatment by artificial extremely low frequency magnetic field on potential quantum yield of photosystem II (Fv/Fm) (**a**), effective quantum yield of photosystem II

under illumination ($\Phi_{PSII}{}^{L}$) (**b**), and time taken for 50% increase of Φ_{PSII} under illumination ($t_{1/2}(\Phi_{PSII})$) (**c**) in wheat seedlings (n = 30). Action of the artificial magnetic field was immediately initiated before dark adaptation; total duration of its action was 30 min. Photosynthetic parameters were measured by the action of this field. Magnitude of the magnetic fields was 18 μT; frequencies were 7.8, 14.3, and 20.8 Hz. Control plants were not treated by this artificial magnetic field. *, difference between the experiment and control plants was significant ($p < 0.05$).

Influence of short-term treatment by the investigated ELFMFs on non-photochemical quenching of chlorophyll fluorescence was more expressive. It was shown (Figure 4a) that the treatment of wheat seedlings by MFs with 14.3 and 20.8 Hz frequency significantly decreased fast-relaxing component of NPQ, which can be considered as an energy-dependent component of the non-photochemical quenching [40,43,44]. Figure 4b shows that treatment by ELFMFs with frequencies equaling 7.8, 14.3, and 20.8 Hz could also decrease the slow-relaxing component of non-photochemical quenching, which was probable to be related to components of NPQ, which were caused by state transition (migration of the light harvesting complex II from photosystem II to photosysthem I) and photodamage of photosystem II [3,40,43,45,46]. Figure 4c shows that all used frequencies of ELFMFs, which were used for short-term treatment, decreased the maximal value of NPQ. Figure 4d shows that ELFMFs with 7.8 and 14.3 Hz frequencies decreased time taken for 50% increase of NPQ under illumination. The 20.8 Hz magnetic field also decreased this NPQ activation time; however, the effect was not significant. Considering the strong relation between light-induced photosynthetic activation and transient increase of NPQ (e.g., see our theoretical work [43]), decreases of both NPQ_{max} and $t_{1/2}(NPQ)$ could reflect stimulation of electron transport through ETC and acceleration of H^+ transport through thylakoid membranes.

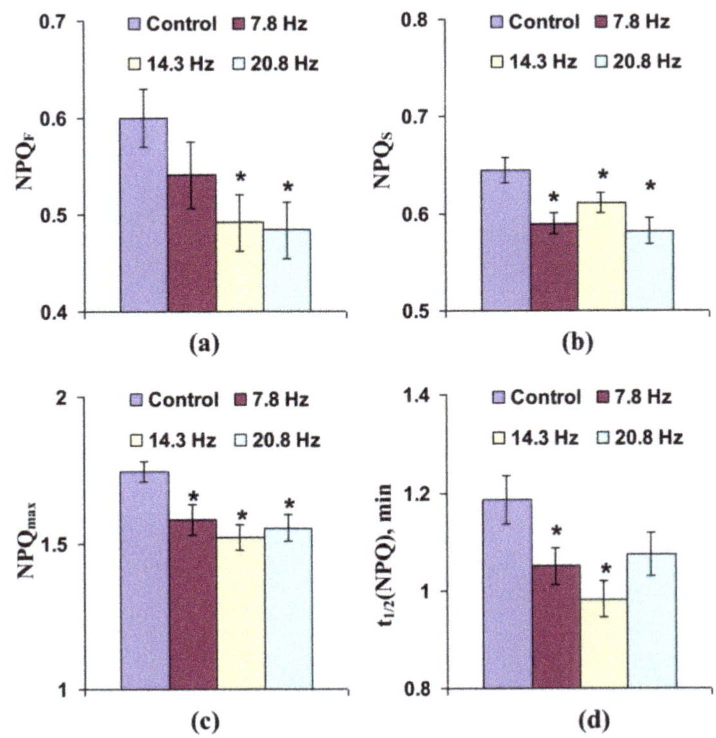

Figure 4. Influence of short-term treatment by artificial extremely low frequency magnetic field on the fast-relaxing component of non-photochemical quenching under illumination (NPQ_F) (**a**),

slow-relaxing component of non-photochemical quenching after this illumination (NPQ_S) (**b**), maximal value of non-photochemical quenching (NPQ_{max}) (**c**), and time taken for 50% increase of NPQ under illumination ($t_{1/2}$(NPQ)) (**d**) in wheat seedlings (n = 30). Action of the artificial magnetic field was immediately initiated before dark adaptation; total duration of its action was 30 min. Photosynthetic parameters were measured by the action of this field. Magnitude of the magnetic field was 18 μT; frequencies were 7.8, 14.3 and 20.8 Hz. Control plants were not treated by this artificial magnetic field. *, difference between the experiment and control plants was significant ($p < 0.05$).

Thus, results of analysis of the influence of short-term treatment by ELFMFs with frequencies equaling 7.8, 14.3, and 20.8 Hz on photosynthetic light reactions in leaves of wheat seedlings showed that these MFs could modify photosystem II parameters (especially, NPQ). This effect was most stable under MFs of 14.3 Hz frequency; as a result, we supposed that ELFMF with this frequency could be used for the analysis of the action of chronic treatment of wheat on photosynthetic light reactions.

Analysis of the influence of short-term treatment by ELFMFs with 7.8, 14.3, and 20.8 Hz frequencies on photosynthetic light reactions in leaves of pea seedlings showed other results. Figure S2 shows light-induced changes in Φ_{PSII} and NPQ, which seemed to be similar in the control plants and those treated by ELFMF. Figures 5 and 6 show that this treatment did not influence all photosynthetic parameters, which were investigated in our work (Fv/Fm, $\Phi_{PSII}{}^L$, $t_{1/2}(\Phi_{PSII})$, NPQ_F, NPQ_S, NPQ_{max}, and $t_{1/2}$(NPQ)); even tendencies to changes were absent. As a result, we could not select an optimal frequency for analysis of the chronic action of ELFMFs on pea seedlings on the basis of these results. Considering the high efficiency of short-term treatment of wheat seedlings by 14.3 Hz MFs, we also used the second harmonic in Schumann resonance frequencies for investigation of the influence of chronic ELFMF on pea seedlings in a further investigation.

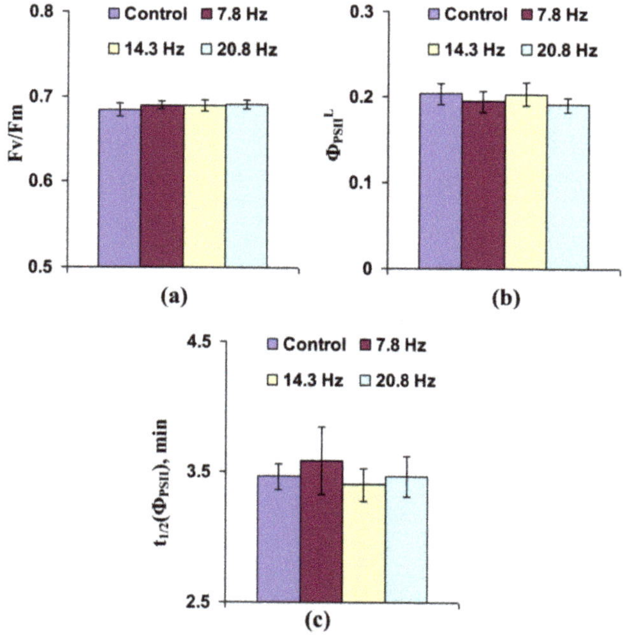

Figure 5. Influence of short-term treatment by artificial extremely low frequency magnetic field on potential quantum yield of photosystem II (Fv/Fm) (**a**), effective quantum yield of photosystem II

under illumination ($\Phi_{PSII}{}^{L}$) (**b**), and time taken for 50% increase of Φ_{PSII} under illumination ($t_{1/2}(\Phi_{PSII})$) (**c**) in pea seedlings (n = 9). Action of the artificial magnetic field was immediately initiated before dark adaptation; total duration of its action was 30 min. Photosynthetic parameters were measured by the action of this field. Magnitude of the magnetic fields was 18 μT; frequencies were 7.8, 14.3 and 20.8 Hz. Control plants were not treated by this artificial magnetic field. Significant differences between the experiment and control plants were absent.

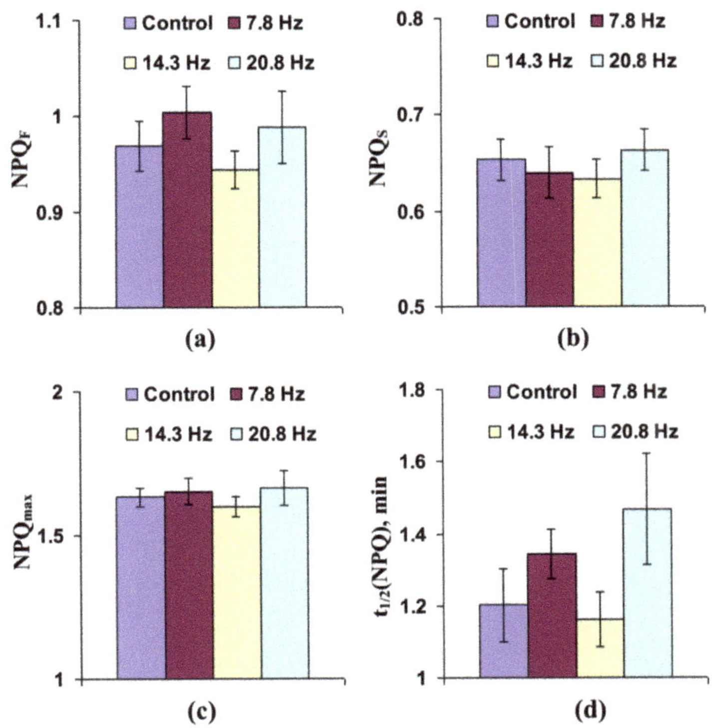

Figure 6. Influence of short-term treatment by artificial extremely low frequency magnetic field on fast-relaxing component of non-photochemical quenching under illumination (NPQ_F) (**a**), slow-relaxing component of non-photochemical quenching after this illumination (NPQ_S) (**b**), maximal value of non-photochemical quenching (NPQ_{max}) (**c**), and time taken for 50% increase of NPQ under illumination ($t_{1/2}$(NPQ)) (**d**) in pea seedlings (n = 9). Action of the artificial magnetic field was immediately initiated before dark adaptation; total duration of its action was 30 min. Photosynthetic parameters were measured by the action of this field. Magnitude of the magnetic fields was 18 μT; frequencies were 7.8, 14.3, and 20.8 Hz. Control plants were not treated by this artificial magnetic field. Significant differences between the experiment and control plants were absent.

3.2. Investigation of the Influence of Chronic Treatment by Magnetic Fields with the Second Harmonic in Schumann Resonance Frequencies on Parameters of Photosynthetic Light Reactions

Analysis of the influence of chronic treatment by ELFMF with 14.3 Hz frequency on photosynthetic light reactions in leaves of wheat seedlings showed that the effect of this treatment was different from the effect of short-term treatments by the investigated ELFMFs (Figure 7 and Figure S3). The potential quantum yield of photosystem II decreased under chronic treatment by 14.3 Hz frequency MF (Figure 7a); in contrast, the effective quantum yield of photosystem II under light conditions was increased under this treatment (Figure 7b). It is important that magnitudes of both changes were small (about 1.5% for

Fv/Fm and 4% for $\Phi_{PSII}{}^{L}$); these magnitudes were lower than photosynthetic changes, which were induced by short-term ELFMFs treatment (mainly, 10–20%). Figure 7c shows that time taken for 50% increase of Φ_{PSII} under illumination decreased under chronic action of 14.3 Hz frequency MF; magnitude of this effect was about 20% and was similar to the magnitudes of photosynthetic changes under short-term action of ELFMFs.

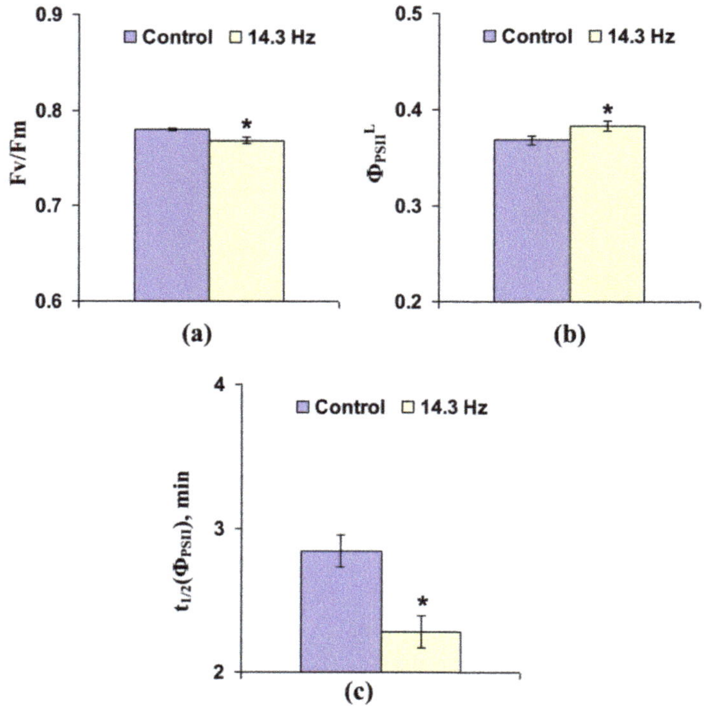

Figure 7. Influence of chronic treatment by artificial extremely low frequency magnetic field on potential quantum yield of photosystem II (Fv/Fm) (**a**), effective quantum yield of photosystem II under illumination ($\Phi_{PSII}{}^{L}$) (**b**), and time taken for 50% increase of Φ_{PSII} under illumination ($t_{1/2}(\Phi_{PSII})$) (**c**) in wheat seedlings (n = 30). Chronic action of the artificial magnetic field was initiated from soaking of seeds. Magnitude of the magnetic field was 18 µT; frequency was 14.3 Hz. Photosynthetic parameters were measured by the action of this field. Control plants were not treated by this artificial magnetic field. *, difference between the experiment and control plants was significant ($p < 0.05$).

Figure 8 shows that chronic treatment by 14.3 Hz frequency ELFMF did not change the fast and slow relaxing components of NPQ and maximal value of non-photochemical quenching in leaves of wheat seedlings; in contrast, time taken for 50% increase of NPQ under illumination was significantly decreased by this chronic MF treatment (more than 20%).

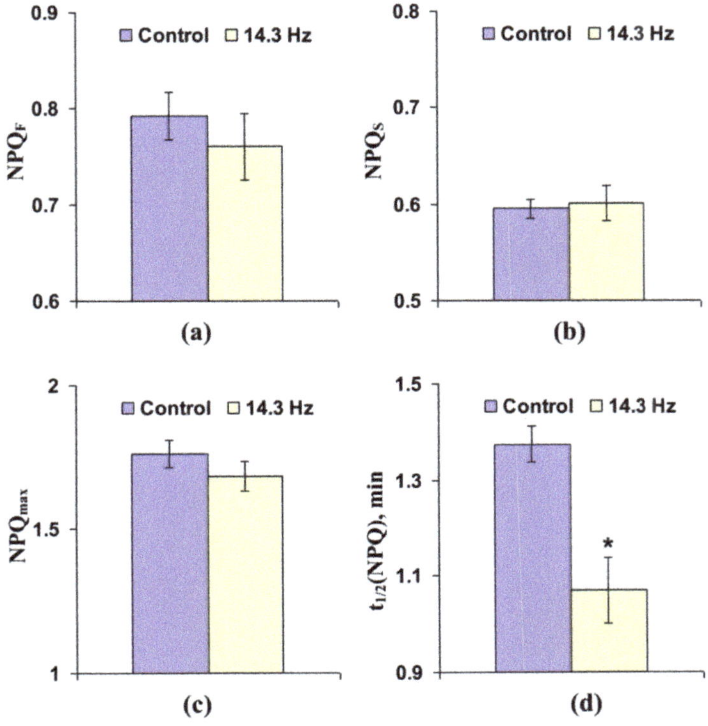

Figure 8. Influence of chronic treatment by artificial extremely low frequency magnetic field on fast-relaxing component of non-photochemical quenching under illumination (NPQ_F) (**a**), slow-relaxing component of non-photochemical quenching after this illumination (NPQ_S) (**b**), maximal value of non-photochemical quenching (NPQ_{max}) (**c**), and time taken for 50% increase of NPQ under illumination ($t_{1/2}$(NPQ)) (**d**) in wheat seedlings (n = 30). Chronic action of the artificial magnetic field was initiated from soaking of seeds. Magnitude of the magnetic field was 18 μT; frequency was 14.3 Hz. Photosynthetic parameters were measured by the action of this field. Control plants were not treated by this artificial magnetic field. *, difference between the experiment and control plants was significant ($p < 0.05$).

Figure 9, Figure 10, and Figure S4 show that chronic action of 14.3 Hz frequency ELFMF did not influence most of the investigated parameters of photosynthetic light reactions in leaves of peas seedlings, excluding the slow-relaxing component of NPQ. This parameter was significantly decreased under chronic action of ELFMF with frequency equaling 14.3 Hz (Figure 10b); the magnitude of this effect was about 14%. It is also interesting that a weak tendency to decrease under chronic treatment by 14.3 Hz frequency ELFMF was observed for $t_{1/2}(\Phi_{PSII})$ and $t_{1/2}$(NPQ). This effect was similar to changes in these parameters in wheat seedlings under short-term and chronic treatment by MFs.

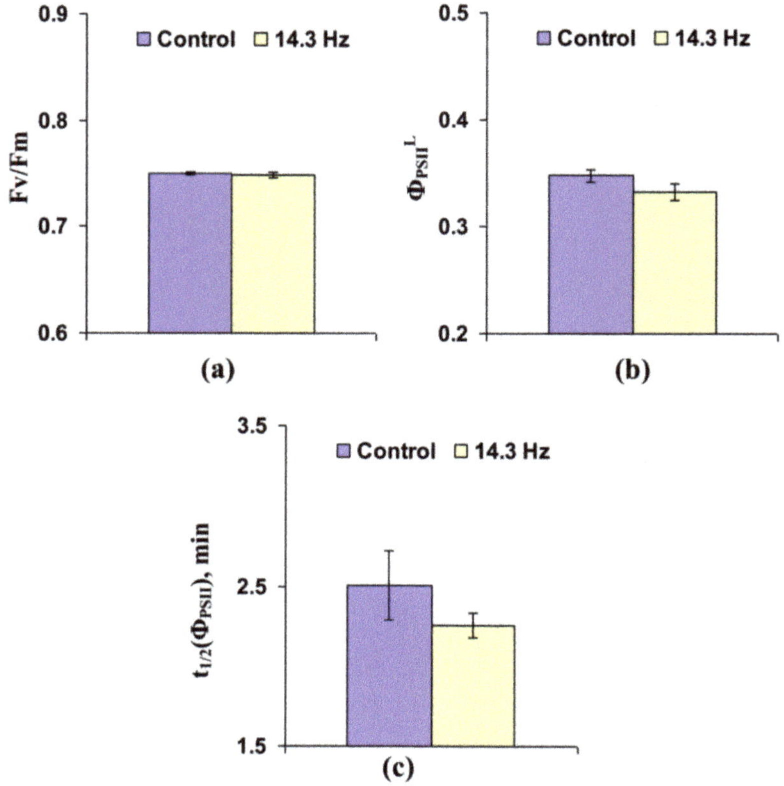

Figure 9. Influence of chronic treatment by the artificial extremely low frequency magnetic field on the potential quantum yield of photosystem II (Fv/Fm) (**a**), effective quantum yield of photosystem II under illumination ($\Phi_{PSII}{}^{L}$) (**b**), and time taken for 50% increase of Φ_{PSII} under illumination ($t_{1/2}(\Phi_{PSII})$) (**c**) in pea seedlings (n = 6). Chronic action of the artificial magnetic field was initiated from soaking of seeds. Magnitude of the magnetic field was 18 μT; frequency was 14.3 Hz. Photosynthetic parameters were measured by the action of this field. Control plants were not treated by this artificial magnetic field. Significant differences between the experiment and control plants were absent.

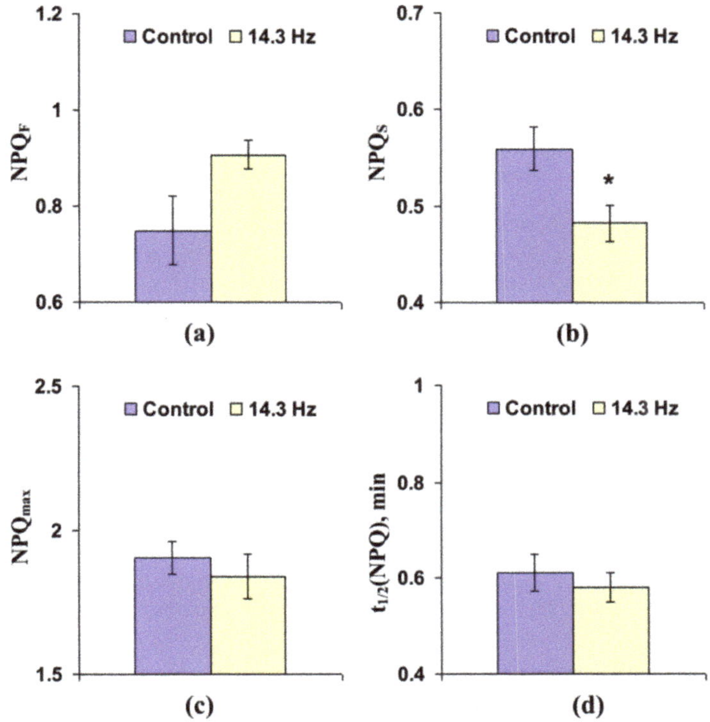

Figure 10. Influence of chronic treatment by the artificial extremely low frequency magnetic field on fast-relaxing component of non-photochemical quenching under illumination (NPQ_F) (**a**), slow-relaxing component of non-photochemical quenching after this illumination (NPQ_S) (**b**), maximal value of non-photochemical quenching (NPQ_{max}) (**c**), and time taken for 50% increase of NPQ under illumination ($t_{1/2}$(NPQ)) (**d**) in pea seedlings (n = 6). Chronic action of the artificial magnetic field was initiated from soaking of seeds. Magnitude of the magnetic field was 18 μT; frequency was 14.3 Hz. Photosynthetic parameters were measured by the action of this field. Control plants were not treated by this artificial magnetic field. *, difference between the experiment and control plants was significant ($p < 0.05$).

4. Discussion

It is known that photosynthesis can be strongly affected by numerous environmental stressors, including light with excess intensity [1–3], low or high temperatures [4–6], drought [7–9], salinity [10–12] or long-distance signals, relating to the action of stressors (e.g., electrical signals [13–15]). However, there are environmental factors with weakly-investigated influence on photosynthetic processes; in particular, ELFMFs that can be produced by direct human activity (mainly, 50 and 60 Hz [22]) and by natural environmental events (mainly, thunderstorm-caused Schumann resonance frequencies [23–25]). Considering the strong relation of photosynthetic activity to electron transport through chloroplast's ETC [16,17] and ion fluxes through thylakoid membranes [20,21], it can be expected that these processes should be affected by EFLMFs, which can likely influence charge transfer in different ways [27].

There is a small amount of work devoted to the investigation of ELFMFs on photosynthesis and relating processes (see reviews [27–30]). Moreover, only few works show the influence of these MFs on photosynthesis in plant seedlings [36,39]; other works investigate the influence on photosynthetic parameters in plants after seed treatments by ELFMFs [35,37,38]. Our current work, which is devoted to the analysis of the influence of

ELFMFs with 7.8, 14.3, and 20.8 Hz frequencies (the first, second and third Schumann resonance frequencies) on parameters of photosynthetic light reactions, shows the following points:

1. ELFMFs with Schumann resonance frequencies can significantly influence photosynthetic parameters in plants; however, this effect is dependent on plant species: changes are observed in wheat seedlings (Figures 3, 4, 7 and 8) and mainly absent in pea seedlings (Figures 5, 6, 9 and 10), excluding NPQ_S decrease under chronic action of MF (Figure 10b).

2. Effects induced by short-term and chronic treatments by ELFMFs are different: short-term action mainly influences NPQ (Figure 4a–c), chronic action mainly modifies quantum yields (Figure 7a,b). However, magnitudes of changes in these yields under chronic action of MFs (1.5–4%) are much lower than magnitudes of changes in NPQ under the short-term action (10–20%).

3. Both short-term and chronic treatments by ELFMFs are likely to decrease the time of light-induced activation of ETC in wheat seedlings, because decreases of $t1/2(\Phi_{PSII})$ (Figures 3c and 7c) and t1/2(NPQ) (Figures 4d and 8d) are shown under short-term and chronic treatments. Relative magnitudes of these decreases are about 20%.

The last point shows some potential mechanisms of revealing photosynthetic changes. t1/2(NPQ) for wheat seedling is about 0.9–1.4 min in our experiments (Figures 4d and 8d). These fast changes are related to the induction of the energy-dependent component of NPQ, which is caused by pH decrease in the chloroplast lumen and protonation of PsbS proteins in the light-harvesting complex [46–48]; i.e., they are dependent on proton transport through a thylakoid membrane. It should be additionally noted that slower mechanisms of NPQ can also be dependent on proton transport: synthesis of zeaxanthin and anteraxanthin from violaxanthin in the xanthophyll cycle [46,49–51] and state transition [52] are activated by decrease of pH in lumen.

It is probable that light-induced increase of Φ_{PSII} (Figure 3c, Figure 7c, and Figures S1 and S3) is also influenced by proton transport through thylakoid membrane and increased pH in the stroma of chloroplasts [53] because stroma alkalization can stimulate enzymes of the Calvin-Benson cycle [54,55] and activate ferredoxin-NADP-reductase through change in its localization [18,19].

If the hypothesis about participation of proton transport in both NPQ and Φ_{PSII} increase under illumination is correct, it should be expected that $t_{1/2}$(NPQ) and $t_{1/2}(\Phi_{PSII})$ should be related. Correlation analysis shows that Pearson's correlation coefficients between these values are 0.71 ($p < 0.05$) and 0.68 ($p < 0.05$) for the control wheat seedlings in short-term and chronic experiments, respectively. Moreover, these coefficients are 0.61–0.78 for different experimental variants with treatment by MFs ($p < 0.05$ for all variants, data not shown). The results support the hypothesis about participation of proton transport in the increase of NPQ and Φ_{PSII} under illumination.

There are different mechanisms of proton transport through thylakoid membranes [20,21,56]: proton influx to lumen at ETC activity, proton efflux to stroma through H^+-ATP-synthase, and proton fluxes through co-transporters of protons and ions (in particular, Ca^{2+}/H^+ antiport in thylakoid membrane [57]). Potentially, increases in rates of both H^+ influx to lumen and H^+ efflux to stroma should decrease time of forming of stationary pH; however, increase in H^+ influx is accompanied by stimulation of lumen acidification under illumination and increase of H^+ efflux is related to weakening of this acidification. Our experimental results show that NPQ_F, NPQ_S, and NPQ_{max} are decreased in wheat seedlings under short-term treatment by ELFMFs, which induces a decrease of $t_{1/2}$(NPQ) and $t_{1/2}(\Phi_{PSII})$ (Figure 4); moreover, similar insignificant decreases are also observed under chronic action of ELFMFs (Figure 8). Considering a strong relation between NPQ formation and decrease of pH in lumen [46–52], the results support weakening of lumen acidification under treatment by ELFMFs; i.e., increase of rate of H^+ efflux is a more probable result of wheat treatment by ELFMFs than increase in H^+ influx rate.

Probable stimulation of H^+ efflux by treatment by ELFMFs with Schumann resonance frequencies can also be supported by literature data. (i) Increase in the membrane permeability for ions (including protons) after plant treatment by ELFMFs [58]; the effect is observed at 10 or 100 μT and 50 or 60 Hz. Similar increase in the membrane permeability can contribute to proton leak and stimulate H^+ efflux to lumen. (ii) ELFMFs can influence Ca^{2+} homeostasis in plants (including increase in free Ca^{2+} concentration) [59–63] that can be related to the direct or indirect effects of cyclotron resonance [27,62]. Considering Ca^{2+}/H^+ antiport in thylakoid membrane [53], increase in Ca^{2+} concentration can stimulate H^+ efflux. (iii) Increase in Ca^{2+} concentration can also modify the activity of the Calvin-Benson cycle [57,64]; in accordance with our previous work, modification of the photosynthetic dark stage can strongly influence the activity of H^+-ATP-synthase and luminal and stromal pH [65].

The influence of ELFMFs on photosynthetic light reactions in wheat seedlings and the absence of the influence in pea plants is another important result of our work. It is in good agreement with the results of others (e.g., [31,34]), which show different MF-induced physiological changes in different plant species. Mechanisms of revealed differences in photosynthetic responses on ELFMFs require future investigations; however, some potential ways can be speculated. Our results show that pea seedlings are more sensitive to light intensity than the wheat ones. In particular, pea leaves have low Φ_{PSII} (Figures 5b and 9b) in comparison to wheat leaves (Figures 3b and 7b) under the used intensity of actinic light. The energy-dependent component of NPQ in pea leaves is larger than this parameter in wheat (average NPQ_F, which is calculated on the basis of all control plants, is 0.881 ± 0.042 in pea and 0.696 ± 0.023 in wheat, $p < 0.05$) and NPQ relaxation under light conditions is slower (Figures S1–S4). Considering the relation between energy-dependent NPQ forming and lumen acidification [46–52], pea chloroplasts can likely have a larger magnitude of proton gradient across the thylakoid membrane than wheat ones. In this case, the influence of the ELFMF-induced increase of proton leak on pH in lumen should be weak (proton concentration in lumen is high). Additionally, the potential effect of ELFMF-induced stimulation of Ca^{2+}/H^+ antiport, which is localized in the thylakoid membrane [53], can also be weak because high proton gradient should induce strong Ca^{2+} flux into the lumen even without the additional stimulation by ELFMFs (i.e., this additional stimulation of antiporter probably weakly influences Ca^{2+} concentrations in lumen and stroma of chloroplast).

As a whole, we reveal changes in the parameters of photosynthetic light reactions in wheat seedlings under short-term and chronic treatment by ELFMFs with Schumann resonance frequencies. The results show that photosynthetic light reactions can be affected by ELFMFs and that light-induced changes in photosynthetic processes (possibly relating to proton transport through thylakoid membrane) are likely to be more sensitive to the action of these magnetic fields than stationary photosynthetic parameters.

5. Conclusions

We investigated the influence of ELFMFs with Schumann resonance frequencies (7.8, 14.3, and 20.8 Hz) on parameters of photosynthetic light reactions in wheat and pea seedlings. These are the following points shown in our current investigation:

ELFMFs with Schumann resonance frequencies can significantly influence photosynthetic parameters in plants; however, this effect is dependent on plant species: changes are observed in wheat seedlings and mainly absent in pea seedlings, excluding NPQ_S decrease under chronic action of MF.

Effects induced by short-term and chronic treatments by ELFMFs are different: short-term action mainly influences NPQ, while chronic action mainly modifies quantum yields. However, magnitudes of changes in these yields under chronic action of MFs are much lower than those in NPQ under the short-term action.

Both short-term and chronic treatments by ELFMFs are likely to decrease the time of light-induced photosynthetic activation in wheat seedlings, because decreases in $t_{1/2}(\Phi_{PSII})$ and $t_{1/2}(NPQ)$ are shown under short-term and chronic treatments.

Supplementary Materials: The following are available online at https://www.mdpi.com/2073-4409/10/1/149/s1, Figure S1: Average dynamics of changes in quantum yield of photosystem II (Φ_{PSII}) (a) and non-photochemical quenching (NPQ) (b) under action of actinic light (its intensity is marked as PAR) in wheat seedlings under short-term action of ELFMFs with different frequencies. Figure S2: Average dynamics of changes in quantum yield of photosystem II (Φ_{PSII}) (a) and non-photochemical quenching (NPQ) (b) under action of actinic light (its intensity is marked as PAR) in pea seedlings under short-term action of ELFMFs with different frequencies. Figure S3: Average dynamics of changes in quantum yield of photosystem II (Φ_{PSII}) (a) and non-photochemical quenching (NPQ) (b) under action of actinic light (its intensity is marked as PAR) in wheat seedlings under chronic action of ELFMF. Figure S4: Average dynamics of changes in quantum yield of photosystem II (Φ_{PSII}) (a) and non-photochemical quenching (NPQ) (b) under action of actinic light (its intensity is marked as PAR) in pea seedlings under chronic action of ELFMF.

Author Contributions: Conceptualization, V.S., V.V., and C.P.; methodology, V.S., V.V., and N.I.; formal analysis, V.S. and E.S.; investigation, Y.S., E.G., and N.M.; writing—original draft preparation, V.S., E.S., and A.R.; writing—review and editing, C.P., and E.M.; visualization, V.S. and E.S..; supervision, C.P.; project administration, V.S, and V.V.; funding acquisition, C.P. and E.M. All authors have read and agreed to the published version of the manuscript.

Funding: The reported study was funded by the Government of the Russian Federation, contract number 075-15-2019-1892.

Institutional Review Board Statement: Not applicable.

Informed Consent Statement: Not applicable.

Data Availability Statement: The data presented in this study are available on request from the corresponding author.

Conflicts of Interest: The authors declare no conflict of interest. The funders had no role in the design of the study; in the collection, analyses, or interpretation of data; in the writing of the manuscript, or in the decision to publish the results.

References

1. Quiles, M.J.; López, N.I. Photoinhibition of photosystems I and II induced by exposure to high light intensity during oat plant growth. Effects on the chloroplast NADH dehydrogenase complex. *Plant Sci.* **2004**, *166*, 815–823. [CrossRef]
2. Kalaji, H.M.; Carpentier, R.; Allakhverdiev, S.I.; Bosa, K. Fluorescence parameters as early indicators of light stress in barley. *J. Photochem. Photobiol. B* **2012**, *112*, 1–6. [CrossRef] [PubMed]
3. Murata, N.; Allakhverdiev, S.I.; Nishiyama, Y. The mechanism of photoinhibition *in vivo*: Re-evaluation of the roles of catalase, α-tocopherol, non-photochemical quenching, and electron transport. *Biochim. Biophys. Acta* **2012**, *1817*, 1127–1133. [CrossRef] [PubMed]
4. Kanervo, E.; Tasaka, Y.; Murata, N.; Aro, E.M. Membrane lipid unsaturation modulates processing of the photosystem II reaction center protein D1 at low temperatures. *Plant Physiol.* **1997**, *114*, 841–849. [CrossRef]
5. Battisti, D.S.; Naylor, R.L. Historical warnings of future food insecurity with unprecedented seasonal heat. *Science* **2009**, *323*, 240–244. [CrossRef]
6. Allakhverdiev, S.I.; Kreslavski, V.D.; Klimov, V.V.; Los, D.A.; Carpentier, R.; Mohanty, P. Heat stress: An overview of molecular responses in photosynthesis. *Photosynth. Res.* **2008**, *98*, 541–550. [CrossRef]
7. Lawlor, D.W.; Tezara, W. Causes of decreased photosynthetic rate and metabolic capacity in water-deficient leaf cells: A critical evaluation of mechanisms and integration of processes. *Ann. Bot.* **2009**, *103*, 561–579. [CrossRef]
8. Zivcak, M.; Brestic, M.; Balatova, Z.; Drevenakova, P.; Olsovska, K.; Kalaji, H.M.; Yang, X.; Allakhverdiev, S.I. Photosynthetic electron transport and specific photoprotective responses in wheat leaves under drought stress. *Photosynth. Res.* **2013**, *117*, 529–546. [CrossRef]
9. Urban, L.; Aarrouf, J.; Bidel, L.P.R. Assessing the effects of water deficit on photosynthesis using parameters derived from measurements of leaf gas exchange and of chlorophyll a fluorescence. *Front. Plant Sci.* **2017**, *8*, 2068. [CrossRef]
10. Stepien, P.; Johnson, G.N. Contrasting responses of photosynthesis to salt stress in the glycophyte *Arabidopsis* and the halophyte *Thellungiella*: Role of the plastid terminal oxidase as an alternative electron sink. *Plant Physiol.* **2009**, *149*, 1154–1165. [CrossRef]
11. Mehta, P.; Kraslavsky, V.; Bharti, S.; Allakhverdiev, S.I.; Jajoo, A. Analysis of salt stress induced changes in Photosystem II heterogeneity by prompt fluorescence and delayed fluorescence in wheat (*Triticum aestivum*) leaves. *J. Photochem. Photobiol. B.* **2011**, *104*, 308–313. [CrossRef] [PubMed]
12. Acosta-Motos, J.R.; Ortuño, M.F.; Bernal-Vicente, A.; Diaz-Vivancos, P.; Sanchez-Blanco, M.J.; Hernandez, J.A. Plant responses to salt stress: Adaptive mechanisms. *Agronomy* **2017**, *7*, 18. [CrossRef]

13. Pavlovič, A. The effect of electrical signals on photosynthesis and respiration. In *Plant Electrophysiology*; Volkov, A., Ed.; Springer: Berlin/Heidelberg, Germany, 2012; pp. 33–62.
14. Szechyńska-Hebda, M.; Lewandowska, M.; Karpiński, S. Electrical signaling, photosynthesis and systemic acquired acclimation. *Front Physiol.* **2017**, *8*, 684. [CrossRef] [PubMed]
15. Sukhov, V.; Sukhova, E.; Vodeneev, V. Long-distance electrical signals as a link between the local action of stressors and the systemic physiological responses in higher plants. *Prog. Biophys. Mol. Biol.* **2019**, *146*, 63–84. [CrossRef] [PubMed]
16. Kalaji, H.M.; Schansker, G.; Ladle, R.J.; Goltsev, V.; Bosa, K.; Allakhverdiev, S.I.; Brestic, M.; Bussotti, F.; Calatayud, A.; Dąbrowski, P.; et al. Frequently asked questions about in vivo chlorophyll fluorescence: Practical issues. *Photosynth. Res.* **2014**, *122*, 121–158. [CrossRef]
17. Cruz, J.A.; Avenson, T.J.; Kanazawa, A.; Takizawa, K.; Edwards, G.E.; Kramer, D.M. Plasticity in light reactions of photosynthesis for energy production and photoprotection. *J. Exp. Bot.* **2005**, *56*, 395–406. [CrossRef]
18. Alte, F.; Stengel, A.; Benz, J.P.; Petersen, E.; Soll, J.; Groll, M.; Bölter, B. Ferredoxin: NADPH oxidoreductase is recruited to thylakoids by binding to a polyproline type II helix in a pH-dependent manner. *Proc. Natl. Acad. Sci. USA* **2010**, *107*, 19260–19265. [CrossRef]
19. Benz, J.P.; Stengel, A.; Lintala, M.; Lee, Y.H.; Weber, A.; Philippar, K.; Gügel, I.L.; Kaieda, S.; Ikegami, T.; Mulo, P.; et al. Arabidopsis Tic62 and ferredoxin-NADP(H) oxidoreductase form light-regulated complexes that are integrated into the chloroplast redox poise. *Plant Cell.* **2010**, *21*, 3965–3983. [CrossRef]
20. Pottosin, I.; Shabala, S. Transport across chloroplast membranes: Optimizing photosynthesis for adverse environmental conditions. *Mol. Plant* **2016**, *9*, 356–370. [CrossRef]
21. Szabò, I.; Spetea, C. Impact of the ion transportome of chloroplasts on the optimization of photosynthesis. *J. Exp. Bot.* **2017**, *68*, 3115–3128. [CrossRef]
22. Price, C.; Williams, E.; Elhalel, G.; Sentman, D. Natural ELF fields in the atmosphere and in living organisms. *Int. J. Biometeorol.* **2020**, *8*, 1–8. [CrossRef] [PubMed]
23. Price, C.; Melnikov, A. Diurnal, seasonal and inter-annual variations in the Schumann resonance parameters. *J. Atmos. Sol. Terr. Phys.* **2004**, *66*, 1179–1185. [CrossRef]
24. Price, C. ELF Electromagnetic waves from lightning: The Schumann resonances. *Atmosphere* **2016**, *7*, 116. [CrossRef]
25. Elhalel, G.; Price, C.; Fixler, D.; Shainberg, A. Cardioprotection from stress conditions by weak magnetic fields in the Schumann Resonance band. *Sci. Rep.* **2019**, *9*, 1645. [CrossRef] [PubMed]
26. Sakurai, T.; Yoshimoto, M.; Koyama, S.; Miyakoshi, J. Exposure a extremely low frequency magnetic fields affects insulin-secreting cells. *Bioelectromagnetics* **2008**, *29*, 118–124. [CrossRef] [PubMed]
27. Maffei, M.E. Magnetic field effects on plant growth, development, and evolution. *Front. Plant Sci.* **2014**, *5*, 445. [CrossRef]
28. Da Silva, J.A.T.; Dobránszki, J. Magnetic fields: How is plant growth and development impacted? *Protoplasma* **2016**, *253*, 231–248. [CrossRef]
29. Radhakrishnan, R. Magnetic field regulates plant functions, growth and enhances tolerance against environmental stresses. *Physiol. Mol. Biol. Plants* **2019**, *25*, 1107–1119. [CrossRef]
30. Sarraf, M.; Kataria, S.; Taimourya, H.; Santos, L.O.; Menegatti, R.D.; Jain, M.; Ihtisham, M.; Liu, S. Magnetic field (MF) applications in plants: An overview. *Plants* **2020**, *9*, 1139. [CrossRef]
31. Davies, M.S. Effects of 60 Hz electromagnetic fields on early growth in three plant species and a replication of previous results. *Bioelectromagnetics* **1996**, *17*, 154–161. [CrossRef]
32. Aksyonov, S.J.; Bulychev, A.A.; Grunina, T.Y.; Goryachev, S.N.; Turovetsky, V.B. Effects of ELF-EMF treatment on wheat seeds at different stages of germination and possible mechanism of their origin. *Electromagn. Biol. Med.* **2001**, *20*, 231–253.
33. Iqbal, M.; Haq, Z.U.; Jamil, Y.; Ahmad, M.R. Effect of presowing magnetic treatment on properties of pea. *Int. Agrophys.* **2012**, *26*, 25–31. [CrossRef]
34. Fischer, G.; Tausz, M.; Köck, M.; Grill, D. Effects of weak 16 3/2 Hz magnetic fields on growth parameters of young sunflower and wheat seedlings. *Bioelectromagnetics* **2004**, *25*, 638–641. [CrossRef] [PubMed]
35. Rochalska, M. Influence of frequent magnetic field on chlorophyll content in leaves of sugar beet plants. *Nukleonika* **2005**, *50*, S25–S28.
36. Aleman, E.I.; Mbogholi, A.; Boix, Y.F.; Gonzalez-Ohnedo, J.; Chalfun, A. Effects of EMFs on some biological parameters in coffee plants (*Coffea arabica* L.) obtained by in vitro propagation. *Polish, J. Environ. Stud.* **2014**, *23*, 95–101.
37. Kornarzyński, K.; Dziwulska-Hunek, A.; Kornarzyńska-Gregorowicz, A.; Sujak, A. Effect of electromagnetic stimulation of amaranth seeds of different initial moisture on the germination parameters and photosynthetic pigments content. *Sci. Rep.* **2018**, *8*, 14023. [CrossRef]
38. De Souza-Torres, A.; Sueiro-Pelegrín, L.; Zambrano-Reyes, M.; Macías-Socarras, I.; González-Posada, M.; García-Fernández, D. Extremely low frequency non-uniform magnetic fields induce changes in water relations, photosynthesis and tomato plant growth. *Int. J. Radiat. Biol.* **2020**, *96*, 951–957. [CrossRef]
39. Yano, A.; Ohashi, Y.; Hirasaki, T.; Fujiwara, K. Effects of a 60 Hz magnetic field on photosynthetic CO_2 uptake and early growth of radish seedlings. *Bioelectromagnetics* **2004**, *25*, 572–581. [CrossRef]
40. Maxwell, K.; Johnson, G.N. Chlorophyll fluorescence—A practical guide. *J. Exp. Bot.* **2000**, *51*, 659–668. [CrossRef]

41. Porcar-Castell, A.; Tyystjärvi, E.; Atherton, J.; van der Tol, C.; Flexas, J.; Pfündel, E.E.; Moreno, J.; Frankenberg, C.; Berry, J.A. Linking chlorophyll a fluorescence to photosynthesis for remote sensing applications: Mechanisms and challenges. *J. Exp. Bot.* **2014**, *65*, 4065–4095. [CrossRef]
42. Kalaji, H.M.; Schansker, G.; Brestic, M.; Bussotti, F.; Calatayud, A.; Ferroni, L.; Goltsev, V.; Guidi, L.; Jajoo, A.; Li, P.; et al. Frequently asked questions about chlorophyll fluorescence, the sequel. *Photosynth. Res.* **2017**, *132*, 13–66. [CrossRef] [PubMed]
43. Sukhova, E.; Khlopkov, A.; Vodeneev, V.; Sukhov, V. Simulation of a nonphotochemical quenching in plant leaf under different light intensities. *Biochim. Biophys. Acta Bioenerg.* **2020**, *1861*, 148138. [CrossRef]
44. Sukhova, E.; Sukhov, V. Analysis of light-induced changes in the photochemical reflectance index (PRI) in leaves of pea, wheat, and pumpkin using pulses of green-yellow measuring light. *Remote Sens.* **2019**, *11*, 810. [CrossRef]
45. Murata, N.; Takahashi, S.; Nishiyama, Y.; Allakhverdiev, S.I. Photoinhibition of photosystem II under environmental stress. *Biochim. Biophys. Acta Bioenerg.* **2007**, *1767*, 414–421. [CrossRef]
46. Ruban, A.V. Evolution under the sun: Optimizing light harvesting in photosynthesis. *J. Exp. Bot.* **2015**, *66*, 7–23. [CrossRef] [PubMed]
47. Li, X.P.; Björkman, O.; Shih, C.; Grossman, A.R.; Rosenquist, M.; Jansson, S.; Niyogi, K.K. A pigment-binding protein essential for regulation of photosynthetic light harvesting. *Nature* **2000**, *403*, 391–395. [CrossRef] [PubMed]
48. Li, X.P.; Gilmore, A.M.; Caffarri, S.; Bassi, R.; Golan, T.; Kramer, D.; Niyogi, K.K. Regulation of photosynthetic light harvesting involves intrathylakoid lumen pH sensing by the PsbS protein. *J. Biol. Chem.* **2004**, *279*, 22866–22874. [CrossRef]
49. Demmig-Adams, B.; Adams III, W.W. The role of xanthophyll cycle carotenoids in the protection of photosynthesis. *Trends Plant Sci.* **1996**, *1*, 21–26. [CrossRef]
50. Müller, P.; Li, X.P.; Niyogi, K.K. Non-photochemical quenching. A response to excess light energy. *Plant Physiol.* **2001**, *125*, 1558–1566. [CrossRef]
51. Jajoo, A.; Mekala, N.R.; Tongra, T.; Tiwari, A.; Grieco, M.; Tikkanen, M.; Aro, E.M. Low pH-induced regulation of excitation energy between the two photosystems. *FEBS Lett.* **2014**, *588*, 970–974. [CrossRef]
52. Singh-Rawal, P.; Jajoo, A.; Mathur, S.; Mehta, P.; Bharti, S. Evidence that pH can drive state transitions in isolated thylakoid membranes from spinach. *Photochem. Photobiol. Sci.* **2010**, *9*, 830–837. [CrossRef] [PubMed]
53. Sukhov, V. Electrical signals as mechanism of photosynthesis regulation in plants. *Photosynth. Res.* **2016**, *130*, 373–387. [CrossRef] [PubMed]
54. Flügge, U.I.; Freisl, M.; Heldt, H.W. The mechanism of the control of carbon fixation by the pH in the chloroplast stroma: Studies with acid mediated proton transfer across the envelope. *Planta* **1980**, *149*, 48–51. [CrossRef] [PubMed]
55. Wolosiuk, R.A.; Ballicora, M.A.; Hagelin, K. The reductive pentose phosphate cycle for photosynthetic CO2 assimilation: Enzyme modulation. *FASEB J.* **1993**, *7*, 622–637. [CrossRef] [PubMed]
56. Shikanai, T.; Yamamoto, H. Contribution of cyclic and pseudo-cyclic electron transport to the formation of proton motive force in chloroplasts. *Mol. Plant.* **2017**, *10*, 20–29. [CrossRef] [PubMed]
57. Ettinger, W.F.; Clear, A.M.; Fanning, K.J.; Peck, M.L. Identification of a Ca^{2+}/H^{+} antiport in the plant chloroplast thylakoid membrane. *Plant Physiol.* **1999**, *119*, 1379–1386. [CrossRef] [PubMed]
58. Stange, B.C.; Rowland, R.E.; Rapley, B.I.; Podd, J.V. ELF magnetic fields increase amino acid uptake into *Vicia faba* L. roots and alter ion movement across the plasma membrane. *Bioelectromagnetics* **2002**, *23*, 347–354. [CrossRef]
59. Pazur, A.; Rassadina, V.; Dandler, J.; Zoller, J. Growth of etiolated barley plants in weak static and 50 Hz electromagnetic fields tuned to calcium ion cyclotron resonance. *Biomagn. Res. Technol.* **2006**, *4*, 1. [CrossRef]
60. Betti, L.; Trebbi, G.; Fregola, F.; Zurla, M.; Mesirca, P.; Brizzi, M.; Borghini, F. Weak static and extremely low frequency magnetic fields affect in vitro pollen germination. *Sci. World J.* **2011**, *11*, 875–890. [CrossRef]
61. Belyavskaya, N.A. Biological effects due to weak magnetic field on plants. *Adv. Space Res.* **2004**, *34*, 1566–1574. [CrossRef]
62. Goldsworthy, A. Effects of electrical and electromagnetic fields on plants and related topics. In *Plant Electrophysiology. Theory and Methods*; Volkov, A.G., Ed.; Springer-Verlag: Berlin/Heidelberg, Germany, 2006.
63. Pazur, A.; Rassadina, V. Transient effect of weak electromagnetic fields on calcium ion concentration in Arabidopsis thaliana. *BMC Plant Biol.* **2009**, *9*, 47. [CrossRef] [PubMed]
64. Krupenina, N.A.; Bulychev, A.A. Action potential in a plant cell lowers the light requirement for non-photochemical energy-dependent quenching of chlorophyll fluorescence. *Biochim. Biophys. Acta* **2007**, *1767*, 781–788. [CrossRef] [PubMed]
65. Sukhov, V.; Surova, L.; Morozova, E.; Sherstneva, O.; Vodeneev, V. Changes in H^{+}-ATP synthase activity, proton electrochemical gradient, and pH in pea chloroplast can be connected with variation potential. *Front Plant Sci.* **2016**, *7*, 1092. [CrossRef] [PubMed]

cells

MDPI

Article

The Effect of Abiotic Factors on Abundance and Photosynthetic Performance of Airborne Cyanobacteria and Microalgae Isolated from the Southern Baltic Sea Region

Kinga Wiśniewska [1], Sylwia Śliwińska-Wilczewska [2,*], Anita Lewandowska [1] and Marta Konik [3]

[1] Division of Marine Chemistry and Environmental Protection, Institute of Oceanography, University of Gdansk, Avenue Piłsudskiego 46, P-81-378 Gdynia, Poland; kinga.wisniewska@phdstud.ug.edu.pl (K.W.); anita.lewandowska@ug.edu.pl (A.L.)

[2] Division of Marine Ecosystems Functioning, Institute of Oceanography, University of Gdansk, Avenue Piłsudskiego 46, P-81-378 Gdynia, Poland

[3] Department of Marine Physics, Institute of Oceanology Polish Academy of Sciences, P-81-779 Sopot, Poland; mk@iopan.gda.pl

* Correspondence: ocessl@ug.edu.pl; Tel.: +48-58-523-68-92

Abstract: Cyanobacteria and microalgae present in the aquatic or terrestrial environment may be emitted into the air and transported along with air masses over long distances. As a result of staying in the atmosphere, these organisms may develop a greater tolerance to stressful factors, but this topic is still relatively unknown. The main aim was to show an autecological characteristic of some airborne microalgae and cyanobacteria strains by a factorial laboratory experiment approach, including changes in irradiance, temperature, and salinity conditions. The additional purpose of this work was also to present part of the Culture Collection of Baltic Algae (CCBA) collection, which consists of airborne algae (AA) isolated from the atmospheric air of the southern Baltic Sea region. Altogether, 61 strains of airborne cyanobacteria and microalgae from the southern Baltic Sea region were isolated from May 2018 to August 2020. Selected microorganisms were tested in controlled laboratory conditions to identify their response to different irradiance (10–190 μmol photons m^{-2} s^{-1}), temperature (13–23 °C), and salinity conditions (0–36 PSU). The highest numbers of cells (above 30×10^5 cell mL^{-1}) were recorded for cyanobacterium *Nostoc* sp., and for diatoms *Nitzschia* sp., *Amphora* sp., and *Halamphora* sp. We found that for cyanobacterium *Nostoc* sp. as well as for green alga *Coccomyxa* sp. the maximum cell concentrations were recorded at the salinity of 0 PSU. Moreover, cyanobacteria *Planktolyngbya contorta*, *Pseudanabaena catenata*, *Leptolyngbya foveolarum*, *Gloeocapsa* sp., and *Rivularia* sp. were able to grow only at a salinity of 0 PSU. On the other hand, in the range of 16–24 PSU, the highest cell numbers of examined diatoms have been identified. Our research provided that deposited airborne microalgae and cyanobacteria showed full colonization potential. The present experiment suggests that the adaptive abilities of microorganisms, in particular those producing toxins, may contribute to the spread in the future. Thus, it may increase human exposure to their negative health effects. Any distinctive adaptations of the genera give them an additional competitive advantage and a greater chance for territorial expansion.

Keywords: environmental stress; abiotic stressors; chlorophyll fluorescence; growth; photosynthetic pigments; plant physiology

check for updates

Citation: Wiśniewska, K.; Śliwińska-Wilczewska, S.; Lewandowska, A.; Konik, M. The Effect of Abiotic Factors on Abundance and Photosynthetic Performance of Airborne Cyanobacteria and Microalgae Isolated from the Southern Baltic Sea Region. *Cells* **2021**, *10*, 103. https://doi.org/10.3390/cells10010103

Received: 16 November 2020
Accepted: 7 January 2021
Published: 8 January 2021

1. Introduction

Cyanobacteria and microalgae present in the aquatic or terrestrial environment can be emitted into the air and transported along with air masses over long distances [1–3]. As of today, it is known that the cyanobacteria and microalgae present in the air play a role in processes occurring in the air and in global climate change [4,5]. Through absorbing and dispersing solar radiation, these species affect the radiation budget of the Earth and may also form ice-nucleating particles and cloud condensation nuclei [3,4,6,7]. There are also scientific reports that cyanobacteria and microalgae can, after the aerosolization process, invade the human respiratory tract and constitute a serious health risk [1,8–11]. Bioaerosols may be emitted from the water surface with the air bubbles and spray, which is correlated with the wind speed and wave conditions. The mechanisms of the emission of cyanobacteria and microalgae to the air are complex processes and were described in detail in several reviews [3–5]. Nevertheless, in both aerobiology and phycology, these microorganisms are the least understood [12,13].

The conditions in the atmosphere are certainly not favorable for the presence of cyanobacteria and microalgae. Bioaerosols, including cyanobacteria and microalgae present in the air, are exposed to stress factors such as temperature, humidity, oxidative stress, nitrogen starvation, radiation, and osmotic stress [2–5]. There is a theory suggesting that this stress is an evolutionary force that causes selection pressure and hence affects the spread and evolution of organisms [2]. The atmospheric survival capability depends on the adaptability of cyanobacteria and microalgae to changing environmental conditions [2,3,5]. The stress response is of great importance to their capability to colonize new habitats and to survive. Certain cyanobacteria can form sheath and mucilage, while microalgae form resistant stages [5,14]. They also tolerate demanding environmental conditions: irradiance, temperature, and humidity gradients [3,5,14,15]. According to Tesson et al. [5] and Jewson et al. [16], there is a possibility that microalgae may change the stage of life e.g., into dormant cells, during their stay in unfavorable weather conditions. Therefore, further investigation of the physiological modifications that affect airborne microalgae as they disperse is essential. Mathematical models exist to help understand the emission and transport of bioaerosols in the atmosphere, but do not provide information on the biological consequences for the organisms concerning survival, vitality, and metabolic activity [2]. Hence, the knowledge about these adaptations is still insufficient and requires environmental research and laboratory experiments.

The emergence of new phytoplankton species and changes in their taxonomical composition indicate changes in biodiversity in the Baltic Sea [17]. Ojaveer et al. [17] emphasize that the causes of biodiversity change are not yet fully understood, and the impact of biodiversity changes on the Baltic Sea ecosystem cannot be determined yet. The authors clarify that the possibility of introduction and acclimatization of new phytoplankton organisms to the Baltic Sea is enhanced by global patterns related to climate change [17]. Thus, the goal of our study was to indicate from which regions of the Baltic Sea the isolated airborne microalgae are originated. In addition, our research was aimed at demonstrating their potential for expansion range and the possibility of an invasion into other regions of the Baltic Sea and even further aquatic ecosystems. The study will extend the knowledge of algal biodiversity and help to preserve the microbial and biological resources of the region. Thus, cyanobacteria and microalgae strains from CCBA AA will help to investigate the biodiversity of the atmospheric environment, and through isolation, collection, and maintenance of new strains they can give information about their potentially negative impact on human health.

2. Results and Discussion

2.1. *Characterization of Airborne Cyanobacteria and Microalgae Strains*

Altogether, 61 strains of airborne cyanobacteria and microalgae belonging to 14 orders and 4 phyla from the southern Baltic Sea region were isolated from May 2018 to August 2020 (Figure 1). In scientific studies, the number of recorded organisms is given, but usually, they are not isolated and cultivated [3]. One of the largest amounts of cyanobacteria and microalgae in the air was recorded by Brown et al. [18] and it was 62 species. The success of the study was attributed to the diverse methods used by the authors to sample by using vehicle, aircraft, and normally-exposed Petri dishes at many sites. According to a review study conducted by Wiśniewska et al. [3], *Chlorella* sp. was the most common Chlorophyte in these bioaerosol studies, while *Chlorococcum* sp. was another that appeared regularly. Among the analyzed phylum, Cyanobacteria *Phormidium* sp. was the most common Cyanobacteria; however, *Lyngbya* sp., *Nostoc* sp. and *Anabaena* sp. were also frequently noted. On the other hand, *Navicula* sp. and *Nitzschia* sp. were the most commonly found Bacillariophyta [3]. A detailed analysis of the number of blue-green algae and microalgae identified in the air on a global scale was described by Wiśniewska et al. [3]. Analyzing the number of isolated strains, it was shown that cyanobacteria (Cyanophyta) occurred in June, July, August, and September 2018–2019 (Figure 1, Table S1). The highest number of cyanobacteria isolates were recorded in August (12 strains) and in June (11 strains). Among the isolated strains, cyanobacteria belonging to the orders Nostocales and Synechococcales dominated in the study period. The green algae (Chlorophyta) were isolated from May 2018 to September 2020. The largest number of green algae strains were isolated in July (9 strains). The isolated green algae were also characterized by the highest diversity (6 orders). Green algae belonging to the orders Chlorellales and Sphaeropleales were the most numerous. In June and September 2018–2019, four diatom species (Bacillariophyta) belonging to 3 orders (Naviculales, Thalassiophysales, and Bacillariales) were also isolated. In addition, one strain belonging to Klebsormidiales (Charophyta) was isolated in September 2019, and one strain belonging to Eustigmatales (Ochrophyta) in November 2019. There have been months when individual strains could not be isolated, e.g., due to not enough organisms in the sample (Figure 1, Table S1).

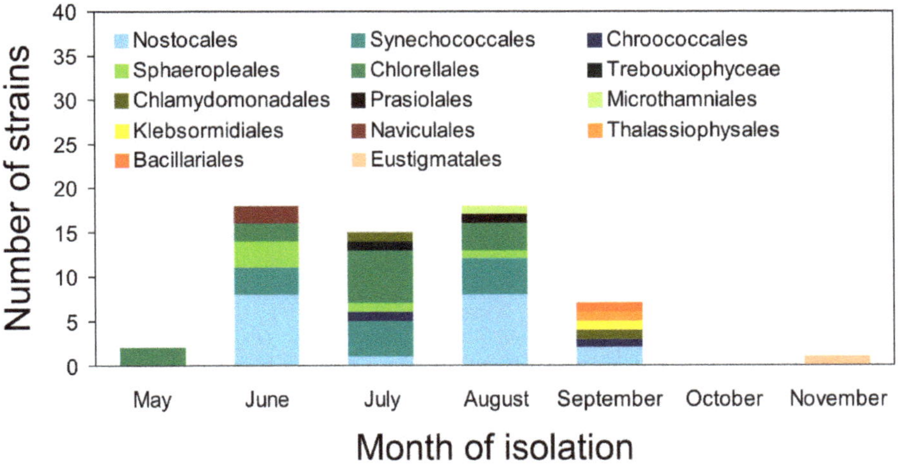

Figure 1. The number of airborne cyanobacteria and microalgae strains belonging to orders isolated in the study period (2018–2020).

As the largest number of isolated strains was recorded in the summer (June–August), it was decided to select three strains of cyanobacteria, green algae, and diatoms belonging to different orders for further research. In this study, we undertook the characterization of three strains of airborne cyanobacteria: *Nostoc* sp. (CCAA 03), *Synechococcus* sp. (CCAA 14), *Aphanothece* sp. (CCAA 48), green algae (B): *Oocystis* sp. (CCAA 20), *Coccomyxa* sp. (CCAA 21), *Kirchneriella* sp. (CCAA 38), and diatoms (C): *Nitzschia* sp. (CCAA 17), *Amphora* sp. (CCAA 34), *Halamphora* sp. (CCAA 47), which were the most abundant in aerosols of the Baltic Sea region. The listed organisms, depending on their preferences, can occur in various environments, from marine, through brackish, to freshwaters. However, the exact classification is often related to the species or strain of the microorganism [19].

It was shown that with the same initial value (OD_{750} = 0.1), the tested strains differed in the absorbance spectrum (Figure 2). Apart from the visible peaks of chlorophyll *a* (Chl *a*), which have approximate absorbance maxima of 430 nm and 662 nm, phycobilin pigments (Phyco) are visible in cyanobacteria. Absorption peaks for phycoerythrin (PE) in the visible light spectrum are measured at 498 and 565 nm. Phycocyanin (PC) has a single absorption peak at 621 nm. In green algae, there is also a peak for chlorophyll *b* (Chl *b*), which has approximate maxima of 453 nm and 642 nm. While in diatoms there is a characteristic peak for chlorophyll *c* (Chl *c*), the absorption maxima of which are around 450, 581, and 629 nm. Cyanobacteria and microalgae have a variety of different pigments that can absorb energy from a wide range of wavelengths [20]. However, when an absorbance spectrum is performed on living organisms, the peaks may overlap, masking the view of the individual pigments. A more adequate way to characterize live strains is to perform a flow cytometer measurement. Thus, it can easily distinguish and identify selected airborne cyanobacteria and microalgae. It has been shown that depending on photosynthetic pigments, small organisms (*Synechococcus* sp. and *Aphanothece* sp.) are visible by the low signal intensity at 640 nm and 480 nm on the cytogram, while larger-sized organisms (*Nitzschia* sp., *Oocystis* sp.) and filamentous *Nostoc* sp. were marked by the high signal intensity at 640 nm and 480 nm on the cytogram (Figure 2). It is worth noticing that based on our observations, only the strains within the groups can be differentiated (although for green algae it would be difficult for accurate separation in a mixed sample). Some organisms from the same group (green algae, cyanobacteria, diatoms) with a different structure and cell size may not be ideal and separated.

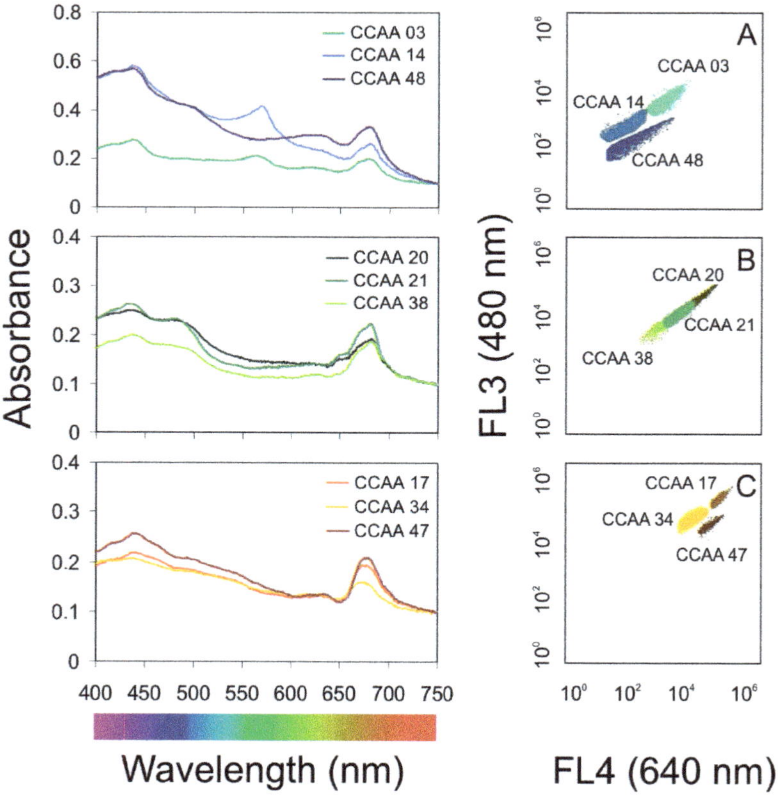

Figure 2. Left-hand panel—PAR absorbance spectra determined for the airborne cyanobacteria (**A**): *Nostoc* sp. (CCAA 03), *Synechococcus* sp. (CCAA 14), *Aphanothece* sp. (CCAA 48), green algae (**B**): *Oocystis* sp. (CCAA 20), *Coccomyxa* sp. (CCAA 21), *Kirchneriella* sp. (CCAA 38), and diatoms (**C**): *Nitzschia* sp. (CCAA 17), *Amphora* sp. (CCAA 34), *Halamphora* sp. (CCAA 47) at an optical density (OD_{750}) = 0.1; right-hand panel—cytograms obtained using a BD Accuri C6 Plus flow cytometer. FL3 detectors (red fluorescence) read fluorescence emissions excited by the blue laser (480 nm), whereas the FL4 detector (orange fluorescence) reads emissions excited by the red laser (640 nm).

Organisms obtained for the algae collections are usually isolated from seas and different inland waters [21,22]. This research focused on the isolation of airborne cyanobacteria and microalgae. At the moment, little data are allowing to determine the differences in the physiology of strains obtained from the air and the water samples [23]. However, it can be assumed that differences within one species may exist [21,24]. A great advantage of isolating cyanobacteria and microalgae from the atmosphere is the fact that it enables us to obtain species unusual for a given water reservoir or area because these organisms can be introduced from remote areas. The transport distance depends on the size and shape of particles. The large and heavy particles are dragging by the wind at a shorter distance. Small cells may be transported on hundreds of kilometers and then be deposited [1,4,5]. Studies are confirming the influence of meteorological parameters such as wind speed, air masses direction, air temperature, and humidity on the diversity of cyanobacteria and microalgae from the air [25,26]. However, the determination of their source of origin focuses on the use of the backward trajectories of air masses and the analysis of the taxonomic composition of potential sources location [3,26]. The isolation of species from the air can significantly enrich collection cultures of cyanobacteria and microalgae with organisms from distant regions. To the best of our knowledge, there is no significant algal culture collection maintaining airborne cyanobacteria and microalgae except for our newly established airborne algae (AA) collection, which is part of the Culture Collection of Baltic Algae (CCBA).

2.2. The Cell Concentration of Airborne Strains under Different Culture Conditions

The Baltic Sea is relatively shallow and small (the average depth of the Baltic Sea is 52.3 m, the maximum—459 m, and the area is approximately 415,266 km^2) with limited water exchange through the narrow and shallow Danish Straits. Photosynthetically active radiation may fluctuate between 389 and 2117 μmol photons m^{-2} s^{-1} during the day [27]. Many different habitats and marine species are affected by pollution, fishing, physical modification, and other human activities, making the sea a harsh place for organisms to live [28,29]. The unique feature of the Baltic Sea is that there are areas where freshwater, brackish, and marine species occur [30]. Climatic diversity within the Baltic Sea shows the wide fluctuation in many physical and chemical parameters such as river runoff, salinity, sea level, and sea ice. Its salinity increases from east to west and north to south, from almost freshwater conditions in the northern Gulf of Bothnia, with salinity fluctuaing between 1–3 PSU, to almost oceanic conditions in the northern Kattegat, with salinity within the range 18–30 PSU. Sea surface temperature in the Proper Baltic varies from 1–2 °C in February up to 17–18 °C in August [30]. In such a variable ecosystem as the Baltic Sea region, autotrophic organisms should show the ability to adapt quickly, which is crucial for their survival and the possibility of settling in new areas. The future climate changes leading to global warming may favor the development of picocyanobacterial—the smallest cell-size cyanobacteria (0.2–2 μm) [31]—which under high temperatures achieve the maximum abundance [32,33]. Our research, conducted between 2018 and 2020 in the southern Baltic Sea region, showed that the increasing intensity of light and temperature had a positive effect on the cell concentration of the studied cyanobacterium *Nostoc* sp., and for diatoms *Nitzschia* sp., *Amphora* sp., and *Halamphora* sp.

In this study, the concentration of cyanobacteria and microalgae cells were determined under different irradiance and temperature conditions (Figure 3). In general, it was shown that both irradiance and temperature significantly affected the number of cells of examined airborne cyanobacteria (ANOVA, $p < 0.001$, $p < 0.001$, $p < 0.01$, for *Nostoc* sp., *Synechococcus* sp., and *Aphanothece* sp., respectively), green algae (ANOVA, $p < 0.001$, $p < 0.05$, $p < 0.001$, for *Oocystis* sp., *Coccomyxa* sp., and *Kirchneriella* sp., respectively), and diatoms (ANOVA, $p < 0.001$, $p < 0.001$, $p < 0.001$, for *Nitzschia* sp., *Amphora* sp., and *Halamphora* sp., respectively; Table S2). ANOVA results also showed that for each airborne strain, the effect of temperature on the cell concentration was higher than the influence of irradiance and the interaction of both factors simultaneously. It is worth mentioning here that irradiance had no significant effect on *Coccomyxa* sp. (ANOVA, $p > 0.5$; Table S2). *Coccomyxa* sp. is resistant to enduring particularly harsh environmental conditions, e.g., grown under a low and high temperature and even high irradiance [34]. Its low variability of cells number indicates a good adaptation of the harsh conditions during emissions from the water reservoir and transport in the atmosphere.

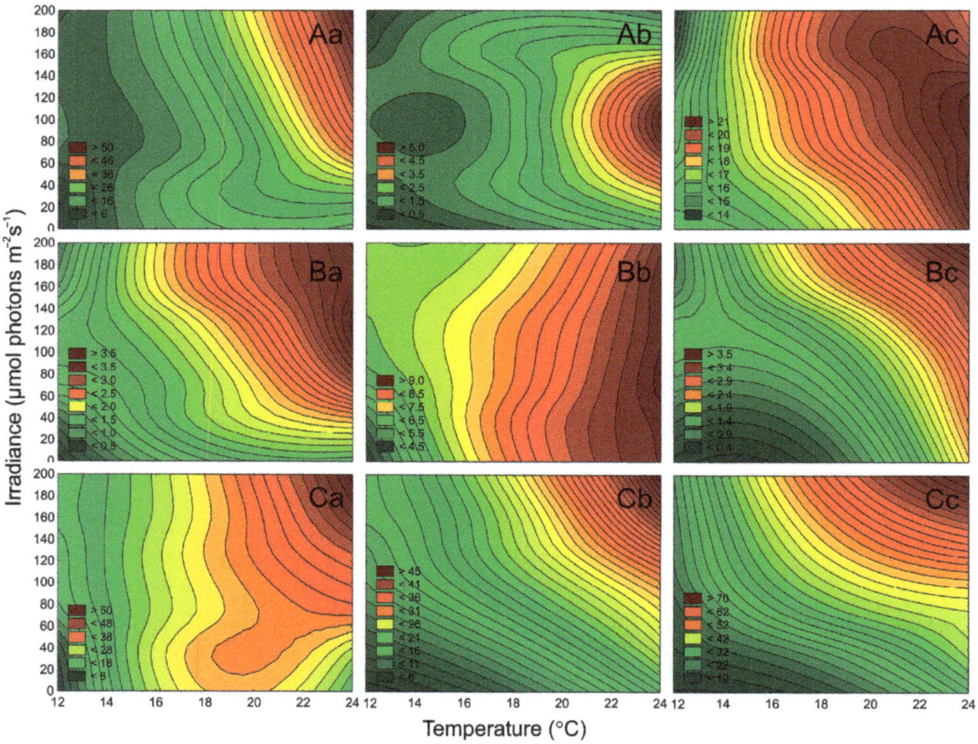

Figure 3. The number of cells ($N \times 10^5$ mL^{-1}) obtained after 7 days of experiment for airborne cyanobacteria: *Nostoc* sp. (**Aa**), *Synechococcus* sp. (**Ab**), and *Aphanothece* sp. (**Ac**); airborne green algae: *Oocystis* sp. (**Ba**), *Coccomyxa* sp. (**Bb**), and *Kirchneriella* sp. (**Bc**); airborne diatoms: *Nitzschia* sp. (**Ca**), *Amphora* sp. (**Cb**), and *Halamphora* sp. (**Cc**) under different irradiance (µmol photons m^{-2} s^{-1}) and temperature (°C) conditions.

It was shown that for cyanobacterium *Nostoc* sp., green algae *Oocystis* sp. and *Kirchneriella* sp., and for all tested diatoms, the increase in light and temperature had a positive effect on the cell numbers in the cultures. In turn, cyanobacterium *Synechococcus* sp. showed growth inhibition in the highest tested irradiance. Moreover, picocyanobacteria *Synechococcus* sp. and *Aphanothece* sp. showed a significant increase in cell concentration with increasing temperature (Figure 3A). This indicates that in the future *Aphanothece* sp. and also *Synechococcus* sp. may become dominant in the Baltic Sea region and other aquatic ecosystems. Thus, if a new organism inhabits a given area, there is a high probability that if it acclimates there, it can show allelopathic effects on organisms already present there. The minimum number of cells was obtained for *Kirchneriella* sp. at 13 °C and 10 µmol photons m^{-2} s^{-1} (0.45×10^5 cell mL^{-1}; Figure 3B). The highest number of cells (71.66×10^5 cell mL^{-1}) was noted for *Halamphora* sp. at 190 µmol photons m^{-2} s^{-1} and 23 °C (Figure 3C). *Halamphora* sp., isolated from the Baltic Sea, has a wide temperature range of 15–25 °C, but a 30 °C-preferring strain has also been isolated. By higher temperatures, the authors noted a decrease in photosynthesis [35]. Moreover, high numbers of cells (above 30×10^5 cell mL^{-1}) were also recorded for diatoms *Nitzschia* sp. and *Amphora* sp. and for cyanobacterium *Nostoc* sp. (Figure 3) The latest research on airborne microalgae proves that these organisms have several adaptations, like better UV tolerance, that distinguish them from microalgae present in water [23]. Better adaptation of these organisms to stress treatment can give them the advantage to settle in new areas and become invasive species. It indicates that, after entering the appropriate environment, these species will easily attain substantial biomass and even become the dominant species in this aquatic ecosystem. It is worth noting here that *Nostoc* sp. is an important CyanoHABs species [36,37], as well as *Nitzschia* sp. as a Red Tide organism [38]. The presence of microorganisms that produce toxins in the air, and their ability to spread and acclimate to new environmental conditions is of particular importance in terms of its impact on human health [3].

Assessing the origin of microorganisms that are present in the air in small amounts is a difficult task. It is often based solely on the analysis of the backward air masses trajectory. Hence it is worth enriching them with laboratory experiments such as the analysis of the number of organisms under different salinity conditions. To estimate the region of origin of airborne cyanobacteria and microalgae, additional experiments have been performed to determine the effect of salinity on the cells' number (Figure 4). In our work, we used different salinity of culture medium to isolate new strains of airborne cyanobacteria and microalgae. Based on our observations, we have shown that most of the CCAA strains grow satisfactorily at the salinity medium of 8 PSU. However, there are several strains (*Planktolyngbya contorta* CCAA11, *Pseudanabaena catenata* CCAA12, *Leptolyngbya foveolarum* CCAA15, *Gloeocapsa* sp. CCAA18, *Pseudanabaena catenata* CCAA19, and *Rivularia* sp. CCAA 49) that grow at 0 PSU salinity.

In general, one-way ANOVA tests showed that salinity significantly affected the number of cells of examined airborne cyanobacteria ($p < 0.001$, for all), green algae ($p < 0.001$, for all), and diatom ($p < 0.001$, for all; Table S3). It was found that for cyanobacterium *Nostoc* sp. (Figure 4A) as well as for green alga *Coccomyxa* sp. (Figure 4B) the maximum cell concentration was recorded at the salinity of 0 PSU (24.15×10^5 cell mL^{-1} and 3.82×10^5 cell mL^{-1}, respectively). It is worth mentioning here that *Coccomyxa* sp. did not grow at salinity above 16 PSU, despite numerous information related to their wide tolerance to harsh environmental conditions including salinity [34]. Moreover, the highest cell numbers of tested diatoms were found in the range of 16–24 PSU (21.14×10^5 cell mL^{-1} for *Nitzschia* sp., 15.52×10^5 cell mL^{-1} for *Amphora* sp., and 20.32×10^5 cell mL^{-1} for *Halamphora* sp.; Figure 4C).

Our measurements have shown that salinity has also a significant impact on the concentration of cells of individual species of algae and cyanobacteria. It was found that for cyanobacterium *Nostoc* sp. as well as for green alga *Coccomyxa* sp. the maximum cell concentrations were recorded at the salinity of 0 PSU (Figure 4A,B). Moreover, cyanobacteria *Planktolyngbya contorta*, *Pseudanabaena catenata*, *Leptolyngbya foveolarum*, *Gloeocapsa* sp., and *Rivularia* sp. belonging to the CCAA were able to grow only at a salinity of 0 PSU (data not shown). It suggests that these species come from freshwater communities. Research conducted by Woelfel et al. [27] indicates that salinity did not strongly influence the growth of diatoms, they also have a high tolerance to low salinity. In contrast, diatoms isolated from the air had the highest cell numbers found in the range of 16–24 PSU; however, all of them had lower tolerance to 0 PSU (Figure 4C). According to Potapova [39], the disparity in salt content between coastal and inland waters was considered a significant barrier that only a few diatom lineages were able to cross in one direction, from marine to freshwater habitat. These findings indicate that the airborne diatoms may come from the western part of the Baltic Sea [17]. Surprisingly, it was found that the picocyanobacteria *Synechococcus* sp. and *Aphanothece* sp. showed a similar abundance in the whole range of tested salinity. Considering their small cell size and their ability to be transported with aerosols over long distances, this property may give them an advantage in colonizing new water areas even with various salinity. The HYSPLIT air mass trajectories model [40,41] let us determine whether the given cyanobacteria and microalgae present in the air were brought from near or far, land or sea source areas [1,3]. We noted that only in the case of *Nitzschia* sp.

and *Aphanothece* sp. Did the analysis of the backward trajectories of air masses indicated their transport from over the land (Figure S1e). When analyzing the salinity preferences of selected strains, with the highest abundance by 4 PSU, it can be also concluded that *Aphanothece* sp. had a terrestrial origin. The preference for *Aphanothece* sp. to freshwater is confirmed by literature data [42]. In addition, Dembowska [43] showed the presence of *Aphanothece* sp. in the waters of lakes (Iławskie Lake District) over which the mass of air passed through.

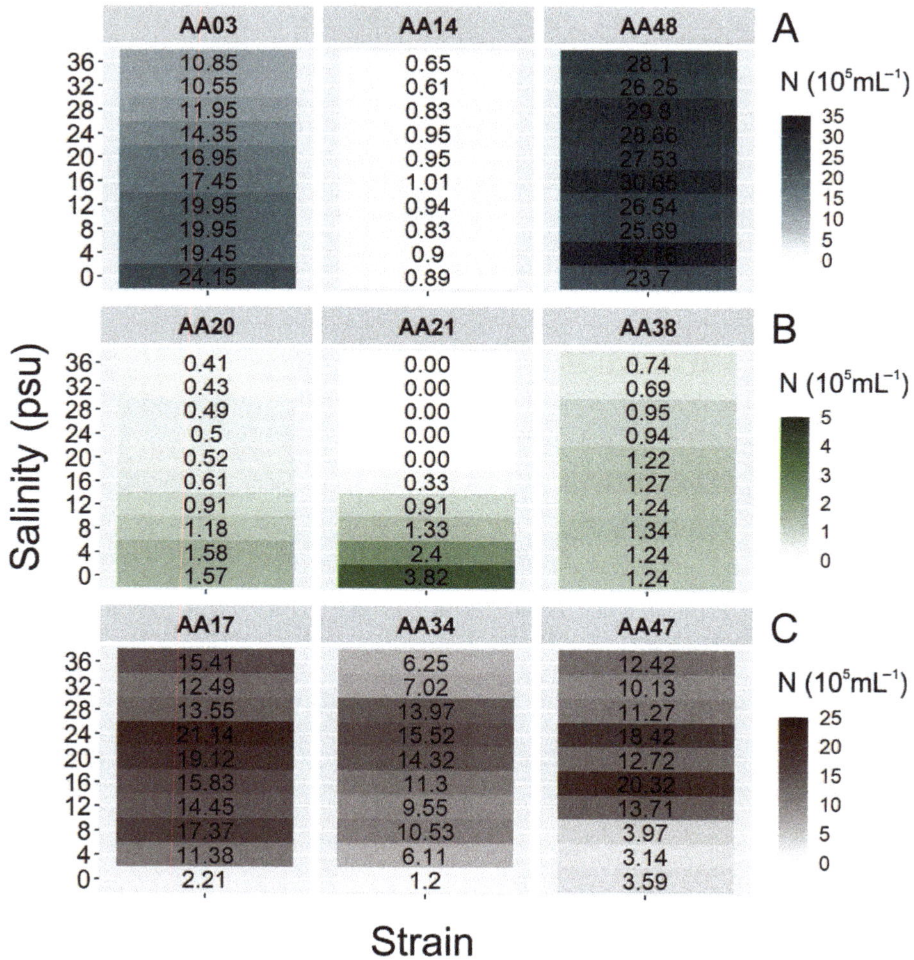

Figure 4. The number of cells ($N \times 10^5$ mL^{-1}) obtained after 7 days of experiment for airborne cyanobacteria (**A**): *Nostoc* sp. (CCAA 03), *Synechococcus* sp. (CCAA 14), *Aphanothece* sp. (CCAA 48), green algae (**B**): *Oocystis* sp. (CCAA 20), *Coccomyxa* sp. (CCAA 21), *Kirchneriella* sp. (CCAA 38), and diatoms (**C**): *Nitzschia* sp. (CCAA 17), *Amphora* sp. (CCAA 34), *Halamphora* sp. (CCAA 47) under different salinity conditions.

In all other cases, the analysis of taxa origin was much more difficult because the air masses passed through the Southern Baltic Proper first and then through the land. In the case of *Synechococcus* sp., *Oocystis* sp., *Coccomyxa* sp., and *Halamphora* sp., the air masses flowed over the measuring station all the way from the North Sea (Figure S1b). Analyzing salinity preferences in combination with the backward trajectories of air masses, it is most likely that only *Halamphora* sp. could have arrived from the distant Kattegat region (Figure S1a).

2.3. Effect of Irradiance and Temperature on Pigments Content

The autecology of airborne cyanobacteria and microalgae was characterized by the changes in composition and proportion of photosynthetic pigments, i.e., Chl *a*, Chl *b*, Chl *c*, and Phyco under different irradiance (µmol photons m^{-2} s^{-1}) and temperature (°C) conditions. As Chl *a* is the main photosynthetic pigment, and Car is responsible for protecting the microorganisms and is a variable when cells are under stress, they are presented in Figures 5 and 6. While additional pigments (Phyco, Chl *b*, Chl *c*) were presented in the Supplementary Materials Figure S2.

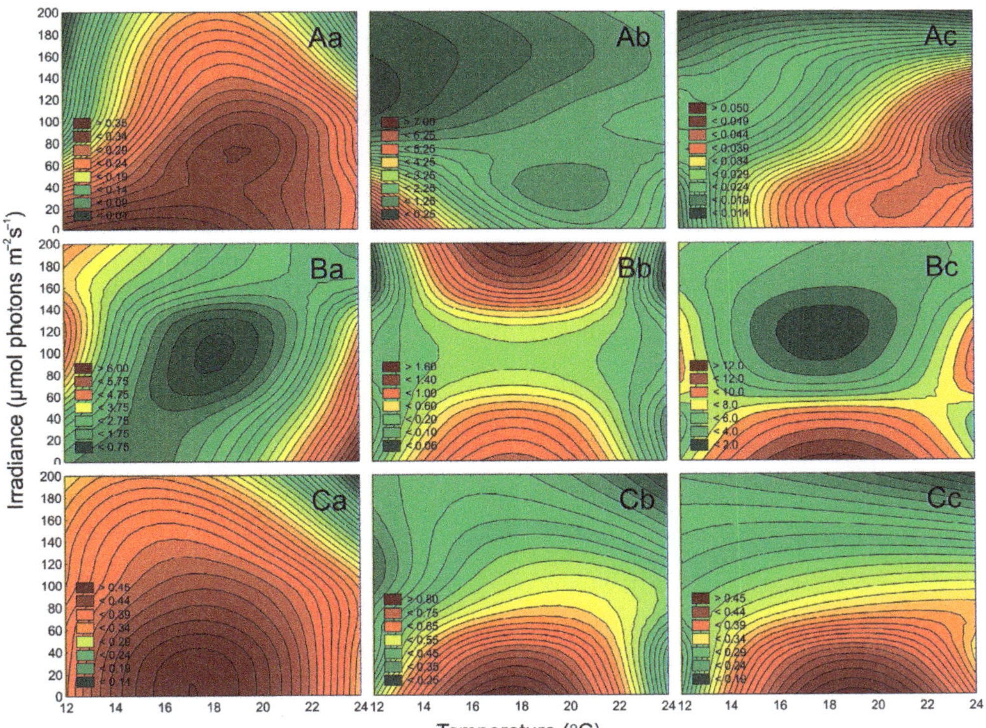

Figure 5. Changes in Chl *a* content (pg $cell^{-1}$) obtained after 7 days of experiment for airborne cyanobacteria: *Nostoc* sp. (Aa), *Synechococcus* sp. (**Ab**), and *Aphanothece* sp. (**Ac**); airborne green algae: *Oocystis* sp. (**Ba**), *Coccomyxa* sp. (**Bb**), and *Kirchneriella* sp. (**Bc**); airborne diatoms: *Nitzschia* sp. (**Ca**), *Amphora* sp. (**Cb**), and *Halamphora* sp. (**Cc**) under different irradiance (µmol photons m^{-2} s^{-1}) and temperature (°C) conditions.

It was found that irradiance and temperature as well as their interaction significantly affected the Chl *a* content in *Nostoc* sp., *Synechococcus* sp., and *Aphanothece* sp. (ANOVA, $p < 0.001$, $p < 0.001$, and $p < 0.001$, respectively), as well as Car content (ANOVA, $p < 0.001$, $p < 0.001$, and $p < 0.001$, respectively). ANOVA indicated that for examined cyanobacteria, the effect of irradiance on cell-specific pigments content was higher than the effect of temperature and the interaction of both factors (Table S4). It was shown that the cell-specific pigments content (Chl *a*, Car) for *Nostoc* sp. and *Synechococcus* sp. decreased with increasing irradiance (Figures 5 and 6). On the other hand, for *Aphanothece* sp. a decrease in Chl *a* (Figure 5(Ac)) and Car content (Figure 6(Ac)) with increasing irradiance was shown. The highest content of the studied pigments was found in *Synechococcus* sp., while the lowest was in *Aphanothece* sp. The highest content of Chl *a* (Figure 5(Ab)) and Car (Figure 6(Ab)) was recorded in *Synechococcus* sp. At the lowest irradiance (10 µmol photons m^{-2} s^{-1}) and the lowest temperature (13 °C), which is respectively 5.77 pg $cell^{-1}$ and 4.69 pg $cell^{-1}$.

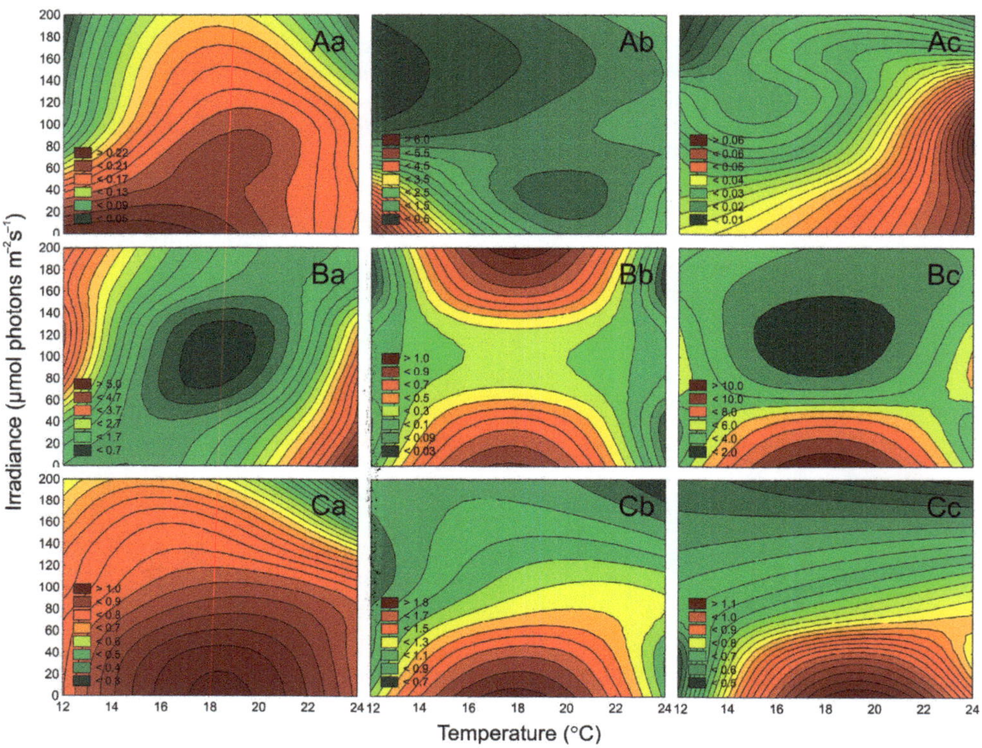

Figure 6. Changes in Car content (pg $cell^{-1}$) obtained after 7 days of experiment for airborne cyanobacteria: *Nostoc* sp. (**Aa**), *Synechococcus* sp. (**Ab**), and *Aphanothece* sp. (**Ac**); airborne green algae: *Oocystis* sp. (**Ba**), *Coccomyxa* sp. (**Bb**), and *Kirchneriella* sp. (**Bc**); airborne diatoms: *Nitzschia* sp. (**Ca**), *Amphora* sp. (**Cb**), and *Halamphora* sp. (**Cc**) under different irradiance (µmol photons m^{-2} s^{-1}) and temperature (°C) conditions.

Irradiance and temperature, as well as their interaction, were found to significantly affect the Chl *a* content in tested strains of green algae (ANOVA, $p < 0.001$, for *Oocystis* sp., $p < 0.001$, for *Coccomyxa* sp., and $p < 0.001$, for *Kirchneriella* sp.), and Car content (ANOVA, $p < 0.001$, for all). For Chl *a*, and Car content, the effect of temperature for *Oocystis* sp. And *Coccomyxa* sp. Was higher than the effect of irradiance and the interaction of both factors (Table S5). In turn, the Chl *a* and Car of *Kirchneriella* sp. Was more affected by irradiance than by temperature. It was shown that the cell-specific pigments content (Chla *a* and Car) for the analyzed green algae decreased with increasing irradiance at 13 °C and 23 °C. On the other hand, at the temperature of 18 °C, the content of the analyzed pigments remained at a similar level. The lowest Chla *a*, and Car content were recorded for *Coccomyxa* sp. On the other hand, it was demonstrated that *Kirchneriella* sp. Had the highest concentration of cell-specific pigments content was compared to examined green algae. The highest content of Chla *a* (Figure 5(Bc)) and Car (Figure 6(Bc)) was recorded at irradiance 10 µmol photons m^{-2} s^{-1} and temperature 18 °C, which were 12.84 pg $cell^{-1}$ and 10.33 pg $cell^{-1}$ respectively.

Both irradiance and temperature significantly affected also the cell-specific Chl *a* content of *Nitzschia* sp., *Amphora* sp., and *Halamphora* sp. (ANOVA, $p < 0.001$, for all), as well as Car content (ANOVA, $p < 0.001$, for all). ANOVA indicated that the effect of irradiance on pigments content was higher than the effect of temperature and of the relationship of both factors (Table S6). It was shown that the cell-specific pigments content for the analyzed diatoms decreased with increasing irradiance (Figures 5C and 6C). In general, the lowest cell-specific pigment content for the analyzed diatoms was recorded at 190 µmol photons $m^{-2} s^{-1}$ and 23 °C. The highest content of Chla *a* and Car was recorded for *Amphora* sp., which in the light of 10 µmol photons $m^{-2} s^{-1}$ and temperature of 18 °C was equal to 0.85 pg $cell^{-1}$ and 1.90 pg $cell^{-1}$ respectively. Moreover, it was also shown that *Amphora* sp. had the highest concentration of cell-specific pigment content compared to other diatoms.

For many cyanobacteria and microalgae, the high light intensity is an unfavorable environmental factor [44], which organisms can respond to by, for example, changing the concentration of pigments in cells [23,32,45]. Generally, the factorial experiments performed in this study showed a negative effect of the increasing irradiance on the content of the cell-specific pigment for the analyzed airborne cyanobacteria and microalgae, obtaining the highest contents at 10 µmol photons $m^{-2} s^{-1}$ and the lowest for 190 µmol photons $m^{-2} s^{-1}$. It was shown that only the cell-specific Phyco content for *Aphanothece* sp. increased with increasing light (Figure S2). High Chl *a* content in low light may indicate that airborne cyanobacteria and microalgae may deposit in highly shaded waters and continue to grow intensively [46]. The increased chlorophyll content found in microalgae grown at low light intensity is possibly due to light/shaded adaptations that enhance the use of light energy [47]. It is also worth noting that in the cases of cyanobacteria and diatoms, irradiance has a stronger effect on pigments than temperature.

Moreover, under low irradiance, the greatest increase of Phyco in cyanobacterial cells was noted (Figure S2), which additionally confirms the ability of these species to appear in shaded waters [48,49]. It can be because photoaclimatization occurs when photosynthetic pigments are reduced with the increase of irradiation intensity [50]. Thus, in the case of isolated in the region of the southern Baltic Sea airborne cyanobacteria and microalgae, except *Aphanothece* sp., the photoaclimatization ability was noted, which proves the ability to adapt to changing environmental conditions. It is also worth mentioning here that stressful factors such as high light intensity favor the accumulation of Car pigments that have a protective function. The latest research on airborne green-algae has shown that these organisms accumulate high levels of carotenoids in such stressful conditions [23]. However, in the present study, this phenomenon was not observed, which may suggest that the isolated microorganisms adapted to these conditions and were not under stress. According to the "Everything small is everywhere" hypothesis, not much can be done to prevent airborne microalgae and cyanobacteria to colonize new regions, when the environmental factors change. Thus, such experiments are especially important for estimating future effects in a changing environment.

2.4. Effect of Irradiance and Temperature on the Maximum PSII Quantum Efficiency

The effect of irradiance and temperature on changes in the maximum quantum efficiency of PSII photochemistry (F_v/F_m parameter) of airborne cyanobacteria and microalgae strains was also determined. It was found that irradiance and temperature, as well as their interaction significantly affected the F_v/F_m of tested cyanobacteria (ANOVA, $p < 0.001$, for all), green algae (ANOVA, $p < 0.001$, for all), and diatoms (ANOVA, $p < 0.001$, for *Amphora* sp., and *Halamphora* sp., respectively). It was also shown that irradiance and temperature had no significant effect on *Nitzschia* sp. (ANOVA, $p > 0.05$; Table S7). Generally, ANOVA indicated that for tested airborne cyanobacteria, green algae, and diatoms the effect of temperature on the F_v/F_m parameter was higher than the effect of irradiance and the interaction of both factors (Table S7).

It was shown that for the tested cyanobacteria, an increase in the F_v/F_m value with increasing temperature was noted. In turn, a decrease in F_v/F_m with increasing irradiance was also recorded. The highest values of F_v/F_m among the examined cyanobacteria, as in the case of the number of cells, were recorded for *Nostoc* sp. The highest value of the Fv/Fm parameter, equal to 0.76, for this cyanobacterium, occurred under 100 µmol photons m^{-2} s^{-1} and 23 °C (Figure 7A). In general, for green algae and diatoms, a decrease in F_v/F_m was recorded with increasing irradiance (Figure 7B,C). The greatest differences were recorded for *Coccomyxa* sp. at 13 °C. In such conditions, the decrease of this parameter from 10 to 190 µmol photons m^{-2} s^{-1} was more than 1.2-fold (Figure 7B). Moreover, green algae were characterized by the highest values of the F_v/F_m parameter compared to cyanobacteria and diatoms. For *Oocystis* sp. and *Kirchneriella* sp. the F_v/F_m always exceeded the value of 0.8 (Figure 7B). In turn, the lowest F_v/F_m value (0.12) among all the analyzed strains was recorded for *Aphanothece* sp. at irradiance 190 µmol photons m^{-2} s^{-1} and temperature 13 °C (Figure 8A).

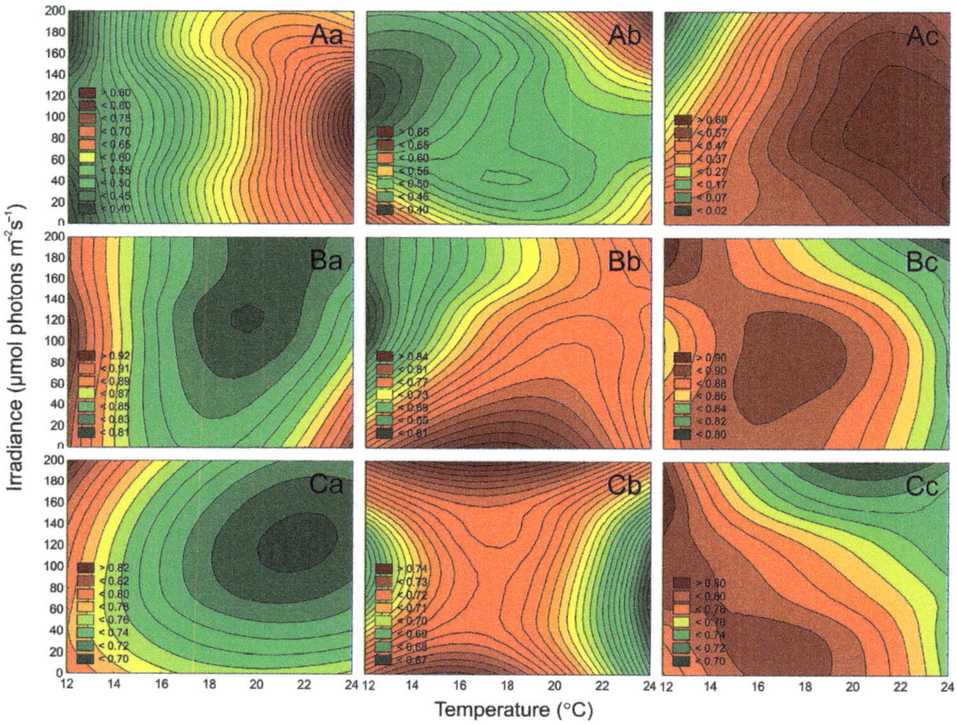

Figure 7. Changes in the maximum PSII quantum efficiency (F_v/F_m) obtained after 7 days of experiment for airborne cyanobacteria: *Nostoc* sp. (**Aa**), *Synechococcus* sp. (**Ab**), and *Aphanothece* sp. (**Ac**); airborne green algae: *Oocystis* sp. (**Ba**), *Coccomyxa* sp. (**Bb**), and *Kirchneriella* sp. (**Bc**); airborne diatoms: *Nitzschia* sp. (**Ca**), *Amphora* sp. (**Cb**), and *Halamphora* sp. (**Cc**) under different irradiance (µmol photons m^{-2} s^{-1}) and temperature (°C) conditions.

The chlorophyll *a* fluorescence measurements may indicate changes in photosystem II (PSII) activity by calculating the maximum quantum yield of PSII (F_v/F_m) [51]. On the basis of our research, we determined that the studied airborne cyanobacteria and microalgae showed differences in the functioning of the maximum quantum yield of PSII, depending on the environmental factors. In general, for cyanobacteria, green algae, and diatoms, a decrease in the maximum quantum yield of PSII was recorded with increasing light. It was shown that for the tested cyanobacteria, an increase in the maximum quantum yield of PSII with increasing temperature was noted. The green algae were characterized by the highest values of the F_v/F_m parameter compared to cyanobacteria and diatoms. In turn, the lowest F_v/F_m value among all the analyzed strains was recorded for *Aphanothece* sp. The appropriate composition of photosynthetic pigments, which is responsible for the proper functioning of the photosynthesis mechanism, enables airborne cyanobacteria and microalgae to grow under the control of changing environmental conditions and to colonize different water bodies [20].

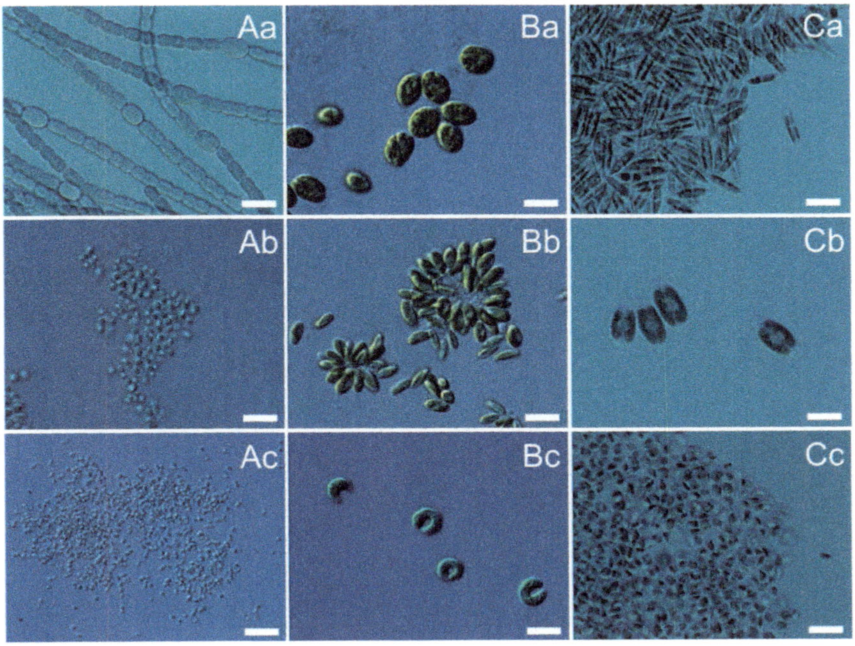

Figure 8. Photographs of airborne cyanobacteria (**A**): *Nostoc* sp. (CCAA 03; (**a**)), *Synechococcus* sp. (CCAA 14; (**b**)), *Aphanothece* sp. (CCAA 48; (**c**)), green algae (**B**): *Oocystis* sp. (CCAA 20; (**a**)), *Coccomyxa* sp. (CCAA 21; (**b**)), *Kirchneriella* sp. (CCAA 38; (**c**)), and diatoms (**C**): *Nitzschia* sp. (CCAA 17; (**a**)), *Amphora* sp. (CCAA 34; (**b**)), *Halamphora* sp. (CCAA 47; (**c**)) used in this study. Scale bar = 10 μm.

3. Materials and Methods

3.1. Culture Conditions

All airborne cyanobacteria and microalgae were collected between 2018 and 2020 at a research station in Gdynia, located 1 km from the coastal zone of the Gulf of Gdańsk, southern Baltic Sea region [1]. Algae have been identified by were identified using keys and world literature [52–58]. The taxonomic composition was analyzed under a Nikon Eclipse 80i microscope at a magnification of 10 and 100×. A list of species isolated in the study period is shown in Table S1 (in Supplementary Material). The factorial experiments were conducted on airborne cyanobacterial strains: *Nostoc* sp. (CCAA 03), *Synechococcus* sp. (CCAA 14), *Aphanothece* sp. (CCAA 48), green algae strains: *Oocystis* sp. (CCAA 20), *Coccomyxa* sp. (CCAA 21), *Kirchneriella* sp. (CCAA 38), and diatoms strains: *Nitzschia* sp. (CCAA 17), *Amphora* sp. (CCAA 34), *Halamphora* sp. (CCAA 47) (Figure 8).

The strains isolated from the air over the Baltic Sea region are maintained as unialgal cultures in the Culture Collection of Baltic Algae (Airborne Algae—AA) at the Institute of Oceanography, University of Gdańsk, Poland. Tests on the "batch cultures" were carried out in 25-mL glass Erlenmeyer flasks containing sterilized F/2 medium [59]. The strains were incubated under a 16:8 h light:dark cycle at three PAR irradiances (10, 100, and 190 μmol photons m^{-2} s^{-1}), at three temperatures (13 °C, 18 °C, and 23 °C), and salinity of 8 PSU. For the experiment determining the abundance of the tested strains under the influence of different salinity, ten different F/2 mediums (in the range of 0 to 36 PSU) were made. These mediums were obtained by dissolving the appropriate amount of Tropic Marin Sea Salt in a specific volume of distilled water. The salinity of the medium was measured by a salinometer (inoLabCond Level 1, Weilheim in Oberbayern, Germany).

The initial number of cells in all conducted experiments was 10^5 cells per mL. The test cultures were grown in three replicates and were incubated for one week. After that time in the exponential growth phase, the cells concentration, pigments content, as well as photosynthesis performance, were measured in each replicate.

3.2. Calculation of Cell Density

Cell density was calculated using linear regression models based on cell concentration (N mL^{-1}) and optical density (OD). Calculation of the cell number was conducted using the procedure described by Śliwińska-Wilczewska et al. [60] with a BD Accuri C6 Plus flow cytometer (BD Biosciences, San Jose, CA, USA) or by Śliwińska-Wilczewska et al. [61] with a light microscope (Nikon Eclipse 80i, Nikon, Tokyo, Japan) and the Bürker counting chamber for filamentous cyanobacteria. This method makes it possible to determine the correlation coefficients and the linear correlations between the number of cells and OD. The cell concentrations in the test cultures were estimated on the basis of calibration curves (Table S8).

3.3. Air Masses Trajectories

For each sampling day, using 6-h intervals, the air mass 48 h backward trajectory was obtained using the HYSPLIT model (NOAA Air Resources Laboratory, College Park, MD, USA) [40,41]. HYSPLIT is a complete system for computing basic trajectories of air parcels for dynamic simulations of dispersion and deposition. The three standard air mass (500, 1000, and 1500 m AGL) arrival heights were used to test the location of air mass overlap. By tracking the path of air masses at a given time 48h before sampling, this information helps to determine the source of both chemical and biological air pollution.

3.4. Determination of the Chlorophyll, Carotenoids, and Phycobiliproteins Content

The concentration of chlorophyll (Chl *a*) for tested cyanobacterial and microalgal strains was calculated according to Jeffrey and Humphrey [62]. The concentration of carotenoids (Car) was calculated using the formula employed by Strickland and Parsons [63]. After 7 days of incubation, 20 mL of culture was filtered using 0.45 μm filters (Macherey-Nagel MN GF-5, Dueren, Germany). Chl *a* and Car were extracted with cold 90% acetone in the dark for 2 h at −20 °C. To remove cell debris and filter particles the pigment extract was centrifuged at 12,000× *g* rpm for 2 min (Sigma 2-16P, Osterode am Harz, Germany). The extinction was determined at 750 nm, 665 nm, and 480 nm with a UV-VIS Multiskan GO spectrophotometer (Thermo Scientific, Waltham, MA, USA) and using 1 cm glass cuvette.

Phycobiliproteins (Phyco) were extracted for tested cyanobacterial strains according to Stewart and Farmer [64]. After 7 days of incubation, 20 mL of culture was filtered using 0.45 μm filters (Macherey-Nagel MN GF-5, Dueren, Germany). Each filter was thoroughly homogenized in a medium consisting of 0.25 M Trizma Base, 10 mM disodium EDTA, and 2 mg/mL lysozyme. The pH of the medium was adjusted to 5.0 through the addition of hydrochloric acid (concentrated HCl). Homogenates were incubated in darkness for 2 h at 37 °C and then for 20 h at 2 °C. To remove cell debris and filter particles, the pigment extract was centrifuged at 12,000× *g* rpm for 2 min (Sigma 2-16P, Osterode am Harz, Germany). The absorbance of the pigment extract was measured at 565 nm, 620 nm, 650 nm, and 750 nm with a Multiskan GO spectrophotometer (Thermo Scientific, Waltham, MA, USA) and using a 1 cm glass cuvette. The concentration of Phyco was calculated according to Tandeau de Marsac and Houmard [65].

3.5. Determination of the Chlorophyll a Fluorescence

The measurement of the chlorophyll fluorescence of airborne cyanobacteria and microalgae was conducted after 7 days of the experiment according to the method described by Śliwińska-Wilczewska et al. [24]. Fluorescence parameter F_v/F_m (where F_v—the difference between the maximum and minimum fluorescence and F_m—the maximum fluorescence) [51] was analyzed using pulse amplitude modulated (PAM) fluorometer (FMS1, Hansatech, King's Lynn, United Kingdom). Illumination was provided by a 594 nm amber modulating beam with 4 step frequency control. Cyanobacterial and microalgal materials were placed in the leaf clip on the 13 mm glass fiber filter (Whatman GF/C, Saint Louis, MO, USA). The actinic light was the same as the one used for the growth of algal cells to provide optimal conditions for photosynthetic activity. Saturation pulses (4500 μmol photons m^{-2} s^{-1}) of 0.7 s duration were used for all airborne cyanobacterial and microalgal species. All samples were dark-adapted for 5 min before measurements.

3.6. Statistical Analyses

To test the influence of a light intensity, temperature, and their interaction on studied parameters the two-way ANOVA was used. One-way ANOVA was used to test the effect of salinity on the number of cells of tested strains on the last day of the experiment. Levels of significance were: * $p < 0.05$; ** $p < 0.01$; *** $p < 0.001$. All data are reported as means ± standard deviations (SD). The statistical analyses were performed using Statistica® 13.1 software (TIBCO Software Inc., StatSoft, Kraków, Poland). The heat maps were performed using RStudio 1.3.1056 software (Boston, MA, USA).

4. Conclusions

Transport of bioaerosols with air masses may play important role in microbial dispersal, and thus maintaining the diversity of water systems [5,66]. To date, the invasiveness of new species has been closely connected to their transfer from the basin to the basin along with the ballast waters. Almost all known harmful algal bloom species have been reported in viable form from this source [67]. However, nowadays research on airborne algae confirms that these organisms can be carried far distances from their source of origin [1,5,68], thus it is worth testing their preferences and backward trajectories in order to estimate their source. Our research has shown that cyanobacteria *Nostoc* sp. as well as green alga *Coccomyxa* sp. may come from freshwater communities. On the other hand, the highest cell numbers of analyzed diatoms were found in the range of 16–24 PSU. These findings indicate that the airborne diatoms may come from the western part of the Baltic Sea. Moreover, analyzing salinity preferences in combination with the backward trajectories of air masses, it is most likely that only *Halamphora* sp. could have arrived from the distant Kattegat region. It was found that the picocyanobacteria *Synechococcus* sp. and *Aphanothece* sp. showed a similar abundance in the whole range of tested salinity. Considering picocyanobacterial small cell size and their ability to be transported with aerosols over long distances, this property may give them an advantage in colonizing new water areas even with various salinity conditions. We also found that cyanobacterium *Nostoc* sp., and diatom *Nitzschia* sp., a well-known bloom-forming species, can quickly reach significant biomass and even become the dominant species in the aquatic ecosystem as they showed the highest numbers of cells under favorable light, temperature, and salinity conditions. The present experiment suggests that the adaptive abilities of microorganisms, in particular those producing toxins, may contribute to the spread in the future. Thus, it may increase human exposure to their negative health effects.

Airborne cyanobacteria and microalgae still remain a group of poorly understood organisms due to difficulties in their extraction and isolation. Moreover, the increase in the number of alien and invasive species is becoming an increasingly serious problem in the world. Therefore, it is necessary to carefully examine the physiology and autecology of airborne organisms and their ability to inhabit distant water bodies. Isolating and creating a collection of cyanobacteria and microalgae has great potential. These organisms can later be used in environmental studies, education, or biotechnology [21]. In addition, the isolation of individual strains allows their use in laboratory conditions exploring the relationships between organisms present in the natural environment [21]. Thanks to this, the complex relationships that occur in nature can be better understood.

Supplementary Materials: The following are available online at https://www.mdpi.com/2073-4409/10/1/103/s1. Figure S1: 48-h air mass backward trajectory analysis from HYSPLIT model (Draxler and Hess, 1998, NOAA Air Resources Laboratory, US) for the date of sampling; Figure S2: Changes in Phyco (A; pg·cell^{-1}), Chl *b* (B; pg·cell^{-1}), and Chl *c* (C; pg·cell^{-1}) content obtained after 7 days of experiment for airborne cyanobacteria; Table S1: A list of species from the southern Baltic Sea region isolated in 2018-2020.; Table S2: Two-way factorial ANOVA of cells concentration measured in tested airborne cyanobacteria and microalgae growing at different temperatures (°C) and irradiance (µmol photons m^{-2} s^{-1}); Table S3: One-way ANOVA of cells concentration measured in tested airborne cyanobacteria and microalgae growing at different salinities (PSU); Table S4: Two-way factorial ANOVA of cell-specific Chl *a*. Car. and Phyco content measured in tested airborne cyanobacteria growing at different temperatures (°C) and irradiance (µmol photons m^{-2} s^{-1}); Table S5: Two-way factorial ANOVA of cell-specific Chl *a*. Car. and Chl *b* content measured in tested airborne green algae growing at different temperatures (°C) and irradiance (µmol photons m^{-2} s^{-1}); Table S6: Two-way factorial ANOVA of cell-specific Chl *a*. Car. and Chl *c* content measured in tested airborne diatoms growing at different temperatures (°C) and irradiance (µmol photons m^{-2} s^{-1}); Table S7: Two-way factorial ANOVA of F_v/F_m parameter measured in tested airborne cyanobacteria and microalgae growing at different temperatures (°C) and irradiance (µmol photons m^{-2} s^{-1}); Table S8: Linear regression and correlation coefficients (r) used to calculate the number (*N*) of studied airborne cyanobacteria, green algae, and diatoms cells in cultures based on optical density (OD) measurements.

Author Contributions: Conceptualization, K.W., S.Ś.-W., M.K., A.L.; methodology, K.W., S.Ś.-W.; formal analysis, K.W., S.Ś.-W.; investigation, K.W., S.Ś.-W.; data curation, K.W., S.Ś.-W., M.K.; writing—original draft preparation, K.W., S.Ś.-W., M.K., A.L.; supervision, A.L. All authors have read and agreed to the published version of the manuscript.

Funding: This research was funded by BMN grant number 539-O160-B432-20 and NCN PRELUDIUM 17 (UMO-2019/33/N/ST10/00585).

Institutional Review Board Statement: Not applicable.

Informed Consent Statement: Not applicable.

Data Availability Statement: Data is contained within the article or supplementary material.

Conflicts of Interest: The authors declare no conflict of interest.

References

1. Lewandowska, A.U.; Śliwińska-Wilczewska, S.; Wozniczka, D. Identification of cyanobacteria and microalgae in aerosols of various sizes in the air over the southern Baltic Sea. *Mar. Pollut. Bull.* **2017**, *125*, 30–38. [CrossRef] [PubMed]
2. Fröhlich-Nowoisky, J.; Kampf, C.J.; Weber, B.; Huffman, J.A.; Pöhlker, C.; Andreae, M.O.; Lang-Yona, N.; Burrows, S.M.; Gunthe, S.S.; Elbert, W.; et al. Bioaerosols in the earth system: Climate, health, and ecosystem interactions. *Atmos. Res.* **2016**, *182*, 346–376. [CrossRef]
3. Wiśniewska, K.; Lewandowska, A.U.; Śliwińska-Wilczewska, S. The importance of cyanobacteria and microalgae present in aerosols to human health and the environment–Review study. *Environ. Int.* **2019**, *131*, 104964. [CrossRef]
4. Després, V.R.; Huffman, J.A.; Burrows, S.M.; Hoose, C.; Safatov, A.S.; Buryak, G.; Fröhlich- Nowoisky, J.; Elbert, W.; Andreae, M.O.; Pöschl, U.; et al. Primary biological aerosol particles in the atmosphere: A review. *Tellus Ser. B Chem. Phys. Meteorol.* **2012**, *64*, 15598–15656. [CrossRef]
5. Tesson, S.V.M.; Skjøth, C.A.; Santl-Temkiv, T.; Londahl, J. Airborne microalgae: Insights, opportunities, and challenges. *Appl. Environ. Microbiol.* **2016**, *82*, 1978–1991. [CrossRef]
6. Hoose, C.; Möhler, O. Heterogeneous ice nucleation on atmospheric aerosols: A review of results from laboratory experiments. *Atmos. Chem. Phys.* **2012**, *12*, 9817–9854. [CrossRef]
7. Tesson, S.V.M.; Šantl-Temkiv, T. Ice nucleation activity and Aeolian dispersal success in airborne and aquatic microalgae. *Front. Microbiol.* **2018**, *9*, 2681. [CrossRef]

8. Franck, U.; Herbarth, O.; Manjarrez, M.; Wiedensohler, A.; Tuch, T.; Holstein, P. Indoor and outdoor fine particles: Exposure and possible health impact. In Proceedings of the Abstracts of the European Aerosol Conference 2003, Madrid, Spain, 31 August–5 September 2003; pp. S1357–S1358.
9. Backer, L.C.; McNeel, S.V.; Barber, T.; Kirkpatrick, B.; Williams, C.; Irvin, M.; Zhou, Y.; Johnson, T.B.; Nierenberg, K.; Aubel, M.; et al. Recreational exposure to microcystins during algal blooms in two California lakes. *Toxicon* **2010**, *55*, 909–921. [CrossRef]
10. May, N.W.; Olson, N.E.; Panas, M.; Axson, J.L.; Tirella, P.S.; Kirpes, R.M.; Craig, R.L.; Gunsch, M.J.; China, S.; Laskin, A.; et al. Aerosol emissions from great lakes harmful algal blooms. *Environ. Sci. Technol.* **2018**, *52*, 397–405. [CrossRef]
11. Facciponte, D.N.; Bough, M.W.; Seidler, D.; Carroll, J.L.; Ashare, A.; Andrew, A.S.; Tsongalis, G.J.; Vaickus, L.J.; Henegan, P.L.; Butt, T.H.; et al. Identifying aerosolized cyanobacteria in the human respiratory tract: A proposed mechanism for cyanotoxin-associated diseases. *Sci. Total Environ.* **2018**, *645*, 1003–1013. [CrossRef]
12. Sharma, N.K.; Rai, A.K.; Singh, S.; Brown, R.M., Jr. Airborne algae: Their present status and relevance. *J. Phycol.* **2007**, *43*, 615–627. [CrossRef]
13. Sahu, N.; Tangutur, A.D. Airborne algae: Overview of the current status and its implications on the environment. *Aerobiologia* **2014**, *31*, 89–97. [CrossRef]
14. Gupta, S.; Agrawal, S.C. Survival of blue-green and green algae under stress conditions. *Folia. Microbiol.* **2006**, *51*, 121–128. [CrossRef] [PubMed]
15. Śliwińska-Wilczewska, S.; Cieszyńska, A.; Konik, M.; Maculewicz, J.; Latała, A. Environmental drivers of bloom-forming cyanobacteria in the Baltic Sea: Effects of salinity, temperature, and irradiance. *Estuar. Coast. Shelf Sci.* **2019**, *219*, 139–150. [CrossRef]
16. Jewson, D.H.; Lowry, S.F.; Bowen, R. Co-existence and survival of diatoms on sand grains. *Eur. J. Phycol.* **2006**, *41*, 131–146. [CrossRef]
17. Ojaveer, H.; Jaanus, A.; Mackenzie, B.R.; Martin, G.; Olenin, S.; Radziejewska, T.; Telesh, I.; Zettler, M.L.; Zaiko, A. Status of biodiversity in the Baltic sea. *PLoS ONE* **2010**, *5*, e12467. [CrossRef]
18. Brown, R.M.; Larson, D.A.; Bold, H.C. Airborne algae: Their abundance and heterogeneity. *Science* **1964**, *143*, 583–585. [CrossRef]
19. Guiry, M.D.; Guiry, G.M. *AlgaeBase World-Wide Electronic Publication*; National University of Ireland: Galway, Ireland, 2020; Available online: http://www.algaebase.org (accessed on 4 November 2020).
20. Roy, S.; Llewellyn, C.; Egeland, E.S.; Johnsen, G. *Phytoplankton Pigments: Characterization, Chemotaxonomy and Applications in OceanOgraphy*; Cambridge University Press: Cambridge, UK, 2011; p. 784.
21. Latała, A.; Jodłowska, S.; Pniewski, F. Culture Collection of Baltic Algae (CCBA) and characteristic of some strains by factorial experiment approach. *Algol. Stud.* **2006**, *122*, 137–154. [CrossRef]
22. Norton, T.A.; Melkonian, M.; Andersen, R.A. Algal biodiversity. *Phycologia* **1996**, *35*, 308–326. [CrossRef]
23. Chiu, C.-S.; Chiu, P.-H.; Yong, T.C.; Tsai, H.-P.; Soong, K.; Huang, H.-E.; Chen, C.-N.N. Mechanisms protect airborne green microalgae during long distance dispersal. *Sci. Rep.* **2020**, *10*, 1–12. [CrossRef]
24. Śliwińska-Wilczewska, S.; Cieszyńska, A.; Maculewicz, J.; Latała, A. Ecophysiological characteristics of red, green and brown strains of the Baltic picocyanobacterium *Synechococcus* sp.—A laboratory study. *Biogeosciences* **2018**, *15*, 6257–6276. [CrossRef]
25. Sharma, K.; Rai, K.A.; Singh, S.A. Meteorological factors affecting the diversity of airborne algae in an urban atmosphere. *Ecography* **2006**, *29*, 766–772. [CrossRef]
26. Singh, H.W.; Wade, R.M.; Sherwood, A.R. Diurnal patterns of airborne algae in the Hawaiian Islands: A preliminary study. *Aerobiologia* **2018**, *34*, 363–373. [CrossRef]
27. Woelfel, J.; Schoknecht, A.; Schaub, I.; Enke, N.; Schumann, R.; Karsten, U. Growth and photosynthesis characteristics of three benthic diatoms from the brackish southern Baltic Sea in relation to varying environmental conditions. *Phycologia* **2014**, *53*, 639–651. [CrossRef]
28. HELCOM. The Baltic Marine Environment 1999–2002. *Baltic Sea Environ. Proc.* **2003**, *87*, 1–48.
29. Neumann, T. Climate-change effects on the Baltic Sea ecosystem: A model study. *J. Mar. Sys.* **2010**, *81*, 213–224. [CrossRef]
30. Walday, M.; Kroglund, T. *The Baltic Sea—The Largest Brackish Sea in the World*; European Environment Agency: Copenhagen, Denmark, 2003; Available online: https://www.eea.europa.eu/publications/report_2002_0524_154909/regional-seas-around-europe/page141.html (accessed on 1 August 2020).
31. Flombaum, P.; Gallegos, J.L.; Gordillo, R.A.; Rincón, J.; Zabala, L.L.; Jiao, N.; Vera, C.S. Present and future global distributions of the marine Cyanobacteria *Prochlorococcus* and *Synechococcus*. *Proc. Natl. Acad. Sci. USA* **2013**, *110*, 9824–9829. [CrossRef]
32. Śliwińska-Wilczewska, S.; Konarzewska, Z.; Wiśniewska, K.; Konik, M. Photosynthetic pigments changes of three phenotypes of picocyanobacteria *Synechococcus* sp. under different light and temperature conditions. *Cells* **2020**, *9*, 2030. [CrossRef]
33. Jodłowska, S.; Śliwińska, S. Effects of light intensity and temperature on the photosynthetic irradiance response curves and chlorophyll fluorescence in three picocyanobacterial strains of Synechococcus. *Photosynthetica* **2014**, *52*, 223–232. [CrossRef]
34. Fuentes, J.L.; Huss, V.A.R.; Montero, Z.; Torronteras, R.; Cuaresma, M.; Garbayo, I.; Vílchez, C. Phylogenetic characterization and morphological and physiological aspects of a novel acidotolerant and halotolerant microalga *Coccomyxa onubensis* sp. nov. (Chlorophyta, Trebouxiophyceae). *J. Appl. Phycol.* **2016**, *28*, 3269–3279. [CrossRef]
35. Prelle, L.R.; Graiff, A.; Gründling-Pfaff, S.; Sommer, V.; Kuriyama, K.; Karsten, U. Photosynthesis and Respiration of Baltic Sea Benthic Diatoms to Changing Environmental Conditions and Growth Responses of Selected Species as Affected by an Adjacent Peatland (Hütelmoor). *Front. Microbiol.* **2019**, *10*, 1500. [CrossRef] [PubMed]

36. Dodds, W.K.; Gudder, D.A.; Mollenhauer, D. The ecology of *Nostoc*. *J. Phycol.* **1995**, *31*, 2–18. [CrossRef]
37. Ploutno, A.; Carmeli, S. Modified peptides from a water bloom of the cyanobacterium *Nostoc* sp. *Tetrahedron* **2002**, *58*, 9949–9957. [CrossRef]
38. Yu, Z.M.; Zou, J.Z.; Ma, X.N. Application of clays to removal of red tide organisms II. Coagulation of different species of red tide organisms with montmorillonite and effect of clay pretreatment. *ChJOL* **1994**, *12*, 316–324.
39. Potapova, M. Patterns of Diatom Distribution In Relation to Salinity. In *The Diatom World. Cellular Origin, Life in Extreme Habitats and Astrobiology*; Seckbach, J., Kociolek, P., Eds.; Springer: Dordrecht, The Nederlands, 2011; pp. 315–332.
40. Draxler, R.R.; Rolph, G.D. *HYSPLIT (Hybrid Single-Particle Lagrangian Integrated Trajectory) Model Access via NOAA ARL READY Website*; NOAA Air Resources Laboratory: Silver Spring, MD, USA, 2003.
41. Rolph, G.D. *Real-Time Environmental Applications and Display System (READY) Website*; NOAA Air Resources Laboratory: Silver Spring, MD, USA, 2003. Available online: www.arl.noaa.gov/ready/hysplit4.html (accessed on 5 September 2020).
42. Jasser, I.; Callieri, C. Picocyanobacteria: The Smallest Cell-Size Cyanobacteria. In *Handbook of Cyanobacterial Monitoring and Cyanotoxin Analysis*; Meriluoto, J., Spoof, L., Codd, G., Eds.; John Wiley & Sons: Hoboken, NJ, USA, 2017; pp. 19–27.
43. Dembowska, E. Cyanobacterial blooms in shallow lakes of the Iławskie Lake District. *Limnol. Rev.* **2011**, *11*, 69–79. [CrossRef]
44. Guillard, R.R.L.; Murphy, L.S.; Foss, P.; Liaaen-Jensen, S. *Synechococcus* spp. as likely zeaxanthin-dominant ultraphytoplankton in the North Atlantic. *Limnol. Oceanogr.* **1985**, *30*, 412–414. [CrossRef]
45. Barlow, R.G.; Alberte, R.S. Photosynthetic characteristic of phycoerythrin-containing marine *Synechococcus* spp. I. Responses to growth photon flux density. *Mar. Biol.* **1985**, *86*, 63–74. [CrossRef]
46. Stal, L.J.; Albertano, P.; Bergman, B.; Bröckel, K.; Gallon, J.R.; Hayes, P.K.; Sivonen, K.; Walsby, A.E. BASIC: Baltic Sea cyanobacteria. An investigation of the structure and dynamics of water blooms of cyanobacteria in the Baltic Sea—Responses to a changing environment. *Cont. Shelf Res.* **2003**, *23*, 1695–1714.
47. Darley, W.M. Phytoplankton: Environmental factorsaffecting growth. In *Algal Biology: Aphysiological Approach*; Darley, W.M., Ed.; Blackwell Scientific Publications: Belgium, Spain, 1982; pp. 921–952.
48. Kana, T.M.; Glibert, P.M. Effect of irradiances up to 2000 µmol $E \cdot m^{-2} \cdot s^{-1}$ on marine *Synechococcus* WH7803-I. Growth, pigmentation, and cell composition. *Deep-Sea Res.* **1987**, *34*, 479–495. [CrossRef]
49. Kana, T.M.; Glibert, P.M. Effect of irradiances up to 2000 µmol $E \cdot m^{-2} \cdot s^{-1}$ on marine *Synechococcus* WH7803-II. Photosynthetic responses and mechanisms. *Deep-Sea Res.* **1987**, *34*, 497–516. [CrossRef]
50. Foy, R.H.; Gibson, C.E. Photosynthetic characteristics of planktonic blue-green algae: Changes in photosynthetic capacity and pigmentation of *Oscillatoria redekei* Van Goor under high and low light. *Br. Phycol. J.* **1982**, *17*, 183–193. [CrossRef]
51. Campbell, D.; Hurry, V.; Clarke, A.K.; Gustafsson, P.; Öquist, G. Chlorophyll fluorescence analysis of cyanobacterial photosynthesis and acclimation. *Microbiol. Mol. Biol. Rev.* **1998**, *62*, 667–683. [CrossRef] [PubMed]
52. Huber-Pestalozzi, G. Das Phytoplankton des Süsswassers, 3 Teil. Cryoptophyceen, Chloromonadien, Peridineen. In *Die Binnengewasser*; Thienemann, A., Ed.; E. Schweizerbart'sche Verlagsbuchhandlung: Stuttgart, Germany, 1950; p. 310.
53. Lind, M.E.; Brook, A.J. *A key to the Commoner Desmids of the English Lake District*; Freswater Biol. Assoc.: Cracow, UK, 1980; p. 123.
54. Komarek, J.; Fott, B. Chlorococcales, 7. Teil. Halfte. In *Das Phytoplankton des Süsswassers*; Elster, J., Ohle, W., Eds.; E. Schweizerbart'sche Verlagsbuchhandlung: Stuttgart, Germany, 1983; p. 1043.
55. Popovski, J.; Pfiester, L.A. Dinophyceae (Dinoflagellida). In *Süsswasserflora von Mitteleuropa*; Ettl, H., Gerloff, J., Heynig, H., Mollenhauer, D., Eds.; Gustav Fishre Verlag: Jena, Germany, 1990; p. 243.
56. Cox, E.J. *Identification of Freshwater Diatoms from Live Material*; Chapman and Hall: London, UK, 1996; p. 158.
57. Komarek, J.; Anagnostidis, K. Cyanoprokaryota 1. Teil: Chroococcales. In *Süsswasserflora von Mitteleuropa, Spektrum Akademischer Verlag*; Ettl, H., Gartner, G., Heynig, H., Mollenhauer, D., Eds.; Springer: Heidelberg, Germany, 1999; p. 548.
58. Hindák, F. *Fotograficky Atlas Mcrsocopickych Sinic*; VEDA, Vydavatelstvo Slovenskey Akademie Vied: Bratislava, Slovakia, 2001; p. 45.
59. Guillard, R.R.L. Culture of phytoplankton for feeding marine invertebrates. In *Culture of Marine Invertebrate Animals*; Smith, W.L., Chanle, M.N., Eds.; Plenum Press: New York, NY, USA, 1975; pp. 29–60.
60. Śliwińska-Wilczewska, S.; Barreiro Felpeto, A.; Maculewicz, J.; Sobczyk, A.; Vasconcelos, V.; Latała, A. Allelopathic activity of the picocyanobacterium Synechococcus sp. on unicellular eukaryote planktonic microalgae. *Mar. Freshw. Res.* **2018**, *69*, 1472–1479. [CrossRef]
61. Śliwińska-Wilczewska, S.; Maculewicz, J.; Barreiro Felpeto, A.; Vasconcelos, V.; Latała, A. Allelopathic activity of the picocyanobacterium *Synechococcus* sp. on filamentous cyanobacteria. *J. Exp. Mar. Biol. Ecol.* **2017**, *496*, 16–21. [CrossRef]
62. Jeffrey, S.T.; Humphrey, G.F. New spectrophotometric equations for determining chlorophylls a, b, c1 and c2 in higher plants, algae and natural phytoplankton. *Biochem. Physiol. Pflanz.* **1975**, *167*, 191–194. [CrossRef]
63. Strickland, I.D.H.; Parsons, T.R. A practical handbook of seawater analysis. *Bull. Fish. Res. Board Can.* **1972**, *167*, 1–310.
64. Stewart, D.E.; Farmer, F.H. Extraction, identification, and quantification of phycobiliprotein pigments from phototrophic plankton. *Limnol. Oceanogr.* **1984**, *29*, 392–397. [CrossRef]
65. Tandeau de Marsac, N.; Houmard, J. Complementary chromatic adaptation: Physiological conditions and action spectra. In *Methods in Enzymology*; Packer, L., Glazer, A.N., Eds.; Academic Press: New York, NY, USA, 1988; pp. 318–328.
66. Mayol, E.; Jimenez, M.A.; Herndl, G.J.; Duarte, C.M.; Arrieta, J.M. Resolving the abundance and air-sea fluxes of airborne microorganisms in the North Atlantic Ocean. *Front. Microbiol.* **2014**, *5*, 557. [CrossRef]

67. Hallegraeff, G. Transport of harmful marine microalgae via ship's ballast water: Management and mitigation with special reference to the Arabian Gulf region. *Aquat. Ecosyst. Health Manag.* **2015**, *18*, 290–298. [CrossRef]
68. Finlay, B.J. Global Dispersal of Free-Living Microbial Eukaryote Species. *Science* **2002**, *296*, 1061–1063. [CrossRef]

cells

MDPI

Article

Complexity of *Brassica oleracea*–*Alternaria brassicicola* Susceptible Interaction Reveals Downregulation of Photosynthesis at Ultrastructural, Transcriptional, and Physiological Levels

Violetta Katarzyna Macioszek [1,*], Magdalena Gapińska [2], Agnieszka Zmienko [3], Mirosław Sobczak [4], Andrzej Skoczowski [5], Jakub Oliwa [6] and Andrzej Kiejstut Kononowicz [7]

1 Laboratory of Plant Physiology, Department of Biology and Plant Ecology, Faculty of Biology, University of Bialystok, 15-245 Bialystok, Poland
2 Laboratory of Microscopy Imaging and Specialized Biological Techniques, Faculty of Biology and Environmental Protection, University of Lodz, 90-237 Lodz, Poland; magdalena.gapinska@biol.uni.lodz.pl
3 Department of Molecular and Systems Biology, Institute of Bioorganic Chemistry, Polish Academy of Sciences, 61-704 Poznan, Poland; akisiel@ibch.poznan.pl
4 Department of Botany, Institute of Biology, Warsaw University of Life Sciences (SGGW), 02-787 Warsaw, Poland; miroslaw_sobczak@sggw.edu.pl
5 Institute of Biology, Pedagogical University in Krakow, 30-084 Krakow, Poland; andrzej.skoczowski@up.krakow.pl
6 Department of Chemistry and Biochemistry, Institute of Basic Sciences, University of Physical Education in Krakow, 31-571 Krakow, Poland; jakub.oliwa@gmail.com
7 Department of Plant Ecophysiology, Faculty of Biology and Environmental Protection, University of Lodz, 90-237 Lodz, Poland; andrzej.kononowicz@biol.uni.lodz.pl
* Correspondence: v.macioszek@uwb.edu.pl; Tel.: +48-85-738-8225

Received: 3 October 2020; Accepted: 18 October 2020; Published: 20 October 2020

check for updates

Abstract: Black spot disease, caused by *Alternaria brassicicola* in *Brassica* species, is one of the most devastating diseases all over the world, especially since there is no known fully resistant *Brassica* cultivar. In this study, the visualization of black spot disease development on *Brassica oleracea* var. *capitata* f. *alba* (white cabbage) leaves and subsequent ultrastructural, molecular and physiological investigations were conducted. Inter- and intracellular hyphae growth within leaf tissues led to the loss of host cell integrity and various levels of organelle disintegration. Severe symptoms of chloroplast damage included the degeneration of chloroplast envelope and grana, and the loss of electron denseness by stroma at the advanced stage of infection. Transcriptional profiling of infected leaves revealed that photosynthesis was the most negatively regulated biological process. However, in infected leaves, chlorophyll and carotenoid content did not decrease until 48 hpi, and several chlorophyll *a* fluorescence parameters, such as photosystem II quantum yield (F_V/F_m), non-photochemical quenching (NPQ), or plant vitality parameter (Rdf) decreased significantly at 24 and 48 hpi compared to control leaves. Our results indicate that the initial stages of interaction between *B. oleracea* and *A. brassicicola* are not uniform within an inoculation site and show a complexity of host responses and fungal attempts to overcome host cell defense mechanisms. The downregulation of photosynthesis at the early stage of this susceptible interaction suggests that it may be a part of a host defense strategy, or, alternatively, that chloroplasts are targets for the unknown virulence factor(s) of *A. brassicicola*. However, the observed decrease of photosynthetic efficiency at the later stages of infection is a result of the fungus-induced necrotic lesion expansion.

Keywords: *Alternaria brassicicola*; chlorophyll a fluorescence; chloroplast ultrastructure; defense response; microarray; photosynthesis; susceptibility

1. Introduction

Brassica oleracea var. *capitata* (head cabbage) from the *Brassicaceae* family is one of eight *Brassica* subspecies and an important vegetable in the human diet due to its nutritional values [1]. Many white, red, and Savoy cabbage cultivars are cultivated widely in Europe, Asia, and North America, and used as a staple diet item, an ingredient of many national cuisine vegetable dishes or an addition to salads. In Northern Europe and New Zealand, cabbage is often used as feed for sheep and cattle [1,2]. As a widespread crop, cabbage is exposed to many bacterial and fungal diseases, which lower its yield all over the world. One of the most devastating fungus-induced diseases to all Brassicas is black spot disease, caused by *Alternaria brassicicola*. Cabbage cultivars show various levels of susceptibility to *A. brassicicola*, ranging from moderately to highly susceptible, but there is still not any known fully resistant cabbage cultivar [3].

Alternaria brassicicola belongs to the large fungal division ascomycota (*Ascomycetes*), order *Pleosporales*. According to the novel taxonomic classification based on the molecular phylogenetic analysis, genus *Alternaria* consists of 27 sections with a distinct Brassicicola one [4]. As a necrotrophic fungus, *A. brassicicola* is seed-transmitted, and thus infects young seedlings causing their damping off [5,6]. In the case of mature plants, the fungus is wind- and insect-spread, infecting preferentially older leaves. This results in black/brownish spreading lesions, which, at advanced stages of infection, can be responsible for the decay of the whole plant [7–9]. As a post-harvest pathogen, *A. brassicicola* has been found on cabbage debris and cabbage heads during storage [10]. The infection cycle of *A. brassicicola* is simple, from a single conidium to a mycelial network, often with overlapping stages typical for necrotrophic fungi [11]. After the initial attachment of the conidia to the leaf surface and germination, the fungus penetrates the host tissues through the appressoria and stomata, or invades them directly, using a preferential mode of penetration (depending on *Brassica* cultivar), or all of them simultaneously [3,11,12]. The fungus forms dome-shaped and usually aseptate appressoria at germ tube tips, similar to appressoria formed by *Botrytis cinerea*. Humidity, temperature, and conidial concentration are factors that influence *A. brassicicola* germination and infection of a host under field and laboratory conditions [13,14]. Prior to and during colonization, the fungus actively kills host cells primarily through the secretion of lipases, cell wall degrading enzymes, small secondary metabolites and was recently identified as the most abundant phytotoxin in *A. brassicicola* cultures—brassicicolin A [15–17]. Many *A. brassicicola* mutants with a depleted production of these secreted compounds, which also exhibit different levels of germination inhibition, were tested *in planta*, showing various degrees of virulence inhibition [16]. However, a primary virulence factor of *A. brassicicola* has not been discovered to date, thus it remains unknown how this fungus kills host plant cells. Moreover, it has been postulated that the possible resistance against *A. brassicicola* in Brassicas might be based on the cumulative effect of several genes rather than monogenicity [18,19].

The knowledge concerning plant defense mechanisms against *A. brassicicola* infection is mostly based on the extensively investigated model pathosystem for *Brassicaceae*—non-host *Arabidopsis thaliana* and *A. brassicicola*. Arabidopsis resistance against the fungus is compromised in *pad3* and *coi1* mutants, indicating that it requires camalexin and jasmonic acid (JA)-dependent signaling, respectively [20]. The hypersensitive host cell death during this interaction is restricted to the inoculation site, and accompanied by generation of reactive oxygen species (ROS) and callose deposition in wild-type Arabidopsis Col-0 [21,22]. On the other hand, camalexin is not metabolized by the fungus and inhibits its germination and development, but Brassicas phytoalexins induced during interaction with *A. brassicicola*, in most cases, are detoxified [23–25]. Although the transcriptome profiling of Arabidopsis Col-0 during *A. brassicicola* infection indicates that the photosynthesis-related genes remain unaffected in the resistant interaction [26]; Zmienko and Macioszek, unpublished data], the microarray analysis of a susceptible *pad3* mutant infected with *A. brassicicola* has demonstrated downregulation of photosynthesis-related genes [27]. Moreover, recent research on susceptible *Brassica juncea* infected with *A. brassicicola* revealed severe changes in chloroplast ultrastructure, and a post-inoculation time- and leaf position-dependent decrease in chlorophyll *a*:*b* ratio and photosynthesis efficiency [9].

Here, we report on the susceptible interaction between *B. oleracea* var. *capitata* f. *alba* (cultivar 'Glory of Enkhuizen') and *A. brassicicola*, both from the fungus and host plant perspective. In our work, we focused on the details of the fungal development and colony formation and the plant cell reactions during infection, at both light and transmission microscopy levels. Moreover, analyses of changes in the host transcriptional profiles with a special attention paid to photosynthesis-related genes and related physiological host responses to the fungus were conducted.

2. Materials and Methods

2.1. Plant Growth, Fungal Strain, and Inoculation

The seeds of *Brassica oleracea* var. *capitata* f. *alba* (white cabbage) early cultivar 'Glory of Enkhuizen' were obtained from a Polish seed company and grown in soil:perlite mixture (15:1) in a plant growth room under fluorescent light (Super TLD Philips 865) at 100 μmol m^{-2} s^{-1}, 16 h day/8 h night photoperiod, at temperature 22 ± 1 °C and relative humidity of approximately 65%. The wild type strain of *A. brassicicola* (ATTC 96836) was grown on potato dextrose agar plates (PDA; Difco, the Netherlands) for 7–10 days under the same conditions as plants, but in the dark. The second leaves of four-leaf mature *B. oleracea* plants were inoculated with the *A. brassicicola* conidial suspension at a concentration of 5×10^5 conidia per ml of distilled water, or just distilled water in case of control plants, by putting one or two 10 μL drops per leaf or spraying the leaves using a Nalgene aerosol spray bottle (Sigma-Aldrich, St. Louis, MO, USA) [9,11]. In all experiments, plants were inoculated 4-6 h after switching on the light (about 10.00 a.m.–12.00 a.m. local time). The inoculated plants were incubated in transparent plastic boxes to maintain high humidity under the same light and temperature conditions as for growing.

2.2. Disease Progression Analysis

Brassica oleracea plants were drop-inoculated, and necrotic spot parameters such as area, perimeter and average radius were measured using the WinDIAS Leaf Image Analysis System (Delta-T Devices, Cambridge, UK). For each time point, the second leaf of at least 6 plants was detached and images were taken immediately at 24, 48 and 72 h post-inoculation (hpi) in 3 independent experiments ($n = 3$).

2.3. Light Microscopy

To investigate the stages of *A. brassicicola* development during leaf infection, aniline blue-lactophenol (Sigma-Aldrich, St. Louis, MO, USA) stain was applied on drop-inoculated leaves at 4, 8, 12, 14, 16, 20, and 24 hpi [11]. The numbers of germinated conidia, germ tubes, and appressoria were counted per 100 conidia at random sites of the inoculation area on the second leaf of 3 plants per each time point.

2.4. Scanning and Transmission Electron Microscopy

To investigate the changes of leaf surface and fungal growth during infection, 1.5 cm diameter leaf discs were cut out from the *B. oleracea* second leaves sprayed or drop-inoculated with the conidial suspension. The samples were examined under a tabletop scanning electron microscope TM-1000 (SEM; Hitachi, Tokyo, Japan) operating at 15 kV, without any pre-processing [11]. Samples from six plants per experiment were harvested at 12, 24, 48, and 72 hpi. The experiment was repeated independently twice.

For the ultrastructural investigations, samples from the second leaf of the control and drop-inoculated plants were collected at 48 hpi and embedded in Epon–Spurr resin mixture [9]. The semi-thin (1 μm thick) and ultra-thin sections (80 nm thick) were obtained using a Reichert Jung ultramicrotome (Leica, Germany). The ultra-thin sections, after staining with uranyl acetate and lead citrate [28], were examined under a transmission electron microscope JEOL 1010 (TEM; JEOL, Japan) operating at 80 kV. The ultrastructure of the plant cells was observed on at least 50 micrographs per each treatment.

2.5. Microarray Experiment

The time-course pattern of gene expression was examined in the second leaves of *B. oleracea* sprayed with *A. brassicicola* conidial suspension by cDNA hybridization to 29 k Arabidopsis Oligonucleotide Microarrays (University of Arizona) (GEO platform GPL7725). Each sample treated as a single biological replicate was collected from 3 control or 3 infected plants. For each time point (0, 12, 24, and 48 hpi), 2 biological replicates were analyzed and a common reference design was applied, where each sample was labeled with Cy5 and hybridized against the Cy3-labeled pool of the control plants.

Each microarray probe sequence (~70-mer) was used as a query in a BLASTN search against *Brassica* nucleotide sequences available in the GenBank database. The best hit was recorded for each probe. Probes with high homology to *Brassica* targets were selected by applying a Bit score cutoff value of 40. For 66% of them, probe:target alignments were >60 nt in length and displayed >75% sequence identity.

The total RNA was extracted with RNeasy Plant Mini Kit (Qiagen, Hilden, Germany) and samples were DNase-digested with a TURBO DNA-free kit (Ambion, Austin, TX, USA). The sample quality was analyzed with a Nanodrop 1000 spectrophotometer (Thermo Scientific, Miami, OK, USA) and 2100 Bioanalyzer (Agilent Technologies, Wood Dale, IL, USA). For each labeling reaction, 15 µg of high-quality RNA ($A_{260}/A280$ and $A_{260}/A_{230} \geq 2$, RIN ≥ 9) was reverse-transcribed using the SuperScript Indirect cDNA Labeling System (Invitrogen, Waltham, MA, USA) according to the manufacturer's protocol. Aminoallyl-modified cDNA was coupled with 5 µL of Cy3 or Cy5 ester dye (Amersham Pharmacia, Buckinghamshire, UK) and unbound dye was removed with a MinElute Reaction Cleanup Kit (Qiagen, Hilden, Germany). The efficiency of the labeling was monitored with NanoDrop measurements. Corresponding Cy3 and Cy5-cDNA were mixed, the volume was adjusted to 10 µL and the samples were heat-denatured at 95 °C for 10 min and added to 115 µL pre-warmed (68 °C) hybridization buffer SlideHyb #3 (Ambion, Austin, TX, USA). Hybridization was carried out in a HybArray 12 workstation (PerkinElmer, Hebron, KY, USA) at 42 °C for 18 h. The microarrays were washed with 2× SSC (Ambion, Austin, TX, USA) supplemented with 0.1% SDS (Sigma-Aldrich, St. Louis, MO, USA) at 42 °C, 0.5× SSC at 30 °C and 0.005× SSC at 25 °C each for 5 cycles (20 s flow, 40 s hold) and dried by centrifuging. Images were collected with a ScanArray Express scanner (PerkinElmer, Hebron, KY, USA) at 100% laser intensity and variable PMT settings.

The images were analyzed with GenePix 6.1 software (Molecular Devices, San Jose, CA, USA) using the morphological opening background subtraction method with default parameters. The data were further analyzed in an R/Bioconductor environment using the limma package [29–31]. Foreground and background intensities were read with "genepix.custom" function. After background subtraction and printtip-loss normalization within arrays, followed by scaling, the log-ratios to have the same median-absolute-deviation (MAD) across arrays, a linear model was constructed for the contrasts of interest. Differential expression of genes for which the probes passed homology-based filtering criteria (see above) was assessed by moderated t-statistics and correcting for multiple testing using the false discovery rate (FDR) [32]. The adjusted p-value 0.1 was chosen as a significance threshold.

For functional analysis and gene annotation, *A. thaliana* gene IDs originally assigned to the microarray probes were used. Only 8014 genes with high homology to *Brassica* were taken into consideration and used as a background in the Gene Ontology (GO) and MapMan analyses.

A GO enrichment analysis of differentially regulated genes was performed separately for each time point against the list of all genes represented on the microarray using the ThaleMine v.4.1.1 online analysis tool [33]. The p-value was calculated using a hypergeometric distribution test and adjusted with the Holm–Bonferroni multiple test correction, and the significance threshold was set at a p-value < 0.05. The visualization of significantly regulated GO terms was performed using REVIGO [34], which reduces long lists of Gene Ontology terms by summarizing and removing redundant GO terms.

The input data were prepared as follows. From the list of all microarray results (Table S1), genes differentially expressed at 48 hpi (FDR < 0.1) were selected. The GO_cellular_process terms and the log fold change expression values of these genes were used as an input for REVIGO. Genes without assigned GO_cellular_process terms as well as genes for which GO: 0008150 (biological_process_unknown) was the only assigned GO term, were excluded from the analysis, resulting in an input list of 275 genes. The output list of REVIGO-summarized GO terms is presented in Table S2 and visualized in Figure 7. The MapMan (v.3.5.0) analysis was performed based on the probe mapping files for *A. thaliana* and pathways downloaded from the MapMan repository [35].

2.6. Photosynthetic Pigment Content

Chlorophyll was extracted in 100% methanol and measured using a spectrophotometer PowerWave XP (BioTek, Crawfordsville, IN, USA). The content of chlorophyll *a*, chlorophyll *b*, and carotenoids was calculated according to Wellburn [36]. The samples were harvested from 3 control and 3 infected plants per time point, and each experiment was repeated independently 3 times ($n = 3$).

2.7. Chlorophyll a Fluorescence

The control and drop-inoculated infected plants were dark-adapted for 30 min, then each second leaf was detached and immediately subjected to analysis of the kinetics of chlorophyll a fluorescence quenching using a Handy FluorCam 1000-H System (Photon Systems Instruments, Czech Republic) according to the manufacturer's built-in protocol. The measurements were performed on the selected area (diameter about 1.5 cm) of control untreated leaves, as well as infected ones containing necrosis and adjacent tissue. The duration of the reading procedure per one leaf was 4 min, and parameters such as F_0 (minimum fluorescence), F_m (maximum fluorescence), F_v (variable fluorescence) as well as F_p (peak fluorescence during the initial phase of Kautsky effect) were measured accordingly during 4 light adaptation (L1–L4), steady-state in light (Lss) and/or 3 dark relaxation periods (D1–D3). Based on the measured chlorophyll *a* fluorescence parameters, automatic calculation of over 45 parameters such as QYmax (maximum photosystem II quantum yield, F_V/F_M), F_v/F_m (photosystem II quantum yield), F_q/F_v (photosystem II efficiency factor), Rfd (fluorescence decline ratio) and qL (fraction of photosystem II open centers) in light adaptation periods and steady-state in light was performed and NPQ (non-photochemical quenching) and qP (photochemical quenching of variable fluorescence) were also calculated in the dark relaxation periods. All the measured and calculated chlorophyll a fluorescence parameters are shown in Table S6. In each experiment, 3–4 control and 3–4 infected plants were used per time point. The experiment was repeated independently 3 times ($n = 3$).

2.8. Statistical Analysis

The statistical analysis of all the data obtained in this work, except the microarray data, was performed using analysis of variance (ANOVA) and post-hoc Duncan's test ($p < 0.05$) using STATISTICA v.13.3.

All the figures were composed using Adobe Photoshop or Corel Software.

3. Results

3.1. Black Spot Disease Development

The areas infected with A. brassicicola became macroscopically visible in the form of small brown necrotic spots on the second leaf of *B. oleracea* plants at 20–24 hpi. The round-shaped necroses expanded beyond the inoculation sites at 48 and 72 hpi, and they were surrounded by a discrete chlorotic ring (Figure 1a). Necrosis parameters such as area, perimeter, and average radius increased in a time-dependent manner, indicating fast and massive fungal development on the susceptible host (Figure 1b). Inoculation of all four fully developed leaves of *B. oleracea* plants revealed that the size of the necrotic spots was leaf position-dependent, the older the leaf, the larger the necrosis, both on *in planta* and detached leaves assays (Figure S1).

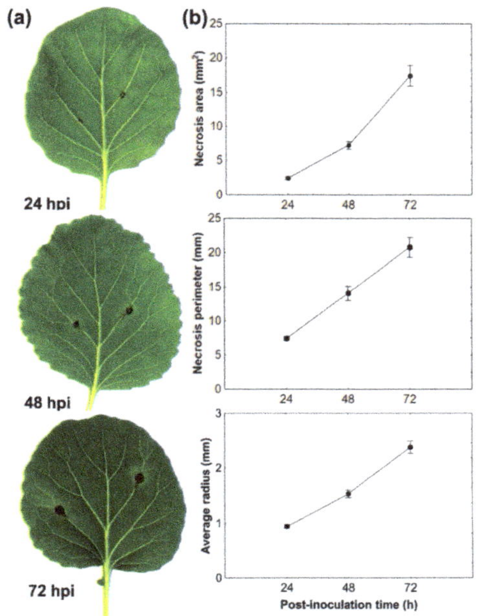

Figure 1. Analyses of necrotic lesion development on the second fully developed leaf of *B. oleracea* infected with *A. brassicicola*: (**a**) time-course images of necrosis development on the drop-inoculated leaves; (**b**) quantitative analysis of necrotic spot size changes. Data were gained using the WinDIAS system. The means ± SE were calculated from at least 18 measurements per time point obtained in 3 independent experiments (n = 3).

3.2. *A. brassicicola Development on Leaf Surface and Host Cell Responses*

The appearance of macroscopically visible necrotic spots at the inoculation sites on *B. oleracea* leaves at 24 hpi was strictly correlated with the overlapping stages of the *A. brassicicola* infection cycle, such as pre-penetration, penetration, colonization, and conidiation, which started before this time point.

Following inoculation of *B. oleracea* leaves, conidia began to germinate at 6–8 hpi, developing germ tubes at subsequent time points, which branched and elongated into hyphae at a later stage of infection (Figure S2). Although the number of both germinating conidia and germ tubes increased gradually with the post-inoculation time flow ($p < 0.001$, r = 0.94), the number of germ tubes did not significantly increase over the number of germinating conidia at any investigated time points (Figure 2). First dome-shaped appressoria were formed at 8 hpi and their number increased in a time-dependent manner ($p < 0.001$, r = 0.86), remaining at a similar level (not exceeding 20%) from 14 hpi (Figure 2). The fungus penetrated host epidermal cells through appressoria, stomata or directly triggering various host cell reactions at the same time (Figure 3). At the beginning of penetration, a bright 'halo' could be observed under a scanning electron microscope (SEM) around the penetration sites, regardless of the mode of penetration (Figure 3a and Figure S3a,b). Direct penetration caused rapid disturbance and collapse of the host epidermal cells, visible as dark areas divided by light cell walls (Figure 3b), but also dissolution of the leaf wax layer (Figure 3c and Figure S3c). Although penetration through stomata caused epidermal cell disturbance and death spreading to adjacent cells (Figure 3d), host epidermal cells with fortified cell walls were also observed in this case (Figure 3e). Regardless of the mode of fungal penetration, the first brownish host epidermal cells, indicating successful penetration sites, appeared as early as 14–16 hpi (Figure S2). However, germinating conidia showing symptomless growth could also be observed on the leaf surface even at 24 hpi (Figure 3f).

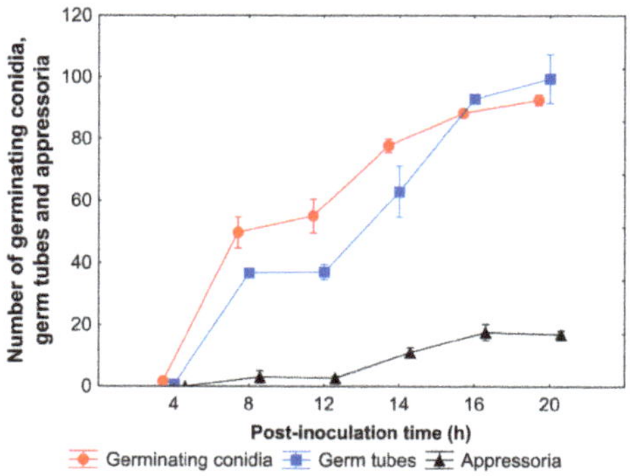

Figure 2. Quantitative data of *A. brassicicola* development on *B. oleracea* leaves. The values are means ± SE of 3 replicates.

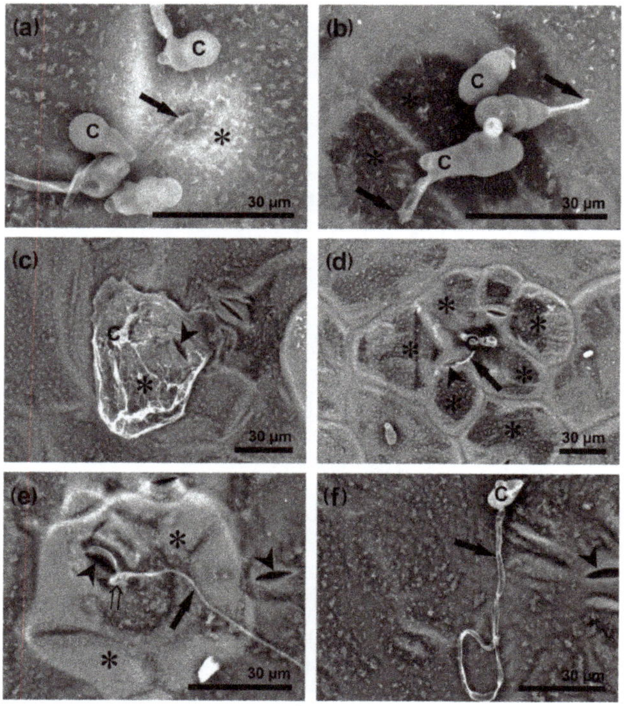

Figure 3. *Alternaria brassicicola* penetration modes and different host cell reactions. The images were capture using SEM at 24 hpi. (**a**) a bright 'halo' (asterisk) around the appressorium (arrow) penetrating the host cell; (**b**) direct penetration (arrows) caused a collapse of epidermal cells (asterisks) underneath germinating conidia; (**c**) dissolved wax layer (asterisk) around germinating conidium (arrowhead points to stoma); (**d**) germination tube (arrow) entering the leaf via stoma (arrowhead) surrounded by collapsed epidermal cells (asterisks); (**e**) germination tube (arrow) forming an appressorium (double tail arrow) over stoma (arrowhead). Neighboring epidermal cells (asterisks) have fortified cell walls; (**f**) symptomless growth of hypha (arrow) on leaf surface (arrowhead points to stoma). Abbreviation: C—conidium.

New young conidia were produced by mature parental conidia (used as an inoculum) during direct conidiation, taking place between 14 and 24 hpi (Figure 4a). The formation of the conidial chain developing on short conidiophores was observed later at 48 hpi.

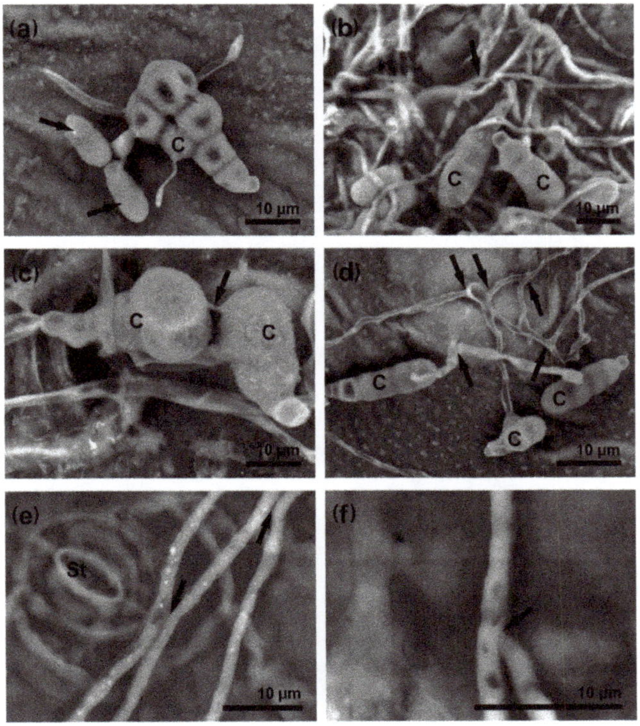

Figure 4. *Alternaria brassicicola* mycelium formation on *B. oleracea* leaves. The images were captured using SEM. (**a**) direct conidiation—new conidia (arrows) bud off from a mature conidium (24 hpi); (**b**) massive elongation and branching of hyphae (arrows) on the leaf surface (24 hpi); (**c**,**d**) conidial anastomosis tubes (CAT, arrows) form connections between conidia (24 hpi); (**e**) anastomoses (arrows) between hyphae (48 hpi); (**f**) fusion of hyphae (arrow) (48 hpi). Abbreviations: C—conidium, St—stoma.

The advanced stage of *A. brassicicola* development—colony formation—began as a tangle of elongating germ tubes developing into hyphae at 16–24 hpi (Figure 4b, Figures S2 and S3d). Connections between different parts of a developing colony were provided by bridges formed directly between two mature conidia or germ tubes, in the form of conidial anastomosis tubes (CATs) (Figure 4c,d) and fusions between hyphae—anastomoses (Figure 4e,f). The expanded colony caused damage and overgrew host tissues from the surface to the bottom side of the infected leaf at 48 hpi (Figure S3e–g) and sometimes chlamydospores could be observed on leaf surface at 72 hpi (Figure S3h).

3.3. Ultrastructure of Infected Host Cells

The ultrastructure of infected B. oleracea leaves revealed that *A. brassicicola* hyphae grew both inter-and intracellularly within the host leaf tissues (Figures S4 and S5a–d). Most of the hyphae had a regular ultrastructure, but a small number of hyphal cells with some level of degradation and osmiophilic granules were also observed (Figure S5e). In the infected mesophyll cells, the changes such as damage to the cell wall, plasmolysis and various levels of organelle disintegration, were dependent on the distance from the necrotic area and presence of hyphae (Figure S4). In the chlorotic area without hyphae, the host cell integrity was retained, although minor ultrastructural changes became evident.

In the control leaves, the mesophyll cells had a classical ultrastructure, with a large central electron-translucent vacuole and a narrow layer of homogeneously electron-dense cytoplasm, containing organelles, located parallel to the cell wall. Cell nuclei were surrounded by a continuous envelope perforated with nuclear pores only. Only few and small electron-dense heterochromatic grains were located next to the nuclear envelope. A well-distinguishable single nucleolus was present inside uniformly electron-dense euchromatin (Figure 5a). Lenticular chloroplasts with electron-dense stroma and a regularly arranged thylakoid and grana system contained up to three small starch grains (Figure 6a).

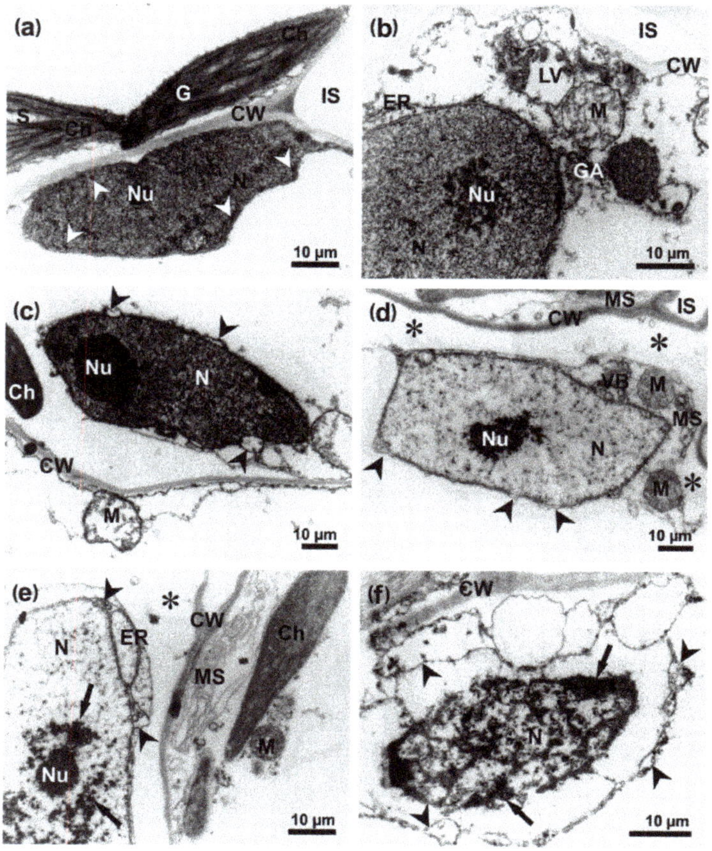

Figure 5. Degradation of mesophyll cells in *B. oleracea* leaf infected with *A. brassicicola* at 48 hpi. The images were captured using TEM. (**a**) a typical nucleus with a nucleolus and small heterochromatin grains (arrowheads), and chloroplasts in control mesophyll cell; (**b**) fragment of cell with degraded protoplast and nucleus with well-preserved nucleolus and fine granular chromatin; (**c**) nucleus with a nucleolus, condensed clumped chromatin and locally dilating nuclear envelope (arrowheads); (**d**) fragment of cell with degraded protoplast detached from the cell walls (asterisks), nucleus with granular chromatin, locally dilated envelope (arrowheads) and electron-dense deteriorating nucleolus; (**e**) fragment of strongly degraded cell with protoplast detached from cell walls (asterisk), pycnotic stage nucleus with osmiophilic remnants of chromatin (arrows) and dilated nuclear envelope (arrowheads); (**f**) terminal stage of nucleus degradation with clumps of chromatin (arrows), broken and dilated envelope (arrowheads). Abbreviations: Ch—chloroplast, CW—cell wall, ER—endoplasmic reticulum, G—granum, GA—Golgi apparatus, IS—intercellular space, LV—lytic vacuole, M—mitochondrion, MS—multilamellar structure, N—nucleus, Nu—nucleolus, VB—vesicular body.

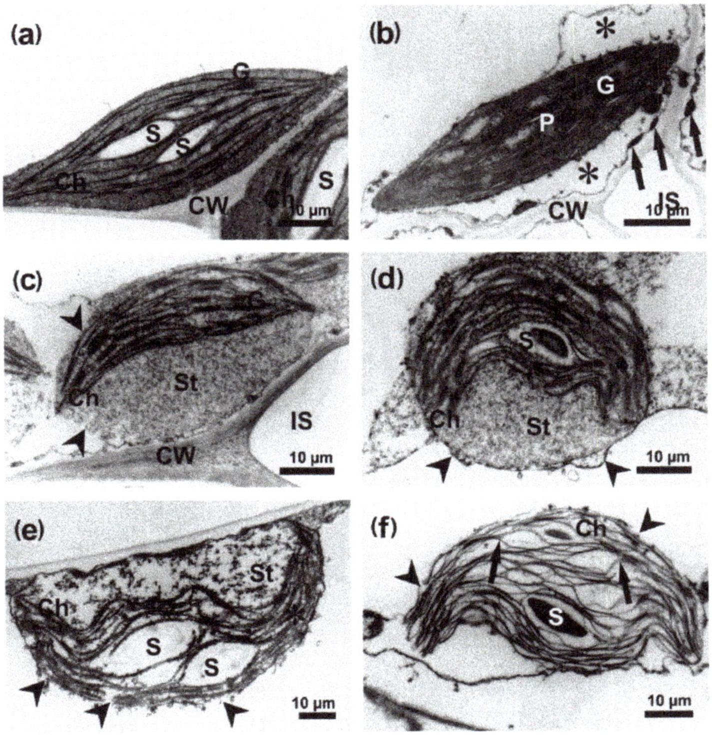

Figure 6. Disintegration of chloroplasts in *B. oleracea* leaf mesophyll cells infected with *A. brassicicola* at 48 hpi. The images were captured using TEM. (**a**) lenticular-shaped chloroplast with typical thylakoid and grana structure, and small starch grains in uninfected leaf; (**b**) fragment of mesophyll cell with degraded cytoplasm (asterisks), osmiophilic granules in plasmalemma (arrows) and a typical well-preserved chloroplast with plastoglobuli; (**c**) round-shaped chloroplast with a still preserved thylakoid and grana system, disintegrated chloroplast envelope (arrowheads) and swollen electron-translucent stroma; (**d**) round-shaped chloroplast with cup-shaped thylakoid system, reduced number of grana, locally swollen envelope (arrowheads) and electron-translucent stroma; (**e**) strongly disintegrated chloroplast with flocculent remnants of stroma, locally disintegrated envelope (arrowheads) and collapsed thylakoid system (asterisks) devoid of grana; (**f**) remnants of chloroplast with destroyed envelope (arrowheads) and thylakoids with widened lumens (arrows). Grana and stroma are missing. Abbreviations: Ch—chloroplast, CW—cell wall, G—granum, P—plastoglobule, S—starch grain, St—stroma.

The mesophyll cells at the border of the chlorotic and necrotic areas, especially in the close proximity of hyphae, showed various degrees of plasmolysis and organelle disintegration. They contained numerous lytic vacuoles, membranous multilamellar structures and vesicular bodies as well as osmiophilic granules close to the cell wall and organelles (Figures 5 and 6). Despite the appearance of symptoms of cell lysis, the ultrastructure of the nuclei was still well-preserved in some cells (Figure 5b), showing only local nuclear envelope swelling. However, the nucleoplasm slowly lost its electron-density and acquired a fine granular and fibrillar appearance (Figure 5c,d). In the more advanced stages of degradation, the nucleoplasm region turned strongly electron-translucent, and remnants of chromatin coalesced, forming electron-dense grains (Figure 5e,f). With the exception of the terminal stages of nucleus degradation, the remnants of nucleoli remained recognizable (Figure 5c,e) and the nuclear envelope was absent only in completely degraded cells (Figure 5f).

The degradation of chloroplasts was more spectacular and distinct, and clear stages of their disintegration could be discriminated. First of all, numerous plastoglobuli appeared in still well-preserved chloroplasts (Figure 6b). Subsequently, chloroplasts started to change their shape from lenticular to round (Figure 6c,d). This phenomenon was accompanied by the swelling of the stroma that lost its electron-density, gradual disintegration of the chloroplast envelope, and formation of the cup-shaped or bowed arrangement of the thylakoid systems (Figure 6c–e). Further degradation of the chloroplasts encompassed the final degeneration of the chloroplast envelope, disappearance of grana, distention, and dilution of the thylakoid lumens and the disintegration of the stroma (Figure 6f). Despite the level of degeneration of the lamellar system, the chloroplasts still contained starch grains or their hydrolyzed remains (Figure 6).

3.4. Microarray Analysis

The transcriptional profiling of *B. oleracea* leaves infected with *A. brassicicola* using 29 k Arabidopsis Oligonucleotide Microarrays was performed. Out of 29,100 microarray oligonucleotide probes, 8654 could be reliably mapped to *B. oleracea* nucleotide sequences, which represented 8014 unique genes. The expression of these genes in infected plants at 12, 24, and 48 hpi was compared to control plants (0 hpi). Significant changes in gene expression (FDR < 0.1) were observed as early as 12 hpi (27 genes) and they intensified while the infection progressed, with 189 genes regulated at 24 hpi and 410 genes at 48 hpi. Overall, 487 distinct genes were significantly affected in infected *B. oleracea* leaves (Table S1).

Based on the results of the statistical analysis, the Gene Ontology (GO) and MapMan analyses were performed to annotate the significantly up- and downregulated genes, and to determine the most affected biological processes and activation of individual genes during *B. oleracea* infection with *A. brassicicola*. Biotic stress response-related processes, such as the cell wall macromolecule catabolic process and defense response to fungus, were among the most extensively upregulated GO terms (Figure 7, Table S2). They were correlated to the microscopic results obtained in this study to uncover the differential responses of host cells to the infection within the inoculation site (Figures 3 and 7, Tables S2 and S3). The upregulation of host genes involved in aging, detection of ethylene stimulus and thiamine biosynthesis indicated general reprogramming of the host transcriptome in response to the infection. Analysis of stress-related genes at 48 hpi revealed differential expression of 167 genes, with 87 of them, mainly involved in cell response signaling, cell wall degradation, protein degradation, auxin-induced metabolism, being downregulated. However, transcriptional activation of 80 stress-related genes, among others *LRR-RLK7* (AT1G09970) encoding leucine-rich repeat transmembrane protein receptor-like kinase 7, *WRKY33* (AT2G38470) encoding transcription factor, *PDF1.2* (AT5G44420) encoding jasmonate-responsive plant defensin or *FAH1* (AT4G36220) encoding ferulate-5-hydroxylase 1 and *CCoAMT* (AT1G67980) encoding caffeoyl-CoA 3-O-methyltransferase involved in lignin biosynthesis, even at 48 hpi, indicated that, despite the successful colonization of the susceptible host by *A. brassicicola*, some of the host cells battled with the fungus and basal defense signaling was still active (Table S4).

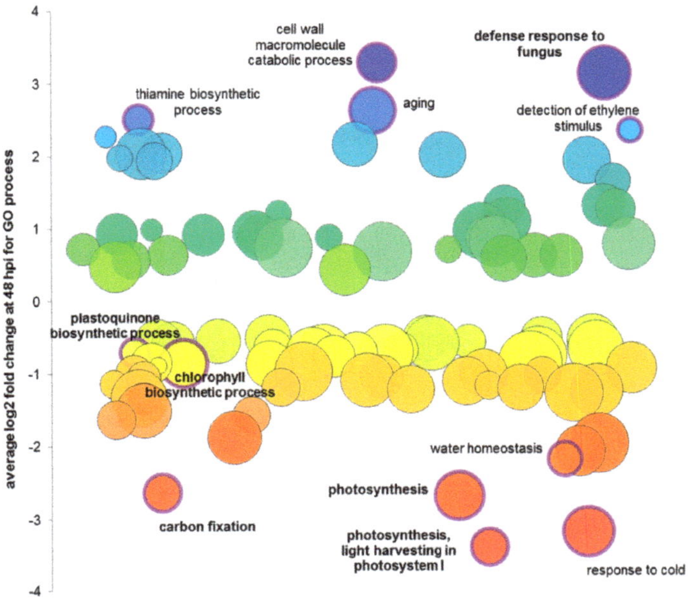

Figure 7. Gene Ontology of Biological Processes in *B. oleracea* leaves during *A. brassicicola* infection.

In turn, photosynthesis-related processes were strongly downregulated (Figure 7, Tables S2 and S3) in agreement with the observed severe changes in the ultrastructure of chloroplasts during infection (Figure 6). Interestingly, gradual downregulation of photosynthesis-related gene expression was observed during the entire course of *B. oleracea*–*A. brassicicola* interaction investigated in this study (Table S5). The majority of the 104 genes that were mapped to photosynthesis-related processes by the MapMan analysis, were negatively regulated at one or more time points (Figure 8a) and the number of significantly downregulated genes (at FDR < 0.1) increased from 6 at 12 hpi, through 32 at 24 hpi, up to 44 genes at 48 hpi (Figure 8b, Table S5). All crucial photosynthesis-related processes, such as light reactions, Calvin cycle, photorespiration, plastid ribosomal protein synthesis and tetrapyrrole synthesis were affected (Figure 8c). The distinctive genes in the photosynthesis light reaction category were LHCb and LHCa, encoding chlorophyll *a*/*b*-binding proteins, which are the part of photosystem II and photosystem I light harvesting complexes, respectively, and thus are important components of the chloroplast membrane system. These genes were downregulated as early as 12 hpi, with an increasing effect at subsequent time points. In the Calvin cycle, constituting the dark reaction of photosynthesis, the genes encoding the RuBisCO small subunit and glyceraldehyde 3-phosphate dehydrogenase were also downregulated at all investigated time points in infected *B. oleracea* leaves. However, changes in the expression of genes involved in photorespiration, plastid ribosomal protein synthesis and tetrapyrrole synthesis started later, at 24 and 48 hpi (Table S5). Additionally, consistent with the negative regulation of photorespiration-related genes, the downregulation of genes involved in water homeostasis was also observed (Figure 7). Remarkably, *AtCLH2* (AT5g43860) encoding chlorophyll-chlorophyllido hydrolase 2 annotated by MapMan as a tetrapyrrole synthesis-related gene was the only photosynthesis-related gene consistently upregulated at all three examined time points of infection (however, it did not pass the criteria of statistical significance) (Table S5).

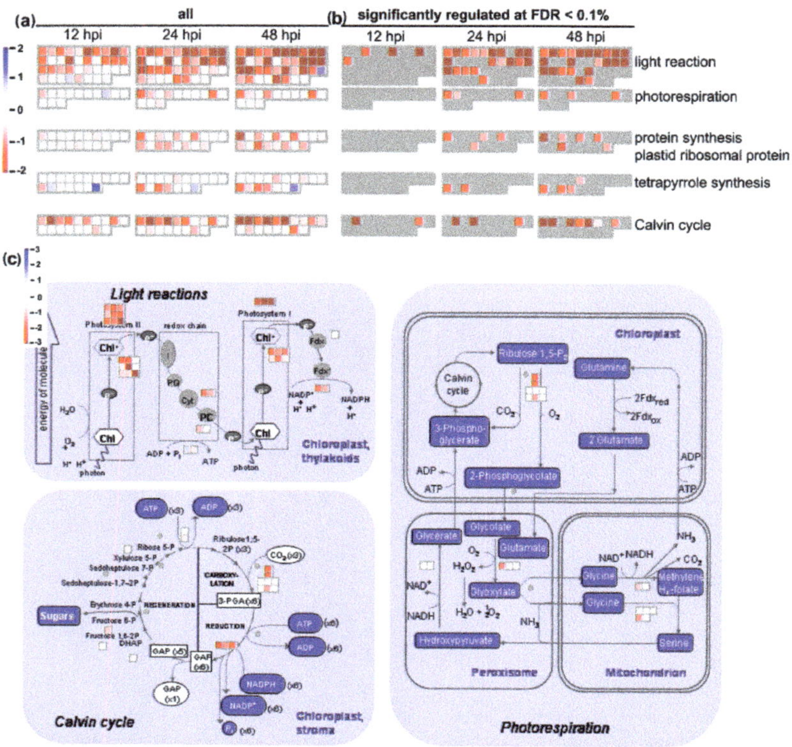

Figure 8. Microarray-based analysis of expression changes of photosynthesis-related *B. oleracea* genes during the course of *A. brassicicola* infection. (**a**) MapMan analysis of all mapped photosynthesis-related genes (104 genes); (**b**) MapMan analysis of photosynthesis-related genes significantly regulated at FDR < 0.1; (**c**) MapMan diagram of significant expression changes of photosynthesis-related genes at 48 hpi.

3.5. Photosynthetic Parameters Analysis

Based on the results of the host cell ultrastructure and microarray data, potential changes in photosynthetic pigment content and chlorophyll a fluorescence were investigated. As suspected, the content of chlorophyll *a*, *b*, *a* + *b* and carotenoids decreased in infected leaves compared to control at 48 and 72 hpi, but no significant differences between the content of any investigated photosynthetic pigments in the control and infected leaves were observed at 24 hpi (Figure 9).

The analysis of chlorophyll *a* fluorescence quenching revealed that the steady-state photosystem II (PSII) quantum yield (F_v/F_m_Lss) and the empiric parameter used to assess plant vitality (Rdf) significantly decreased in infected *B. oleracea* leaves in a time-dependent manner (Figure 10, Table S6). The significant decrease of variable fluorescence photochemical quenching (qP), PSII open centers fraction (qL), and PSII efficiency factor (F_q/F_v) in light adaptation periods and steady-state light during infection indicated partial blocking of electron transport to PSII reaction centers (Figure 10). The temporary increase in the value of non-photochemical quenching (NPQ) in infected leaves at 48 hpi was only visible during one of the dark relaxation periods (NPQ_D3). In turn, NPQ in light adaptation periods (NPQ_L1-L3) showed a significant decrease in infected leaves compared to the control at 24 hpi and 48 hpi (Figure 10, Table S6). Despite this, the maximum quantum yield of PSII significantly decreased in infected *B. oleracea* leaves compared to the control at 48 hpi, as evidenced by the values of QY_max and QY_Lss (Figure 10).

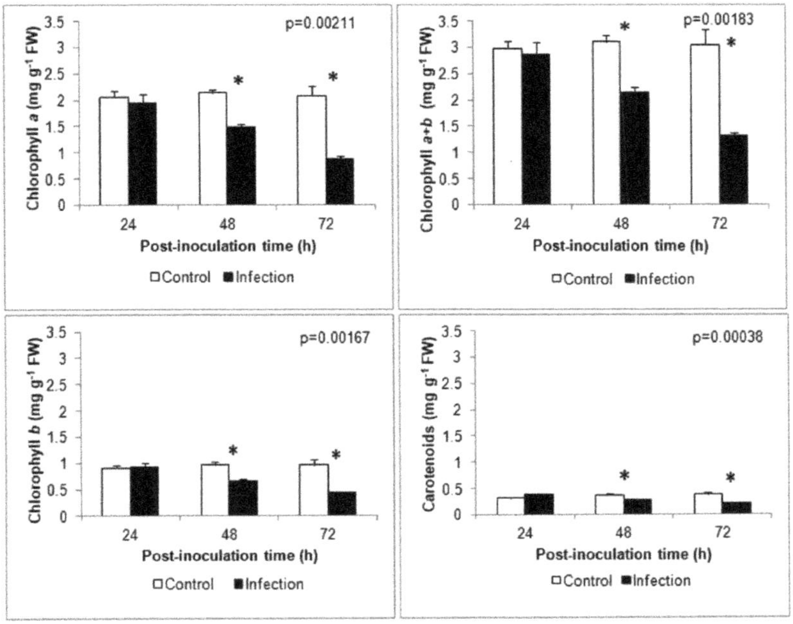

Figure 9. Time-course analysis of photosynthetic pigment content. The means ± SE were obtained in 3 independent experiments ($n = 3$). Asterisks indicate significant differences between control and infection at each time point according to the Duncan test (* $p < 0.05$).

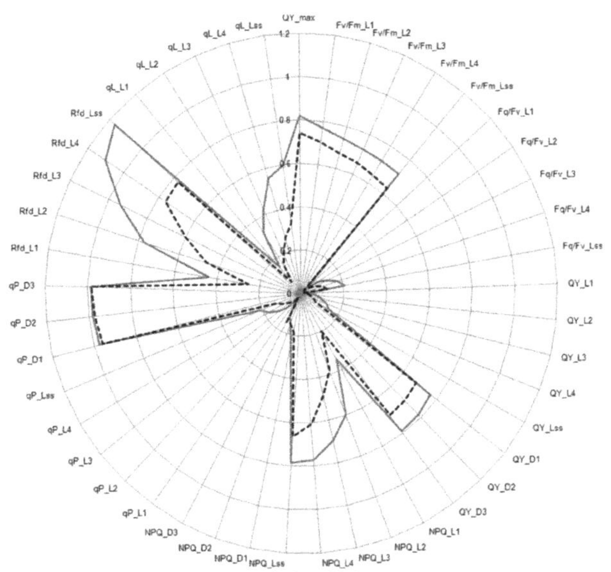

Figure 10. Analysis of chlorophyll *a* fluorescence parameters in *B. oleracea* leaves infected with *A. brassicicola* at 48 hpi. Values are the means obtained in 3 independent experiments ($n = 3$; detailed analysis is available in Table S6).

4. Discussion

Infections caused by necrotrophic fungi in host plants during susceptible interactions are related to unrestricted cell death visible as spreading lesions, the appearance of which is strictly correlated to a fungal infection cycle and the production of a wide set of virulence factors [37,38]. The cabbage cultivar 'Glory of Enkhuizen' used in the presented study has been considered to be susceptible to *A. brassicicola* infection under laboratory conditions and moderately resistant in the field [3]. Macroscopically visible necrotic lesions on leaves during *A. brassicicola* infection spread gradually from small brown spots at 24 hpi to larger necroses surrounded by a chlorotic area in a leaf position-dependent manner similarly as described for other Brassicas [3,9,39].

4.1. Timing of *A. brassicicola* Infection Cycle Depends on the Host Leaf Surface

In our study, the *A. brassicicola* infection cycle was delayed on the second leaf of 'Glory of Enkhuizen' compared to the susceptible white, red and Savoy cabbage cultivars infected with the same strain. The fungal germination began about 2 h later than on the white cabbage cultivar 'Stone Head'; however, the appearance of the first appressoria at 8 hpi was in accordance with the previous study [11]. The delayed germination of *A. brassicicola* conidia may be related to the thicker wax layer on the 'Glory of Enkhuizen' leaf surface that might have caused difficulties in conidial adherence to the leaf surface and/or a delayed recognition of the host surface signals by the conidia [40–42]. Following germination, the fungus penetrated host epidermal cells through appressoria, stomata or directly without any preferential mode of penetration, but, on other susceptible cabbage cultivars, *A. brassicicola* penetrated the leaf surface mainly through appressoria and rarely through stomata [11]. Nowakowska et al. [3] have claimed that the cabbage cultivars were penetrated mainly directly and through appressoria, and only rarely through stomata or without any preferential mode of penetration independently of a cultivar.

4.2. Host Cells Respond Differentially to Penetration

Using a SEM, we found that the fungus attempts to penetrate leaf surface, regardless of the mode of penetration, were accompanied by a bright 'halo' formation as early as 12 hpi (Figure 3a and Figure S3), indicating that, at first, the host cell reaction to penetration was defensive. Such clear 'halos' around penetration sites have been observed during a resistant interaction of a biotrophic fungus *Blumeria graminis* f. sp. *tritici* on a wheat cultivar carrying effective resistance genes and also on a susceptible one, using cryoscanning electron microscopy [43]. Possibly, these 'halos' were described as papillae formed around penetration sites of *A. brassicicola* on the leaves of both susceptible and moderately resistant cabbage cultivars after double staining with trypan blue and aniline blue using confocal laser scanning microscopy [3]. The biochemical analysis revealed that the 'halo' contains callose, phenolic compounds and an elevated level of calcium ions [43]. The 'halo' phenomenon indicates that the host cells are trying to combat fungal invasion at this point of the susceptible interaction and stay alive. It also shows that the *A. brassicicola* infection cycle probably contains a very short biotrophic phase, although there is a general agreement that a necrotrophic fungus first kills a host cell by secreting toxins, and then invades it. Interestingly, it has been postulated that another necrotrophic fungus, *B. cinerea*, should be considered as a hemibiotroph having a short biotrophic phase. *Botrytis cinerea* suppresses early host defense reactions by the secretion of small RNAs (sRNAs), thus leading to the silencing of host genes, but, to achieve this, the host cell must be alive [44]. It has been also shown that during invasion and the establishment of a necrotrophic interaction, *B. cinerea* 'sacrifices' many of its invading hyphae by subjecting them to cell death induced by plant-secreted cell death inducing factors, to get a chance to release fungal cell death inducing factors by the surviving hyphae into plant cells [45]. It could also be a mechanism used by *A. brassicicola* and an explanation for the presence of hyphae with deteriorating protoplasts (Figure S5e).

At later stages of an *A. brassicicola* infection (from 12–16 hpi), the successful penetration sites appeared on SEM images as electron-dense collapsed epidermal cells (Figure 3b) or brownish possibly dead cells under bright field light microscopy (Figure S2), which have been previously described during *A. brassicicola* infection [11]. Brownish cells are also frequently found at penetration sites of other necrotrophic as well as biotrophic fungi [46]. Browning of successfully infected epidermal cells may have been caused by the activation of peroxidases and phenolics within the cell wall and/or a production of ROS by dying host cells, indicating the suppression of host cell defense reactions and a highly susceptible response at this stage of infection [11,22,47,48].

4.3. *A. brassicicola Colony Formation Is Triggered by Successful Penetration*

Successful penetration of the susceptible host epidermal cells by *A. brassicicola* was a signal triggering colony formation at inoculation sites. Connections between different regions of an expending mycelial network were established via CATs and anastomoses (Figure 4). Their formation is a typical feature of many filamentous fungi [49,50], but these phenomena are mostly described when fungi grow in media rather than *in planta* during infection. CATs are usually formed during the initiation of a colony, and facilitate the transport of water and nutrients as well as horizontal gene transfer [51,52]. Similarly to CATs, the formation of fusions between hyphae (anastomoses), originating from the same or different conidia allows them a proper distribution of nutrients, transduction of chemical signals, and even exchange of genetic material [53]. Moreover, anastomosis formation is a prerequisite for *A. brassicicola* virulence, as indicated by the development of an *A. brassicicola aso1* mutant that is unable to form anastomoses, and, when tested in planta, appears to be unable to spread beyond the inoculation site [54].

4.4. *Changes in Host Cell Ultrastructure and Transcriptome Reprogramming*

The appearance of necrotic spots indicated successful invasion and colonization of the host tissues by *A. brassicicola*, although host cells were differentially affected within the inoculation site depending on the distance from the invading hyphae, as evidenced by the gradual degradation of organelles. Similar individual changes in the ultrastructure of infected plant host cells, such as cell lysis, disintegration of the nuclei and chloroplasts, or the presence of osmiophilic granules have been observed in other pathosystems during infection with viruses, bacteria, and fungi [55–58]. It has to be mentioned that rounded chloroplasts were also observed in *B. juncea* cells in response to *A. brassicicola* infection, although the cup-shaped thylakoid system was not found [9]. Changes of chloroplast shape and location of thylakoids could be related to cell lysis and, as a consequence, changes of osmotic pressure of stroma could have occurred. However, the underlying processes behind chloroplast degradation during infection are complex and require further extensive investigation. In the case of infected *B. oleracea* mesophyll cells, changes of organelle ultrastructure could be a result of the action of toxins and secondary metabolites secreted by *A. brassicicola* [16] and general reprogramming of the host transcriptome and metabolome in response to the infection [59,60]. *Alternaria* species secrete various non-host-(nHSTs) and host-selective toxins (HSTs) which are responsible for the degradation of different organelles within an infected cell in over 200 plant species. For example, *Alternaria alternata* pathotypes produce different HSTs depending on the plant species, such as the AM-toxin that degrades the plasma membrane and chloroplasts in apple; the ACR-toxin that causes damage of the mitochondria in lemon; the AK-toxin, AF-toxin and ACT-toxin that target the plasma membrane in susceptible cultivars of Japanese pear, strawberry and tangerine, respectively [61–63]. However, a possible effect of identified *A. brassicicola* secreted compounds during infection on plant cell structures or their influence on the host transcriptome has not been described yet. It has to be emphasized that visualized differential *B. oleracea* cell responses to *A. brassicicola* infection, from defensive to highly susceptible, were confirmed by our microarray results (Table S4), and were also concordant with the previous studies on plant gene expression during infection of susceptible *B. oleracea* and *B. napus* with *A. brassicicola* [64,65]. The most upregulated categories revealed in the analysis of Gene Ontology of Biological Processes were aging, the detection of ethylene stimulus, thiamine biosynthetic processes, cell wall macromolecule catabolic processes and defense response to fungus. Additionally, analysis of stress-related genes revealed upregulation of the individual genes associated with hormone metabolism (brassinosteroids, ethylene and salicylic acid biosynthesis), peroxidases, glutathione S-transferases and secondary metabolism connected to flavonoid biosynthesis (Table S4). Many of these genes have been previously described as important factors in general plant cell response to biotic stress and identified in transcriptome profile analyses of other susceptible hosts infected with *Alternaria* species [66,67]. Some of these genes undoubtedly are characteristic for the *B. oleracea* response to infection by *A. brassicicola* [59,64], albeit confirmation of their roles requires further investigations based on comparative transcriptomics and proteomic and metabolomic approaches. It has to be emphasized that the up- and downregulation of the host genes in response to *A. brassicicola* revealed in our microarray analysis is concordant with a general scheme of transcriptomics response of a host to necrotrophic pathogens. Perception of necrotrophic fungi activates hormone-dependent and hormone-independent signaling pathways [68]. In resistant interactions, a hormone-dependent response to necrotrophs is mainly based on induction of genes associated with JA and ET-dependent signaling pathways to restrict pathogen spread. In susceptible interactions, genes involved in abscisic acid (ABA), brassinosteroid (BR) and auxin-dependent signaling are also often induced [67,69,70]. Regardless of the mode of interaction, hormone-independent signaling is mediated by induction of genes involved in biosynthesis of phytoalexins and pathogenesis-related proteins (PR) [68]. If a pathogen is unable to detoxify host-specific phytoalexins, they often restrict its growth as in the case of Arabidopsis and *A. brassicicola* interaction [23]. In addition, PR proteins are plant species-specific and play important roles in a pathogen recognition processes by a host as well as a host defense response. Transcriptomics analysis of tomato genotypes contrasting in response to *A. solani* infection revealed differential expression of PR protein-encoding genes. In resistant tomato genotype, the most of identified PR genes was extremely upregulated, whereas, in a susceptible genotype, the majority of these genes was downregulated [66].

4.5. Defense-Related Genes Are Activated Also in Susceptible Interaction

The upregulation of several stress-related genes involved in the defense response to fungus in the examined susceptible cultivar of *B. oleracea* even at 48 hpi, when necrotic lesions were fully developed, indicates that some of the host cells still attempted to combat the invasion of the fungus. The fungal hyphae spreading beyond inoculation site, even at a later stage of infection, apparently induced host defense response through PTI (pattern-triggered immunity) [59]. Probably, the perception of DAMP (damage-associated molecular pattern, e.g., products of the *A. brassicicola* cell wall degrading enzymes or plant secreted peptides) or PAMP (pathogen-associated molecular pattern, e.g., chitin) through a PRR (pattern recognition receptor) triggered the PTI-associated signaling cascade (albeit transcription of MEK1 was negatively regulated) and activated WRKY33 in infected *B. oleracea* cells, similarly to the situation observed in Arabidopsis during fungal infection [71]. Interestingly, the only upregulated gene encoding PRR in our microarray analysis was *RLK7* (Table S4). RLK7 belongs to the category XI of RLKs and acts as a receptor for PIP1 secreted peptide. Elevated expression of *PIP1* has been described in Arabidopsis guard cells during *Pseudomonas*-induced PTI, and thus the peptide plays a role in stomatal immunity [72]. Moreover, RLK7 also contributes to Arabidopsis resistance to *B. cinerea* [73] and *Phytium irregulare* [74]. However, *LYK5*, the other gene encoding RLK, which is an important receptor engaged in chitin perception in Arabidopsis [75], was downregulated. The upregulated genes involved in defense signaling such as *WRKY33* [76] and *PDF1.2* [77] have been identified in *A. thaliana* signaling pathways and are required for signaling resistance to necrotrophic fungi *A. brassicicola* and *B. cinerea*. In Arabidopsis, WRKY33 transcription factor is responsible for the activation of camalexin biosynthesis [78], which efficiently inhibits development of necrotrophic fungi. Possibly, WRKY33 activates *de novo* biosynthesis of *B. oleracea* fungus-induced phytoalexins, such as brassinin, which can be metabolized by *A. brassicicola*, and thus the fungus suppresses the first line of host defense [79]. Subsequently, JA-dependent signaling, which is characteristic for host cells infected by fungal necrotrophs, activates transcription of JA-responsive defense genes encoding PR (pathogenesis-related) proteins such as PDF1.2 involved in resistance against *B. cinerea* [80] and *A. brassicicola* [23]. However, the downregulation of *LOX2* chloroplast lipooxygenase required for JA accumulation may be also responsible for the impaired defense of the host and, in turn, salicylic acid-dependent genes become activated, which promote host cell death [81]. Moreover, local fortification of the host cell walls microscopically observed within inoculation site in our study was reflected in the upregulation of several stress-responsive cell wall and secondary metabolism-related genes, such as *FAH1* involved in lignin biosynthesis and an important component of *A. thaliana* resistance mechanism to *B. cinerea* [82] and *CCoAMT*, which is involved in the biosynthesis of cell wall-bound phenolics and lignin [83]. The strengthening of host cell walls through lignin biosynthesis and its deposition at a pathogen entry site also constitutes the first line of the host defense reaction in order to slow down or restrict pathogen development, especially as lignin is not metabolized by most pathogens [84,85]. Thus, even successful invasion of host tissues within inoculation sites does not signify that every single host cell immediately surrenders to a fungal invader. However, downregulation of numerous stress-related genes found in our study indicates that *A. brassicicola* effectively overcomes the host arsenal of defenses (Table S4). Although it has been reported that *A. brassicicola* secretes phytotoxin brassicicolin A, histone deacetylase inhibitor depudicin, siderophore N,N-dimethylcoproge [15,17], a proteinaceous host-specific toxin—AB-toxin [86,87] and many low molecular weight secondary metabolites e.g., brassicenes A to F [88], there is not yet any known putative *A. brassicicola* effector(s) targeting also still unknown *Brassica* receptor(s) and triggering ETS (effector-triggered susceptibility) or ETI (effector-triggered immunity), albeit many research groups all over the world work on *A. brassicicola* pathogenicity factors [16].

4.6. Downregulation of Photosynthesis Is Probably Not Only a Part of Susceptible Interaction

Furthermore, we have focused on changes in the ultrastructure of chloroplasts in infected mesophyll cells, mostly due to the observed clear stages of their degradation (Figure 6) and the fact that analysis of our microarray data pointed out photosynthesis as the most negatively regulated process during infection of *B. oleracea* leaves with *A. brassicicola* (Table S3). Chloroplasts are energy and carbon source organelles, and play an important role in plant immunity as a compartment for ROS generation and the production of phytohormones, secondary metabolites, and their precursors [89]. As sensors of environmental changes, chloroplasts can shape nuclear gene expression and activate defense responses through redox flux [90]. Moreover, many pathogen-derived effectors target chloroplast-localized proteins, including components of the photosynthetic electron transport chain [91]. Therefore, changes in chloroplast ultrastructure are often used as good indicators of biotic/abiotic stress [92].

Gradual degradation of the chloroplast membrane system, such as widening of the thylakoid lumens and the disappearance of grana observed in infected *B. oleracea* leaves, indicated the suppression of photosynthesis light reactions and potential damage of light harvesting complexes and reaction centers. As a result, chlorophyll and carotenoid content and photosynthesis efficiency at a physiological level were decreased. Ultrastructural changes of the stroma, from its swelling to disintegration, suggested downregulation of the photosynthesis dark reaction and a decrease of carotenoid content. In severely degraded cells, the thylakoid system was still to some extent preserved, whereas the chloroplast envelope and stroma were totally disintegrated (Figure 6f). The observed changes in the chloroplast ultrastructure were associated with changes in the expression of photosynthesis-related genes and the physiological response of the host cells. Early downregulation of photosynthesis-related genes could be a result of the host cell defense strategy related to the shift from photosynthesis to non-assimilatory metabolism, as observed during plant infection by various pathogens, or an effect of action of an unknown *A. brassicicola*-secreted effector protein [93]. At later stages of infection (24 and 48 hpi), more genes involved in light reactions were downregulated. Fluorescence decline ratio in steady-state (Rfd) and a significant decrease in steady-state PSII quantum yield (F_v/F_mLss) in infected *B. oleracea* leaves indicate inhibition of electron transport in light-dependent reactions. Confirmation of this was the decrease in the value of F_q/F_v, qL, as well as the qP parameter, which describes the level of energy transferred to the reaction centers and informs about the proportion of open PSII reaction centers (Figure 10) [94,95]. The reduction of photosynthesis, determined by a decrease in the effective quantum yield of PSII, has been previously observed in the interaction of *B. juncea–A. brassicicola* [9] and in other plants infected with necrotrophic fungi [96,97]. In addition, in wheat plants infected with *Bipolaris sorokiniana*, a decrease in F_v/F_m correlating with the loss of chlorophyll has been noted [98].

In severe stress caused by *A. alternata* in rice, a decrease in the electron transport rate was correlated with an increase in non-photochemical quenching (NPQ) [99]. An early increase in NPQ took place following a decline in photosynthetic electron transport activity. A similar PSII response was also observed in plants infected with viruses, e.g., Pepper mild mottle virus (PMMoV) infecting pepper leaves [100]. However, such an increase can be the result of both: protective processes of PSII and damage to the photosystem [84]. In infected *B. oleracea* leaves, we generally observed a decrease in NPQ (Figure 10), which indicates a reduction in the dissipation efficiency of excess excitation energy as heat. However, it should be mentioned that not all processes associated with non-photochemical quenching lead to an increase in NPQ [101]. The decrease in NPQ in infected *B. oleracea* leaves may be due to reduced light absorption, e.g., as a result of the destruction of chloroplasts and a decrease in chlorophyll content (Figure 9), rather than a lack of thermal dissipation. In addition, in other plants (i.e., rice, tomato) exposed to necrotrophic fungi, a significant decrease in NPQ was observed both in the necrotic zone and in adjacent areas of the leaf blade, without significantly reducing the photosynthesis efficiency (measured as F_v/F_m) [97,102].

In the case of progressive degradation of chloroplasts at the inoculation site, it is difficult to expect discrete changes in the photosynthesis light reactions. Therefore, the significant decrease in the values of chlorophyll a fluorescence quenching parameters is not surprising. However, a slight decrease in QY_max and QY_Lss (Figure 10) suggests that the destruction of the photosynthetic apparatus did not completely block the electron transfer in PSII in the analyzed leaf area. It is probable that the chloroplasts of cells that have not been invaded by the fungus allow photochemical reactions to occur, although it is known that even *Alternaria* spp. metabolites alone can cause a reduction or complete inhibition of the electron transport chain from QA to QB [103,104]. This is due to competition between the toxin and QB for a binding site in the D1 protein on the thylakoid PSII membrane. In general, the photosynthetic yield of plants infected with necrotrophic fungi presents complex spatial and temporal patterns, depending on the degree of colonization the individual regions of a leaf blade by the pathogen [105].

5. Conclusions

Our results show that initial stages of a susceptible interaction between *B. oleracea* and *A. brassicicola* are complex, not uniform within an inoculation site, and host cell defenses are still active even at later stages of infection (48 hpi). Analysis of the microarray data suggested the possible involvement of putative genes encoding LRR receptor-like kinases in PTI in this pathosystem. Ultrastructural, molecular, and physiological analysis of infected leaves revealed photosynthesis as the most downregulated process from the onset of the infection. This finding should be taken into consideration in further research and work on a strategy for the management of black spot disease.

Supplementary Materials: The following are available online at http://www.mdpi.com/2073-4409/9/10/2329/s1, Figure S1: Leaf position-dependent necrosis formation on *B. oleracea* leaves during *A. brassicicola* infection, Figure S2: Bright field light microscopy images depicting germination and development of *A. brassicicola* on *B. oleracea* leaf surface, Figure S3: Scanning electron microscopy (SEM) images depicting formation of *A. brassicicola* mycelium on abaxial side of *B. oleracea* leaf, Figure S4: Selected light microscopy images taken from semi-thin sections of control and infected *B. oleracea* leaves, Figure S5: Transmission electron microscopy micrographs depicting ultrastructural details of *A. brassicicola* hyphae growing inside the host leaf mesophyll, Table S1: Statistical analysis of microarray data, Table S2: Gene Ontology of Biological Processes at 48 hpi significant with FDR<0.1 expressed in *B. oleracea* during *A. brassicicola* infection, Table S3: Gene Ontology enrichment analysis, Table S4: Microarray-based analysis of stress-related genes significantly expressed during *B. oleracea* infection by *A. brassicicola* at 48 hpi (MapMan analysis), Table S5: Microarray-based analysis of photosynthesis-related *B. oleracea* genes differentially expressed during *A. brassicicola* infection (MapMan analysis), Table S6: Detailed statistical analysis of measured and calculated chlorophyll a fluorescence parameters in *B. oleracea* leaves infected with *A. brassicicola*. The microarray gene expression data generated during the current study is available in the Gene Expression Omnibus (GEO) repository under experiment number GSE155051 (https://www.ncbi.nlm.nih.gov/geo/query/acc.cgi?acc=GSE155051).

Author Contributions: Conceptualization, V.K.M.; methodology, V.K.M., M.G., and A.Z.; formal analysis, V.K.M. and A.Z.; investigation, V.K.M., M.G., and A.Z.; data curation, V.K.M., M.G., and A.Z.; writing—original draft preparation, V.K.M.; writing—review and editing, V.K.M., M.S., A.S., J.O., and A.K.K.; visualization, V.K.M., A.Z., and M.S.; supervision, V.K.M. and A.K.K.; funding acquisition, V.K.M. and A.K.K. All authors have read and agreed to the published version of the manuscript.

Funding: This research was funded by the Ministry of Science and Higher Education, Grant No. N N302 3188 33 and in part by the National Center of Research and Development, Grant No. ERA-CAPS II/1/2015. The APC was funded by the University of Bialystok, Poland.

Acknowledgments: The authors thank Paweł Jedyński for excellent technical assistance and Michal Goralski for help with the microarray data analysis.

Conflicts of Interest: The authors declare no conflict of interest. The funders had no role in the design of the study; in the collection, analyses, or interpretation of data; in the writing of the manuscript, or in the decision to publish the results.

References

1. Babula, D.; Kaczmarek, M.; Ziółkowski, P.A.; Sadowski, J. Brassica oleracea. In *Genome Mapping and Molecular Breeding in Plants*; Kole, C., Ed.; Springer: Berlin/Heidelberg, Germany, 2007; Volume 5, pp. 227–285.
2. Ryder, E.J. Cabbage. In *Leafy Salad Vegetables*; Ryder, E.J., Ed.; The AVI Publishing Company, Springer: Berlin, Germany, 1979; pp. 127–169.
3. Nowakowska, M.; Wrzesińska, M.; Kamiński, P.; Szczechura, W.; Lichocka, M.; Tartanus, M.; Kozik, E.U.; Nowicki, M. Alternaria brassicicola–Brassicaceae pathosystem: Insights into the infection process and resistance mechanisms under optimized artificial bio-assay. *Eur. J. Plant Pathol.* **2019**, *153*, 131–151. [CrossRef]
4. Lawrence, D.P.; Rotondo, F.; Gannibal, P.B. Biodiversity and taxonomy of the pleomorphic genus *Alternaria*. *Mycol. Prog.* **2016**, *15*, 3. [CrossRef]
5. Maude, R.B.; Humpherson-Jones, F.M. Studies on the seed-borne phases of dark leaf spot *Alternaria brassicicola* and grey leaf spot *Alternaria brassicae* of brassicas. *Ann. Appl. Biol.* **1980**, *95*, 311–319. [CrossRef]
6. Humpherson-Jones, F.M.; Maude, R.B. Studies on the epidemiology of *Alternaria brassicicola* in *Brassica oleracea* seed production crops. *Ann. Appl. Biol.* **1982**, *100*, 61–71. [CrossRef]
7. Dillard, H.R.; Cobb, A.C.; Lamboy, J.S. Transmission of *Alternaria brassicicola* to cabbage by flea beetles (*Phyllotreta cruciferae*). *Plant Dis.* **1998**, *82*, 153–157. [CrossRef]
8. Nowicki, M.; Nowakowska, M.; Niezgoda, A.; Kozik, E.U. *Alternaria* black spot of *Crucifers*: Symptoms, importance of disease, and perspectives of resistance breeding. *Veg. Crop. Res. Bull.* **2012**, *76*, 5–19. [CrossRef]
9. Macioszek, V.K.; Wielanek, M.; Morkunas, I.; Ciereszko, I.; Kononowicz, A.K. Leaf position-dependent effect of *Alternaria brassicicola* development on host cell death, photosynthesis and secondary metabolites in *Brassica juncea*. *Physiol. Plant* **2020**, *168*, 601–616. [CrossRef]
10. Humpherson-Jones, F.M. Survival of *Alternaria brassicae* and *Alternaria brassicicola* on crop debris of oilseed rape and cabbage. *Ann. Appl. Biol.* **1989**, *115*, 45–50. [CrossRef]
11. Macioszek, V.K.; Lawrence, C.B.; Kononowicz, A.K. Infection cycle of *Alternaria brassicicola* on *Brassica oleracea* leaves under growth room conditions. *Plant Pathol.* **2018**, *67*, 1088–1096. [CrossRef]
12. Sharma, P.; Deep, S.; Bhati, D.S.; Sharma, M.; Chowdappa, P. Penetration and infection processes of *Alternaria brassicicola* on cauliflower leaf and *Alternaria brassicae* on mustard leaf: A histopathological study. *Plant. Pathol. J.* **2014**, *13*, 100–111. [CrossRef]
13. Humpherson-Jones, F.M.; Phelps, K. Climatic factors influencing spore production in *Alternaria brassicae* and *Alternaria brassicicola*. *Ann. Appl. Biol.* **1989**, *114*, 449–458. [CrossRef]
14. Chen, L.Y.; Price, T.V.; Park-Ng, Z. Conidial dispersal by *Alternaria brassicicola* on Chinese cabbage (*Brassica pekinensis*) in the field and under simulated conditions. *Plant Pathol.* **2003**, *52*, 536–545. [CrossRef]
15. Pedras, M.S.C.; Chumala, P.B.; Jin, W.; Islam, M.S.; Hauck, D.W. The phytopathogenic fungus *Alternaria brassicicola*: Phytotoxin production and phytoalexin elicitation. *Phytochemistry* **2009**, *70*, 394–402. [CrossRef] [PubMed]

16. Cho, Y. How the necrotrophic fungus *Alternaria brassicicola* kills plant cells remains an enigma. *Eukaryot. Cell* **2015**, *14*, 335–344. [CrossRef] [PubMed]
17. Pedras, M.S.C.; Park, M.R. The biosynthesis of brassicicolin A in the phytopathogen *Alternaria brassicicola*. *Phytochemistry* **2016**, *132*, 26–32. [CrossRef] [PubMed]
18. Aneja, J.K.; Agnihotri, A. Alternaria blight of oilseed Brassicas: Epidemiology and disease control strategies with special reference to use of biotechnological approaches for attaining host resistance. *J. Oilseed Brassica* **2013**, *4*, 1–10.
19. Saharan, G.S.; Mehta, N.; Meena, P.D. Resistance. In *Alternaria Diseases of Crucifers: Biology, Ecology and Disease Management*; Saharan, G.S., Mehta, N., Meena, P.D., Eds.; Springer Science+Business Media: Singapore, 2016; pp. 175–210.
20. van Wees, S.C.M.; Chang, H.-S.; Zhu, T.; Glazebrook, J. Characterization of the early response of Arabidopsis to *Alternaria brassicicola* infection using expression profiling. *Plant Physiol.* **2003**, *132*, 606–617. [CrossRef]
21. Su'udi, M.; Kim, M.G.; Park, S.R.; Hwang, D.J.; Bae, S.C.; Ahn, I.P. Arabidopsis cell death in compatible and incompatible interactions with *Alternaria brassicicola*. *Mol. Cells* **2011**, *31*, 593–601. [CrossRef]
22. Kámán-Tóth, E.; Dankó, T.; Gullner, G.; Bozsó, Z.; Palkovics, L.; Pogány, M. Contribution of cell wall peroxidase- and NADPH oxidase-derived reactive oxygen species to *Alternaria brassicicola*-induced oxidative burst in Arabidopsis. *Mol. Plant Pathol.* **2019**, *20*, 485–499. [CrossRef]
23. Thomma, B.P.; Nelissen, I.; Eggermont, K.; Broekaert, W.F. Deficiency in phytoalexin production causes enhanced susceptibility of *Arabidopsis thaliana* to the fungus *Alternaria brassicicola*. *Plant J.* **1999**, *19*, 163–171. [CrossRef]
24. Pedras, M.S.; Minic, Z.; Sarma-Mamillapalle, V.K. Substrate specificity and inhibition of brassinin hydrolases, detoxifying enzymes from the plant pathogens *Leptosphaeria maculans* and *Alternaria brassicicola*. *FEBS J.* **2009**, *276*, 7412–7428. [CrossRef]
25. Pedras, M.S.C.; Abdoli, A. Pathogen inactivation of cruciferous phytoalexins: Detoxification reactions, enzymes and inhibitors. *RSC Adv.* **2017**, *7*, 23633. [CrossRef]
26. Bilgin, D.D.; Zavala, J.A.; Zhu, J.; Clough, S.J.; Ort, D.R.; DeLucia, E.H. Biotic stress globally downregulates photosynthesis genes. *Plant Cell Environ.* **2010**, *33*, 1597–1613. [CrossRef]
27. Narusaka, Y.; Narusaka, M.; Seki, M.; Ishida, J.; Nakashima, M.; Kamiya, A.; Enju, A.; Sakurai, T.; Satoh, M.; Kobayashi, M.; et al. The cDNA microarray analysis using an *Arabidopsis pad3* mutant reveals the expression profiles and classification of genes induced by *Alternaria brassicicola* attack. *Plant Cell Physiol.* **2003**, *44*, 377–387. [CrossRef] [PubMed]
28. Reynolds, E.S. The use of lead citrate at high pH as an electron-opaque stain in electron microscopy. *J. Cell Biol.* **1963**, *17*, 208–212. [CrossRef] [PubMed]
29. Smyth, G.K.; Speed, T. Normalization of cDNA microarray data. *Methods* **2003**, *31*, 265–273. [CrossRef]
30. Smyth, G.K. Linear models and empirical bayes methods for assessing differential expression in microarray experiments. *Stat. Appl. Genet. Mol. Biol.* **2004**, *3*, 3. [CrossRef]
31. Ritchie, M.E.; Silver, J.; Oshlack, A.; Holmes, M.; Diyagama, D.; Holloway, A.; Smyth, G.K. A comparison of background correction methods for two-colour microarrays. *Bioinformatics* **2007**, *23*, 2700–2707. [CrossRef]
32. Benjamini, Y.; Hochberg, Y. Controlling the false discovery rate: A practical and powerful approach to multiple testing. *J. R. Stat. Soc. Ser. B* **1995**, *57*, 289–300. [CrossRef]
33. Krishnakumar, V.; Hanlon, M.R.; Contrino, S.; Ferlanti, E.S.; Karamycheva, S.; Kim, M.; Rosen, B.D.; Cheng, C.-Y.; Moreira, W.; Mock, S.A.; et al. Araport: The arabidopsis information portal. *Nucleic Acids Res.* **2015**, *43*, D1003–D1009. [CrossRef]
34. Supek, F.; Bošnjak, M.; Škunca, N.; Šmuc, T. REVIGO summarizes and visualizes long lists of Gene Ontology terms. *PLoS ONE* **2011**, *6*, e21800. [CrossRef] [PubMed]
35. Thimm, O.; Bläsing, O.; Gibon, Y.; Nagel, A.; Meyer, S.; Krüger, P.; Selbig, J.; Müller, L.A.; Rhee, S.Y.; Stitt, M. MAPMAN: A user-driven tool to display genomics data sets onto diagrams of metabolic pathways and other biological processes. *Plant J.* **2004**, *37*, 914–939. [CrossRef]
36. Wellburn, A.R. Spectral determination of chlorophylls *a* and *b*, as well as total carotenoids, using various solvents with spectrophotometers of different resolution. *J. Plant Physiol.* **1994**, *144*, 307–313. [CrossRef]
37. van Kan, J.A.L. Licensed to kill: The lifestyle of a necrotrophic plant pathogen. *Trends Plant Sci.* **2006**, *11*, 247–253. [CrossRef]
38. Mengiste, T. Plant immunity to necrotrophs. *Annu. Rev. Phytopathol.* **2012**, *50*, 267–294. [CrossRef] [PubMed]

39. Doullah, M.A.U.; Meah, M.B.; Okazaki, K. Development of an effective screening method for partial resistance to *Alternaria brassicicola* (dark leaf spot) in *Brassica rapa*. *Eur. J. Plant Pathol.* **2006**, *116*, 33–43. [CrossRef]
40. Łaźniewska, J.; Macioszek, V.K.; Kononowicz, A.K. Plant-fungus interface: The role of surface structures in plant resistance and susceptibility to pathogenic fungi. *Physiol. Mol. Plant Pathol.* **2012**, *78*, 24–30. [CrossRef]
41. Kou, Y.; Naqvi, N.I. Surface sensing and signaling networks in plant pathogenic fungi. *Semin. Cell Dev. Biol.* **2016**, *57*, 84–92. [CrossRef]
42. Ziv, C.; Zhao, Z.; Gao, Y.G.; Xia, Y. Multifunctional roles of plant cuticle during plant-pathogen interactions. *Front. Plant Sci.* **2018**, *9*, 1088. [CrossRef]
43. Babosha, A.V.; Ryabchenko, A.S.; Avetisyan, G.A.; Avetisyan, T.V. Visualization of the halo region in plant–powdery mildew interactions by cryoscanning electron microscopy. *J. Plant Pathol.* **2020**, *102*, 103–111. [CrossRef]
44. Veloso, J.; van Kan, J.A.L. Many shades of grey in *Botrytis*-host plant interactions. *Trends Plant Sci.* **2018**, *23*, 613–622. [CrossRef] [PubMed]
45. Shlezinger, N.; Minz, A.; Gur, J.; Hatam, I.; Dagdas, Y.F.; Talbot, N.J.; Sharon, A. Anti-apoptotic machinery protects the necrotrophic fungus *Botrytis cinerea* from host-induced apoptotic-like cell death during plant infection. *PLoS Pathog.* **2011**, *7*, e1002185. [CrossRef] [PubMed]
46. Fung, R.W.M.; Gonzalo, M.; Fekete, C.; Kovacs, L.G.; He, Y.; Marsh, E.; McIntyre, L.M.; Schachtman, D.P.; Qiu, W. Powdery mildew induces defense-oriented reprogramming of the transcriptome in a susceptible but not in a resistant grapevine. *Plant Physiol.* **2008**, *146*, 236–249. [CrossRef]
47. Nokthai, P.; Lee, V.S.; Shank, L. Molecular modeling of peroxidase and polyphenol oxidase: Substrate specificity and active site comparison. *Int. J. Mol. Sci.* **2010**, *11*, 3266–3276. [CrossRef]
48. Mandal, S.; Kar, I.; Mukherjee, A.K.; Acharya, P. Elicitor-induced defense responses in *Solanum lycopersicum* against *Ralstonia solanacearum*. *Sci. World J.* **2013**, *2013*, 561056. [CrossRef] [PubMed]
49. Roca, M.G.; Read, N.D.; Wheals, A.E. Conidial anastomosis tubes in filamentous fungi. *FEMS Microbiol. Lett.* **2005**, *249*, 191–198. [CrossRef] [PubMed]
50. Read, N.D.; Lichius, A.; Shoji, Y.-J.; Goryachev, A.B. Self-signalling and self-fusion in filamentous fungi. *Curr. Opin. Microbiol.* **2009**, *12*, 608–615. [CrossRef] [PubMed]
51. Kurian, S.M.; Di Pietro, A.; Read, N.D. Live-cell imaging of conidial anastomosis tube fusion during colony initiation in *Fusarium oxysporum*. *PLoS ONE* **2011**, *13*, e0195634. [CrossRef]
52. Read, N.D.; Goryachev, A.B.; Lichius, A. The mechanistic basis of self-fusion between conidial anastomosis tubes during fungal colony initiation. *Fungal Biol. Rev.* **2012**, *26*, 1–11. [CrossRef]
53. Shahi, S.; Fokkens, L.; Houterman, P.M.; Rep, M. Suppressor of fusion, a *Fusarium oxysporum* homolog of Ndt80, is required for nutrient-dependent regulation of anastomosis. *Fungal Genet. Biol.* **2016**, *95*, 49–57. [CrossRef]
54. Craven, K.D.; Vélëz, H.; Cho, Y.; Lawrence, C.B.; Mitchell, T.K. Anastomosis is required for virulence of the fungal necrotroph *Alternaria brassicicola*. *Eukaryot. Cell* **2008**, *7*, 675–683. [CrossRef] [PubMed]
55. Kozieł, E.; Otulak-Kozieł, K.; Bujarski, J.J. Modifications in tissue and cell ultrastructure as elements of immunity-like reaction in *Chenopodium quinoa* against prune dwarf virus (PDV). *Cells* **2020**, *9*, 148. [CrossRef] [PubMed]
56. Otulak, K.; Chouda, M.; Bujarski, J.; Garbaczewska, G. The evidence of tobacco rattle virus impact on host plant organelles ultrastructure. *Micron* **2015**, *70*, 7–20. [CrossRef]
57. Krzymowska, M.; Konopka-Postupolska, D.; Sobczak, M.; Macioszek, V.K.; Ellis, B.E.; Hennig, J. Infection of tobacco with different *Pseudomonas syringae* pathovars leads to distinct morphotypes of programmed cell death. *Plant J.* **2007**, *50*, 253–264. [CrossRef]
58. Gabara, B.; Kuźniak, E.; Skłodowska, M.; Surówka, E.; Miszalski, Z. Ultrastructural and metabolic modifications at the plant-pathogen interface in *Mesembryanthemum crystallinum* leaves infected by *Botrytis cinerea*. *Environ. Exp. Bot.* **2012**, *77*, 33–43. [CrossRef]
59. Lai, Z.; Mengiste, T. Genetic and cellular mechanisms regulating plant responses to necrotrophic pathogens. *Curr. Opin. Plant Biol.* **2013**, *16*, 505–512. [CrossRef] [PubMed]
60. Chen, F.; Ma, R.; Chen, X.-L. Advances of metabolomics in fungal pathogen–plant interactions. *Metabolites* **2019**, *9*, 169. [CrossRef]

61. Tsuge, T.; Harimoto, Y.; Akimitsu, K.; Ohtani, K.; Kodama, M.; Akagi, Y.; Egusa, M.; Yamamoto, M.; Otani, H. Host-selective toxins produced by the plant pathogenic fungus *Alternaria alternata*. *FEMS Microbiol. Rev.* **2013**, *37*, 44–66. [CrossRef] [PubMed]
62. Akimitsu, K.; Tsuge, T.; Kodama, M.; Yamamoto, M.; Otani, H. *Alternaria* host-selective toxins: Determinant factors of plant disease. *J. Gen. Plant Pathol.* **2014**, *80*, 109–122. [CrossRef]
63. Meena, M.; Samal, S. *Alternaria* host-specific (HSTs) toxins: An overview of chemical characterization, target sites, regulation and their toxic effects. *Toxicol. Rep.* **2019**, *6*, 745–758. [CrossRef]
64. Cramer, R.A.; La Rota, C.M.; Cho, Y.; Thon, M.; Craven, K.D.; Knudson, D.L.; Mitchell, T.K.; Lawrence, C.B. Bioinformatic analysis of expressed sequence tags derived from a compatible *Alternaria brassicicola-Brassica oleracea* interaction. *Mol. Plant Pathol.* **2006**, *7*, 113–124. [CrossRef] [PubMed]
65. Schenk, P.M.; Thomas-Hall, S.R.; Nguyen, A.V.; Manners, J.M.; Kazan, K.; Spangenberg, G. Identification of plant defence genes in canola using Arabidopsis cDNA microarrays. *Plant Biol.* **2008**, *10*, 539–547. [CrossRef] [PubMed]
66. Upadhyay, P.; Rai, A.; Kumar, R.; Singh, M.; Sinha, B. Differential expression of pathogenesis related protein genes in tomato during inoculation with *A. solani*. *J. Plant Pathol. Microb.* **2014**, *5*, 1. [CrossRef]
67. Zhu, L.; Ni, W.; Liu, S.; Cai, B.; Xing, H.; Wang, S. Transcriptomics analysis of apple leaves in response to *Alternaria alternata* apple pathotype infection. *Front. Plant Sci.* **2017**, *8*, 22. [CrossRef]
68. Pandey, D.; Rajendran, S.R.C.K.; Gaur, M.; Sajeesh, P.K.; Kumar, A. Plant defense signaling and responses against necrotrophic fungal pathogens. *J. Plant Growth Regul.* **2016**, *35*, 1159–1174. [CrossRef]
69. Glazebrook, J. Contrasting mechanisms of defense against biotrophic and necrotrophic pathogens. *Annu. Rev. Phytopathol.* **2005**, *43*, 205–227. [CrossRef]
70. Malik, N.A.A.; Kumar, I.S.; Nadarajah, K. Elicitor and receptor molecules: Orchestrators of plant defense and immunity. *Int. J. Mol. Sci.* **2020**, *21*, 963. [CrossRef]
71. Gong, B.-Q.; Wang, F.-Z.; Li, J.-F. Hide-and-seek: Chitin-triggered plant immunity and fungal counterstrategies. *Trends Plant Sci.* **2020**, *25*, 805–816. [CrossRef]
72. Hou, S.; Shen, H.; Shao, H. PAMP-induced peptide 1 cooperates with salicylic acid to regulate stomatal immunity in *Arabidopsis thaliana*. *Plant Signal. Behav.* **2019**, *14*, 1666657. [CrossRef]
73. Liu, Z.; Wu, Y.; Yang, F.; Zhang, Y.; Chen, S.; Xie, Q.; Tian, X.; Zhou, J.-M. BIK1 interacts with PEPRs to mediate ethylene-induced immunity. *Proc. Natl. Acad. Sci. USA* **2013**, *110*, 6205–6210. [CrossRef]
74. Yamaguchi, Y.; Huffaker, A.; Bryan, A.C.; Tax, F.E.; Ryan, C.A. PEPR2 is a second receptor for the Pep1 and Pep2 peptides and contributes to defense responses in Arabidopsis. *Plant Cell* **2010**, *22*, 508–522. [CrossRef]
75. Erwig, J.; Ghareeb, H.; Kopischke, M.; Hacke, R.; Matei, A.; Petutschnig, E.; Lipka, V. Chitin-induced and chitin elicitor receptor kinase1 (CERK1) phosphorylation-dependent endocytosis of Arabidopsis thaliana lysin motif-containing receptor-like kinase5 (LYK5). *New Phytol.* **2017**, *215*, 382–396. [CrossRef] [PubMed]
76. Zheng, Z.; Qamar, S.A.; Chen, Z.; Mengiste, T. Arabidopsis WRKY33 transcription factor is required for resistance to necrotrophic fungal pathogens. *Plant J.* **2006**, *48*, 592–605. [CrossRef] [PubMed]
77. Manners, J.M.; Penninckx, I.A.; Vermaere, K.; Kazan, K.; Brown, R.L.; Morgan, A.; Maclean, D.J.; Curtis, M.D.; Cammue, B.P.; Broekaert, W.F. The promoter of the plant defensin gene *PDF1.2* from *Arabidopsis* is systemically activated by fungal pathogens and responds to methyl jasmonate but not to salicylic acid. *Plant Mol. Biol.* **1998**, *38*, 1071–1080. [CrossRef]
78. Zhou, J.; Wang, X.; He, Y.; Sang, T.; Wang, P.; Dai, S.; Zhang, S.; Meng, X. Differential phosphorylation of the transcription factor WRKY33 by the protein kinases CPK5/CPK6 and MPK3/MPK6 cooperatively regulates camalexin biosynthesis in Arabidopsis. *Plant Cell* **2020**. [CrossRef] [PubMed]
79. Srivastava, A.; Cho, I.K.; Cho, Y. The Bdtf1 gene in *Alternaria brassicicola* is important in detoxifying brassinin and maintaining virulence on *Brassica* species. *Mol. Plant Microbe Interact.* **2013**, *26*, 1429–1440. [CrossRef] [PubMed]
80. Meng, X.; Xu, J.; He, Y.; Yang, K.-Y.; Mordorski, B.; Liu, Y.; Zhang, S. Phosphorylation of an ERF transcription factor by Arabidopsis MPK3/MPK6 regulates plant defense gene induction and fungal resistance. *Plant Cell* **2013**, *25*, 1126–1142. [CrossRef]
81. Vidhyasekaran, P. *Plant. Hormone Signalling Systems in Plant Innate Immunity*; Springer Science+Business Media: Dordrecht, The Netherlands, 2015.

82. Lloyd, A.J.; Allwood, J.W.; Winder, C.L.; Dunn, W.B.; Heald, J.K.; Cristescu, S.M.; Sivakumaran, A.; Harren, F.J.M.; Mulema, J.; Denby, K.; et al. Metabolomic approaches reveal that cell wall modifications play a major role in ethylene-mediated resistance against *Botrytis cinerea*. *Plant J.* **2011**, *67*, 852–868. [CrossRef]
83. Vidhyasekaran, P. Cell wall degradation and fortification. In *Fungal Pathogenesis in Plants and Crop Molecular Biology of Host Defence Mechanism*, 2nd ed.; Vidhyasekaran, P., Ed.; CRC Press, Taylor and Francis Group: New York, NY, USA, 2008; pp. 275–320.
84. Bhuiyan, N.H.; Selvaraj, G.; Wei, Y.; King, J. Role of lignification in plant defense. *Plant Signal. Behav.* **2009**, *4*, 158–159. [CrossRef]
85. Laluk, K.; Mengiste, T. Necrotroph attacks on plants: Wanton destruction or covert extortion? *Arab. Book* **2010**, *8*, e0136. [CrossRef]
86. Otani, H.; Kohnobe, A.; Kodama, M.; Kohmoto, K. Production of a host-specific toxin by germinating spores of *Alternaria brassicicola*. *Physiol. Mol. Plant Pathol.* **1998**, *52*, 285–295. [CrossRef]
87. Oka, K.; Akamatsu, H.; Kodama, M.; Nakajima, H.; Kawada, T.; Otani, H. Host-specific AB-toxin production by germinating spores of *Alternaria brassicicola* is induced by a host-derived oligosaccharide. *Physiol. Mol. Plant Pathol.* **2005**, *66*, 12–19. [CrossRef]
88. MacKinnon, S.L.; Keifer, P.; Ayer, W.A. Components from the phytotoxic extract of *Alternaria brassicicola*, a black spot pathogen of canola. *Phytochemistry* **1999**, *51*, 215–221. [CrossRef]
89. Kretschmer, M.; Damoo, D.; Djamei, A.; Kronstad, J. Chloroplasts and plant immunity: Where are the fungal effectors? *Pathogens* **2020**, *9*, 19. [CrossRef] [PubMed]
90. Rossi, F.R.; Krapp, A.R.; Bisaro, F.; Maiale, S.J.; Pieckenstain, F.L.; Carrillo, N. Reactive oxygen species generated in chloroplasts contribute to tobacco leaf infection by the necrotrophic fungus *Botrytis cinerea*. *Plant J.* **2017**, *92*, 761–773. [CrossRef] [PubMed]
91. Xu, Q.; Tang, C.; Wang, X.; Sun, S.; Zhao, J.; Kang, Z.; Wang, X. An effector protein of the wheat stripe rust fungus targets chloroplasts and suppresses chloroplast function. *Nat. Commun.* **2019**, *10*, 5571. [CrossRef] [PubMed]
92. Zechmann, B. Ultrastructure of plastids serves as reliable abiotic and biotic stress marker. *PLoS ONE* **2019**, *14*, e0214811. [CrossRef] [PubMed]
93. Kangasjärvi, S.; Neukermans, J.; Li, S.; Aro, E.-M.; Noctor, G. Photosynthesis, photorespiration, and light signalling in defence responses. *J. Exp. Bot.* **2012**, *63*, 1619–1636. [CrossRef]
94. Maxwell, K.; Johnson, G.N. Chlorophyll fluorescence—A practical guide. *J. Exp. Bot.* **2000**, *51*, 659–668. [CrossRef]
95. Nosek, M.; Kornaś, A.; Kuźniak, E.; Miszalski, Z. Plastoquinone redox state modifies plant response to pathogen. *Plant Physiol. Biochem.* **2015**, *96*, 163–170. [CrossRef]
96. Chaerle, L.; Hagenbeek, D.; De Bruyne, E.; Valcke, R.; Van Der Straeten, D. Thermal and chlorophyll-fluorescence imaging distinguish plant-pathogen interactions at an early stage. *Plant Cell Physiol.* **2004**, *45*, 887–896. [CrossRef] [PubMed]
97. Pérez-Bueno, M.L.; Pineda, M.; Barón, M. Phenotyping plant responses to biotic stress by chlorophyll fluorescence imaging. *Front. Plant Sci.* **2019**, *10*, 1135. [CrossRef] [PubMed]
98. Rios, J.A.; Aucique-Pérez, C.E.; Debona, D.; Cruz Neto, L.B.M.; Rios, V.S.; Rodrigues, F.A. Changes in leaf gas exchange chlorophyll *a* fluorescence and antioxidant metabolism within wheat leaves infected by *Bipolaris sorokiniana*. *Ann. Appl. Biol.* **2017**, *170*, 189–203. [CrossRef]
99. Yang, Z.-X.; Yang, Y.-F.; Yu, S.-Z.; Wang, R.-G.; Wang, Y.; Chen, H.-L. Photosynthetic photochemical and osmotic regulation changes in tobacco resistant and susceptible to *Alternaria alternata*. *Trop. Plant Pathol.* **2018**, *43*, 413–421. [CrossRef]
100. Rys, M.; Juhász, C.; Surówka, E.; Janeczko, A.; Saja, D.; Tóbiás, I.; Skoczowski, A.; Barna, B.; Gullner, G. Comparison of a compatible and an incompatible pepper-tobamovirus interaction by biochemical and non-invasive techniques: Chlorophyll *a* fluorescence, isothermal calorimetry and FT-Raman spectroscopy. *Plant Physiol. Biochem.* **2014**, *83*, 267–278. [CrossRef]
101. Malnoë, A. Photoinhibition or photoprotection of photosynthesis? Update on new sustained quenching component, qH. *Environ. Exp. Bot.* **2018**, *154*, 123–233. [CrossRef]
102. Ghosh, S.; Kanwar, P.; Jha, G. Alterations in rice chloroplast integrity photosynthesis and metabolome associated with pathogenesis of *Rhizoctonia solani*. *Sci. Rep.* **2017**, *7*, 41610. [CrossRef]

103. Chen, S.; Dai, X.; Qiang, S.; Tang, Y. Effect of a nonhost-selective toxin from *Alternaria alternata* on chloroplast-electron transfer activity in *Eupatorium adenophorum*. *Plant Pathol.* **2005**, *54*, 671–677. [CrossRef]
104. Chen, S.; Kim, C.; Lee, J.M.; Lee, H.-A.; Fei, Z.; Wang, L.; Apel, K. Blocking the QB-binding site of photosystem II by tenuazonic acid, a non-host-specific toxin of *Alternaria alternata*, activates singlet oxygen-mediated and EXECUTER-dependent signalling in Arabidopsis. *Plant Cell Environ.* **2015**, *38*, 1069–1080. [CrossRef]
105. Barón, M.; Pineda, M.; Pérez-Bueno, M.L. Picturing pathogen infection in plants. *Z. Nat. C* **2016**, *71*, 355–368. [CrossRef]

cells MDPI

Article

Photosynthetic Pigments Changes of Three Phenotypes of Picocyanobacteria *Synechococcus* sp. under Different Light and Temperature Conditions

Sylwia Śliwińska-Wilczewska [1,*], Zofia Konarzewska [1], Kinga Wiśniewska [2] and Marta Konik [3]

[1] Division of Marine Ecosystems Functioning, Institute of Oceanography, University of Gdansk, Avenue Piłsudskiego 46, P-81-378 Gdynia, Poland; zofia.konarzewska@gmail.com

[2] Division of Marine Chemistry and Environmental Protection, Institute of Oceanography, University of Gdansk, Avenue Piłsudskiego 46, P-81-378 Gdynia, Poland; kinga.wisniewska@phdstud.ug.edu.pl

[3] Department of Marine Physics, Institute of Oceanology Polish Academy of Sciences, P-81-779 Sopot, Poland; mk@iopan.gda.pl

* Correspondence: s.sliwinska-wilczewska@ug.edu.pl; Tel.: +48-58-523-68-92

Received: 16 August 2020; Accepted: 1 September 2020; Published: 3 September 2020

Abstract: It is estimated that the genus *Synechococcus* is responsible for about 17% of net primary production in the Global Ocean. Blooms of these organisms are observed in tropical, subtropical and even temperate zones, and they have been recorded recently even beyond the polar circle. The long-term scenarios forecast a growing expansion of *Synechococcus* sp. and its area of dominance. This is, among others, due to their high physiological plasticity in relation to changing environmental conditions. Three phenotypes of the genus *Synechococcus* sp. (Type 1, Type 2, and Type 3a) were tested in controlled laboratory conditions in order to identify their response to various irradiance (10, 55, 100 and 145 μmol photons m^{-2} s^{-1}) and temperature (15, 22.5 and 30 °C) conditions. The highest total pigment content per cell was recorded at 10 μmol photons m^{-2} s^{-1} at all temperature variants with the clear dominance of phycobilins among all the pigments. In almost every variant the highest growth rate was recorded for the Type 1. The lowest growth rates were observed, in general, for the Type 3a. However, it was recognized to be less temperature sensitive in comparison to the other two types and rather light-driven with the highest plasticity and adaptation potential. The highest amounts of carotenoids were produced by Type 2 which also showed signs of the cell stress even around 55 μmol photons m^{-2} s^{-1} at 15 °C and 22.5 °C. This may imply that the Type 2 is the most susceptible to higher irradiances. Picocyanobacteria *Synechococcus* sp. require less light intensity to achieve the maximum rate of photosynthesis than larger algae. They also tolerate a wide range of temperatures which combined together make them gain a powerful competitive advantage. Our results will provide key information for the ecohydrodynamical model development. Thus, this work would be an important link in forecasting future changes in the occurrence of these organisms in the context of global warming.

Keywords: abiotic stressors; environmental stress; growth; light intensity; photosynthetic pigments; picocyanobacteria; plant physiology

1. Introduction

The discovery of autotrophic picoplankton in the late 1970s [1,2] has contributed to numerous scientific studies on these organisms and demonstrated their significant role as a missing link in the carbon cycle and a major producer in oceanic waters [3]. Many researchers proved that picoplankton also plays an important role in more productive waters, often exceeding the abundance and biomass of other phytoplankton species [4]. The genus *Synechococcus* is a polyphyletic group of picoplanktonic cyanobacteria that constitutes one of the major contributors to oceanic primary production [5,6] and is a key worldwide distributed component of marine planktonic communities [7]. It is estimated that for about 17% of net primary production in the Global Ocean is responsible solely the genus *Synechococcus* [8]. Blooms of these organisms are observed in tropical, subtropical and even temperate zones [9]. The present global warming causes temperature rise which was recognized as a main cause of the massive shift of species northwards [10]. Furthermore, *Synechococcus* has been recorded far beyond the polar circle, e.g., dragged with a strong Atlantic inflow in 2014, as far as 82.5° N [11]. In the future ocean scenarios, a growing expansion of *Synechococcus* sp. and its area of dominance is forecasted [8,12]. A significant increase in the frequency of their blooms has already been detected [9]. This is, among others, due to their high physiological plasticity in relation to changing environmental conditions [13]. Organisms from the genus *Synechococcus* are represented by three phenotypes that complement each other and fill tightly the ecological niche due to varying photosynthetic pigment profiles and high chromatic adaptation potential.

The photosynthetic pigment observed in cells of picoplanktonic cyanobacteria is chlorophyll *a* (Chl *a*), carotenoid (Car) pigments, and phycobiliproteins (Phyco) [14]. Chl *a* is the most important pigment because it controls photosynthesis and this transformation of the absorbed energy from sunlight into chemical compounds determines the biomass growth rates [14]. The most dominant Car pigment is zeaxanthin (Zea), representing 40% to 80%. The presence of cell-specific Zea content in *Synechococcus* sp. and high Zea/Chl *a* ratios may be regarded as a diagnostic feature [15]. Besides Zea, β-carotene (β-Car) is also present among Car pigments [16]. Car pigments play an important photoprotective role against damage to the photosystem [17]. Furthermore, cells of picocyanobacteria contain accessory phycobilin pigments instead of the additional chlorophylls that are common among other phytoplankton organisms. There are three types of Phyco containing: phycoerythrin (PE), phycocyanin (PC), and allophycocyanin (APC), which absorb green, yellow-orange, and red light, respectively [18]. In cyanobacterial cells, Phyco are organized into aggregates consisting of many subunits called phycobilisomes, which are connected in regular rows to the surface of thylakoid membranes. The main component of the core complex is APC while PE is located in the peripheral parts of these formations [19]. Phyco absorb light in the 500–650 nm range and provide additional energy to photosynthetic centers. The transfer process is highly efficient and reaches 80–90% of the energy absorbed by phycobilin pigments. Their role is vital, especially in case of any light shortages to maintain high photosynthesis rate which guarantees cyanobacteria competitive advantage in low-light conditions. The red PE absorbs the blue-green light that penetrates the deepest into the water column. It enables conducting photosynthesis even at the bottom of the euphotic zone. The deeper live an organism, the more PE it contains and the higher is the PE to Chl *a* pigment ratio. In the cells of cyanobacteria living in the upper layers of the ocean the dominant pigments are the blue PC and APC [19].

The distinction between the three main identified phenotypes of the genus *Synechococcus* is based on the phycobilin pigments composition [20,21]. Six et al. [22] in their research presented a classification that divides marine *Synechococcus* to Type 1, Type 2, and Type 3. Organisms with the dominance of PC were classified as Type 1. Type 2 incorporates phenotypes with a dominance of PE, more specifically PEI, while Type 3 consists of organisms in which PC, as well as PEI and PEII, dominates in phycobilisomes. Furthermore, Six et al., [22] divided Type 3 into four subcategories from a to d, according to the increasing phycoerythrobilin (PEB) and phycocyanobilin (PCB) ratios. Organisms with high levels of PE are found mainly in oligotrophic oceans, while green (PC-rich) phenotypes prefer turbid freshwater [23,24]. In general, picocyanobacteria prefer lower irradiance intensity to reach the maximum rate of photosynthesis than larger algae [25]. Furthermore, studies have shown that the reduction of radiation intensity does not change the efficiency of carbon incorporation during photosynthesis, as is the case with larger plant organisms that exceed 3 μm. Marine *Synechococcus* sp. is able to saturate photosynthesis and growth rates at very low radiation [26]. Under culture conditions, some strains of picoplankton have shown the ability to survive and grow again after periods of total darkness [27,28]. Platt et al. [29] observed photosynthetic picoplankton at a depth of 1000 m in the depths of the eastern Pacific Ocean and Cai et al. [30] confirmed the presence of small populations of *Synechococcus* sp. in the Chesapeake Bay during winter months. Furthermore, Ernst [31] isolated *Synechocystis* sp. (Maple BO 8402) from the Lake Constance with a different type of pigmentation than any described so far. This strain contained Phyco similar to the PC, characterized by very strong red fluorescence occurring after stimulation of the cells with wavelengths of 600 nm but also with wavelengths of 436 and 546 nm [32]. Most cyanobacteria, especially those living all year round in coastal ocean waters, contain PE [23,33,34].

The main aim of this study was to determine the acclimatization capacity of three Baltic phenotypes of *Synechococcus* sp.: Type 1, Type 2, Type 3a. Furthermore, the study focused on the effect of irradiance, temperature, and their mutual interactions on the content and proportions of cell-specific photosynthetic pigments of the examined cyanobacterial phenotypes. The cell-specific Chl *a* and Car content was determined by the HPLC method, whereas the content of Phyco was determined by the spectrophotometric method. The detailed characterization of the quantitative and qualitative composition of pigments is important to determine the level of acclimatization of the examined phenotypes of cyanobacteria to specific environmental conditions. The knowledge of biology and especially the physiology of these organisms by capturing their reactions to various environmental factors is important for forecasting the possible expansion of these organisms.

2. Results

2.1. The Cell Concentration and the Growth Rate of Three Synechococcus sp. Phenotypes under Different Culture Conditions

In this study, the concentration of picocyanobacterial cells as well as the growth rate under different irradiance and temperature conditions were determined for the three studied phenotypes of *Synechococcus* sp. (Type 1, Type 2, and Type 3a). In general, factorial tests showed that both irradiance and temperature significantly affected the number of cells of three *Synechococcus* sp. phenotypes (ANOVA, $p < 0.001$, $p < 0.01$, $p < 0.01$, for Type 1, Type 2, and Type 3a, respectively; Table S1). Moreover, ANOVA results indicated that for each picocyanobacteria phenotype the effect of temperature on the culture concentration was higher than the influence of irradiance and the interaction of both factors (Table S1). The highest picocyanobacterial cell numbers (59.5×10^7 and 60.2×10^7 cell mL^{-1}) was noted for *Synechococcus* sp. Type 1 at 10 μmol photons m^{-2} s^{-1} and 55 μmol photons m^{-2} s^{-1}, respectively and 30 °C (Figure 1Aa), and it was about 4-fold higher that the minimum values observed in 15 °C and 145 μmol photons m^{-2} s^{-1} (15.2×10^7 cell mL^{-1}). For *Synechococcus* sp. Type 2 (Figure 1Ba) and Type 3a (Figure 1Ca) the maximum cell concentration were recorded at the temperature of 22.5 °C and 30 °C, respectively. Moreover, the highest picocyanobacterial cell numbers for Type 2 was found at irradiance 55 μmol photons m^{-2} s^{-1} (49.4×10^7 cell mL^{-1}), whereas for Type 3a at 10 μmol photons m^{-2} s^{-1} (25.8×10^7 cell mL^{-1}). For both phenotypes, similar to Type 1, the minimum number of cells were obtained at 15 °C and 145 μmol photons m^{-2} s^{-1} (about 9.7×10^7 and 6.5×10^7 cell mL^{-1}, respectively).

It was found that analyzed phenotypes of *Synechococcus* sp. showed different growth rates (μ) under different temperature and light conditions. For *Synechococcus* sp. Type 1, Type 2, and Type 3a the highest growth rate was recorded at the highest temperature (30 °C). Moreover, the highest growth rate for Type 1 (Figure 1Ab) and Type 2 (Figure 1Bb) was noted at 55 μmol photons m^{-2} s^{-1} (0.457, 0.443, respectively) whereas for Type 3a at 10 μmol photons m^{-2} s^{-1} (0.396; Figure 1Cb). On the other hand, for Type 1, Type 2, and Type 3a, the shortest growth rate (0.359, 0.327, 0.298, respectively) was obtained at 15 °C and 145 μmol photons m^{-2} s^{-1}.

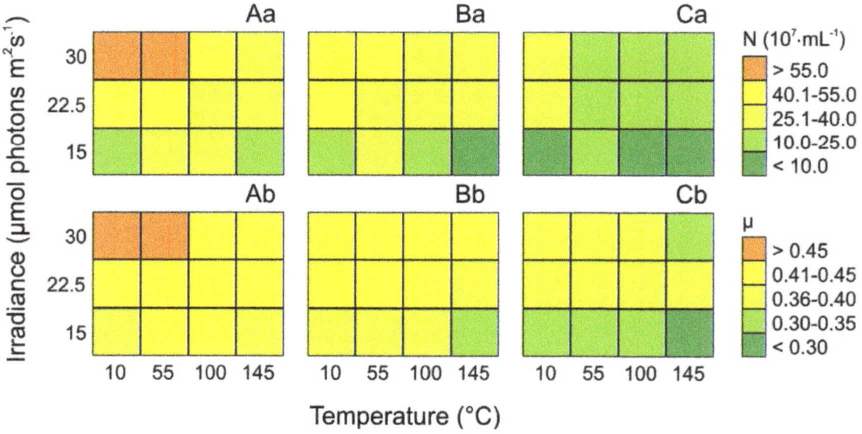

Figure 1. Changes in the number of cells (N × 10^7 mL^{-1}; **a**) and the growth rate (µ; **b**) obtained after 14 days of experiment for three phenotypes of *Synechococcus* sp.: Type 1 (**A**), Type 2 (**B**), Type 3a (**C**) under different irradiance (µmol photons m^{-2} s^{-1}) and temperature (°C) conditions.

2.2. The Total Pigments Content for Three Phenotypes of the Genus Synechococcus

The acclimation mechanisms of three *Synechococcus* sp. phenotypes was characterized by the concentration of changes in composition and proportion of photosynthetic pigments i.e., chlorophyll *a* (Chl *a*), zeaxanthin (Zea), β-carotene (β-Car), phycoerythrin (PE), phycocyanin (PC), and allophycocyanin (APC) under different light (µmol photons m^{-2} s^{-1}) and temperature (°C) conditions. In this work, the composition and proportions of Chl *a* and Car pigments (Zea and β-Car) of three *Synechococcus* sp. phenotypes were determined by HPLC method, while the content of phycobilins (Phyco) were determined by spectrophotometric method.

Both light and temperature significantly affected the cell-specific Chl *a* content of *Synechococcus* sp. Type 1, Type 2, and Type 3a (ANOVA, $p < 0.001$, for all) and Phyco content (ANOVA, $p < 0.001$, $p < 0.001$, and $p < 0.001$, for Type 1, Type 2, and Type 3a, respectively). Moreover, these factors significantly affected the cell-specific Car content of *Synechococcus* sp. phenotypes (ANOVA, $p < 0.001$, $p < 0.001$, $p < 0.001$ for Type 1, Type 2, and Type 3a, respectively; Table S2). Generally, ANOVA results indicated that the effect of irradiance on the Chl *a* and Phyco concentration for picocyanobacteria phenotypes was higher than the influence of temperature and the interaction of the two factors (Table S2). In contrast, the cell-specific Car content of *Synechococcus* sp. Type 1, Type 2, and Type 3a was more affected by temperature and the interaction of the two factors than by irradiance (Table S2).

The maximum cell-specific concentration of Chl *a* (about 8.11 pg·cell^{-1}) was noted for *Synechococcus* sp. Type 3a at 10 µmol photons m^{-2} s^{-1} light intensity and 15 °C, and it was about 5.5 times higher than the minimum at 145 µmol photons m^{-2} s^{-1} and 30 °C (Figure 2Ca). For *Synechococcus* sp. Type 1 and Type 2 the maximum cell-specific Chl *a* concentrations (4.51 pg·cell^{-1} and 4.82 pg·cell^{-1}, respectively) were recorded at 10 µmol photons m^{-2} s^{-1} and 15 °C for Type 1 and 30 °C for Type 2. On the other hand, the minimum values for these phenotypes were obtained at 145 µmol photons m^{-2} s^{-1} and 30 °C (0.68 pg·cell^{-1} and 0.67 pg·cell^{-1}, respectively; Figure 2Aa–Ba).

On the basis of the results obtained in this study, it was found that the analyzed phenotypes were characterized by a similar maximum cell-specific Car content. It was also shown that cell-specific Car content was the lowest among all analyzed photosynthetic pigments. The total Car content for *Synechococcus* sp. Type 1, Type 2, and Type 3a constituted approximately 7%, 11%, and 12% of the sum of Chl *a* and Phyco, respectively. It was also found that for Type 2 (Figure 2Bb) and Type 3a (Figure 2Cb) the maximum cell-specific Car content (2.01 pg·cell^{-1} and 2.25 pg·cell^{-1}, respectively) were recorded at 190 µmol photons m^{-2} s^{-1} and 30 °C. By contrast, the minimum values of cell-specific Car content were obtained at 100 μmol photons m^{-2} s^{-1} and 22.5 °C (1.20 pg·$cell^{-1}$, for Type 2 and 0.60 pg·$cell^{-1}$, for Type 3a). On the other hand, for *Synechococcus* sp. Type 1, the reported maximum value of cell-specific Car content (1.74 pg·$cell^{-1}$) at 100 μmol photons m^{-2} s^{-1} and 15 °C was approximately 4-fold higher compared to the recorded minimum values at 10 μmol photons m^{-2} s^{-1} and 30 °C (Figure 2Ab).

It was noted that the total Phyco pigments were always greater than cell-specific Chl *a* and Car content of the three examined *Synechococcus* sp. phenotypes. The study found that the total Phyco content for *Synechococcus* sp. Type 1, Type 2, and Type 3a constituted about 80%, 75%, and 65% of the sum of Chl *a* and Car, respectively. The highest cell-specific Phyco content was measured in *Synechococcus* sp. Type 2 (45.90 pg·$cell^{-1}$) at 10 μmol photons m^{-2} s^{-1} and 30 °C (Figure 2Bc) while the minimum values of these pigments was noted at 55 μmol photons m^{-2} s^{-1} and 15 °C (2.70 pg·$cell^{-1}$). The greatest decrease in the cell-specific Phyco content was noted for *Synechococcus* sp. Type 1 (Figure 2Ac), which under minimal conditions (100 μmol photons m^{-2} s^{-1} and 15 °C) was about 30 times lower than the recorded under maximum values at 10 μmol photons m^{-2} s^{-1} and 30 °C (33.56 pg·$cell^{-1}$). In turn, *Synechococcus* sp. Type 3a showed the highest resistance to light and temperature, and its decrease in the cell-specific Phyco content under minimal conditions (145 μmol photons m^{-2} s^{-1} and 15 °C) was about 12.7 times lower (2.25 pg·$cell^{-1}$) than the recorded under maximum values (10 μmol photons m^{-2} s^{-1} and 22.5 °C; Figure 2Cc).

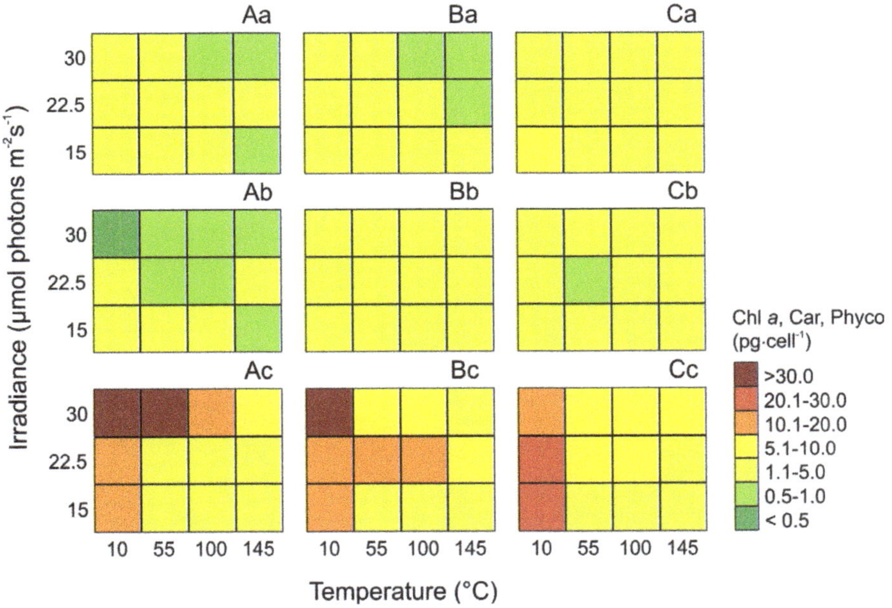

Figure 2. Changes in content (pg·$cell^{-1}$) of Chl *a* (**a**), sum of total Car (**b**), and sum of total Phyco (**c**) obtained after 14 days of experiment for three phenotypes of *Synechococcus* sp.: Type 1 (**A**), Type 2 (**B**), Type 3a (**C**) under different irradiance (μmol photons m^{-2} s^{-1}) and temperature (°C) conditions.

2.3. *Effect of Irradiance and Temperature on Phycocyanin, Phycoerythrin, and Allophycocyanin Content*

The presence of phycoerythrin (PE), phycocyanin (PC), and allophycocyanin (APC) was demonstrated for all picocyanobacterial phenotypes by spectrophotometric analysis. It was found that irradiance and temperature as well as their interaction significantly affected the cell-specific PE content of *Synechococcus* sp. (ANOVA, $p < 0.001$, for Type 1, Type 2, and Type 3a), PC content (ANOVA, $p < 0.001$, fot Type 1, $p < 0.001$, for Type 2, and $p < 0.001$, for Type 3a) and APC content (ANOVA,

$p < 0.001$, $p < 0.01$, and $p < 0.05$, for Type 1, Type 2, and Type 3a, respectively; Table S3). ANOVA indicated that for most of *Synechococcus* sp. phenotypes, the effect of irradiance on PE was higher than the effect of temperature. In contrast, the PC and APC content of analyzed phenotypes was more affected by temperature than by irradiance and by the interaction of both factors (Table S3).

In all the phenotypes, the cell-specific (pg·cell^{-1}) PE, PC, and APC pigment contents were environmentally driven (Figure 3). The cell-specific PE content increased with decrease of irradiance and increase of the temperature, reaching the highest values at the intensity of 10 μmol photons m^{-2} s^{-1} and temperature 22.5 °C (21.16 pg·cell^{-1} for Type 3a; Figure 3Ca) and 30 °C (8.59 pg·cell^{-1} for Type 1 and 40.35 pg·cell^{-1} for Type 2; Figure 3Aa,Ba). Under these conditions, the PE in the cells of the tested picocyanobacteria increased approximately 20.0-fold, 19.7-fold, and 13.6-fold, for Type 1, Type 2, and Type 3a, respectively, compared with the observed minimum values at 100–145 μmol photons m^{-2} s^{-1} and 15 °C.

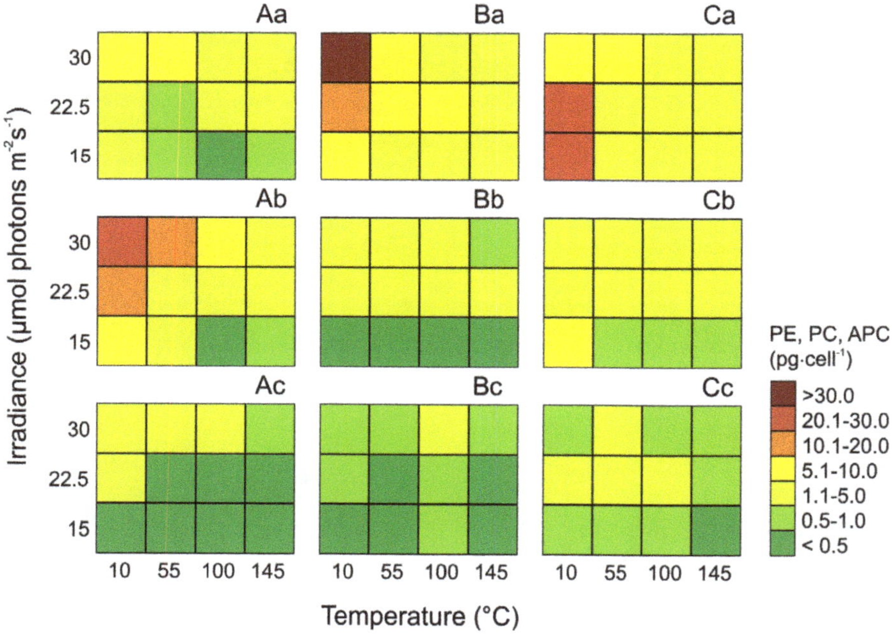

Figure 3. Changes in content (pg·cell^{-1}) of PE (**a**), PC (**b**), and APC (**c**) obtained after 14 days of experiment for three phenotypes of *Synechococcus* sp.: Type 1 (**A**), Type 2 (**B**), Type 3a (**C**) under different irradiance (μmol photons m^{-2} s^{-1}) and temperature (°C) conditions.

On the basis of the conducted analyzes, it was found that the conditions under which the *Synechococcus* sp. Type 1 and Type 2 achieved the highest concentrations of the cell-specific PC were the low light intensity of 10 μmol photons m^{-2} s^{-1} and a high temperature of 30 °C. On the other hand, for Type 3a the maximal value of this pigment was noted at 10 μmol photons m^{-2} s^{-1} and 15 °C. The highest concentration value of PC pigments under optimal conditions was observed for *Synechococcus* Type 1 (20.95 pg·cell^{-1}; Figure 3Ab), and the lowest for *Synechococcus* Type 2 (4.64 pg·cell^{-1}; Figure 3Bb). The greatest decrease in cell-specific PC (about 64-fold) was noted for *Synechococcus* Type 1. However, the least susceptible to analyzed factors was *Synechococcus* Type 3a, with a 10-fold decrease in PC pigments (Figure 3Cb).

The highest cell-specific APC content (4.34 pg·cell^{-1}) was recorded for *Synechococcus* sp. Type 1 in the 55 μmol photons m^{-2} s^{-1} and 30 °C (Figure 3Ac). For these light and temperature conditions, over 18-fold increase was observed in relation to the lowest recorded values at 10 μmol photons m^{-2} s^{-1}

and 15 °C. For *Synechococcus* sp. Type 2 and Type 3a the maximum cell-specific APC concentrations (1.09 pg·cell^{-1} and 1.98 pg·cell^{-1}, respectively) were recorded at 55–100 µmol photons m^{-2} s^{-1} and 22.5–30 °C. On the other hand, the minimum values for these phenotypes were obtained at 145 µmol photons m^{-2} s^{-1} and 15 °C (0.28 pg·cell^{-1} for Type 2 and 0.44 pg·cell^{-1}, for Type 3a; Figure 3Bc,Cc).

2.4. Effect of Irradiance and Temperature on Zeaxanthin and β-carotene

On the basis of the results, the effect of irradiance and temperature on changes in individual Car pigments in the cells of the picocyanobacterial phenotypes was determined. In all the *Synechococcus* sp. phenotypes, the cell-specific (pg·cell^{-1}) pigment contents were environmentally driven (Figure 4). In the most of three tested phenotypes, the cell-specific concentrations of Zea (ANOVA, $p < 0.001$, $p < 0.001$, $p < 0.001$ for Type 1, Type 2, and Type 3a, respectively) and β-Car (ANOVA, $p < 0.001$, $p < 0.01$, $p > 0.05$ for Type 1, Type 2, and Type 3a, respectively) were affected by irradiance and temperature (Table S4). ANOVA indicated that in Type 1 and Type 3a, the effect of temperature on Zea was higher than the effect of irradiance. In contrast, the Zea content of Type 2 was more affected by irradiance than by temperature and by the interaction of both factors. It was also noted that for all tested phenotypes, effect of irradiance on β-Car was not statistically significant (Table S4).

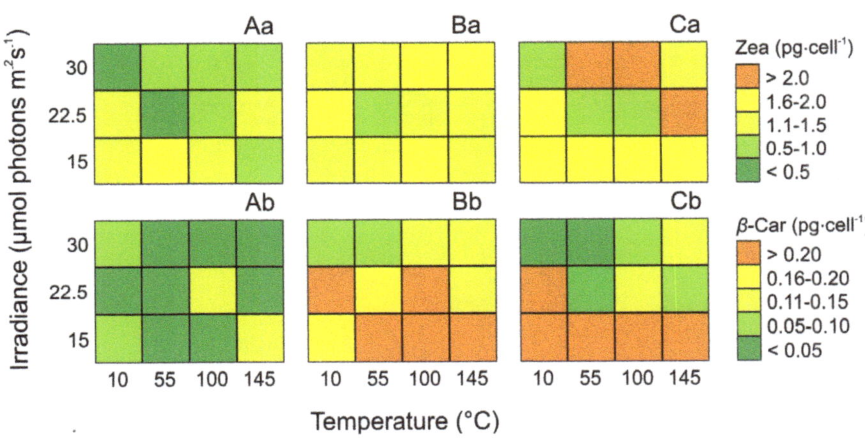

Figure 4. Changes in content (pg·cell^{-1}) of Zea (**a**) and β-Car (**b**) obtained after 14 days of experiment for three phenotypes of *Synechococcus* sp.: Type 1 (**A**), Type 2 (**B**), Type 3a (**C**) under different irradiance (µmol photons m^{-2} s^{-1}) and temperature (°C) conditions.

The highest Zea content for *Synechococcus* sp. Type 2 and Type 3a (1.85 pg·cell^{-1} and 2.11 pg·cell^{-1}, respectively) was noted at 100 µmol photons m^{-2} s^{-1} and 30 °C while the lowest value of this pigment were 1.02 pg·cell^{-1} for Type 2 and 0.53 pg·cell^{-1} for Type 3a at 55 µmol photons m^{-2} s^{-1} and 22.5 °C (Figure 4Ba,Ca). Moreover, the highest value of Zea content for Type 1 was found at irradiance 55 µmol photons m^{-2} s^{-1} and 15 °C (1.68 pg cell^{-1}) while the minimum Zea content was obtained at 30 °C and 10 µmol photons m^{-2} s^{-1} (0.37 pg·cell^{-1}; Figure 4Aa). The highest values of β-Car in Type 2 and Type 3a were noted at 55 µmol photons m^{-2} s^{-1} and 15 °C and 30 °C (0.32 pg·cell^{-1} and 0.40 pg·cell^{-1}, respectively; Figure 4Bb,Cb). In turn, the lowest content of β-Car being found in Type 1 (0.12 pg·cell^{-1}) at 145 µmol photons m^{-2} s^{-1} and 15 °C (Figure 4Ab).

2.5. Effect of Irradiance and Temperature on Pigments Ratios

Light and temperature as well as their interaction were found to significantly affect the Zea/Chl *a* ratio only in *Synechococcus* sp. Type 2 (ANOVA, $p < 0.001$) and the effect of light was higher than the effect of temperature and the interaction of both factors (Table S5). On the other hand, irradiance and temperature as well as their interaction significantly affected the β-Car/Chl *a* ratio in three

Synechococcus sp. phenotypes (ANOVA, $p < 0.001$, $p < 0.01$, and $p < 0.001$, for Type 1, Type 2, and Type 3, respectively). ANOVA indicated that in Type 1 and Type 2, the effect of light on β-Car/Chl *a* ratio was higher than the effect of temperature. In contrast, the β-Car/Chl *a* ratio of Type 3a was more affected by temperature than by irradiance and by the interaction of both factors (Table S5). The highest values of Zea/Chl *a* ratio in *Synechococcus* sp. Type 2, at the 145 µmol photons m^{-2} s^{-1} and the temperature of 30 °C (2.3; Figure 5Ba) was about 11 times higher than the lowest values observed at the light intensity of 10 µmol photons m^{-2} s^{-1} and 30 °C. In turn, the lowest value of Zea/Chl *a* ratio was noted in Type 3a under the same light and temperature conditions (0.8; Figure 5Ca). Besides, the highest β-Car/Chl *a* ratio was also observed for *Synechococcus* sp. Type 2, which at the irradiance of 145 µmol photons m^{-2} s^{-1}, and the temperature of 15 °C was 0.19 (Figure 5Bb). On the other hand, the lowest pigments ratio was recorded for *Synechococcus* sp. Type 3a, which under the same light and temperature conditions was 0.14 (Figure 5Cb).

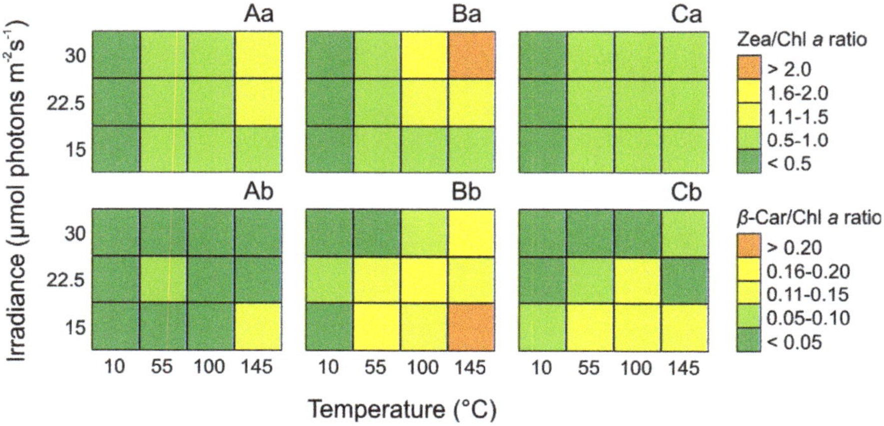

Figure 5. Changes in Zea/Chl *a* ratio (**a**) and β-Car/Chl *a* ratio (**b**) obtained after 14 days of experiment for three phenotypes of *Synechococcus* sp.: Type 1 (**A**), Type 2 (**B**), Type 3a (**C**) under different irradiance (µmol photons m^{-2} s^{-1}) and temperature (°C) conditions.

Since Phyco pigments participate in the transfer of excitation energy to Chl *a* in photosystems, the analysis of changes in these pigments in relation to Chl *a* and Car was also performed (Table S6). It was found that irradiance and temperature as well as their interaction significantly affected the Phyco/Chl *a* ratio in *Synechococcus* sp. Type 1, Type 2, and Type 3 (ANOVA, $p < 0.001$, $p < 0.001$, and $p < 0.01$, respectively) and Phyco/Car ratio (ANOVA, $p < 0.001$, $p < 0.001$, and $p < 0.001$ for Type 1, Type 2, and Type 3a, respectively). ANOVA indicated that in Type 1 and Type 2, the effect of temperature on Phyco/Chl *a* ratio was higher than the effect of irradiance and the interaction of both factors. In turn, the Phyco/Chl *a* ratio of Type 3a was more affected by irradiance than by temperature. For Phyco/Car ratio the effect of temperature for three analyzed phenotypes was higher than the effect of irradiance and the interaction of both factors (Table S7).

The highest Phyco/Chl *a* ratio and Phyco/Car ratio were observed for *Synechococcus* sp. Type 1, which at the light intensity of 55 µmol photons m^{-2} s^{-1} and 10 µmol photons m^{-2} s^{-1} and the temperature of 30 °C was 16.5 and 62.5, respectively. Moreover, the highest values of these pigment ratio in Type 1 was about 33 times and 125 times, respectively higher than the lowest values observed at the light intensity of 100 µmol photons m^{-2} s^{-1} and 15 °C. Conversely, for *Synechococcus* sp. Type 3a the lowest values of Phyco/Chl *a* ratio as well as Phyco/Car ratio were found at 10 µmol photons m^{-2} s^{-1} and 22.5 °C (5.0 and 21.1, respectively) and were about 7 and 21 times higher, respectively, than the minimums obtained at PAR 100 µmol photons m^{-2} s^{-1} and 15 °C (Table S6).

3. Discussion

3.1. Occurrence and Abundance of Picocyanobacteria under Changing Irradiance and Temperature Conditions

Changes in the number of cells of photoautotrophic organisms inhabiting surface waters are the result of the interaction of several physical and chemical environmental factors [35]. Light and temperature play a key role in the occurrence of autotrophic picoplankton [32] and are the main factors causing the appearance of cyanobacteria both at depths and in coastal waters [36,37]. Additionally, light and temperature may be more important abiotic factors influencing the occurrence of picocyanobacteria than the availability of nutrients [36]. In spring, the number of autotrophic picoplankton cells begin to increase which is triggered by the temperature increase due to more intensive insolation of the surface water layers. Their growth reaches its maximum values during summer [36]. Gławdel et al. [38] showed that in the coastal waters of the southern Baltic Sea during the summer period, the autotrophic picoplankton, composed mainly of cyanobacteria in the total biomass exceeded even bacterioplankton. Three phenotypes of picocyanobacteria of the genus *Synechococcus* (Type 1, Type 2, and Type 3a) were isolated from the southern Baltic Sea. This area is characterized by large changes of environmental conditions. Autotrophic organisms living in such a variable ecosystem show the ability to quickly adapt which is essential for their survival. In this work, the influence of temperature and PAR irradiance on the autecology of the investigated phenotypes of *Synechococcus*: Type 1, Type 2, and Type 3a were demonstrated.

It was found that the increasing intensity of light had a negative effect on the cell concentration of the three studied phenotypes of *Synechococcus* sp. The number of picocyanobacteria cells increased as the PAR irradiance decreased, reaching the maximum value in the range of 10–55 μmol photons m^{-2} s^{-1} and the minimum value at 145 μmol photons m^{-2} s^{-1}. Besides, it was shown that *Synechococcus* sp. Type 2 was the most susceptible to high light intensity. Its number of cells was more than 5-fold lower in high light compared to low light. On the other hand, the cell number decreased about 4-fold in the high light compared to low light for both *Synechococcus* sp. Type 1 and Type 3a. Literature data also indicated that picocyanobacteria of the genus *Synechococcus* in natural aquatic communities are adapted to low light and show maximum growth in the deeper layers of the euphotic zone [26,29,33,39,40]. The high abundance of autotrophic picoplankton was recorded even at a depth of 90 m [33]. This may indicate the ability of these organisms to survive seasonal changes and their fall into the aphotic zone. Besides, it is considered that *Synechococcus* sp. found in natural surface water layers may show photoinhibition of growth under high light [29,39,41] as well as the low rate of photosynthesis in the surface layer compared to greater depths [33,39]. On the other hand, Śliwińska-Wilczewska et al. [13] showed that the number of cells of green and brown phenotypes of *Synechococcus* sp. increased with the increase in light and was the highest in 280 μmol photons m^{-2} s^{-1}. Furthermore, Kana and Glibert [42,43] showed that *Synechococcus* sp. could occur and grow in the irradiance reaching even 2000 μmol photons m^{-2} s^{-1}. These studies confirmed that *Synechococcus* sp. can grow in maximally coastal waters due to their adaptation to high light intensities. Thus, picocyanobacteria of the genus *Synechococcus* can occur both at the near-surface layers and deeper waters. Furthermore, the ability of *Synechococcus* to grow in low light intensities and their low photoinhibition in exposure to high irradiance could give picocyanobacteria an advantage in changeable light-limited waters.

Temperature is also a very important factor controlling picocyanobacteria abundance in aquatic ecosystems [7,8,37]. Based on the conducted experiments, the influence of increasing temperature on the number of cells of the studied *Synechococcus* sp. phenotypes was found. The most favorable temperature conditions for the growth of *Synechococcus* sp. Type 1 and Type 2 were at 30 °C, while the highest number of cells for Type 3a was recorded at 22.5 °C. The most susceptible to high temperature was *Synechococcus* sp. Type 2. Its abundance was more than 5 times higher at 30 °C compared to the abundance recorded at 15 °C. On the other hand, for both *Synechococcus* sp. Type 1 and Type 3a, the increase in cell numbers along with the increase in temperature was about 4 times greater than that recorded at the lowest temperature. In laboratory studies, Jodłowska and Śliwińska [44] also found that increasing temperatures from 15 °C to 30 °C increased picocyanobacterial abundances. Similar observations were made by Śliwińska-Wilczewska et al. [13] who showed that with an increase in temperature from 10 °C to 25 °C, the number of cells of the green, red and brown *Synechococcus* sp. phenotype. was increased. Picocyanobacteria prefer high temperature for growth and their temperature optimum is higher than for eukaryotic phytoplankton organisms [37]. Furthermore, Paerl and Huisman [45] explained that the global temperature rise would stabilize or even inhibit the eukaryotic phytoplankton while the number of cyanobacteria would increase. Many cyanobacteria species demonstrate the highest increase in growth at 30–35 °C [46]. Noaman et al. [47] also demonstrated that the optimum temperature for growth of *Synechococcus leopoliensis* was 35 °C. An increase in temperature causes an increase in the number of picocyanobacteria cells, and their maximum occurrence was in the summer period when the water temperature is the highest [48]. This relationship is also apparent for the entire autotrophic picoplankton [49] and was confirmed by numerous studies [36,50,51]. Regarding climate change, picocyanobacteria of the genus *Synechococcus* achieves maximal growth rates at high temperatures and thus can be promoted by future global warming [7,8].

This study also showed that the analyzed phenotypes of *Synechococcus* sp.: Type 1, Type 2, and Type 3a has different growth rates. The highest growth rate was recorded for *Synechococcus* sp. Type 1. It was related to the smallest size obtained by these picocyanobacteria [44]. On the other hand, the lowest growth rate was observed for *Synechococcus* sp. Type 3a. Additionally, it was shown that this phenotype reached the largest cell size in cultures [44]. The research conducted by Stal et al. [52] on PE-rich and PC-rich phenotypes of *Synechococcus* also showed differences in the rate of cell growth depending on their size and picocyanobacteria with a larger cell size grew slower. Small cell size of *Synechococcus* Type 1 resulting in faster nutrient uptake allows picocyanobacteria to compete effectively with larger phytoplankton organisms in surface waters. On the other hand, increasing the cell volume of *Synechococcus* Type 3a may result in better light absorption at greater depths.

3.2. Changes in Pigments Content and Pigment Ratios under Different Irradiance and Temperature Conditions

Cyanobacteria living in coastal waters are often exposed to changes in light and temperature conditions. These factors influence the content of cyanobacterial photosynthetic pigments in aquatic ecosystems [53–57]. The factorial experiments performed in this study showed a negative effect of the increasing intensity of light on the cell-specific Chl *a* content for the three examined phenotypes of picocyanobacteria, obtaining the highest content at 10 µmol photons m^{-2} s^{-1} and the lowest for 145 µmol photons m^{-2} s^{-1}. The conducted factorial experiments also showed a statistically significant influence of temperature on the cell-specific Chl *a* content for the examined phenotypes. The highest concentration of this pigment was observed at 30 °C for *Synechococcus* sp. Type 2 and at 15 °C for *Synechococcus* sp. Type 1 and Type 3a. The greatest decrease in the cell-specific Chl *a* content was noted for *Synechococcus* sp. Type 2, which under minimal conditions was about 7 times lower than the recorded under maximum values. On the other hand, *Synechococcus* sp. Type 3a showed the highest resistance to high values of irradiance, and its decrease in the content of Chl *a* in cells under minimal conditions was about 5.5 times higher than the recorded maximum values. Kana and Glibert [43] also showed that the concentration of this pigment was the highest for *Synechococcus* cells adapted to low light. On the other hand, the greatest decrease in Chl *a* content was recorded in the light greater than 700 µmol photons m^{-2} s^{-1} [42]. High content of Chl *a* in low light may indicate that picocyanobacteria of the genus *Synechococcus* may occur in highly shaded waters [52] and even under conditions of extreme radiation deficiency [58].

High light intensity is an unfavorable environmental factor for many photoautotrophic organisms [59]. However, cyanobacteria living in an environment with a high light intensity developed a defense strategy consisting of special pigmentation of the cells [39,60,61]. Convergence between the accumulation of Car pigments under the influence of high light intensity allows them to be assigned a protective role. The highest content of Zea and β-Car was recorded for *Synechococcus* sp. Type 3a. Zea is an accessory pigment at low light intensities but becomes dominant for cells growing under higher ones [16]. Our research showed that for the examined cyanobacteria cells the amount of Zea was much higher than that of β-Car. The study found that the Zea content for *Synechococcus* sp. Type 1, Type 2, and Type 3a was 93%, 89%, and 87% of the sum of Car pigments, respectively. Guillard et al. [62] observed that Zea may constitute as much as 50–81% of Car pigments for cyanobacteria of the genus *Synechococcus*. The high cell-specific Zea content in the *Synechococcus* sp. is related to the existence of these organisms in surface sea waters and places of exposure to high levels of solar radiation [62,63]. The cell-specific Car content of the tested picocyanobacteria phenotypes changed significantly in response to irradiance increase, which suggests that these organisms reorganize their pigments in order to protect against the unfavorable environmental conditions.

In this study, the factorial experiments carried out showed a negative effect of irradiance on the cell-specific PE, PC, and APC as well as the total sum of Phyco pigments content for the three studied phenotypes of the genus *Synechococcus*. Moreover, it was shown that the cell-specific content of these pigments increased with increasing temperature for Type 1 and Type 2. In turn, for Type 3a, a negative effect of increasing temperature on Phyco content was noted. On the basis of the conducted analyzes, it was found that the conditions under which the examined phenotypes of picocyanobacteria achieved the highest concentrations of the total sum of cell-specific Phyco content were at low light intensity of 10 μmol photons m^{-2} s^{-1} and high temperatures ranging between 22.5 and 30 °C. The greatest decrease in Phyco pigments (about 30-fold) in cyanobacteria cells under the influence of increasing light intensity was noted for *Synechococcus* Type 1. However, the least susceptible to high irradiance was *Synechococcus* Type 3a, with a 13-fold decrease in Phyco pigments. Among all Phyco pigments present in picocyanobacteria cells, the highest content of PE was observed for *Synechococcus* Type 2, whereas for *Synechococcus* Type 1 PC was the dominant pigment. A study by Kana and Glibert [42,43] also showed that the concentration of PE and PC were dependent on the intensity of light. The concentration of PC is related to the number of phycobilisomes [42]. The greatest increase of PC in cells was observed in low light, suggesting a change in phycobilisome numbers in growth-limiting light [42]. Cyanobacteria of the genus *Synechococcus*, depending on the light, can change their number and size of phycobilisomes and this may be associated with acclimatization to different light levels [42]. Photoaclimatization is visible when there is a reduction in photosynthetic pigments with increasing irradiance [64–66]. Hence, it may be concluded that the studied *Synechococcus* sp. phenotypes have a high ability to photoacclimatize to changing environmental conditions.

Based on conducted experiments, the highest Zea/Chl *a* ratio and β-Car/Chl *a* ratio was noted for *Synechococcus* sp. Type 2. On the other hand, the lowest ratios of the discussed pigments were recorded for *Synechococcus* sp. Type 3a. Tang and Vincent [67] showed that the content of Car and Chl *a* increases with increasing temperature. However, carotenoids grow more slowly with temperature, therefore the Car/Chl *a* ratio decreases with temperature [67]. Most cyanobacteria show photoinhibition at low temperatures [68], and an increase in the Car/Chl *a* ratio at low temperature may result in an increase in photoprotective pigments such as carotenoids [69,70]. Studies have shown that a high Car/Chl *a* ratio is characteristic for surface water populations [16]. In addition, Paerl et al. [71] and Paerl [72] suggested that a high Car/Chl *a* ratio has a dual role in cells as it maintains high photosynthetic light absorption capacity and protects cells from photooxidation which may explain why the deeper-living PE-rich *Synechococcus* sp. Type 2 had the highest Zea/Chl *a* ratio and β-Car/Chl *a* ratio of all studied phenotypes. This study also showed an increase in the Phyco/Chl *a* ratio and Phyco/Car ratio in the cells of the investigated cyanobacterial phenotypes with a decrease of irradiance and an increase of temperature. It is related to the advantage of Phyco pigments over Chl *a* and Car pigments for the tested picocyanobacteria phenotypes at low light intensity. Furthermore, a change in color from green, red and brown at low irradiances to bright yellow at high light levels was also observed for three phenotypes of cyanobacteria of the genus *Synechococcus* (Type 1, Type 2, and type 3a, respectively). A clear difference in the color of picocyanobacteria was associated with a change of the proportions between the pigments. At low light intensity, picocyanobacteria phenotypes showed the maximum content of Phyco and Chl *a* pigments. At the highest irradiance, the share of the Car pigments, mainly Zea, increased significantly in picocyanobacterial cells. Similar tendencies were observed by Kana and Glibert [16,42] for picocyanobacteria of the genus *Synechococcus*. Picocyanobacteria can acclimate to different light intensities by changing the content of pigments, especially Phyco and Chl *a* [73–75]. In this work, we observed the effect of light intensity and temperature on the cell-specific pigment content of all studied picocyanobacterial phenotypes. The concentration of Phyco and Chl *a* was the highest for picocyanobacteria cells acclimated to low light and decreased with increasing irradiance. Inverse relationships were noted for the cell-specific Car content. The high content of Phyco pigments and Chl *a* observed in our work indicated that the tested picocyanobacteria phenotypes are well adapted to low light conditions and high temperatures. Besides, the highest differences in the Phyco/Chl *a* ratio and Phyco/Car ratio were observed in *Synechococcus* sp. Type 1, which may confirm that this phenotype showed the best photoaclimatization abilities of all analyzed organisms. Because this PC-rich phenotype occurs in more productive waters [18,34,76], this observation may be important in the era of climate change and the associated mass occurrence of *Synechococcus* sp. in many places around the world [8,9]. It should be emphasized that Flombaum et al. [8] predicted that the number of *Synechococcus* sp. cells would increase by 14% at the end of the 21st century.

4. Materials and Methods

4.1. Culture Conditions

Three different phenotypes of picocyanobacteria from the genus *Synechococcus* were examined: BA-120 (Type 2), BA-124 (Type 1), and BA-132 (Type 3a). The strains were isolated from the coastal zone of the Gulf of Gdansk (the southern Baltic Sea) and maintained as unialgal cultures in the Culture Collection of Baltic Algae (CCBA) at the Institute of Oceanography, University of Gdańsk, Poland. Cyanobacteria were cultured on the BG-11 mineral medium [77], which was prepared with water from the Baltic Sea (salinity 8), which was filtered using 0.45 μm filters (Macherey-Nagel MN GF-5, Dueren, Germany) and autoclaved.

The cultures of cyanobacteria were acclimatized to the new conditions corresponding to the incubation conditions of the proper culture. After a week, the culture, which was in the logarithmic growth phase, was used to establish the proper, experimental culture. After the acclimatization time, proper cultures with known initial cell numbers were prepared. For this purpose, a specific volume of inoculum was taken from the actively growing acclimatization culture and added to the sterile media. The optimal number of the initial proper culture was set at 10^7 cells in 1 mL of the medium. The inoculum selected in this way allowed for a constant increase in the number of cyanobacterial cells without inhibiting logarithmic population growth. The incubation of cultures lasted 14 days. After that time, for three phenotypes of cyanobacteria of the genus *Synechococcus* the cell concentration, the growth rate and photosynthetic pigments were determined. Each variant of the experiment was conducted in three repetitions and the results of the experiments were presented as an average of three measurements.

The cultures of the examined cyanobacterial strains were carried out in thermostat under the following temperature conditions (°C): 15, 22.5, and 30. The effect of PAR irradiance was tested in photoperiod (16 h of light and 8 h of darkness) at the following values (μmol photons m^{-2} s^{-1}): 10, 55, 100, and 145. 36 W Philips fluorescent lamps (Philips Lighting, Amsterdam, The Netherlands) were used as light sources and two additional 120 W halogen lamps by OSRAM (Osram Licht AG, Berlin, Germany) were used for the highest irradiance (145 μmol photons m^{-2} s^{-1}). Measurements of PAR irradiance were made with Li-Cor (Lincoln, NE, USA), model LI-189 with cosine collector.

It is worth mentioning here that a change in the color of the cultures of three phenotypes of picocyanobacteria *Synechococcus* sp. under different light was observed. The phenotypes were shown to be dark green, red and brown at low irradiance (for Type 1, Type 2, and type 3a, respectively), while in the high light their color turned to bright yellow. It was also shown that the examined phenotypes showed differences in PAR absorption spectra when exposed to low and high light (Figure 6).

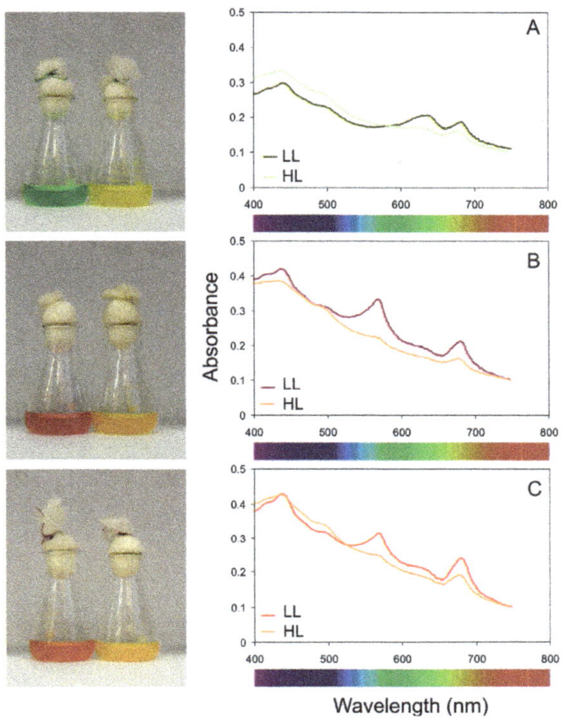

Figure 6. Left-side panel—photographs of the picocyanobacterial phenotypes in 100 mL glass Erlenmeyer flasks: Type 1 (**A**), Type 2 (**B**), and type 3a (**C**), obtained from low (left) and high (right) light; right-side panel—Absorbance spectra measured in the PAR range determined for the picocyanobacterial phenotypes at an optical density (OD_{750}) = 0.1, obtained from low light (LL) and high light (HL).

4.2. Calculation of Cell Density and Growth Rates

Cell density was calculated using linear regression models based on cell concentration (N mL^{-1}) and optical density (OD) at 750 nm [44]. Calculation of the cell number was conducted using the procedure described by Guillard and Sieracki [78], with a light microscope (Nikon Eclipse 80i, Nikon, Tokyo, Japan) and the Bürker counting chamber. To determine the growth rate of cyanobacteria, cell counts were conducted in cultures at two-day intervals from inoculation to the 14th day of culture. Based on these data the parameters characterizing the growth of cyanobacterial cells in the logarithmic phase: growth rate coefficient and cell doubling time were determined [78].

4.3. Determination of the Chlorophyll and Carotenoids Content

The concentration of photosynthetic pigments of analyzed picocyanobacteria was measured by the HPLC method. After 14 days of incubation, 40 mL of culture was filtered using 0.45 μm filters (Macherey-Nagel MN GF-5) to separate the picocyanobacteria cells from the medium. Chl *a* and Car were extracted from the picocyanobacteria cells with 90% acetone (V = 5 mL) and sonicated for 2 min. Then, the test-tube with the extract was held in the dark for 2 h at −80 °C. After 2 h, the pigment extract was centrifuged at 10,000 rpm for 5 min to remove filter particles (Sigma 2-16P, Osterode am Harz, Germany).

Chromatographic analyses were carried out using HPLC equipment of Waters company (Waters Chromatography Europe BV, Etten-Leur, The Netherlands) equipped with: Spectro Vision FD-300 fluorescence detector, Waters 486 absorption detector, Pharmacia autosampler LKB 2157, Waters

Millennium Chromatography software. Measurements of pigment absorption were taken at 440 nm. Pigment separation was carried out according to a method proposed by Llewellyn and Mantour [79], with modifications [80] at room temperature on Vydac 201TP (C18) column 250 mm long. As an eluent A; 0.5 M ammonium acetate/methanol (20/80) was used and as eluent B; acetone/methanol (20/80) was used. Before injection of pigments extract (40 μL) the column was conditioned using an isocratic flow of eluents (40% A and 60% B) for 15 min. The analysis was performed at a flow rate of 1.0 mL min^{-1}. Chl *a*, Zea, and *β*-Car standards were used for the qualitative and quantitative determination of pigments (The International Agency for 14C Determination, VKI, Hørsholm, Denmark). The pigments present in the cells of cyanobacteria strains of the genus *Synechococcus* were identified based on retention times and absorbance spectrum, which were compared with the standards. Calibration curves were plotted for each standard used to quantify assimilation pigments.

4.4. Determination of the Phycobiliproteins Content

The 40 mL of the test material was filtered through a 0.45 μm filter (Macherey-Nagel MN GF-5) and stored in −80 °C. Reagent for phycobiliprotein extraction contained 0.25 M Trizma Base, 10 mM binary EDTA and 2 mg mL^{-1} lysozyme. A pH of 5.5 was obtained by acidifying with concentrated HCl. The filters were homogenized in 5 mL of reagent, sonicated for 5 min and incubated first in the dark at 37 °C for about 2 h, then at 1.5 °C for about 20 h. After this time the pigment extract was centrifuged in experimental flasks for 10 min, at 10,000 rpm. Absorption measurements in 1 cm glass cuvettes on Beckman spectrophotometer (Indianapolis, IN, USA), model DU 530, at wavelengths (nm): 565, 620, 650 and 750, were conducted. The pigment contents: PE, PC, and APC were calculated based on Bennett and Bogorad [81] and Bryant et al. [82].

4.5. Statistical Analyses

To test the influence of a single factor as well as an interplay of factors on studied parameters the two-way ANOVA was used. Moreover, to determine the significance of treatment levels a post hoc test (Tukey's HSD) was conducted. The impact of every environmental agent, as well as an interplay of factors on studied parameters, were measured using the method of orthogonal polynomial tables as described by Fisher and Yates [83]. Furthermore, to describe the connection of the factors and studied parameters regression equations were generated. Data are described as the mean ± standard deviation (SD). Levels of significance were * $p < 0.05$, ** $p < 0.01$, and *** $p < 0.001$. The statistical analyses were executed using the Statistica® 13.1 software (StatSoft Polska, Cracow, Poland).

5. Conclusions

In this work, we found that the three analyzed phenotypes of the genus *Synechococcus* have diverse irradiance and temperature preferences. This, coupled with their high photoacclimation capabilities give them powerful tools to win the competition for the marine resources and provide them opportunity to dominate the area, at least as long as sufficient nutrient amounts are available. In almost all conditions the highest rate of growth was recorded for the *Synechococcus* sp. Type 1 which is the most competitive type. It prefers warmer waters −22.5 °C and above, but it produces the least nominal amounts of Car which is a probable cause of equalisation of the growth rates between the Type 1 and Type 2 at the highest irradiances and at the mentioned temperatures over 22 °C. The lowest growth rates were observed for the Type 3a for all variants. However, Type 3a was recognized to be less temperature sensitive and rather light-driven. Moreover, at low light and low temperature the highest pigment content was observed within the cells Type 3a which may suggest higher tolerance for colder waters such as tested here 15 °C or even below. The highest total pigment content per cell was recorded at 10 μmol photons m^{-2} s^{-1} at all temperature variants with the clear dominance of phycobilins among all the pigments. The high pigment content observed in picocyanobacteria cells proves that they may adapt and live in the deeper layers of the euphotic zone. The highest amounts of carotenoids were produced by Type 2. This may imply lower tolerance of this type to higher irradiance. Our results showed that the best photoaclimation abilities of all analyzed *Synechococcus* sp. types is Type 1 with the highest differences in the Phyco/Chl *a* and Phyco/Car ratios. One of our striking observations is a significant difference between the physiological responses of different *Synechococcus* sp. phenotypes to changeable environmental conditions. Thus, this work would be an important link in forecasting future changes in the occurrence of these organisms in the context of global warming.

Supplementary Materials: The following are available online at http://www.mdpi.com/2073-4409/9/9/2030/s1, Table S1: Two-way factorial ANOVA of cells concentration measured in Synechococcus sp. Type 1, Type 2, and Type 3a growing at different temperatures (°C) and irradiance (μmol photons m^{-2} s^{-1}): df—degrees of freedom; F—Fisher's F-test statistic; Mss—mean sum of squares; Ss—sum of squares. Levels of significance were: * $p < 0.05$; ** $p < 0.01$; *** $p < 0.001$, Table S2: Two-way factorial ANOVA of cell-specific Chl *a*, Phyco, and Car content measured in *Synechococcus* sp. Type 1, Type 2, and Type 3a growing at different temperatures (°C) and irradiance (μmol photons m^{-2} s^{-1}): df—degrees of freedom; F—Fisher's F-test statistic; Mss—mean sum of squares; Ss—sum of squares. Levels of significance were: * $p < 0.05$; ** $p < 0.01$; *** $p < 0.001$, Table S3: Two-way factorial ANOVA of cell-specific PE, PC, and APC content measured in *Synechococcus* sp. Type 1, Type 2, and Type 3a growing at different temperatures (°C) and irradiance (μmol photons m^{-2} s^{-1}): df—degrees of freedom; F—Fisher's F-test statistic; Mss—mean sum of squares; Ss—sum of squares. Levels of significance were: * $p < 0.05$; ** $p < 0.01$; *** $p < 0.001$, Table S4: Two-way factorial ANOVA of cell-specific Zea and β-Car content measured in *Synechococcus* sp. Type 1, Type 2, and Type 3a growing at different temperatures (°C) and irradiance (μmol photons m^{-2} s^{-1}): df—degrees of freedom; F—Fisher's F-test statistic; Mss—mean sum of squares; Ss—sum of squares. Levels of significance were: * $p < 0.05$; ** $p < 0.01$; *** $p < 0.001$. Table S5: Two-way factorial ANOVA of Zea/Chl *a* ratio and β-Car/Chl *a* ratio measured in *Synechococcus* sp. Type 1, Type 2, and Type 3a growing at different temperatures (°C) and irradiance (μmol photons m^{-2} s^{-1}): df—degrees of freedom; F—Fisher's F-test statistic; Mss—mean sum of squares; Ss—sum of squares. Levels of significance were: * $p < 0.05$; ** $p < 0.01$; *** $p < 0.001$, Table S6: The Phyco/Chl *a* ratios and Phyco/Car ratios obtained after 14 days of experiment for three phenotypes of *Synechococcus* sp.: Type 1 (A), Type 2 (B), Type 3a (C) under different temperature (°C) and light (μmol photons m^{-2} s^{-1}) conditions, Table S7: Two-way factorial ANOVA of Phyco/Chl *a* ratio and Phyco/Car ratio measured in *Synechococcus* sp. Type 1, Type 2, and Type 3a growing at different temperatures (°C) and irradiance (μmol photons m^{-2} s^{-1}): df—degrees of freedom; F—Fisher's F-test statistic; Mss—mean sum of squares; Ss—sum of squares. Levels of significance were: * $p < 0.05$; ** $p < 0.01$; *** $p < 0.001$.

Author Contributions: Conceptualization, S.Ś.-W., Z.K., K.W. and M.K.; methodology, S.Ś.-W., Z.K., K.W. and M.K.; formal analysis, S.Ś.-W., Z.K., K.W. and M.K.; investigation, S.Ś.-W., Z.K., K.W. and M.K.; data curation, S.Ś.-W., Z.K., K.W. and M.K.; writing—original draft preparation, S.Ś.-W., Z.K., K.W. and M.K.; supervision, S.Ś.-W. All authors have read and agreed to the published version of the manuscript.

Funding: This research was funded by BMN grant number 539-O140-B416-20 and NCN PRELUDIUM 17 (UMO-2019/33/N/ST10/00585).

Acknowledgments: The authors would like to thank the Editor and anonymous Reviewers for their valuable comments and suggestions to improve the quality of the paper.

Conflicts of Interest: The authors declare no conflict of interest. The funders had a role in the design of the study; in the collection, analyses, and interpretation of data, and in the decision to publish the results.

References

1. Johnson, P.W.; Sieburth, J.M. Chroococcoid cyanobacteria in the sea: A ubiquitous and diverse phototrophic biomass. *Limnol. Oceanogr.* **1979**, *24*, 928–935. [CrossRef]
2. Waterbury, J.B.; Watson, S.W.; Guillard, R.R.L.; Brand, L.E. Widespread occurrence of a unicellular, marine, planktonic cyanobacterium. *Nature* **1979**, *277*, 293–294. [CrossRef]
3. Stockner, J.G. Phototrophic picoplankton: An overview from marine and freshwater ecosystems. *Limnol. Oceanogr.* **1988**, *33*, 765–775. [CrossRef]
4. Jasser, I.; Lehtovaara, A.; Arvola, L. Seasonality and coexistence of autotrophic pico- and nanoplankton and zooplankton in three boreal lakes. *Internationale Vereinigung für Theoretische und Angewandte Limnologie Verhandlungen* **2006**, *29*, 1413–1416. [CrossRef]
5. Mühling, M.; Woolven-Allen, J.; Murrell, J.C.; Joint, I. Improved group-specific PCR primers for denaturing gradient gel electrophoresis analysis of the genetic diversity of complex microbial communities. *ISME J.* **2008**, *2*, 379–392. [CrossRef]

6. Guidi, L.; Chaffron, S.; Bittner, L.; Eveillard, D.; Larhlimi, A.; Roux, S.; Darzi, Y.; Audic, S.; Berline, L.; Brum, J.R.; et al. Plankton networks driving carbon export in the oligotrophic ocean. *Nature* **2016**, *532*, 465–470. [CrossRef]
7. Worden, A.Z.; Wilken, S. A plankton bloom shifts as the ocean warms. *Science* **2016**, *354*, 287–288. [CrossRef]
8. Flombaum, P.; Gallegos, J.L.; Gordillo, R.A.; Rincon, J.; Zabala, L.L.; Jiao, N.; Karl, D.M.; Li, W.K.; Lomas, M.W.; Veneziano, D. Present and future global distributions of the marine Cyanobacteria *Prochlorococcus* and *Synechococcus*. *Proc. Natl. Acad. Sci. USA* **2013**, *110*, 9824–9829. [CrossRef]
9. Li, J.; Chen, Z.; Jing, Z.; Zhou, L.; Li, G.; Ke, Z.; Jiang, X.; Liu, J.; Liu, H.; Tan, Y. *Synechococcus* bloom in the Pearl River Estuary and adjacent coastal area–With special focus on flooding during wet seasons. *Sci. Total Environ.* **2019**, *692*, 769–783. [CrossRef]
10. Oziel, L.; Neukermans, G.; Ardyna, M.; Lancelot, C.; Tison, J.-L.; Wassmann, P.; Sirven, J.; Ruiz-Pino, D.; Gascard, J.-C. Role for Atlantic inflows and sea ice loss on shifting phytoplankton blooms in the Barents Sea. *J. Geophys. Res. Oceans.* **2017**, *122*, 5121–5139. [CrossRef]
11. Paulsen, M.L.; Doré, H.; Garczarek, L.; Seuthe, L.; Müller, O.; Sandaa, R.A.; Bratbak, G.; Larsen, A. *Synechococcus* in the Atlantic Gateway to the Arctic Ocean. *Front. Mar. Sci.* **2016**, *3*, 191. [CrossRef]
12. Dutkiewicz, S.; Morris, J.J.; Follows, M.J.; Scott, J.; Levitan, O.; Dyhrman, S.T.; Berman-Frank, I. Impact of ocean acidification on the structure of future phytoplankton communities. *Nat. Clim. Change* **2015**, *5*, 1002–1006. [CrossRef]
13. Śliwińska-Wilczewska, S.; Cieszyńska, A.; Maculewicz, J.; Latała, A. Ecophysiological characteristics of red, green, and brown strains of the Baltic picocyanobacterium *Synechococcus* sp.-a laboratory study. *Biogeosciences* **2018**, *15*, 6257–6276. [CrossRef]
14. Stransky, H.; Hager, A. Das Carotinoid-muster und die Verbreitung des lichtinduzierten Xanthophyllcyclus in verschiedenen Algenklassen. Cyanophyceae und Rhodophyceae. *Arch. Mikrobiol.* **1970**, *72*, 84–96. [CrossRef] [PubMed]
15. Schlüter, L.; Lauridsen, T.L.; Krogh, G.; Jørgensen, T. Identification and quantification of phytoplankton groups in lakes using new pigment ratios—a comparison between pigment analysis by HPLC and microscopy. *Freshw. Biol.* **2006**, *51*, 1474–1485. [CrossRef]
16. Kana, T.M.; Glibert, P.M. Zeaxanthin and B-carotene in *Synechococcus* WH7803 respond differently to irradiance. *Limnol. Oceanogr.* **1988**, *33*, 1623–1627.
17. Ostrowska, M.; Woźniak, B.; Dera, J. Modelled quantum yields and energy efficiency of fluorescence, photosynthesis and heat production by phytoplankton in the World Ocean. *Oceanologia* **2012**, *54*, 565–610. [CrossRef]
18. Hauschild, C.A.; McMurter, H.J.G.; Pick, F.R. Effect of spectral quality on growth and pigmentation of picocyanobacteria. *J. Phycol.* **1991**, *27*, 698–702. [CrossRef]
19. Cornejo, J.; Beale, S.I. Phycobilin biosynthetic reactions in extracts of cyanobacteria. *Photosynth. Res.* **1997**, *51*, 223–230. [CrossRef]
20. Śliwińska-Wilczewska, S.; Maculewicz, J.; Barreiro Felpeto, A.; Latała, A. Allelopathic and bloom-forming picocyanobacteria in a changing world. *Toxins* **2018**, *10*, 48. [CrossRef]
21. Konarzewska, Z.; Śliwińska-Wilczewska, S.; Barreiro Felpeto, A.; Vasconcelos, V.; Latała, A. Assessment of the Allelochemical Activity and Biochemical Profile of Different Phenotypes of Picocyanobacteria from the Genus *Synechococcus*. *Mar. Drugs* **2020**, *18*, 179. [CrossRef] [PubMed]
22. Six, C.; Thomas, J.C.; Garczarek, L.; Ostrowski, M.; Dufresne, A.; Blot, N.; Scanlan, D.J.; Partensky, F. Diversity and evolution of phycobilisomes in marine *Synechococcus* spp.: A comparative genomics study. *Genome Biol.* **2007**, *8*, R259. [CrossRef] [PubMed]
23. Haverkamp, T.H.; Schouten, D.; Doeleman, M.; Wollenzien, U.; Huisman, J.; Stal, L.J. Colorful microdiversity of *Synechococcus* strains (picocyanobacteria) isolated from the Baltic Sea. *ISME J.* **2009**, *3*, 34–408. [CrossRef] [PubMed]
24. Callieri, C. *Synechococcus* plasticity under environmental changes. *FEMS Microbiol. Lett.* **2017**, *364*, fnx229. [CrossRef]
25. Roy, S.; Llewellyn, C.; Egeland, E.S.; Johnsen, G. *Phytoplankton Pigments: Characterization, Chemotaxonomy and Applications in Oceanography*; Cambridge University Press: Cambridge, UK, 2011; p. 784.
26. Morris, I.; Glover, H. Physiology of photosynthesis by marine coccoid Cyanobacteria -some ecological implications. *Limnol. Oceanogr.* **1981**, *26*, 957–961. [CrossRef]

27. Antia, N.J.; Cheng, J.Y. The survival of axenic cultures of marine planktonic algae from prolonged exposure to darkness at 20 °C. *Phycologia* **1970**, *9*, 179–183. [CrossRef]
28. Antia, N.J. Effects of temperature on the darkness survival of marine microplanktonic algae. *Microb. Ecol.* **1976**, *3*, 41–54. [CrossRef]
29. Platt, T.; Subba-Rao, D.V.; Irwin, B. Photosynthesis of picoplankton in the oligotrophic ocean. *Nature* **1983**, *301*, 702–704. [CrossRef]
30. Cai, H.; Wang, K.; Huang, S.; Jiao, N.; Chen, F. Distinct patterns of picocyanobacterial communities in winter and summer in the Chesapeake Bay. *Appl. Environ. Microbiol.* **2010**, *76*, 2955–2960. [CrossRef]
31. Ernst, A. Cyanobacterial picoplankton from Lake Constance. Isolation by fluorescence characteristics. *J. Plankton Res.* **1991**, *13*, 1307–1312. [CrossRef]
32. Jasser, I.; Callieri, C. Picocyanobacteria. In *Handbook on Cyanobacterial Monitoring and Cyanotoxin Analysis*; Meriluoto, J., Spoof, L., Codd, G.A., Eds.; John Wiley & Sons, Ltd.: Hoboken, NJ, USA, 2017; pp. 19–27.
33. Glover, H.E. The physiology and ecology of marine Cyanobacteria, *Synechococcus* spp. In *Advances in Applied Microbiology*; Jannasch, H.W., Williams Leb, P.J., Eds.; Academic Press: New York, USA, 1985; pp. 49–107.
34. Vörös, L.; Callieri, C.; Katalin, V.; Bertoni, R. Freshwater picocyanobacteria along a trophic gradient and light quality range. *Hydrobiologia* **1998**, *369*, 117–125. [CrossRef]
35. Wang, C.; Wang, Z.; Wang, P.; Zhang, S. Multiple effects of environmental factors on algal growth and nutrient thresholds for harmful algal blooms: Application of response surface methodology. *Environ. Model. Assess.* **2016**, *21*, 247–259. [CrossRef]
36. Jasser, I.; Arvola, L. Potential effects of abiotic factors on the abundance of autotrophic picoplankton in four boreal lakes. *J. Plankton Res.* **2003**, *25*, 873–883. [CrossRef]
37. Jasser, I. The relationship between autotrophic picoplankton (APP)-the smallest autotrophic component of food web and the trophic status and depth of lakes. *Ecohydrol. Hydrobiol.* **2006**, *6*, 69–77. [CrossRef]
38. Gławdel, M.; Mackiewicz, T.; Witek, Z. Composition and abundance of picoplankton in the coastal zone of the Gulf of Gdańsk. *Oceanol. Stud.* **1999**, *28*, 17–30.
39. Barlow, R.G.; Alberte, R.S. Photosynthetic characteristic of phycoerythrin-containing marine *Synechococcus* spp. I. Responses to growth photon flux density. *Mar. Biol.* **1985**, *86*, 63–74. [CrossRef]
40. Glover, H.E.; Campbell, L.; Prézelin, B.B. Contribution of *Synechococcus* spp. to size-fraction primary productivity in three waters masses in the Northwest Atlantic Ocean. *Mar. Biol.* **1986**, *91*, 193–203. [CrossRef]
41. Campbell, L.; Carpenter, E.J. Diel patterns of cell division in marine *Synechococcus* spp. (Cyanobacteria): Use of the frequency of dividing cell technique to measure growth rate. *Mar. Ecol. Prog. Ser.* **1986**, *32*, 139–148. [CrossRef]
42. Kana, T.M.; Glibert, P.M. Effect of irradiances up to 2000 μmol E·m^{-2}·s^{-1} on marine *Synechococcus* WH7803-I. Growth, pigmentation, and cell composition. *Deep-Sea Res.* **1987**, *34*, 479–495. [CrossRef]
43. Kana, T.M.; Glibert, P.M. Effect of irradiances up to 2000 μmol E·m^{-2}·s^{-1} on marine *Synechococcus* WH7803-II. Photosynthetic responses and mechanisms. *Deep-Sea Res.* **1987**, *34*, 497–516. [CrossRef]
44. Jodłowska, S.; Śliwińska, S. Effects of light intensity and temperature on the photosynthetic irradiance response curves and chlorophyll fluorescence in three picocyanobacterial strains of *Synechococcus*. *Photosynthetica* **2014**, *52*, 223–232. [CrossRef]
45. Paerl, H.W.; Huisman, J. Climate change: A catalyst for global expansion of harmful cyanobacterial blooms. *Eviron. Microb. Rep.* **2009**, *1*, 27–37. [CrossRef] [PubMed]
46. Wilde, E.W.; Tilly, L.J. Structural characteristics of algal communities in thermally altered artificial streams. *Hydrobiologia* **1981**, *76*, 57–63. [CrossRef]
47. Noaman, N.H.; Fattah, A.; Khaleafa, M.; Zaky, S.H. Factors affecting antimicrobial activity of *Synechococcus leopoliensis*. *Microbiol. Res.* **2004**, *159*, 395–402. [CrossRef]
48. Robarts, R.D.; Zohary, T. Temperature effects on photosynthetic capacity, respiration, and growth rates of bloom-forming cyanobacteria. *N. Z. J. Mar. Freshwat. Res.* **1987**, *21*, 391–399. [CrossRef]
49. Weisse, T. The microbial food-web and its sensitivity to eutrophication and contaminant enrichment: Across-system overview. *Int. Rev. Ges. Hydrobiol.* **1991**, *76*, 327–338. [CrossRef]
50. Kuosa, H. Picoplanktonic algae in the northern Baltic Sea: Seasonal dynamics and flagellate grazing. *Mar. Ecol. Prog. Ser.* **1991**, *73*, 269–276. [CrossRef]

51. Kononen, K.; Kuparinen, J.; Mäkelä, K.; Laanemets, J.; Pavelson, J.; Nômmann, S. Initiation of cyanobacterial blooms in a frontal region at the entrance to the Gulf of Finland, Baltic Sea. *Limnol. Ocean.* **1996**, *41*, 98–112. [CrossRef]
52. Stal, L.J.; Albertano, P.; Bergman, B.; Bröckel, K.; Gallon, J.R.; Hayes, P.K.; Sivonen, K.; Walsby, A.E.; BASIC: Baltic Sea cyanobacteria. An investigation of the structure and dynamics of water blooms of cyanobacteria in the Baltic Sea—responses to a changing environment. *Cont. Shelf Res.* **2003**, *23*, 1695–1714.
53. Hanelt, D.; Wiencke, C.; Bischof, K. Photosynthesis in Algae. In *Advances in Photosynthesis and Respiration*; Larkum, A.W.D., Douglas, S.E., Raven, J.A., Eds.; Springer Science & Business Media: Berlin, Germany, 2003; Volume 14, p. 417.
54. Stomp, M.; Huisman, J.; Stal, L.J.; Matthijs, H.C. Colorful niches of phototrophic microorganisms shaped by vibrations of the water molecule. *ISME J.* **2007**, *1*, 271–282. [CrossRef]
55. Millie, D.F.; Ingram, D.A.; Dionigi, C.P. Pigment and photosynthetic responses of *Oscillatoria agardhii* (Cyanophyta) to photon flux density and spectral quality. *J. Phycol.* **1990**, *26*, 660–666. [CrossRef]
56. Prézelin, B.B. Light reactions in photosynthesis. In *Physiological Bases of Phytoplankton Ecology*; Platt, T., Ed.; Canadian Bulletin of Fisheries and Aquatic Sciences: Ottawa, ON, Canada, 1981; pp. 1–43.
57. Paerl, H.W. Nuisance phytoplankton blooms in coastal, estuarine, and inland waters. *Limnol. Oceanogr.* **1988**, *33*, 823–847. [CrossRef]
58. Carter, J.R. *Diatom from the Devil's Hole Cave, Fife*; Nova Hedwigia: Scotland, UK, 1971; pp. 657–681.
59. Lund, I.W.G. The ecology of the freshwater phytoplankton. *Biol. Rev. Cambridge Philosoph. Soc.* **1965**, *40*, 231–293. [CrossRef]
60. Foy, R.H.; Gibson, C.E. Photosynthetic characteristics of planktonic blue-green algae: Changes in photosynthetic capacity and pigmentation of *Oscillatoria redekei* Van Goor under high and low light. *Br. Phycol. J.* **1982**, *17*, 183–193. [CrossRef]
61. Post, A.F. Transient state characteristics of adaptation to changes in light conditions for the cyanobacterium *Oscillatoria agardhii*. I. Pigmentation and photosynthesis. *Arch. Microbiol.* **1986**, *145*, 353–357. [CrossRef]
62. Guillard, R.R.L.; Murphy, L.S.; Foss, P.; Liaaen-Jensen, S. *Synechococcus* spp. as likely zeaxanthin-dominant ultraphytoplankton in the North Atlantic. *Limnol. Oceanogr.* **1985**, *30*, 412–414. [CrossRef]
63. Gieskes, W.W.; Kraay, G.W. Floristic and physiological differences between the shallow and the deep nanophytoplankton community in the euphotic zone of the open tropical Atlantic revealed by HPLC analysis of pigments. *Mar. Biol.* **1986**, *91*, 567–576. [CrossRef]
64. Dubinsky, Z.; Falkowski, P.J.; Wyman, K. Light harvesting and utilization by phytoplankton. *Plant. Cell Physiol.* **1986**, *27*, 1335–1349. [CrossRef]
65. Neale, P.J.; Melis, A. Algal photosynthetic membrane complex and the photosynthesis-irradiance curve: A comparison of light-adaptation responses in *Chlamydomonas reinhardtii* (Chlorophyta). *J. Phycol.* **1986**, *22*, 531–538. [CrossRef]
66. Sukenik, A.; Bennett, J.; Falkowski, P.G. Light saturated photosynthesis-limitation by electron transport or carbon-fixation? *Biochim. Biophys. Acta.* **1987**, *891*, 205–215. [CrossRef]
67. Tang, E.P.Y.; Vincent, W.F. Effects of daylength and temperature on the growth and photosynthesis of an Arctic cyanobacterium, *Schizothrix calcicola* (Oscillatoriaceae). *Eur. J. Phycol.* **2000**, *35*, 263–272. [CrossRef]
68. Rae, R.; Vincent, W.F. Phytoplankton production in subarctic lake and river ecosystems: Development of a photosynthesis-temperature-irradiance model. *J. Plankton Res.* **1998**, *20*, 1293–1312. [CrossRef]
69. Young, A.J. Factors that affect the carotenoids composition of higher plants and algae. In *Carotenoids in Photosynthesis*; Young, A.J., Britton, G., Eds.; Chapman and Hall: London, UK, 1993; pp. 160–205.
70. Young, A.J.; Frank, H.A. Energy transfer reactions involving carotenoids: Quenching of chlorophyll fluorescence. *J. Photochem. Photobiol. B* **1996**, *36*, 3–15. [CrossRef]
71. Paerl, H.W.; Tucker, J.; Bland, P.T. Carotenoid enhancement and its role in maintaining blue-green algal (*Microcystis aeruginosa*) surface blooms. Limnol. *Oceanogr.* **1983**, *28*, 847–857.
72. Paerl, H.W. Cyanobacterial carotenoids: Their roles in maintaining optimal photosynthetic production among aquatic bloom-forming genera. *Oecologia* **1984**, *61*, 143–149. [CrossRef]
73. Jones, L.W.; Myers, J. Pigment variations in *Anacystis nidulans* induced by light of selected wavelengths. *J. Phycol.* **1965**, *1*, 7–15. [CrossRef]
74. Glover, H.E.; Keller, M.D.; Spinrad, R.W. The effects of light quality on photosynthesis and growth of marine eukaryotic and prokaryotic phytoplankton clones. *J. Exp. Mar. Biol. Ecol.* **1987**, *105*, 137–159. [CrossRef]

75. Tilzer, M.M. Light-dependence of photosynthesis and growth in cyanobacteria: Implications for their dominance in eutrophic lakes. *New Zeal. J. Mar. Fresh. Res.* **1987**, *21*, 401–412. [CrossRef]
76. Fahnenstiel, G.L.; Carrick, H.J.; Rogers, C.E.; Sicko-Goad, L. Red fluorescing phototrophic picoplankton in the Laurentian Great Lakes: What are they and what are they doing? *Int. Rev. Ges. Hydrobiol.* **1991**, *76*, 603–616. [CrossRef]
77. Stanier, R.Y.; Cohen-Bazire, G.; Kunisawa, R.; Mandel, M. Purification and properties of unicellular blue-green algae (order Chroococcales). *Bact. Rev.* **1971**, *35*, 171–205. [CrossRef]
78. Guillard, R.R.; Sieracki, M.S. Counting cells in cultures with the light microscope. In *Algal Culturing Techniques*; Andersen, R.R., Ed.; Elsevier Academic Press: Burlington, VT, USA, 2005; pp. 239–252.
79. Llewellyn, C.A.; Mantoura, R.F.C. The rapid determination of algal chlorophyll and carotenoid pigments and their breakdown products in natural waters by reverse-phase high-performance liquid chromatography. *Anal. Chem. Acta.* **1983**, *151*, 297–314.
80. Jodłowska, S.; Latała, A. Simultaneous separation of chlorophylls and carotenoids by RP-HPLC in some algae and cyanobacteria from the Southern Baltic. *Oceanol. Hydrobiol. Stud.* **2003**, *32*, 81–89.
81. Bennett, A.; Bogorad, L. Complementary chromatic adaptation in freshwater blue-green alga. *J. Cell Biol.* **1973**, *58*, 419–435. [CrossRef] [PubMed]
82. Bryant, D.A.; Guglielmi, G.; Tandeau de Marsac, N.; Casteta, A.-M.; Cohen-Bazire, G. The structure of cyanobacterial phycobilisomes: A model. *Arch. Microbiol.* **1979**, *123*, 113–127. [CrossRef]
83. Fisher, R.A.; Yates, F. *Statistical Tables for Biological, Agricultural and Medical Research*, 6th ed.; Olivier and Boyd: Edinburgh, UK, 1963; p. 146.

cells

MDPI

Review

Raman Spectroscopy and Its Modifications Applied to Biological and Medical Research

Elvin S. Allakhverdiev [1,2], Venera V. Khabatova [3], Bekzhan D. Kossalbayev [4,5], Elena V. Zadneprovskaya [3], Oleg V. Rodnenkov [1], Tamila V. Martynyuk [1], Georgy V. Maksimov [2,6], Saleh Alwasel [7], Tatsuya Tomo [8] and Suleyman I. Allakhverdiev [3,7,9,*]

1 Russian National Medical Research Center of Cardiology, 3rd Cherepkovskaya St., 15A, 121552 Moscow, Russia; elvin21128@gmail.com (E.S.A.); rodnenkov@mail.ru (O.V.R.); trukhiniv@mail.ru (T.V.M.)
2 Biology Faculty, Lomonosov Moscow State University, Leninskie Gory 1/12, 119991 Moscow, Russia; gmaksimov@mail.ru
3 K.A. Timiryazev Institute of Plant Physiology, RAS, Botanicheskaya Str., 35, 127276 Moscow, Russia; venera.khabatova@ifr.moscow (V.V.K.); zadneprovskaya@ifr.moscow (E.V.Z.)
4 Geology and Oil-Gas Business Institute Named after K. Turyssov, Satbayev University, Satpaeva, 22, Almaty 050043, Kazakhstan; kossalbayev.bekzhan@gmail.com
5 Department of Biotechnology, Faculty of Biology and Biotechnology, Al-Farabi Kazakh National University, Al-Farabi Avenue 71, Almaty 050038, Kazakhstan
6 Department of Physical Materials Science, Technological University "MISiS", Leninskiy Prospekt 4, Office 626, 119049 Moscow, Russia
7 Zoology Department, College of Science, King Saud University, Riyadh 12372, Saudi Arabia; salwasel@ksu.edu.sa
8 Department of Biology, Faculty of Science, Tokyo University of Science, 1-3 Kagurazaka, Shinjuku-ku, Tokyo 162-8601, Japan; tomo@rs.tus.ac.jp
9 Institute of Basic Biological Problems, RAS, Pushchino, 142290 Moscow, Russia
* Correspondence: suleyman.allakhverdiev@gmail.com

Abstract: Nowadays, there is an interest in biomedical and nanobiotechnological studies, such as studies on carotenoids as antioxidants and studies on molecular markers for cardiovascular, endocrine, and oncological diseases. Moreover, interest in industrial production of microalgal biomass for biofuels and bioproducts has stimulated studies on microalgal physiology and mechanisms of synthesis and accumulation of valuable biomolecules in algal cells. Biomolecules such as neutral lipids and carotenoids are being actively explored by the biotechnology community. Raman spectroscopy (RS) has become an important tool for researchers to understand biological processes at the cellular level in medicine and biotechnology. This review provides a brief analysis of existing studies on the application of RS for investigation of biological, medical, analytical, photosynthetic, and algal research, particularly to understand how the technique can be used for lipids, carotenoids, and cellular research. First, the review article shows the main applications of the modified Raman spectroscopy in medicine and biotechnology. Research works in the field of medicine and biotechnology are analysed in terms of showing the common connections of some studies as caretenoids and lipids. Second, this article summarises some of the recent advances in Raman microspectroscopy applications in areas related to microalgal detection. Strategies based on Raman spectroscopy provide potential for biochemical-composition analysis and imaging of living microalgal cells, in situ and in vivo. Finally, current approaches used in the papers presented show the advantages, perspectives, and other essential specifics of the method applied to plants and other species/objects.

Keywords: carotenoids; lipid droplets; microalgae; Raman spectroscopy; Surface-enhanced Raman Spectroscopy

Citation: Allakhverdiev, E.S.; Khabatova, V.V.; Kossalbayev, B.D.; Zadneprovskaya, E.V.; Rodnenkov, O.V.; Martynyuk, T.V.; Maksimov, G.V.; Alwasel, S.; Tomo, T.; Allakhverdiev, S.I. Raman Spectroscopy and Its Modifications Applied to Biological and Medical Research. *Cells* **2022**, *11*, 386. https://doi.org/10.3390/cells11030386

Academic Editor: Alexander E. Kalyuzhny

Received: 13 December 2021
Accepted: 22 January 2022
Published: 24 January 2022

1. Introduction

In recent decades, Raman spectroscopy (RS) has been used in several studies on animal cells [1–3]. The method is popular among biophysicists and life science researchers. RS allows for the study of living cells in their natural conditions without any damage [4,5]. Nowadays, we can see a significant increase in the use of RS in plants, and especially in algae research.

RS is a well-known approach used in many of biomedical studies. Since biomolecules are involved, the main obstacle to the use of such methods in life science research is the low signal of Raman scattering. There are a number of modifications of RS that allow Raman scattering to be improved. There are existing approaches to detect Raman signals not only on the surface of human skin, but also inside the vasculature and various organs of patients.

In this review, we have attempted to cover the list of modifications of RS applied to biological and medical research; moreover, algal research, and especially to understand more detailed mechanisms related to the biosynthesis and transport of lipid droplets/fatty acids and carotenoids. It is important to note that algal research might represent an interest in terms of carotenoids production and further application of carotenoids for medical treatment of human diseases.

We have also covered studies on human cells [1,2] and microalgae [6,7] that we thought would be useful to introduce to the reader, especially in view of future studies of the use of algae. We have tried to cover the variety of algal species that have been used in different studies with the application of RS and its modifications. Figure 1 provides a schematic of the different research areas in which RS can be used. Biotechnological, biomedical, photosynthetic, and analytical research applications of RS are the main interest of this particular review.

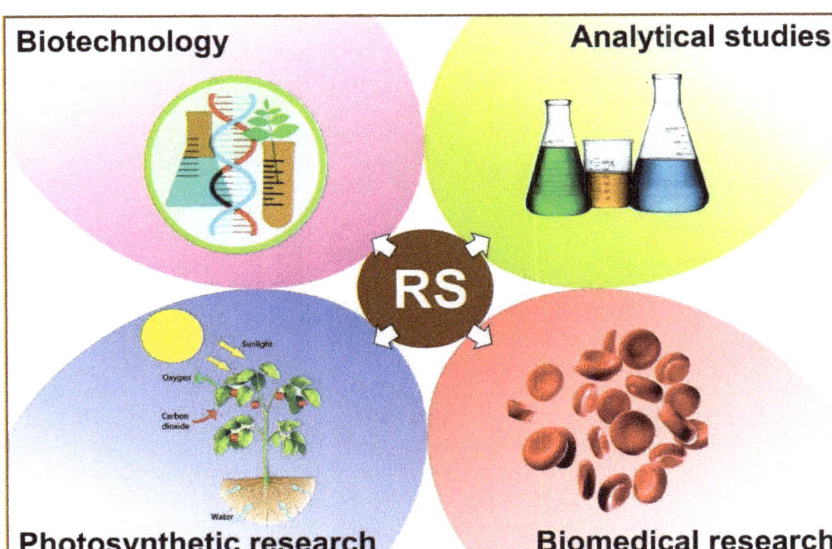

Figure 1. Application of RS in different research.

2. The Principles of the Method of Raman Spectroscopy

Figure 2 shows the energy transition of Rayleigh and Raman scattering. The former (Rayleigh) is based on the principle that the frequency of the absorbed and scattered photon does not change—elastic scattering. In the latter (Raman), on the other hand, there is a shift in frequency of the scattered photon (change in energy or change in wavelength)—inelastic scattering. It is also essential to note that Raman scattering is divided into Stokes and anti-Stokes shifts. The Stokes shift is more likely because it is associated with the shift of the binding maxima to the longer wavelengths (the energy and frequency of the scattered photon are correspondingly lower than those of the absorbed photon—see Figure 2). The anti-Stokes shift, on the other hand, is less likely. It results from the shift of the binding maxima to the shorter wavelengths (energy and frequency of the scattered photon are correspondingly higher than those of the absorbed photon—see Figure 2).

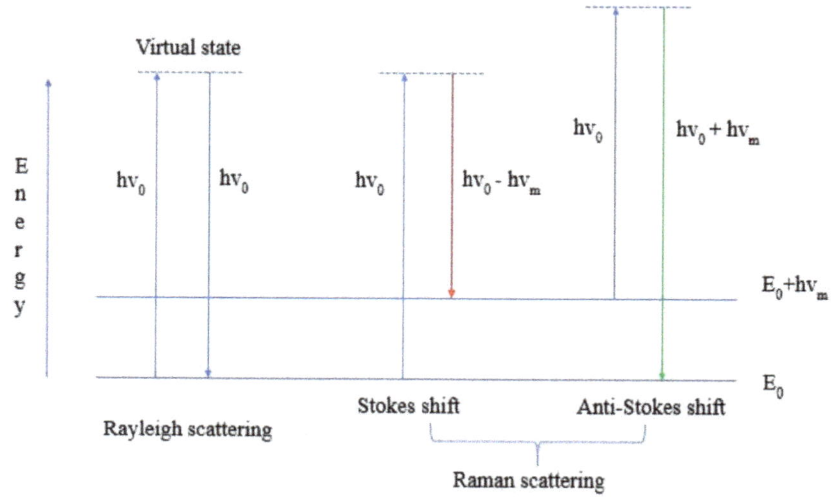

Figure 2. Energy transition of Rayleigh and Raman scattering.

In order to increase the sensitivity and resolution capability of the method, scientists use the approach based on the surface plasmon resonance effect. The electrons on the surface of the metal (silver and/or gold nanoparticles) oscillate. At a certain point, the frequency of the photon resonates with the frequency of the surface electrons. The surface plasmons substantially enhance the local electric field of the incident light (on the molecules near the surface's vicinity of the metal nanoparticles) [8]. In recent decades, this effect has been used to modify the method of RS to apply it in biological and medical research.

3. Raman Spectroscopy and Its Modifications: Advantages and Use

There are many variants of Raman spectroscopy, all of which use the phenomenon of Raman scattering in different ways (Figure 3). The choice of which variant to use for a particular measurement depends on inherent factors, such as the complexity of the sample and/or the concentrations of the target analyses.

The popularity of the RS method among biophysicists around the world is explained by a list of advantages of the method. RS is a non-invasive, rapid, and sensitive method for in vitro investigations [10–12]. Usually we face an obstacle—the biomolecules are present in the cell in very low concentrations, so the Raman scattering/signal is very low. To enhance the Raman scattering signal, the modifications of the method can be used (see Table 1), such as surface-enhanced Raman spectroscopy (SERS), coherent anti-Stokes Raman scattering (CARS), surface-enhanced Raman Spectroscopy (SERCS) [13], and micro-Raman spectroscopy [14].

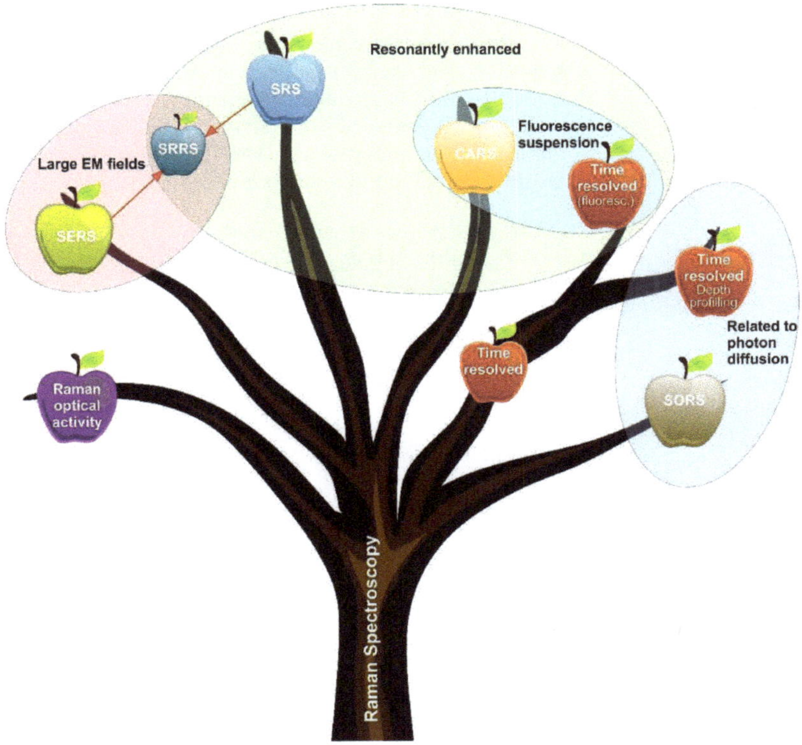

Figure 3. "Family tree" of the RS. The relatively simple RS is the root of the complex surface-enhanced, resonance-enhanced time—and spatially-resolved techniques. Abbreviations: SERS, Surface-enhanced Raman Spectroscopy; CARS, coherent anti-Stokes Raman spectroscopy; RRS, resonance-enhanced Raman scattering; SORS, spatially offset Raman spectroscopy. Modified from Buckley and Ryder [9].

Table 1. The list of modifications of RS with details of the objects and molecules of interest with reference numbers of the papers used in the article.

Modification of Method	Object	Biomolecules	Reference/Link Number
Coherent anti-Stokes Raman scattering (CARS) and microscopy	microalgae	lipids, carotenoids	[4,15–17]
Confocal Raman microscopy	microalgae, algae	lipids	[18,19]
Raman micro spectroscopy	algae, animals	lipids, carotenoids	[2,5,20]
Resonance Raman spectroscopy (RRS)	bacteria, microalgae	carotenoids	[7,19,21]
Single-cell Raman spectroscopy (SCRS)	microalgae	lipids	[22,23]
Surface-enhanced Raman spectroscopy (SERS)	animals, bacteria, microalgae	lipids, carotenoids, proteins	[1,2,6,10,14,24]

As mentioned earlier, there is a special mechanism of Raman scattering (plasmon resonance) enhancement for modifications such as SERS. Moreover, each modification of the method can solve a specific task/objective.

The following sections describe the most commonly used types of RS.

3.1. Surface-Enhanced Raman Spectroscopy (SERS)

There are several clinical trials in progress with SERS. Sample types include blood [25,26], saliva [27], and tears. As reported in articles from two studies, SERS had a susceptibility of 80.7% and 84.1% in detecting squamous cell carcinoma of the oral cavity by analysing blood [28]. Therefore, the tests show that SERS biofluid is suitable as a sample and relies on metal nanoparticles for signal amplification. The status of clinical trials is important for the understanding and future prospect of SERS and Raman spectroscopy [29].

3.2. Coherent Anti-Stokes Raman Scattering (CARS)

Conventional Raman spectroscopy uses only a CW laser to generate spectra, while CARS and SRS use two pulsed lasers with different wavelengths to enable nonlinear optical motions. CARS microscopy can form an optical contrast of endogenous chemical structures, which is popular in various fields of biomedicine as it can provide a high-resolution image. For example, CARS microscopy has been used to visualise tissue structures, skin [30], lung, kidney, and retina [31]. Consequently, CARS has been able to obtain micron-level images of brain slices, which has worked well in cancer diagnosis [32]. Concrete prostatectomy is considered the most popular method in civilised countries for curing members of the stronger sex with clinically localised prostate cancer. In this surgery, the entire prostate is removed, but the urinary ball is reunited with the urethra [33].

3.3. Resonance Raman Spectroscopy (RRS)

One of the drawbacks of Raman spectroscopy is the low signal intensity. This drawback can be corrected with RRS. By matching the wavelength of laser excitation to the electrical absorption maximum of a particular chemical, the Raman signal of certain bands is enhanced. The study was used for a multifaceted study of haemoglobin, and the release of cytochrome-c from mitochondria during apoptosis was also studied. Okada et al. [34] as well as other scientists have used RRS to perform unlabelled studies of molecular dynamics in apoptotic cells. Observation of mitochondrial membrane stained with the dye JC-1 using RRS confirmed that the observed release of cytochrome was due to apoptosis.

3.4. Spatially Offset Raman Spectroscopy (SORS)

Although SORS technology may not be approved in any way for uniform diagnosis of patients in our time, there are significant prospects for the eventual use of this technique in the clinic. Recent advances in medicine have shown SORS can be used in blood testing, such as assessing the quality of erythrocytes during blood transfusion in patients. Vardaki et al. [35] have shown that SORS is able to profile changes in oxygenation when stored for 6 weeks. It is well known that in blood transfusions, the chemical composition of blood units changes differently from time to time. For this reason, a few units over the years are by no means determinative of the relationship between erythrocytes properties. Feng et al. [36] used SORS to measure subcortical bone and biochemical changes with increasing depth in intact mouse bone.

4. Application of Raman Spectroscopy in Biomedical Research

Due to the non-invasive, fast, and highly sensitive advantages of RS and its modifications, there is considerable demand for its use in biomedical research, such as in studying the structure and conformation of molecules of interest and investigating the mechanisms of the drug action [10,11]. Nowadays, RS is used in vivo and ex vivo to solve various biomedical issues, such as early cancer detection, monitoring the effects of different drugs on the skin, determining the composition of atherosclerotic plaques, and rapid identification of pathogenic microorganisms. Detailed information about the RS application can be found in Figure 4.

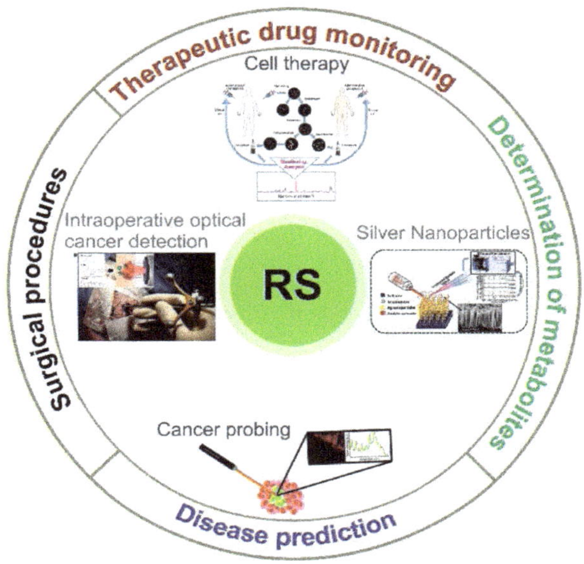

Figure 4. Schematic view of biomedical RS application. Adapted from Desroches et al. [37].

4.1. Disease Prediction

Early diagnosis of diseases, such as those that are life-threatening, is essential to prevent their spread. Based on Raman light scattering, a diagnostic tool has been developed to study important molecules and events in real-time using new RS technologies with high sensitivity to biomolecular changes. In the last decades, there have been many publications showing that RS is used to define several diseases [38]. RS is used to collect biochemical information, such as biomolecules, cells, tissues, and organs, whose biomarkers are various biological fluids, such as urine, saliva, blood, and tears. Selected biomarkers are analysed and evaluated, revealing the relationship between Raman light scattering spectral indicators and clinical condition [39].

Cancer is one of the most common causes of death worldwide and RS makes it possible to diagnose undetected precancerous lesions in various organs, such as breast, skin, brain, gastrointestinal tract, heart, urinary, and reproductive tracts [40]. Hsu et al. [41] investigated that confocal Raman microscopy distinguishes intestinal tumours from adenocarcinomas and normal, healthy organs. RS provides simple and immediate tissue identification during surgery, which allows for cancerous organs to be distinguished from healthy tissue. Using RS, the mechanism of malignant transformation of breast tissue has been studied with great success [42].

The properties of blood vessels in a tumour mass of breast tissue were investigated by Kopeć and Abramczyk [43] using a combination of Raman and atomic force microscopy (AFM) imaging to determine biochemical composition. They found that individuals with breast cancer had higher concentrations of glycogen and lactic acid as well as an increase in the collagen–fibroblast network. An excellent, recent study by Winnard et al. [44] demonstrated the potential of RS in characterising organ-specific metastatic lesions at the molecular level to gain insight into metastatic progression. In this study, they used the combinatorial approach of RS and metabolomics. The stromal adjustments that occur in pre-metastatic lungs caused by breast cancer were analysed using RS. This work was performed with mouse lines in which mice were implanted with breast cancer cells with different metastatic potential. Changes in the extracellular matrix of the congested lungs, such as an increase in collagen and proteoglycan, were examined, and this was directly related to the metastatic potential of the breast cancer cells used [45].

Ryzhikova et al. [46] have shown that RS can be used effectively for the diagnosis of Alzheimer's disease (AD). The novelty of this work is that it is based on the analysis of cerebrospinal fluid (CSF), while the other research has focused on different body fluids for the detection of AD. It is important to emphasise that CSF is the most relevant body fluid for detecting AD. The group of Ryzhikova et al. [46] suggests that early detection of AD is potentially possible using RS. It is expected that the method will be repeated on a larger subject population.

Lednev group [47,48] diagnosed early AD in saliva and serum with potential biomarkers using RS in combination with machine learning. This project aimed to use Raman hyper-spectroscopy in combination with machine learning. New methods were developed to diagnose AD based on the analysis of biological material, such as saliva. The group used biological material from saliva samples from a normal person with AD and mild cognitive impairment. In the end, it turned out that Raman hyper-spectroscopic analysis of saliva could be effective for an accurate diagnostic method in the early stages of AD. It is also possible to diagnose lung cancer with high accuracy at an early stage, as shown by the studies of Shin et al. [49] using a combination of SERS spectra and deep learning diseases. The advantage of this method is that tissue can be seen in the near-infrared region of the electromagnetic spectrum, which in combination with the RS instrument as well as multivariate data analysis, has become an accurately reproducible and non-invasive method for studying tissue pathology.

Barnas et al. [50] used Fourier transform infrared (FTIR) and RS to study endometrial hyperplasia and cancer. The study was performed on tissues from three groups of patients: normal control patients, patients with atypical hyperplasia, and patients with endometrial cancer. It has been revealed that both methods are complementary in terms of tissue examination. The results of the research suggest that the peaks of FTIR and Raman spectra and the changes in the specific peaks (absence of the peak or shift of the peak) can be used to distinguish cancer and atypical hyperplasia from normal endometrial tissue. Further studies are needed to understand whether RS is indeed a practical approach to study carcinogenesis.

SERS immunoassays have labelled/indirect or unlabelled configuration. Without labels, the Raman measurement is based on the fingerprint of the bioanalyte, and the labelled ones are identified by the spectrum of the Raman label. Therefore, labels without labels are not as complex as the labels that make up the labels on metallic nanostructures. Two systems were used to detect proteins, nucleotides, and fatty acids of lipids. The changes that occurred in the bioassay were recorded and diagnosed with infectious and non-infectious diseases [51]. Table 2 shows some examples of bioanalytes or diseases that were detected using SERS.

Table 2. Bioanalytes/diseases detected using SERS.

Bioanalyte/Disease	RS Substrate	Reference
Cancer (blood plasma protein)	Ag NPs	[51]
Quantification of hepatitis B DNA	Ag NPs	[52]
Breast cancer tissue	Ag NPs	[53]
Sjogren's syndrome from saliva	Cl-Ag NPs	[54]
Human tear uric acid	SiO_2 and Au	[55]
Creatinine	Nano-Au	[56]
Mouse IgG	Au NPs	[57]
Single prostate cancer cells	Au NPs	[58]
Plasmodium falciparum DNA	Magnetic beads	[59]
HeLa cells	Au NPs	[60]
Gastritis	Au NPs	[61]

In addition, RS is used to analyse the serum of patients with AD, patients with other types of dementia, and individuals from the control group. The results were analysed using multivariate statistics for differential identification of patients with AD. The study was a confirmation of the concept; this proves that RS and artificial neural network classification were able to differentiate patients with sensitivity and specificity of more than 90%, which shows that a combination test can become a blood test that can support clinical evaluation for effective and accurate differential diagnosis of AD [62].

4.2. Surgical Procedures

In medicine, much attention is paid to optical instruments based on RS, which consist of intraoperative procedures in real-time. The benchmark for surgical guidance is histopathology, which also involves surgical removal of tissue followed by staining and examination under a microscope. This procedure takes a long time and in some cases, results in multiple biopsies, which causes a great deal of discomfort and suffering in patients [49]. Therefore, a sensory system that can provide results during surgery is needed. For example, endoscopic pain analysis before surgery, delineation of the sides of the lesion during surgery, and changes of the single biopsy using RS can contribute to the absolute removal of the affected tissue and reduce the cost of secondary assessments of the disease and surgery [63].

Motz et al. [64] have developed a small diameter Raman probe with integrated filters and a spherical lens to minimise low priority signals. In as little as one second, the probe can show the spectra of arteries and breast tissue at different stages of pathology, which is clinically useful. Jermyn et al. [65] studied cancers of multiple human organs during surgery with 97% accuracy using a trimodal optical imaging system combining Raman. Thus, the method demonstrated that molecular imaging with high sensitivity could dramatically impact such areas of surgical and non-invasive oncological procedures for tumour detection to reduce cancer risk and improve quality of life. Kircher et al. [66] investigated the ternary status of magnetic resonance imaging—photoacoustic Raman imaging of nanoparticles, which revealed brain tumour boundaries and visualised tumour margins using RS. They used Raman imaging to ensure monitoring of intraoperative tumour resection, and histologic interdependence proved that Raman imaging delineated brain tumour boundaries. This latest trimodal aspect using nanoparticles can ensure the clearest visualisation of even the resection of a brain tumour.

In addition, significant steps are being taken to integrate RS with other wide-field and spectroscopic methods to provide additional data to support RS measurements. It has been shown that cancer cells can be diagnosed in less than one second using the broadband fluorescence method together with a Raman micro-spectrometer [67]. The trimodal optical imaging system is a combination of Raman scattering, diffuse reflectance, and intrinsic fluorescence spectroscopy, in which various cancer organs were detected during surgery with 93%, 100%, and 97% accuracy [65]. Kircher et al. [66] also synthesised a nanoparticle with three-component magnetic resonance imaging—photoacoustic and Raman imaging with the aim of preoperative and intraoperative separation of the sides of the leading brain tumour; the presence of this RS was used to visualise the sides of the tumour.

Nowadays, RS is increasingly used for cancer diagnosis and monitoring. As mentioned above, the improvement of algorithms for processing Raman signals, as well as the development of new methods SERS and fibre-optic probes, may make it possible to obtain results with high sensitivity and specificity and to apply RS approaches to cancer diagnosis.

4.3. Therapeutic Drug Monitoring (TDM)

Therapeutic drug monitoring (TDM) is an important method in clinical pharmacology and clinical chemistry that aims to measure drug concentrations in human blood [54]. TDM has been used in medical practice since the 1960s and mainly focuses on drugs with narrow therapeutic targets [68–70].

TDM is more commonly referred to in clinical practice as the observation of drug concentrations in biological fluids over time [69,71]. Karine is important for drugs with a limited therapeutic index, where a low dose is prescribed when the difference in dosage may lead to serious therapeutic consequences such, as drug toxicity and side effects [72].

There is also increasing advocacy in the field of personalised medicine that will be useful for measuring the plasma concentration of a drug at individual doses. Individualised therapy planning in personalised medicine has become a great challenge for clinicians as it is very successful in improving patient care, so that each patient can reduce drug costs while receiving optimal treatment with minimal side effects.

Fei et al. [73] performed the synchronous pharmacokinetics of 6-mercaptopurine and methimazole in the HeLa cell directions using the automated micro-fluid concept in the SERS database, which is convenient for synergistic tumour targeting. Valves and gradient generators can be used to adjust each of their chambers to deliver the required amount of different active cells and drugs at specific concentrations. The aforementioned examples also show that SERS is a robust method to determine the number of entrapped substances.

Nowadays, there is a crucial problem of how to distinguish expired and non-expired drugs using a quick and non-invasive method. Current methods, such as HPLC, thin-layer chromatography, are time-consuming and complicated. In contrast, RS is a rapid and non-invasive method that can be used to examine expired and non-expired drugs. This is a significant problem for the medical world [72,74].

The combination of AFM and RS can distinguish the characteristics of the nucleus and cytoplasm in living cells. By combining both methods, a modification of RS can be developed: tip-enhanced Raman spectroscopy (TERS). Intracellular imaging with TERS has been applied to HeLa cells. It has been shown that the regions of the nucleus and cytoplasm can be effectively distinguished using this method, for which the local information within the cell was obtained. Crucially, scientists have shown that the viability of the cell membrane is very high (about 100%) after the AFM tip penetrates the cell membrane. The method has significant potential for future use in studies where it is necessary to investigate the various organelles and biomolecules within the cell [75].

4.4. *Determination of Metabolites*

Molecularly specific RS is well suited for profiling cellular metabolites, including neurotransmitters, amino acids, lipids, glucose, and nucleic acids, as well as in biofluids. Among all cellular metabolites, lipids are one of the most studied classes of biomolecules because they have large Raman scattering cross-sections. Lipids rich in intracellular bodies are referred to as lipid droplets, making their relationship to the physiological state of the cell increasingly apparent [76]. RS exploits the promise of detecting and imaging lipid droplets for quantification in cancer cells, such as HuH7 and colorectal cancer stem cells [77].

RS and its modifications are widely used in cell research. A variety of microalgae [6] and human [1–3] cells are studied using SERS. It has been shown that information about membrane lipids can be obtained, especially the conformation of membrane lipids and the molecular environment [1].

RS has been used to study nerve myelin during excitation or the effect of neurotransmitter on nerve fibre. It has been shown that the changes in the C–C bonds of the fatty acids can be detected as well as the changes in the conformation [78]. This knowledge will improve our understanding of the mechanisms of lipid–lipid interactions in myelin and many processes associated with various diseases, such as multiple sclerosis, trauma and AD.

It is important to add that human skin is exposed to ultraviolet and infrared radiation, which is the cause of a number of diseases and ageing of human skin. Carotenoids are considered antioxidants that can support the antioxidant status of the human epidermis [79]. RS is one of the most popular methods to study carotenoids. More knowledge about the mechanisms of the photoprotective function of carotenoids is important for biomedical research and the development of commercial products. Similarly, Gellermann et al. [80] used resonant Raman scattering spectroscopy (RRSS) as a novel, non-invasive, in vivo optical technique to measure the concentration of the carotenoid pigments, lutein and zeaxanthin, in the human retina of adolescents and adults. Using RRSS, they found an apparent decrease in macular pigment concentration during the normal ageing process. They suggested the use of RS to measure macular carotenoids as a promising technology.

Ermakov et al. [81] reported that the RS technique has the potential for novel, rapid screening for carotenoid antioxidants in the largest populations at risk of vision loss due to age-related macular degeneration, an important precondition for blindness in the elderly in mature societies.

The method of RS is becoming increasingly popular in biological research. Valpapuram et al. [82] proposed a new technique combining optical biosensing and Raman micro spectroscopy. The particular advantage of this method is its ability to reduce the background signal and thus improve the signal-to-noise ratio. The researchers have shown that the combination of optical bio-sensing and Raman micro spectroscopy is a far more informative method than the conventional RS.

SERS has been used to spatially localise neurotransmitters on living cells and to study protein–neurotransmitter interactions [83,84]. Although it offers the best detection limits, the toxicity of metal nanoparticles in vivo limits its use [85]. Manciu et al. [86] have also demonstrated the usefulness of confocal RS for rapid detection of neurotransmitter predictions, but their studies were limited by in vitro spiked material. They propose real-time detection of serotonin, adenosine, and dopamine in vitro, but in addition, diffusion of dopamine in a heterogeneous base gel used as a surrogate for neural tissue. Raman mapping was performed using alpha 300 WITec confocal Raman system to obtain non-overlapping spectral data of neurotransmitters. Their work demonstrates the power of Raman spectroscopy in the biological sciences and likely provides a novel mechanism for testing the adaptability and kinetics that stimulate the brain [86].

A rapid, non-invasive, label-free approach to biological studies is currently essential for scientific purposes. However, RS has some limitations—it requires longer acquisition times and it is not possible to optically slice the collected signal. This makes it difficult to use RS for tissue research alone. Therefore, Marchetti et al. [87] combined three methods: multiphoton microscopy, fluorescence lifetime imaging microscopy, and RS to perform an efficient study of tissues ex vivo. The mentioned tailored technique is a promising approach to expand the application of RS in biological research.

The use of the RS method in medical studies is becoming more common. For example, RS is used in dentistry. The short- and long-term effects of demineralization can be studied using the RS tool. The major advantage of RS is that it is non-invasive while providing a high degree of sensitivity. In the study of Marin et al. [88], quantitative information on the crystalline structure of the phosphate groups and the loss of the mineral fraction in the organic collagen matrix was discovered.

Nowadays, a precise, fast, and direct analysis tool is needed. The capillary sensor SERS, developed by the group of Arabi et al. [89], is proposed as an ultrasensitive tool and used for protein analysis. Trypsin is a protein that can be used as a biomarker (in urine) for the diagnosis of pancreatitis. The idea is that this approach can be effectively used for early diagnosis of the disease. In addition, and to test the feasibility of the tool, other biological fluids such as saliva and sweat have also been measured. The microsensors are relatively quick and inexpensive to produce.

Another application of RS in biological studies is high-throughput screening Raman spectroscopy (HTS-RS)—presented in the work of Arend et al. [90]. This application is a customised platform for single-cell analysis. In the study in which the group of Arend et al. [90] examined the different types of neutrophils, both infected and uninfected, it has been shown that this type of platform can potentially help to speed up the diagnosis of pathogens. Currently, the routine for such analyses takes 1 working day.

It has been discovered that RS can be used effectively in chronic renal failure (CRF) to differentiate patients with this disease from healthy patients. The group of Chen et al. [74] conducted a study on 47 samples from patients with CRF and 54 samples from control subjects. There is a prospect that the application used, which can be effectively utilised as a rapid diagnostic method for CRF. The plasma RS has been effectively used to study giant unilamellar vesicles (GUV)—simplified models of cellular plasma membranes [91]. The group of Collard et al. [91] has applied the modification of RS in combination with holographic optical tweezers (HOT): HOT-Raman microscopy for the study of curvature gradients on lipid order and cholesterol segregation in GUVs. The RS provides an overall estimate of cholesterol concentration for both leaflets of the bilayer. Importantly, the proposed method also allows for obtaining multiple Raman spectra from different regions of the lipid vesicle.

Raman spectroscopy is a promising high-sense diagnostic method for assessing the oxygen transport function of erythrocytes. Haemoglobin accounts for >95% of the dry weight of erythrocytes and is a suitable subject for RS to study the conformation of globin and haeme. To assess the conformational state of the active site of haemoglobin, we use special Raman spectra to study the conformation of deoxyhaemoglobin (d- Hb) and oxyhaemoglobin (o- Hb), as well as the ability to release oxygen. This approach is important for monitoring changes in the ability of haemoglobin in erythrocytes to carry oxygen and, accordingly, characterising the presence or development of hypoxia in patient tissues. RS has been successfully used to analyse the properties of haemoglobin from healthy donors and patients with various cardiovascular diseases [92,93], diabetes [94], and astronauts after a long space flight [95], as well as for the analysis of animal models of cerebral ischemia and reperfusion, haemorrhagic shock, etc. In addition, RS has been successfully used in experiments to alter the properties of erythrocytes under in vitro conditions. A promising application of RS is the study of the molecular mechanisms of the development of pulmonary hypertension. In patients with IPAH with a typical hemodynamic picture, changes in the ability of hematoporphyrin of Hb to bind O_2 have been detected [96].

5. Biotechnology Application of Raman Spectroscopy

Biotechnology has become one of the most popular areas of research due to the demand for a number of molecules that are important for different types of practical applications [97]. Wang et al. provided a comprehensive and critical review of the most recent advances in the application of RS-enabled technologies, with focus on biomolecular applications in environmental and biotechnological fields [98].

Bioprospecting and mutagenesis are two important strategies that have been studied in the development of algae-based biofuels [18]. Considering these two strategies, there is a need to optimise biofuel production. The ability to rapidly characterise the accumulating algal lipids is essential for algal bioproduction. Confocal Raman microscopy can accomplish this task and it is also possible to localise lipid-rich regions within microalgal cells with high spatial resolution [18]. Among the existing methods, RS is the one that does not require additional long preparations of the research object.

It has been investigated that the chemical composition of lipid droplets can be obtained by using RS. It is claimed that the combination of CARS (coherent) and Raman micro-spectroscopy would allow accurate determination of the harvesting times for algae [4]. This is supposed to be one of the valuable interests in modern biotechnology.

It has been revealed that another modification of RS—single-cell Raman spectroscopy (SCRS) is applicable for gathering information about the lipid content of the cells and the degree of lipid unsaturation [22].

RS is one of the techniques that can fill some gaps in our knowledge of the synthesis and storage of biomolecules in algae [21].

5.1. Application of Raman Spectroscopy in Algae Studies

In algal research, RS has been used to analyse pigments, proteins, carbohydrates, and lipids [98–102] (Figure 5). Huang et al. [14] investigated the composition of microalgae to analyse them using confocal Raman microscopy. The scientists collected Raman spectra while using a 532 nm laser and found a strong background of fluorescence as a function of temporal behaviour. They also used RS to analyse the resolution of individual cells of microalgae. Kaczor et al. [103] published a study on the state of nutrients of single cells of microalgae and the visualisation of astaxanthin in a cell of microalgae, visualising Raman in situ with a 1064 nm laser. Laser capture Raman spectroscopy and a combination of laser capture and micro-Raman spectroscopy were developed by researchers Wu et al. [23] in which the lipid composition of microalgae was analysed using a single cell. For example, Hosokawa et al. [104] used the confocal Raman microscopy method to quantitatively monitor lipids based on a single cell. A brief report on the prospects for algal research based on RS was given in two published review articles [105,106]. Wei et al. [106] have reviewed some of the recent advances in RS applications in areas related to microalgae. Strategies based on RS provide tremendous potential for non-invasive biochemical-composition analysis and imaging of living microalgal cells. The analysis of lipids carried out by the ratiometric method has provided a solid basis, but to improve the quality of data collection and to obtain an accurate analysis, one hundred percent adjustment of the data collection parameters is required.

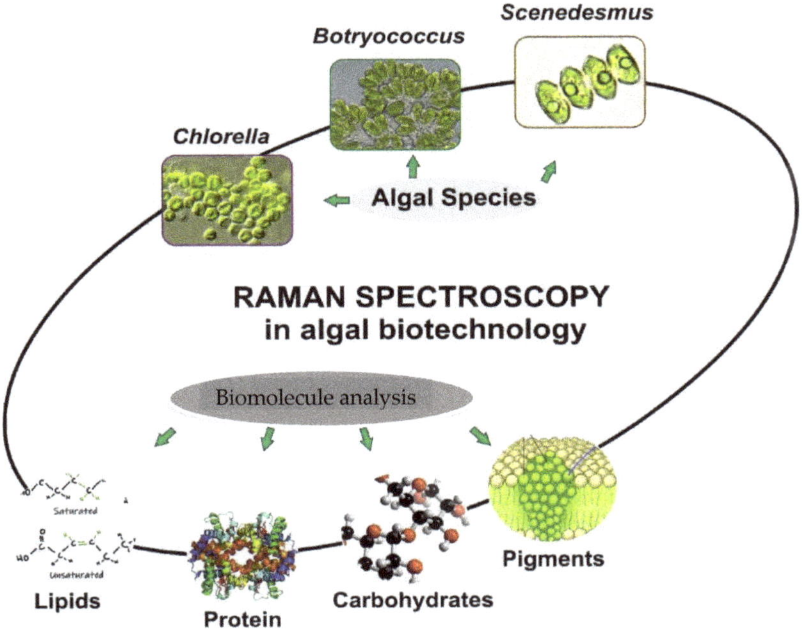

Figure 5. A schematic of the experimental set-up of a typical Raman spectrometer and schematic showing applicability of RS to different aspects of algae.

In addition, the bioactive components of many microalgae species have been studied using this Raman method. Micro-Raman spectroscopy is the most effective method for studying biologically active additives. The most commonly used modifications are macro-Raman spectrometry, single-cell micro-Raman spectrometry, and Surface-enhanced Raman Spectroscopy (Table 3).

Table 3. Summary of the bands observed in the Raman spectra of microalgae and contributing bioactive compounds.

Bioactive Compounds	Microalgal Strain Name	Type of RS	Wavenumber	Ref.
α-helix protein	*Arthrospira platensis*	Macro-Raman spectrometry	1574 cm^{-1}	[107]
Amide bonds	*Arthrospira platensis*	Macro-Raman spectrometry	1400 cm^{-1}	[107]
Antioxidant protein enzyme	*Arthrospira platensis*	Macro-Raman spectrometry	1030 and 1120 cm^{-1}	[107]
Polyphosphates	*Phaeodactylum tricornutum*	Single-cell micro-Raman spectrometry	1160 cm^{-1}	[107]
ß-carotene	*Phaeodactylum tricornutum*	Single-cell micro-Raman spectrometry	1522 cm^{-1}	[107]
ß-carotene	*Dunaliella tertiolecta*	Resonance Raman spectrometry	1158 and 1527 cm^{-1}	[108]
Triglyceride	*Chlorella sorokiniana*	Micro-Raman spectroscopy	2800 and 3000 cm^{-1}	[14]
ß-carotenoid	*Neochloris oleoabundans*	Micro-Raman spectroscopy	1505 and 1535 cm^{-1}	
ß-carotene	*Chlorella sorokiniana*	Raman micro-spectroscopy	1515 and 1157 cm^{-1}	[109]
Astaxanthin	*Chlainomonas* sp.	Micro-Raman spectroscopy	1520 and 1156 cm^{-1}	[20]
Astaxanthin	*Chlamydomonas nivalis*	Micro-Raman spectroscopy	1520 and 1156 cm^{-1}	
Violaxanthin	*Chloromonas nivalis*	Micro-Raman spectroscopy	1525 cm^{-1}	
Antheraxanthin	*Chloromonas nivalis*	Micro-Raman spectroscopy	1523 cm^{-1}	
Myxoxanthophyll	*Botrydiopsis alpina*	Micro-Raman spectroscopy	1527 cm^{-1}	[110]
Neoxanthin	*Dunaliella parva*	Micro-Raman spectroscopy	1525 and 1530 cm^{-1}	
Chlorophyll c	*Dunaliella tertiolecta*	Micro-Raman spectroscopy	1670 cm^{-1}	
Lipid	*Botryococcus brauniiis*	Micro-Raman spectroscopy	1640 and 1674 cm^{-1}	[111]
FAME	*Scenedemus quadricauda*	Surface-enhanced Raman spectroscopy	1430, 1157, 1544, 1257, 1307, 961 and 596 cm^{-1}	[112]

As shown in Table 1, two species of microalgae were studied using single-cell micro- and macro- RS methods. *Arthrospira platensis* was the highest with 1118, 1403, and 1576 cm^{-1}. The significant changes (1118, 1403, and 1576 cm^{-1}) are explained by considerable differences in the structure of secondary proteins, especially amide bonds (1400 cm^{-1}) and the α-helix (1574 cm^{-1}) [113,114]. Furthermore, Venkatesan et al. [115] indicated that the wavenumbers of *Arthrospira platensis* (1030–1120 cm^{-1}) are responsible for the presence of antioxidant protein enzymes. *Phaeodactylum tricornutum* was found to have high wavenumbers of 1522 and 1160 cm^{-1}. Moudříková et al. [116] found that 1160 cm^{-1} corresponds to the polyphosphate-positive groups of microalgae. The authors used RS for quantification as well as for localisation of polyphosphate reserves within an algal cell. These authors improved the method by extracting polyphosphate with phenol–chloroform and by purifying the extract through ethanol precipitation. Wei et al. [106] noted that they (1518–1525 cm^{-1}) correspond to the C–C regions of β-carotene, which are mainly present in the non-polar phase of microalgae. Brahma et al. [108] observed the strongest peaks around 1527 and 1158 cm^{-1} in *Dunaliella tertiolecta* associated with carotenoid pigments. Huang et al. [14] studied *Chlorella sorokiniana* and *Neochloris oleoabundans*. *Chlorella sorokiniana* had strong peaks at 2800 and 3000 cm^{-1}. Further support for the use of a broad band of wavenumbers for lipid identification is provided by several previous CARS studies that used peak positions at 2840 or 2845 cm^{-1} for lipid identification [106]. Osterrothová et al. [20] tested the possibilities of Raman micro-spectroscopy for the determination of carotenoid pigments, both primary (lutein, β-carotene) and secondary (astaxanthin) carotenoids. The components of the xanthophyll cycle, violaxanthin (1525 cm^{-1}) and antheraxanthin (1523 cm^{-1}), can also contribute to a shift in the position of the ν1 (C=C) band. Strong carotenoid signals were observed by Jehlička et al. [117] in *Botrydiopsis alpina* at 1527 cm^{-1}. In *Dunaliella parva*, the secondary carotenoid neoxanthin was found at 1525 and 1530 cm^{-1}.

B. brauniii is a classic producer of liquid hydrocarbons known as botryococci, which can be used as fuel. Weiss et al. [111] have obtained Raman spectra of hydrocarbons isolated from *B. brauniicells*. By comparison with the Raman spectra of pure squalene and computer analysis, they found that the Raman bands at 1640 cm^{-1} and 1647 cm^{-1} were related to the stretching of the C=C bond in the botryococcus branch and the methylene C=C bond. produced by methylation, respectively.

Compared with the standard fatty acid mixture, the main SERS peaks obtained from the cells of *S. quadricauda* are at 1430, 1157, 1544, 1257, 1307, 961, and 596 cm^{-1}, which is in good agreement with the literature data [22,112,118].

5.1.1. Raman Spectroscopy Applied to Lipid

Normally, lipid droplets are assumed to be cellular structures that function only as static lipid storage depots. Recently, however, it has become apparent that lipid droplets may be multifunctional organelles [119,120]. Considering the information available so far, there is a need for deeper research into lipid droplets—their structures and functions in cells. Nowadays, there is a hypothesis that lipid droplets may be involved in the biosynthesis and transport of carotenoids in the cell [121]. It is known that the biosynthesis of the carotenoid astaxanthin is accompanied by a massive accumulation of lipids [122].

Another interesting finding is that the membrane fluidity of cyanobacteria can change when the temperature is lowered [123]. This is thought to be due to the desaturation of fatty acids in the membranes [124]. RS is an attractive alternative for lipid detection that has not yet been sufficiently exploited in microalgae. Mostly this is because RS is hampered in photosynthetic organisms by strong autofluorescence of the pigments, which obscures the characteristic Raman spectral features.

The intensities of the Raman spectral peaks correspond to the saturated and unsaturated C–C bonds in lipid molecules. This information is used to estimate the degree of unsaturation in lipid bodies/droplets [4,5,15,16,18]. It has been shown that Raman micro spectrometry can be used to study the triacylglycerols (TAG) content and accumulation [21], providing a new view of the biosynthesis of fatty acids in the microalgae.

Several authors have shown that the RM methods are suitable for biodiesel production from microalgae to determine the FAs content. Wu et al. [23] demonstrated a method for direct quantitative and in vivo lipid profiling of oil-producing microalgae using single-cell RS laser capture. This approach shows that lipids in microalgae determine the quantitative degree of unsaturation and the transition temperature. As the authors stated, the above factors can be measured on a single living cell of a microalga held in place with an optical trap while Raman data is collected. Raman is used in the study of FAs from microalgae for biofuel production and has shown analytical capabilities and quantification algorithms to be useful in many different organisms and lipidomics (Figure 6).

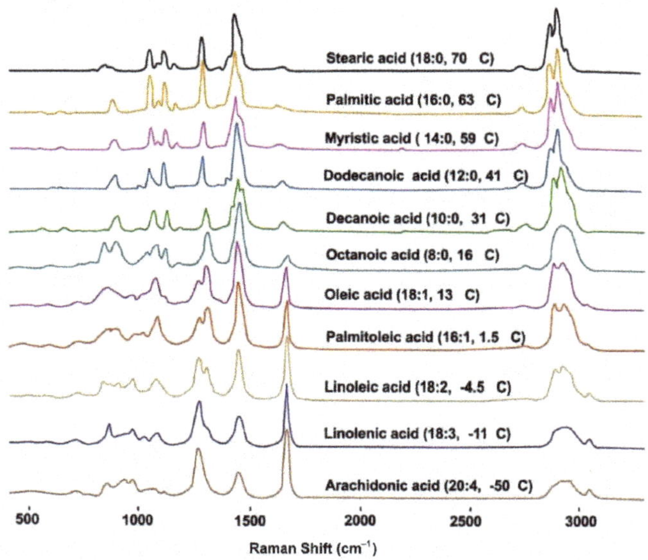

Figure 6. Raman spectra of various lipid molecules of *Botryococcus braunii* [23].

Another ability of Raman microscopy reported by Samek et al. [5] showed how the useful iodine number in lipid bodies in *Chlamydomonas* sp. CCALA can be determined from living algal cells. At the same time, the characteristic peaks in the Raman light spectra at 1656 cm^{-1} and 1445 cm^{-1} were used as markers for fatty acids in algae lipids, indicating the ratio of unsaturated and saturated carbon-carbon bonds [5].

He et al. [19] investigated the accumulation of separation of TAG in *Coccomyxa subellipsoidea* cells in the presence of N-depletion using the broadband CARS concept. Compared to simple Raman imaging, CARS microscopy showed intrinsic advantages in detection speed and spatial resolution, but the concept of CARS imaging was limited by overlapping signals, such as two-photon-excited fluorescence.

Yan et al. reviewed several RS methods in cell sorting to understand the metabolic interactions between bacteria in natural habitat. This review shows current knowledge about the research progress of recognition and assessment of single microorganism cell. The group summarised that Raman-activated cell sorting can be suitable method for cell recognition in application [125].

Thus, Raman spectrometer is used in microalgal biotechnology to screen species for highly concentrated fatty acid mutants [18]. There are two major strategies in algal biofuels development: bioprospecting and mutagenesis. This requires precise sorting and analysis of a large number of algal isolates containing TAG; FACS and ratiometric Raman analysis are the most suitable. Isolation of new algae from field samples can be carried out by UV mutagenesis to increase lipid production. FACS can be used to sample mutant populations and strains that alter lipid production during UV mutagenesis. Central to this workflow is confocal Raman microscopy, which allows for characterisation of the lipids produced by the algae in situ and rapid extraction of lipids from the cells. Raman hyper-spectroscopy is used to localise the lipid-rich region with a low pixel density, allowing faster Raman hyper-spectroscopy imaging. Confocal Raman microscopy characterises the lipid content (Figure 7).

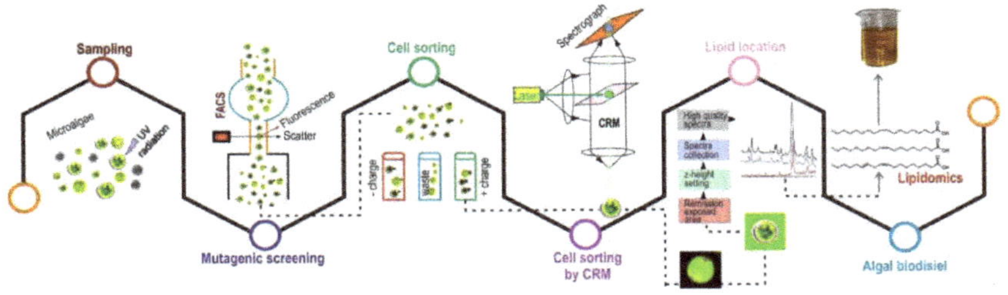

Figure 7. The schematic view in lipid characterisation of microalgae. Bioprospecting of *C. reinhardtii* is performed to generate algal samples with lipid content. The mutagens are sorted by FACS based on the fluorescence of a dye to select cells with high lipid content. The selected cells and mutants are then screened using CRM. This method allows for rapid characterisation of lipids. The spectra yield depends on the number of C=C bonds and the length of the hydrocarbon chains of the lipid molecules. This workflow enables rapid characterisation of cells for molecular traits that are important for the production of biodiesel. Modified from Sharma et al [18].

5.1.2. Application of Raman Spectroscopy on Pigment Investigation in Microalgae

Thus, considering the spectra of the pigments, they are very sensitive to the excitation energy and contribute to a large extent to the Raman spectra of many algae [108,125–128]. Chen et al. showed that when a long excitation wavelength of 488 nm is used, the strongest and most abundant peaks of chlorophyll-d coincide with the peaks of chlorophyll a and chlorophyll b [129]. β-Carotene has intense peaks at 1150 cm^{-1}, 1520 cm^{-1}, and 1008 cm^{-1}, and the most important overtone peaks at 2320 cm^{-1} and 2667 cm^{-1} [130–133]. In addition, due to the identical chemical structure, it is expected that a large number of compounds of both chlorophyll and carotenoids will have similar spectra.

Furthermore, because of the similar chemical structure, we would expect different chlorophyll compounds to give similar spectra and different carotenoids to give similar spectra. Therefore, the major Raman peaks associated with chlorophyll d and β-carotene can be used to represent common chlorophylls and carotenoids. The standard spectra of chlorophyll d and β-carotene are shown together with the experimental spectra of algae (Figure 8).

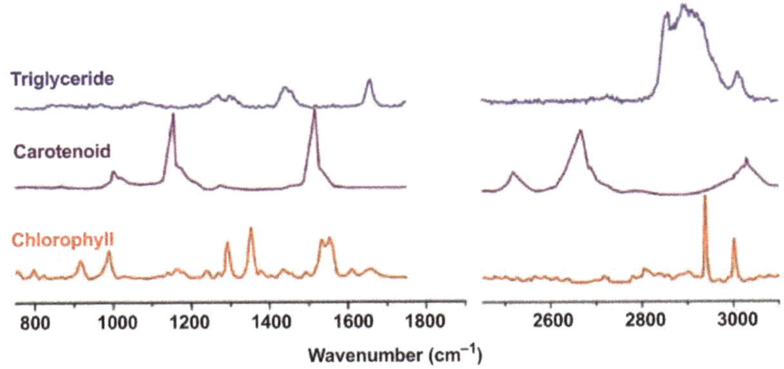

Figure 8. The Raman spectrum of carotenoid [132], chlorophyll [129], and triglyceride and the mean spectra acquired for starved *C. sorokiniana* and starved *N. oleoabundans* in the wavenumber regions of 750–1750 cm^{-1} and 2450–3150 cm^{-1}. Modified from Shutova et al. [134].

Several researchers have shown that RS methods are also suitable for the production of carotenoids by microalgae. Carotenoids are extremely important for human health [135]. Carotenoids are popular biomolecules for biomedical applications. Carotenoids are known to play the role of photoprotection molecules in the cells of phototrophs. Secondary carotenoids are also of interest as secondary carotenogenesis is thought to be the stress response of the cell [121].

It is also believed that carotenogenic microalgae can survive in a wide range of environmental conditions [136]. Nowadays, the analysis of carotenoids in algae is performed using high-performance liquid chromatography (HPLC). However, it is important to emphasise that RS can be used for a more detailed analysis of carotenoids in algal cells [17,20,24]. In addition, the development of RS, which will be applied to algae in the future, can be used for real-time research of algal combination in nature.

Jehlička et al. [7] studied how RS can be used to identify various carotenoids as well as probable biomarkers in algae. A number of laboratory grown algae with different taxonomic groups were studied. The results showed that RS is considered an optimal tool for assessing the presence of carotenoids in a given organism. The comparison was made with the HPLC method to examine the pigments in the concentrates. In summary Raman spectroscopy can be used for the detection of carotenoids and other pigments in algae.

Osterrothová et al. [20] tested the abilities of RS to determine carotenoid pigments—both basic (lutein, β-carotene) and secondary (astaxanthin) carotenoids—in different species of *Chlamydomonadales* algae. They also compared the performance of RS with a standard biological pigment analysis method, such as HPLC. They described the carotenoids of algae using a combination of resonance RS and HPLC, also creating a spectral library for different stages of the algal life cycle. A comprehensive study to find pigments in biomass can show results with HPLC. However, this method requires the extraction of pigments from the biomass, which can lead to data loss (e.g., protein/lipid interactions). Raman macro–microscopy, however, makes it possible to quickly reveal the pigments of single cells, which is another advantage, especially when the heterogeneous nature of the cells is taken into account.

To map the changes in the composition of β-carotene and AXT in different cellular morphotypes of *H. pluvialis*, Collins et al. [137] used a confocal Raman microscope at 532 nm laser excitation. Using a multivariate curve, several readable spectral components were extracted from the data describing Raman scattering and fluorescence of active *H. pluvialis* cells at different life stages. Based on the results, they were able to determine the arrangement of the different pigments in the cells at different time periods. They also concluded that β-carotene can be considered as an ancestor of AXT and a site for the synthesis of AXT. Their study shows that Raman micro-spectroscopy is an important method for studying in vivo changes stimulated by the environment in the life cycle of microalgae.

In their study, Chiu et al. [138] demonstrated for the first time that RS can be used to quantify starch in addition to lipids in algal cells. Because RS is so simple and non-destructive, it is ideal for further investigation of the starch–lipid shift mechanism.

RS provides information about the vibrations of bonds in molecules. This approach is used for the study of carotenoids. It has been shown that changes in the molecular environment (such as pH change) affect the specific bands (Figure 8) in the carotenoids' spectra [134].

This is particularly attractive for applied sciences, such as biotechnology and biomedicine.

6. Raman Spectroscopy for Photosynthetic Studies

Photosynthesis is the most basic and important process on earth. It is the natural way of synthesising carbohydrates using solar energy. Scientists from all over the world are exploring it with a number of applications, one of which is Raman spectroscopy. Findings from a number of recent studies on RS applied to photosynthetic organisms are shared below.

Mishra et al. [139] recently conducted their study on Antarctic lichens using RS. Antarctic lichens are organisms that can change their metabolism and photosynthetic activity in response to changing environmental conditions. Hydration and dehydration are the investigated triggers for the activation/deactivation of photosynthetic processes in the lichens. It has been revealed that photosynthetic activity is activated quite rapidly, which contributes to the hypothesis that the photosynthetic apparatus and carotenoids are not synthesised de novo in the early stages of photosynthesis. Another important discovery made using RS, the bands/features of the pigment scytonemin, are present in the Raman spectra of one of the lichens studied. There is a hypothesis that this pigment plays a photoprotective role in the photobionts of algae and cyanobacteriae.

7. Raman Spectroscopy for Analytical Studies

There is an ongoing need for fast and accurate detection of melamine in dairy products. Melamine is a compound that can be toxic above a certain level when added to food. Therefore, it is important to propose an approach that allows accurate detection of the toxic compound in food products. Liu et al. [120] proposed the SERS method using silver nanoparticles (AgNPs). Not only SERS but also a colourimetric method was used for this idea. The results showed that the colourimetric method can lead to false-positives in detecting the presence of different compounds (AgNPs). The SERS method, on the other hand, can overcome this limitation [120]. Importantly, the scientists suggested using both methods in tandem to achieve accurate and rapid detection of melamine in dairy products.

Fentanyl is one of the most commonly used opioids. However, fentanyl and its analogues caused numerous fatal drug overdose incidents. The problem raised by the group of Mirsafavi et al. [140] is the need for novel analytical methods to effectively distinguish fentanyl from its precursors. The vibrational spectra of this family of analytes are quite similar, so it is difficult to solve the problem using conventional methods. The SERS method enables the distinguishing of fentanyl and its precursors. This approach would be an efficient and effective aid in the field of forensics.

8. Future Perspectives

In recent decades, RS has successfully emerged as a clinical tool for diagnostic, surgical, and pathological applications. The creation of probes in conjunction with modern methods of studying information has led to a surge in studies based on combinatorial light scattering. However, when introducing RS into clinics, there are the major difficulties described in the previous section, which should be overcome by close collaboration between clinicians, material scientists, biomedical engineers, and spectroscopists. Artificial intelligence algorithms are expected to be used to solve complex clinical issues, which will accelerate the work of RS. In addition, the probes must be resistant to disinfection for daily use. Further advances in scientific and technical research, which also guarantee a high signal-to-noise ratio with the lowest laser excitation power in a short time, are accordingly worth examining in order to use RS for intraoperative procedures. Nevertheless, introducing new technology into the clinic remains a challenge, even though recent successes and prospects represent meaningful ideas for us and inspire us to solve certain complex problems, opening the door to an appreciable goal.

In addition, Raman spectroscopy is a common tool for detecting carotenoids in various biological materials, including prokaryotic bacteria, aquatic plants, and lichens. The resonant Raman amplification of the signals enables the detection of carotenoids at low concentrations. In other cases, however, microorganisms also synthesise other pigments, and the examples studied included their composition. In this case, the combinatorial scattering ranges do not at all include a series of sudden bands corresponding to this carotene, nor was there any significant broadening of the bands. The predominant carotenoid can be seen in the spectra. However, it was not possible to use a unique excitation wavelength from a range of microorganisms.

Raman spectroscopy can be used to detect the presence of carotenoids and other pigments in cyanobacteria, microorganisms, and algae. The occurrence of colour combinations in this organism is capable of producing small or moderately significant changes in saturation and in the number of combinatorial scatter bands, which interferes with the likely unambiguous identification of carotenoid c due to shifts in power, particularly in the position of the combinatorial scatter bands.

In recent years, RS has been widely used in research, including photosynthesis and analytical research. However, very little research has been conducted on plant photosynthesis using RS. Most research has been carried out to determine the composition of pigments in cyanobacteria.

9. Conclusions

RS and a spectrum of different modifications of the RS method are increasingly used in biological and medical research. RS is gradually becoming more popular among algae experts. In the list of studies, it has been revealed that we can successfully detect and analyse the Raman scattering signal of algae. This is important not only for biotechnology, but also for a better understanding of the mechanisms of the biomolecule synthesis and storage in algal cells. In this review, we have analysed and presented a number of existing studies in biological, medical, analytical, photosynthetic, and algal research using RS. RS is effectively and widely used for a variety of studies in animals and human research. We have attempted to highlight that a greater focus on the application of RS in algal research will be beneficial for biotechnological purposes and general knowledge of the mechanisms of the biomolecule interactions in algae under natural/environmental conditions. It is worth emphasising that RS is a very attractive and promising approach for algal research, especially because of its advantages.

Author Contributions: Investigation, formal analysis, and data curation: E.S.A., V.V.K., B.D.K. and E.V.Z.; drawing of main conclusions and draft preparation of the manuscript: E.S.A., V.V.K., E.V.Z., O.V.R. and T.V.M.; draft preparation of the manuscript: O.V.R., T.V.M., G.V.M., S.A., T.T. and S.I.A. All authors have read and agreed to the published version of the manuscript.

Funding: S.A. would like to thank the Distinguished Scientists Fellowship Program, King Saud University, Saudi Arabia, for their support. This work was supported by JSPS KAKENHI Grant Numbers 20H05114, J21K06101 to T.T.; V.V.K., E.V.Z., S.I.A. was supported by the grant from Russian Science Foundation (No: 22-44-08001); The results of part (Sections 5 and 6) were obtained within the state assignment of Ministry of Science and Higher Education of the Russian Federation (project No. 121033000136-4). G.V.M. is supported by the Russian Science Foundation (grant No:19-79-30062) as well as the Interdisciplinary Scientific and Educational School of Moscow University "Molecular Technologies of Living Systems and Synthetic Biology".

Institutional Review Board Statement: Not applicable.

Informed Consent Statement: Not applicable.

Data Availability Statement: The data presented in this study are available on request from the corresponding authors. The data are not public.

Conflicts of Interest: The authors declare no conflict of interest.

References

1. Nikelshparg, E.I.; Grivennikova, V.G.; Baizhumanov, A.A.; Semenova, A.A.; Sosnovtseva, V.; Goodilin, E.A.; Maksimov, G.V.; Brazhe, N.A. Probing Lipids in Biological Membranes Using SERS. *Mendeleev Commun.* **2019**, *29*, 635–637. [CrossRef]
2. Brazhe, N.A.; Evlyukhin, A.B.; Goodilin, E.A.; Semenova, A.A.; Novikov, S.M.; Bozhevolnyi, S.I.; Chichkov, B.N.; Sarycheva, A.S.; Baizhumanov, A.A.; Nikelshparg, E.I.; et al. Probing Cytochrome c in Living Mitochondria with Surface-Enhanced Raman Spectroscopy. *Sci. Rep.* **2015**, *5*, 13793. [CrossRef] [PubMed]
3. Subramanian, B.; Thibault, M.-H.; Djaoued, Y.; Pelletier, C.; Touaibia, M.; Tchoukanova, N. Chromatographic, NMR and Vibrational Spectroscopic Investigations of Astaxanthin Esters: Application to "Astaxanthin-Rich Shrimp Oil" Obtained from Processing of Nordic Shrimps. *Analyst* **2015**, *140*, 7423–7433. [CrossRef] [PubMed]
4. Jaeger, D.; Pilger, C.; Hachmeister, H.; Oberländer, E.; Wördenweber, R.; Wichmann, J.; Mussgnug, J.H.; Huser, T.; Kruse, O. Label-Free in vivo Analysis of Intracellular Lipid Droplets in the Oleaginous Microalga Monoraphidium Neglectum by Coherent Raman Scattering Microscopy. *Sci. Rep.* **2016**, *6*, 35340. [CrossRef]
5. Samek, O.; Jonáš, A.; Pilát, Z.; Zemánek, P.; Nedbal, L.; Tříska, J.; Kotas, P.; Trtílek, M. Raman Microspectroscopy of Individual Algal Cells: Sensing Unsaturation of Storage Lipids in vivo. *Sensors* **2010**, *10*, 8635–8651. [CrossRef] [PubMed]

6. Fu, S.; Wang, X.; Wang, T.; Li, Z.; Han, D.; Yu, C.; Yang, C.; Qu, H.; Chi, H.; Wang, Y.; et al. A Sensitive and Rapid Bacterial Antibiotic Susceptibility Test Method by Surface Enhanced Raman Spectroscopy. *Braz. J. Microbiol.* **2020**, *51*, 875–881. [CrossRef]
7. Jehlička, J.; Edwards, H.G.M.; Osterrothová, K.; Novotná, J.; Nedbalová, L.; Kopecký, J.; Němec, I.; Oren, A. Potential and Limits of Raman Spectroscopy for Carotenoid Detection in Microorganisms: Implications for Astrobiology. *Philos. Trans. R. Soc. Math. Phys. Eng. Sci.* **2014**, *372*, 20140199. [CrossRef]
8. Butler, H.J.; Ashton, L.; Bird, B.; Cinque, G.; Curtis, K.; Dorney, J.; Esmonde-White, K.; Fullwood, N.J.; Gardner, B.; Martin-Hirsch, P.L.; et al. Using Raman Spectroscopy to Characterize Biological Materials. *Nat. Protoc.* **2016**, *11*, 664–687. [CrossRef]
9. Buckley, K.; Ryder, A.G. Applications of Raman Spectroscopy in Biopharmaceutical Manufacturing: A Short Review. *Appl. Spectrosc.* **2017**, *71*, 1085–1116. [CrossRef]
10. Alvarez-Puebla, R.A.; Liz-Marzán, L.M. SERS-Based Diagnosis and Biodetection. *Small* **2010**, *6*, 604–610. [CrossRef]
11. Hudson, S.D.; Chumanov, G. Bioanalytical Applications of SERS (Surface-Enhanced Raman Spectroscopy). *Anal. Bioanal. Chem.* **2009**, *394*, 679–686. [CrossRef] [PubMed]
12. Cialla, D.; März, A.; Böhme, R.; Theil, F.; Weber, K.; Schmitt, M.; Popp, J. Surface-Enhanced Raman Spectroscopy (SERS): Progress and Trends. *Anal. Bioanal. Chem.* **2012**, *403*, 27–54. [CrossRef] [PubMed]
13. Asiala, S.M.; Schultz, Z.D. Surface Enhanced Raman Correlation Spectroscopy of Particles in Solution. *Anal. Chem.* **2014**, *86*, 2625–2632. [CrossRef] [PubMed]
14. Huang, Y.Y.; Beal, C.M.; Cai, W.W.; Ruoff, R.S.; Terentjev, E.M. Micro-Raman Spectroscopy of Algae: Composition Analysis and Fluorescence Background Behavior. *Biotechnol. Bioeng.* **2010**, *105*, 889–898. [CrossRef]
15. Cavonius, L.; Fink, H.; Kiskis, J.; Albers, E.; Undeland, I.; Enejder, A. Imaging of Lipids in Microalgae with Coherent Anti-Stokes Raman Scattering Microscopy. *Plant Physiol.* **2015**, *167*, 603–616. [CrossRef]
16. He, X.N.; Allen, J.; Black, P.N.; Baldacchini, T.; Huang, X.; Huang, H.; Jiang, L.; Lu, Y.F. Coherent Anti-Stokes Raman Scattering and Spontaneous Raman Spectroscopy and Microscopy of Microalgae with Nitrogen Depletion. *Biomed. Opt. Express* **2012**, *3*, 2896–2906. [CrossRef]
17. Legesse, F.B.; Rüger, J.; Meyer, T.; Krafft, C.; Schmitt, M.; Popp, J. Investigation of Microalgal Carotenoid Content Using Coherent Anti-Stokes Raman Scattering (CARS) Microscopy and Spontaneous Raman Spectroscopy. *ChemPhysChem* **2018**, *19*, 1048–1055. [CrossRef]
18. Sharma, S.K.; Nelson, D.R.; Abdrabu, R.; Khraiwesh, B.; Jijakli, K.; Arnoux, M.; O'Connor, M.J.; Bahmani, T.; Cai, H.; Khapli, S.; et al. An Integrative Raman Microscopy-Based Workflow for Rapid in Situ Analysis of Microalgal Lipid Bodies. *Biotechnol. Biofuels* **2015**, *8*, 164. [CrossRef]
19. He, S.; Xie, W.; Zhang, P.; Fang, S.; Li, Z.; Tang, P.; Gao, X.; Guo, J.; Tlili, C.; Wang, D. Preliminary Identification of Unicellular Algal Genus by Using Combined Confocal Resonance Raman Spectroscopy with PCA and DPLS Analysis. *Spectrochim. Acta Part A Mol. Biomol. Spectrosc.* **2018**, *190*, 417–422. [CrossRef]
20. Osterrothová, K.; Culka, A.; Němečková, K.; Kaftan, D.; Nedbalová, L.; Procházková, L.; Jehlička, J. Analyzing Carotenoids of Snow Algae by Raman Microspectroscopy and High-Performance Liquid Chromatography. *Spectrochim. Acta Part A Mol. Biomol. Spectrosc.* **2019**, *212*, 262–271. [CrossRef]
21. Kish, E.; Wang, K.; Llansola-Portoles, M.J.; Ilioaia, C.; Pascal, A.A.; Robert, B.; Yang, C. Probing the Pigment Binding Sites in LHCII with Resonance Raman Spectroscopy: The Effect of Mutations at S123. *Biochim. Biophys. Acta BBA- Bioenerg.* **2016**, *1857*, 1490–1496. [CrossRef] [PubMed]
22. Wang, T.; Ji, Y.; Wang, Y.; Jia, J.; Li, J.; Huang, S.; Han, D.; Hu, Q.; Huang, W.E.; Xu, J. Quantitative Dynamics of Triacylglycerol Accumulation in Microalgae Populations at Single-Cell Resolution Revealed by Raman Microspectroscopy. *Biotechnol. Biofuels* **2014**, *7*, 58. [CrossRef] [PubMed]
23. Wu, H.; Volponi, J.V.; Oliver, A.E.; Parikh, A.N.; Simmons, B.A.; Singh, S. In Vivo Lipidomics Using Single-Cell Raman Spectroscopy. *Proc. Natl. Acad. Sci. USA* **2011**, *108*, 3809–3814. [CrossRef] [PubMed]
24. Deng, Y.-L.; Juang, Y.-J. Black Silicon SERS Substrate: Effect of Surface Morphology on SERS Detection and Application of Single Algal Cell Analysis. *Biosens. Bioelectron.* **2014**, *53*, 37–42. [CrossRef]
25. Wang, Y. Construction of Artificial Intelligence-Assisted Prostate Tumor Early Diagnosis System Based on Surface Enhanced Raman Spectroscopy. 2020. Available online: http://www.chictr.org.cn/showproj.aspx?proj=60141 (accessed on 12 December 2021).
26. U.S. National Library of Medicine Non-Invasive Assessment of Mechano-Chemical Properties of Urine Proteins by Hybrid Brillouin-Raman Spectroscopy. Available online: https://clinicaltrials.gov/ct2/show/NCT04311684 (accessed on 12 December 2021).
27. U.S. National Library of Medicine. Raman Analysis of Saliva as Biomarker of COPD (BIO-RAnCh). Available online: https://clinicaltrials.gov/ct2/show/NCT04628962 (accessed on 12 December 2021).
28. Zavaleta, C.L.; Garai, E.; Liu, J.T.C.; Sensarn, S.; Mandella, M.J.; de Sompel, D.V.; Friedland, S.; Dam, J.V.; Contag, C.H.; Gambhir, S.S. A Raman-Based Endoscopic Strategy for Multiplexed Molecular Imaging. *Proc. Natl. Acad. Sci. USA* **2013**, *110*, E2288–E2297. [CrossRef]
29. Jiang, C.; Wang, Y.; Song, W.; Lu, L. Delineating the Tumor Margin with Intraoperative Surface-Enhanced Raman Spectroscopy. *Anal. Bioanal. Chem.* **2019**, *411*, 3993–4006. [CrossRef]
30. Evans, C.L.; Xie, X.S. Coherent Anti-Stokes Raman Scattering Microscopy: Chemical Imaging for Biology and Medicine. *Annu. Rev. Anal. Chem.* **2008**, *1*, 883–909. [CrossRef]

31. Evans, C.L.; Potma, E.O.; Puoris'haag, M.; Côté, D.; Lin, C.P.; Xie, X.S. Chemical Imaging of Tissue in vivo with Video-Rate Coherent Anti-Stokes Raman Scattering Microscopy. *Proc. Natl. Acad. Sci. USA* **2005**, *102*, 16807–16812. [CrossRef]
32. Cheng, J.-X.; Xie, X.S. Coherent Anti-Stokes Raman Scattering Microscopy: Instrumentation, Theory, and Applications. *J. Phys. Chem. B* **2004**, *108*, 827–840. [CrossRef]
33. Su, L.-M. Robot-Assisted Radical Prostatectomy: Advances since 2005. *Curr. Opin. Urol.* **2010**, *20*, 130–135. [CrossRef]
34. Okada, M.; Smith, N.I.; Palonpon, A.F.; Endo, H.; Kawata, S.; Sodeoka, M.; Fujita, K. Label-Free Raman Observation of Cytochrome c Dynamics during Apoptosis. *Proc. Natl. Acad. Sci. USA* **2012**, *109*, 28–32. [CrossRef]
35. Vardaki, M.Z.; Atkins, C.G.; Schulze, H.G.; Devine, D.V.; Serrano, K.; Blades, M.W.; Turner, R.F.B. Raman Spectroscopy of Stored Red Blood Cell Concentrate within Sealed Transfusion Blood Bags. *Analyst* **2018**, *143*, 6006–6013. [CrossRef] [PubMed]
36. Feng, G.; Ochoa, M.; Maher, J.R.; Awad, H.A.; Berger, A.J. Sensitivity of Spatially Offset Raman Spectroscopy (SORS) to Subcortical Bone Tissue. *J. Biophotonics* **2017**, *10*, 990–996. [CrossRef]
37. Desroches, J.; Jermyn, M.; Pinto, M.; Picot, F.; Tremblay, M.-A.; Obaid, S.; Marple, E.; Urmey, K.; Trudel, D.; Soulez, G.; et al. A New Method Using Raman Spectroscopy for in vivo Targeted Brain Cancer Tissue Biopsy. *Sci. Rep.* **2018**, *8*, 1792. [CrossRef] [PubMed]
38. Choo-Smith, L.-P.; Edwards, H.G.M.; Endtz, H.P.; Kros, J.M.; Heule, F.; Barr, H.; Robinson, J.S., Jr.; Bruining, H.A.; Puppels, G.J. Medical Applications of Raman Spectroscopy: From Proof of Principle to Clinical Implementation. *Biopolymers* **2002**, *67*, 1–9. [CrossRef]
39. Pandey, R.; Paidi, S.K.; Valdez, T.A.; Zhang, C.; Spegazzini, N.; Dasari, R.R.; Barman, I. Noninvasive Monitoring of Blood Glucose with Raman Spectroscopy. *Acc. Chem. Res.* **2017**, *50*, 264–272. [CrossRef]
40. Cui, S.; Zhang, S.; Yue, S. Raman Spectroscopy and Imaging for Cancer Diagnosis. *J. Healthc. Eng.* **2018**, *2018*, e8619342. [CrossRef] [PubMed]
41. Hsu, C.-W.; Huang, C.-C.; Sheu, J.-H.; Lin, C.-W.; Lin, L.-F.; Jin, J.-S.; Chen, W. Differentiating Gastrointestinal Stromal Tumors from Gastric Adenocarcinomas and Normal Mucosae Using Confocal Raman Microspectroscopy. *J. Biomed. Opt.* **2016**, *21*, 075006. [CrossRef]
42. Zhang, H.; Wang, X.; Ding, R.; Shen, L.; Gao, P.; Xu, H.; Xiu, C.; Zhang, H.; Song, D.; Han, B. Characterization and Imaging of Surgical Specimens of Invasive Breast Cancer and Normal Breast Tissues with the Application of Raman Spectral Mapping: A Feasibility Study and Comparison with Randomized Single-point Detection Method. *Oncol. Lett.* **2020**, *20*, 2969–2976. [CrossRef]
43. Kopeć, M.; Abramczyk, H. Angiogenesis—A Crucial Step in Breast Cancer Growth, Progression and Dissemination by Raman Imaging. *Spectrochim. Acta Part A Mol. Biomol. Spectrosc.* **2018**, *198*, 338–345. [CrossRef]
44. Winnard, P.T.; Zhang, C.; Vesuna, F.; Kang, J.W.; Garry, J.; Dasari, R.R.; Barman, I.; Raman, V. Organ-Specific Isogenic Metastatic Breast Cancer Cell Lines Exhibit Distinct Raman Spectral Signatures and Metabolomes. *Oncotarget* **2017**, *8*, 20266–20287. [CrossRef] [PubMed]
45. Paidi, S.K.; Rizwan, A.; Zheng, C.; Cheng, M.; Glunde, K.; Barman, I. Label-Free Raman Spectroscopy Detects Stromal Adaptations in Premetastatic Lungs Primed by Breast Cancer. *Cancer Res.* **2017**, *77*, 247–256. [CrossRef] [PubMed]
46. Ryzhikova, E.; Ralbovsky, N.M.; Sikirzhytski, V.; Kazakov, O.; Halamkova, L.; Quinn, J.; Zimmerman, E.A.; Lednev, I.K. Raman Spectroscopy and Machine Learning for Biomedical Applications: Alzheimer's Disease Diagnosis Based on the Analysis of Cerebrospinal Fluid. *Spectrochim. Acta Part A Mol. Biomol. Spectrosc.* **2021**, *248*, 119188. [CrossRef] [PubMed]
47. Ralbovsky, N.M.; Halámková, L.; Wall, K.; Anderson-Hanley, C.; Lednev, I.K. Screening for Alzheimer's Disease Using Saliva: A New Approach Based on Machine Learning and Raman Hyperspectroscopy. *J. Alzheimers Dis.* **2019**, *71*, 1351–1359. [CrossRef] [PubMed]
48. Ralbovsky, N.M.; Fitzgerald, G.S.; McNay, E.C.; Lednev, I.K. Towards Development of a Novel Screening Method for Identifying Alzheimer's Disease Risk: Raman Spectroscopy of Blood Serum and Machine Learning. *Spectrochim. Acta Part A Mol. Biomol. Spectrosc.* **2021**, *254*, 119603. [CrossRef]
49. Shin, H.; Oh, S.; Hong, S.; Kang, M.; Kang, D.; Ji, Y.; Choi, B.H.; Kang, K.-W.; Jeong, H.; Park, Y.; et al. Early-Stage Lung Cancer Diagnosis by Deep Learning-Based Spectroscopic Analysis of Circulating Exosomes. *ACS Nano* **2020**, *14*, 5435–5444. [CrossRef]
50. Barnas, E.; Skret-Magierlo, J.; Skret, A.; Kaznowska, E.; Depciuch, J.; Szmuc, K.; Łach, K.; Krawczyk-Marć, I.; Cebulski, J. Simultaneous FTIR and Raman Spectroscopy in Endometrial Atypical Hyperplasia and Cancer. *Int. J. Mol. Sci.* **2020**, *21*, 4828. [CrossRef]
51. Lin, J.; Chen, R.; Feng, S.; Pan, J.; Li, Y.; Chen, G.; Cheng, M.; Huang, Z.; Yu, Y.; Zeng, H. A Novel Blood Plasma Analysis Technique Combining Membrane Electrophoresis with Silver Nanoparticle-Based SERS Spectroscopy for Potential Applications in Noninvasive Cancer Detection. *Nanomed. Nanotechnol. Biol. Med.* **2011**, *7*, 655–663. [CrossRef]
52. Batool, F.; Nawaz, H.; Majeed, M.I.; Rashid, N.; Bashir, S.; Akbar, S.; Abubakar, M.; Ahmad, S.; Ashraf, M.N.; Ali, S.; et al. SERS-Based Viral Load Quantification of Hepatitis B Virus from PCR Products. *Spectrochim. Acta Part A Mol. Biomol. Spectrosc.* **2021**, *255*, 119722. [CrossRef]
53. Shen, L.; Du, Y.; Wei, N.; Li, Q.; Li, S.; Sun, T.; Xu, S.; Wang, H.; Man, X.; Han, B. SERS Studies on Normal Epithelial and Cancer Cells Derived from Clinical Breast Cancer Specimens. *Spectrochim. Acta Part A Mol. Biomol. Spectrosc.* **2020**, *237*, 118364. [CrossRef]
54. Moisoiu, V.; Badarinza, M.; Stefancu, A.; Iancu, S.D.; Serban, O.; Leopold, N.; Fodor, D. Combining Surface-Enhanced Raman Scattering (SERS) of Saliva and Two-Dimensional Shear Wave Elastography (2D-SWE) of the Parotid Glands in the Diagnosis of Sjögren's Syndrome. *Spectrochim. Acta Part A Mol. Biomol. Spectrosc.* **2020**, *235*, 118267. [CrossRef] [PubMed]

55. Narasimhan, V.; Siddique, R.H.; Park, H.; Choo, H. Bioinspired Disordered Flexible Metasurfaces for Human Tear Analysis Using Broadband Surface-Enhanced Raman Scattering. *ACS Omega* **2020**, *5*, 12915–12922. [CrossRef] [PubMed]
56. Su, X.; Xu, Y.; Zhao, H.; Li, S.; Chen, L. Design and Preparation of Centrifugal Microfluidic Chip Integrated with SERS Detection for Rapid Diagnostics. *Talanta* **2019**, *194*, 903–909. [CrossRef] [PubMed]
57. Frimpong, R.; Jang, W.; Kim, J.-H.; Driskell, J.D. Rapid Vertical Flow Immunoassay on AuNP Plasmonic Paper for SERS-Based Point of Need Diagnostics. *Talanta* **2021**, *223*, 121739. [CrossRef] [PubMed]
58. Willner, M.R.; McMillan, K.S.; Graham, D.; Vikesland, P.J.; Zagnoni, M. Surface-Enhanced Raman Scattering Based Microfluidics for Single-Cell Analysis. *Anal. Chem.* **2018**, *90*, 12004–12010. [CrossRef] [PubMed]
59. Ngo, H.T.; Gandra, N.; Fales, A.M.; Taylor, S.M.; Vo-Dinh, T. Sensitive DNA Detection and SNP Discrimination Using Ultrabright SERS Nanorattles and Magnetic Beads for Malaria Diagnostics. *Biosens. Bioelectron.* **2016**, *81*, 8–14. [CrossRef]
60. Lu, F.-K.; Basu, S.; Igras, V.; Hoang, M.P.; Ji, M.; Fu, D.; Holtom, G.R.; Neel, V.A.; Freudiger, C.W.; Fisher, D.E.; et al. Label-Free DNA Imaging in Vivo with Stimulated Raman Scattering Microscopy. *Proc. Natl. Acad. Sci. USA* **2015**, *112*, 11624–11629. [CrossRef]
61. Harmsen, S.; Rogalla, S.; Huang, R.; Spaliviero, M.; Neuschmelting, V.; Hayakawa, Y.; Lee, Y.; Tailor, Y.; Toledo-Crow, R.; Kang, J.W.; et al. Detection of Premalignant Gastrointestinal Lesions Using Surface-Enhanced Resonance Raman Scattering–Nanoparticle Endoscopy. *ACS Nano* **2019**, *13*, 1354–1364. [CrossRef]
62. Ryzhikova, E.; Kazakov, O.; Halamkova, L.; Celmins, D.; Malone, P.; Molho, E.; Zimmerman, E.A.; Lednev, I.K. Raman Spectroscopy of Blood Serum for Alzheimer's Disease Diagnostics: Specificity Relative to Other Types of Dementia. *J. Biophotonics* **2015**, *8*, 584–596. [CrossRef]
63. Kong, K.; Rowlands, C.J.; Varma, S.; Perkins, W.; Leach, I.H.; Koloydenko, A.A.; Williams, H.C.; Notingher, I. Diagnosis of Tumors during Tissue-Conserving Surgery with Integrated Autofluorescence and Raman Scattering Microscopy. *Proc. Natl. Acad. Sci. USA* **2013**, *110*, 15189–15194. [CrossRef]
64. Motz, J.T.; Hunter, M.; Galindo, L.H.; Gardecki, J.A.; Kramer, J.R.; Dasari, R.R.; Feld, M.S. Optical Fiber Probe for Biomedical Raman Spectroscopy. *Appl. Opt.* **2004**, *43*, 542–554. [CrossRef] [PubMed]
65. Jermyn, M.; Mercier, J.; Aubertin, K.; Desroches, J.; Urmey, K.; Karamchandiani, J.; Marple, E.; Guiot, M.-C.; Leblond, F.; Petrecca, K. Highly Accurate Detection of Cancer In Situ with Intraoperative, Label-Free, Multimodal Optical Spectroscopy. *Cancer Res.* **2017**, *77*, 3942–3950. [CrossRef] [PubMed]
66. Kircher, M.F.; de la Zerda, A.; Jokerst, J.V.; Zavaleta, C.L.; Kempen, P.J.; Mittra, E.; Pitter, K.; Huang, R.; Campos, C.; Habte, F.; et al. A Brain Tumor Molecular Imaging Strategy Using a New Triple-Modality MRI-Photoacoustic-Raman Nanoparticle. *Nat. Med.* **2012**, *18*, 829–834. [CrossRef] [PubMed]
67. Lukic, A.; Dochow, S.; Bae, H.; Matz, G.; Latka, I.; Messerschmidt, B.; Schmitt, M.; Popp, J. Endoscopic Fiber Probe for Nonlinear Spectroscopic Imaging. *Optica* **2017**, *4*, 496–501. [CrossRef]
68. Jaworska, A.; Fornasaro, S.; Sergo, V.; Bonifacio, A. Potential of Surface Enhanced Raman Spectroscopy (SERS) in Therapeutic Drug Monitoring (TDM). A Critical Review. *Biosensors* **2016**, *6*, 47. [CrossRef]
69. Neef, C.; Touw, D.J.; Stolk, L.M. Therapeutic Drug Monitoring in Clinical Research. *Pharm. Med.* **2008**, *22*, 235–244. [CrossRef]
70. Kang, J.-S.; Lee, M.-H. Overview of Therapeutic Drug Monitoring. *Korean J. Intern. Med.* **2009**, *24*, 1–10. [CrossRef]
71. Lennard, L. Therapeutic Drug Monitoring of Antimetabolic Cytotoxic Drugs. *Br. J. Clin. Pharmacol.* **1999**, *47*, 131–143. [CrossRef]
72. Zhang, S.; Chen, H.; Li, R.; Yu, Z.; Lu, F. Raman Spectroscopy and Mapping Technique for the Identification of Expired Drugs. *Spectrochim. Acta Part A Mol. Biomol. Spectrosc.* **2020**, *224*, 117407. [CrossRef]
73. Fei, J.; Wu, L.; Zhang, Y.; Zong, S.; Wang, Z.; Cui, Y. Pharmacokinetics-on-a-Chip Using Label-Free SERS Technique for Programmable Dual-Drug Analysis. *ACS Sens.* **2017**, *2*, 773–780. [CrossRef]
74. Chen, C.; Yang, L.; Li, H.; Chen, F.; Chen, C.; Gao, R.; Lv, X.; Tang, J. Raman Spectroscopy Combined with Multiple Algorithms for Analysis and Rapid Screening of Chronic Renal Failure. *Photodiagnosis Photodyn. Ther.* **2020**, *30*, 101792. [CrossRef] [PubMed]
75. Shibata, T.; Furukawa, H.; Ito, Y.; Nagahama, M.; Hayashi, T.; Ishii-Teshima, M.; Nagai, M. Photocatalytic Nanofabrication and Intracellular Raman Imaging of Living Cells with Functionalized AFM Probes. *Micromachines* **2020**, *11*, 495. [CrossRef] [PubMed]
76. Ohsaki, Y.; Cheng, J.; Fujita, A.; Tokumoto, T.; Fujimoto, T. Cytoplasmic Lipid Droplets Are Sites of Convergence of Proteasomal and Autophagic Degradation of Apolipoprotein, B. *Mol. Biol. Cell* **2006**, *17*, 2674–2683. [CrossRef] [PubMed]
77. Samuel, A.Z.; Miyaoka, R.; Ando, M.; Gaebler, A.; Thiele, C.; Takeyama, H. Molecular Profiling of Lipid Droplets inside HuH7 Cells with Raman Micro-Spectroscopy. *Commun. Biol.* **2020**, *3*, 1–10. [CrossRef]
78. Rodionova, N.N.; Allakhverdiev, E.S.; Maksimov, G.V. Study of Myelin Structure Changes during the Nerve Fibers Demyelination. *PLoS ONE* **2017**, *12*, e0185170. [CrossRef]
79. Akhalaya, M.Y.; Maksimov, G.V.; Rubin, A.B.; Lademann, J.; Darvin, M.E. Molecular Action Mechanisms of Solar Infrared Radiation and Heat on Human Skin. *Ageing Res. Rev.* **2014**, *16*, 1–11. [CrossRef]
80. Gellermann, W.; Ermakov, I.V.; Ermakova, M.R.; McClane, R.W.; Zhao, D.-Y.; Bernstein, P.S. In Vivo Resonant Raman Measurement of Macular Carotenoid Pigments in the Young and the Aging Human Retina. *JOSA A* **2002**, *19*, 1172–1186. [CrossRef]
81. Ermakov, I.V.; Ermakova, M.R.; Gellermann, W.; Lademann, J. Noninvasive Selective Detection of Lycopene and Beta-Carotene in Human Skin Using Raman Spectroscopy. *J. Biomed. Opt.* **2004**, *9*, 332–338. [CrossRef]
82. Valpapuram, I.; Candeloro, P.; Coluccio, M.L.; Parrotta, E.I.; Giugni, A.; Das, G.; Cuda, G.; Di Fabrizio, E.; Perozziello, G. Waveguiding and SERS Simplified Raman Spectroscopy on Biological Samples. *Biosensors* **2019**, *9*, 37. [CrossRef]

83. Dijkstra, R.J.; Scheenen, W.J.J.M.; Dam, N.; Roubos, E.W.; ter Meulen, J.J. Monitoring Neurotransmitter Release Using Surface-Enhanced Raman Spectroscopy. *J. Neurosci. Methods* **2007**, *159*, 43–50. [CrossRef]
84. Silwal, A.P.; Lu, H.P. Mode-Selective Raman Imaging of Dopamine–Human Dopamine Transporter Interaction in Live Cells. *ACS Chem. Neurosci.* **2018**, *9*, 3117–3127. [CrossRef] [PubMed]
85. Bahadar, H.; Maqbool, F.; Niaz, K.; Abdollahi, M. Toxicity of Nanoparticles and an Overview of Current Experimental Models. *Iran. Biomed. J.* **2016**, *20*, 1–11. [CrossRef] [PubMed]
86. Manciu, F.S.; Lee, K.H.; Durrer, W.G.; Bennet, K.E. Detection and Monitoring of Neurotransmitters—A Spectroscopic Analysis. *Neuromodulation Technol. Neural Interface* **2013**, *16*, 192–199. [CrossRef] [PubMed]
87. Marchetti, M.; Baria, E.; Cicchi, R.; Pavone, F.S. Custom Multiphoton/Raman Microscopy Setup for Imaging and Characterization of Biological Samples. *Methods Protoc.* **2019**, *2*, 51. [CrossRef] [PubMed]
88. Marin, E.; Hiraishi, N.; Honma, T.; Boschetto, F.; Zanocco, M.; Zhu, W.; Adachi, T.; Kanamura, N.; Yamamoto, T.; Pezzotti, G. Raman Spectroscopy for Early Detection and Monitoring of Dentin Demineralization. *Dent. Mater.* **2020**, *36*, 1635–1644. [CrossRef]
89. Arabi, M.; Ostovan, A.; Zhang, Z.; Wang, Y.; Mei, R.; Fu, L.; Wang, X.; Ma, J.; Chen, L. Label-Free SERS Detection of Raman-Inactive Protein Biomarkers by Raman Reporter Indicator: Toward Ultrasensitivity and Universality. *Biosens. Bioelectron.* **2021**, *174*, 112825. [CrossRef]
90. Arend, N.; Pittner, A.; Ramoji, A.; Mondol, A.S.; Dahms, M.; Rüger, J.; Kurzai, O.; Schie, I.W.; Bauer, M.; Popp, J.; et al. Detection and Differentiation of Bacterial and Fungal Infection of Neutrophils from Peripheral Blood Using Raman Spectroscopy. *Anal. Chem.* **2020**, *92*, 10560–10568. [CrossRef]
91. Collard, L.; Sinjab, F.; Notingher, I. Raman Spectroscopy Study of Curvature-Mediated Lipid Packing and Sorting in Single Lipid Vesicles. *Biophys. J.* **2019**, *117*, 1589–1598. [CrossRef]
92. Luneva, O.G.; Brazhe, N.A.; Maksimova, N.V.; Rodnenkov, O.V.; Parshina, E.Y.; Bryzgalova, N.Y.; Maksimov, G.V.; Rubin, A.B.; Orlov, S.N.; Chazov, E.I. Ion Transport, Membrane Fluidity and Haemoglobin Conformation in Erythrocyte from Patients with Cardiovascular Diseases: Role of Augmented Plasma Cholesterol. *Pathophysiology* **2007**, *14*, 41–46. [CrossRef]
93. Maksimov, G.V.; Maksimova, N.V.; Churin, A.A.; Orlov, S.N.; Rubin, A.B. Study on Conformational Changes in Hemoglobin Protoporphyrin in Essential Hypertension. *Biochem. Mosc.* **2001**, *66*, 295–299. [CrossRef]
94. Maksimov, G.V.; Luneva, O.G.; Maksimova, N.V.; Matettuchi, E.; Medvedev, E.A.; Pashchenko, V.Z.; Rubin, A.B. Role of Viscosity and Permeability of the Erythrocyte Plasma Membrane in Changes in Oxygen-Binding Properties of Hemoglobin during Diabetes Mellitus. *Bull. Exp. Biol. Med.* **2005**, *140*, 510–513. [CrossRef] [PubMed]
95. Brazhe, N.A.; Baizhumanov, A.A.; Parshina, E.Y.; Yusipovich, A.I.; Akhalaya, M.Y.; Yarlykova, Y.V.; Labetskaya, O.I.; Ivanova, S.M.; Morukov, B.V.; Maksimov, G.V. Studies of the Blood Antioxidant System and Oxygen-Transporting Properties of Human Erythrocytes during 105-Day Isolation. *Hum. Physiol.* **2014**, *40*, 804–809. [CrossRef]
96. Allakhverdiev, E.S.; Slatinskaya, O.V.; Rodnenkov, O.V.; Maksimov, G.V.; Martynyuk, T.V. Evaluation of Hemoglobin Conformation State in Patients with Pulmonary Hypertension. *Russ. J. Cardiol.* **2021**, *25*, 23.
97. Skjånes, K.; Rebours, C.; Lindblad, P. Potential for Green Microalgae to Produce Hydrogen, Pharmaceuticals and Other High Value Products in a Combined Process. *Crit. Rev. Biotechnol.* **2013**, *33*, 172–215. [CrossRef] [PubMed]
98. Schenk, P.M.; Thomas-Hall, S.R.; Stephens, E.; Marx, U.C.; Mussgnug, J.H.; Posten, C.; Kruse, O.; Hankamer, B. Second Generation Biofuels: High-Efficiency Microalgae for Biodiesel Production. *BioEnergy Res.* **2008**, *1*, 20–43. [CrossRef]
99. Sadvakasova, A.K.; Kossalbayev, B.D.; Zayadan, B.K.; Kirbayeva, D.K.; Alwasel, S.; Allakhverdiev, S.I. Potential of Cyanobacteria in the Conversion of Wastewater to Biofuels. *World J. Microbiol. Biotechnol.* **2021**, *37*, 140. [CrossRef]
100. Bolatkhan, K.; Sadvakasova, A.K.; Zayadan, B.K.; Kakimova, A.B.; Sarsekeyeva, F.K.; Kossalbayev, B.D.; Bozieva, A.M.; Alwasel, S.; Allakhverdiev, S.I. Prospects for the Creation of a Waste-Free Technology for Wastewater Treatment and Utilization of Carbon Dioxide Based on Cyanobacteria for Biodiesel Production. *J. Biotechnol.* **2020**, *324*, 162–170. [CrossRef]
101. Zayadan, B.; Usserbayeva, A.; Bolatkhan, K.; Akmukhanova, N.; Kossalbayev, B.; Baizhigitova, A.; Los, D. Screening of Isolated and Collection Strains of Cyanobacteria on Productivity for Determining Their Biotechnological Potential. *Eur. J. Entomol.* **2018**, *55*. [CrossRef]
102. Breuer, G.; de Jaeger, L.; Artus, V.G.; Martens, D.E.; Springer, J.; Draaisma, R.B.; Eggink, G.; Wijffels, R.H.; Lamers, P.P. Superior Triacylglycerol (TAG) Accumulation in Starchless Mutants of Scenedesmus Obliquus: (II) Evaluation of TAG Yield and Productivity in Controlled Photobioreactors. *Biotechnol. Biofuels* **2014**, *7*, 70. [CrossRef]
103. Kaczor, A.; Turnau, K.; Baranska, M. In Situ Raman Imaging of Astaxanthin in a Single Microalgal Cell. *Analyst* **2011**, *136*, 1109–1112. [CrossRef]
104. Hosokawa, M.; Ando, M.; Mukai, S.; Osada, K.; Yoshino, T.; Hamaguchi, H.; Tanaka, T. In Vivo Live Cell Imaging for the Quantitative Monitoring of Lipids by Using Raman Microspectroscopy. *Anal. Chem.* **2014**, *86*, 8224–8230. [CrossRef] [PubMed]
105. Pořízka, P.; Prochazková, P.; Prochazka, D.; Sládková, L.; Novotný, J.; Petrilak, M.; Brada, M.; Samek, O.; Pilát, Z.; Zemánek, P.; et al. Algal Biomass Analysis by Laser-Based Analytical Techniques—A Review. *Sensors* **2014**, *14*, 17725–17752. [CrossRef] [PubMed]
106. Wei, X.; Jie, D.; Cuello, J.J.; Johnson, D.J.; Qiu, Z.; He, Y. Microalgal Detection by Raman Microspectroscopy. *TrAC Trends Anal. Chem.* **2014**, *53*, 33–40. [CrossRef]

107. Gouda, M.; Chen, K.; Li, X.; Liu, Y.; He, Y. Detection of Microalgae Single-Cell Antioxidant and Electrochemical Potentials by Gold Microelectrode and Raman Micro-Spectroscopy Combined with Chemometrics. *Sens. Actuators B Chem.* **2021**, *329*, 129229. [CrossRef]
108. Brahma, S.K.; Hargraves, P.E.; Howard, W.F.; Nelson, W.H. A Resonance Raman Method for the Rapid Detection and Identification of Algae in Water. *Appl. Spectrosc.* **1983**, *37*, 55–58. [CrossRef]
109. Vítek, P.; Jehlička, J.; Edwards, H.G.M.; Hutchinson, I.; Ascaso, C.; Wierzchos, J. The Miniaturized Raman System and Detection of Traces of Life in Halite from the Atacama Desert: Some Considerations for the Search for Life Signatures on Mars. *Astrobiology* **2012**, *12*, 1095–1099. [CrossRef] [PubMed]
110. Jehlička, J.; Edwards, H.G.M.; Oren, A. Raman Spectroscopy of Microbial Pigments. *Appl. Environ. Microbiol.* **2014**, *80*, 3286–3295. [CrossRef]
111. Weiss, T.L.; Chun, H.J.; Okada, S.; Vitha, S.; Holzenburg, A.; Laane, J.; Devarenne, T.P. Raman Spectroscopy Analysis of Botryococcene Hydrocarbons from the Green Microalga Botryococcus Braunii*. *J. Biol. Chem.* **2010**, *285*, 32458–32466. [CrossRef]
112. Ramya, A.N.; Ambily, P.S.; Sujitha, B.S.; Arumugam, M.; Maiti, K.K. Single Cell Lipid Profiling of Scenedesmus Quadricauda CASA-CC202 under Nitrogen Starved Condition by Surface Enhanced Raman Scattering (SERS) Fingerprinting. *Algal Res.* **2017**, *25*, 200–206. [CrossRef]
113. Byler, D.M.; Susi, H. Examination of the Secondary Structure of Proteins by Deconvolved FTIR Spectra. *Biopolymers* **1986**, *25*, 469–487. [CrossRef]
114. Barka, A.; Amira, A.B.; Francis, F.; Blecker, C. Physicochemical Characterization of Colored Soluble Protein Fractions Extracted from Spirulina (Spirulina Platensis). *Food Sci. Technol. Int.* **2018**, *24*, 651–663. [CrossRef] [PubMed]
115. Venkatesan, S.; Pugazhendy, K.; Sangeetha, D.; Vasantharaja, C.; Prabakaran, S.; Meenambal, M. Fourier Transform Infrared (FT-IR) Spectoroscopic Analysis of Spirulina. *Int. J. Pharm. Biol. Arch.* **2012**, *3*, 969–972.
116. Moudříková, Š.; Sadowsky, A.; Metzger, S.; Nedbal, L.; Mettler-Altmann, T.; Mojzeš, P. Quantification of Polyphosphate in Microalgae by Raman Microscopy and by a Reference Enzymatic Assay. *Anal. Chem.* **2017**, *89*, 12006–12013. [CrossRef]
117. Jehlička, J.; Culka, A.; Nedbalová, L. Colonization of Snow by Microorganisms as Revealed Using Miniature Raman Spectrometers—Possibilities for Detecting Carotenoids of Psychrophiles on Mars? *Astrobiology* **2016**, *16*, 913–924. [CrossRef] [PubMed]
118. Javee, A.; Sulochana, S.B.; Pallissery, S.J.; Arumugam, M. Major Lipid Body Protein: A Conserved Structural Component of Lipid Body Accumulated during Abiotic Stress in *S. quadricauda* CASA-CC202. *Front. Energy Res.* **2016**, *4*. [CrossRef]
119. Zhang, C.; Liu, P. The New Face of the Lipid Droplet: Lipid Droplet Proteins. *Proteomics* **2019**, *19*, 1700223. [CrossRef] [PubMed]
120. Liu, S.; Kannegulla, A.; Kong, X.; Sun, R.; Liu, Y.; Wang, R.; Yu, Q.; Wang, A.X. Simultaneous Colorimetric and Surface-Enhanced Raman Scattering Detection of Melamine from Milk. *Spectrochim. Acta Part A Mol. Biomol. Spectrosc.* **2020**, *231*, 118130. [CrossRef] [PubMed]
121. Solovchenko, A.; Neverov, K. Carotenogenic Response in Photosynthetic Organisms: A Colorful Story. *Photosynth. Res.* **2017**, *133*, 31–47. [CrossRef]
122. Chekanov, K.; Lukyanov, A.; Boussiba, S.; Aflalo, C.; Solovchenko, A. Modulation of Photosynthetic Activity and Photoprotection in Haematococcus Pluvialis Cells during Their Conversion into Haematocysts and Back. *Photosynth. Res.* **2016**, *128*, 313–323. [CrossRef]
123. Maksimov, E.G.; Mironov, K.S.; Trofimova, M.S.; Nechaeva, N.L.; Todorenko, D.A.; Klementiev, K.E.; Tsoraev, G.V.; Tyutyaev, E.V.; Zorina, A.A.; Feduraev, P.V.; et al. Membrane Fluidity Controls Redox-Regulated Cold Stress Responses in Cyanobacteria. *Photosynth. Res.* **2017**, *133*, 215–223. [CrossRef]
124. Los, D.A.; Murata, N. Structure and Expression of Fatty Acid Desaturases. *Biochim. Biophys. Acta BBA- Lipids Lipid Metab.* **1998**, *1394*, 3–15. [CrossRef]
125. Kubo, Y.; Ikeda, T.; Yang, S.-Y.; Tsuboi, M. Orientation of Carotenoid Molecules in the Eyespot of Alga: In Situ Polarized Resonance Raman Spectroscopy. *Appl. Spectrosc.* **2000**, *54*, 1114–1119. [CrossRef]
126. Heraud, P.; Wood, B.R.; Beardall, J.; McNaughton, D. Effects of Pre-Processing of Raman Spectra on in Vivo Classification of Nutrient Status of Microalgal Cells. *J. Chemom.* **2006**, *20*, 193–197. [CrossRef]
127. Heraud, P.; Beardall, J.; McNaughton, D.; Wood, B.R. In Vivo Prediction of the Nutrient Status of Individual Microalgal Cells Using Raman Microspectroscopy. *FEMS Microbiol. Lett.* **2007**, *275*, 24–30. [CrossRef]
128. Wood, B.R.; Heraud, P.; Stojkovic, S.; Morrison, D.; Beardall, J.; McNaughton, D. A Portable Raman Acoustic Levitation Spectroscopic System for the Identification and Environmental Monitoring of Algal Cells. *Anal. Chem.* **2005**, *77*, 4955–4961. [CrossRef] [PubMed]
129. Chen, M.; Zeng, H.; Larkum, A.W.D.; Cai, Z.-L. Raman Properties of Chlorophyll d, the Major Pigment of Acaryochloris Marina: Studies Using Both Raman Spectroscopy and Density Functional Theory. *Spectrochim. Acta. A. Mol. Biomol. Spectrosc.* **2004**, *60*, 527–534. [CrossRef]
130. Cannizzaro, C.; Rhiel, M.; Marison, I.; von Stockar, U. On-Line Monitoring of Phaffia Rhodozyma Fed-Batch Process with in Situ Dispersive Raman Spectroscopy. *Biotechnol. Bioeng.* **2003**, *83*, 668–680. [CrossRef]
131. Marshall, C.P.; Leuko, S.; Coyle, C.M.; Walter, M.R.; Burns, B.P.; Neilan, B.A. Carotenoid Analysis of Halophilic Archaea by Resonance Raman Spectroscopy. *Astrobiology* **2007**, *7*, 631–643. [CrossRef]

132. Parker, S.F.; Tavender, S.M.; Dixon, N.M.; Herman, H.; Williams, K.P.J.; Maddams, W.F. Raman Spectrum of β-Carotene Using Laser Lines from Green (514.5 Nm) to Near-Infrared (1064 Nm): Implications for the Characterization of Conjugated Polyenes. *Appl. Spectrosc.* **1999**, *53*, 86–91. [CrossRef]
133. Mostaert, A.S.; Giordani, C.; Crockett, R.; Karsten, U.; Schumann, R.; Jarvis, S.P. Characterisation of Amyloid Nanostructures in the Natural Adhesive of Unicellular Subaerial Algae. *J. Adhes.* **2009**, *85*, 465–483. [CrossRef]
134. Shutova, V.V.; Tyutyaev, E.V.; Churin, A.A.; Ponomarev, V.Y.; Belyakova, G.A.; Maksimov, G.V. IR and Raman Spectroscopy in the Study of Carotenoids of Cladophora Rivularis Algae. *Biophysics* **2016**, *61*, 601–605. [CrossRef]
135. Eggersdorfer, M.; Wyss, A. Carotenoids in Human Nutrition and Health. *Arch. Biochem. Biophys.* **2018**, *652*, 18–26. [CrossRef] [PubMed]
136. Chekanov, K.; Fedorenko, T.; Kublanovskaya, A.; Litvinov, D.; Lobakova, E. Diversity of Carotenogenic Microalgae in the White Sea Polar Region. *FEMS Microbiol. Ecol.* **2020**, *96*, fiz183. [CrossRef] [PubMed]
137. Collins, A.M.; Jones, H.D.T.; Han, D.; Hu, Q.; Beechem, T.E.; Timlin, J.A. Carotenoid Distribution in Living Cells of Haematococcus Pluvialis (Chlorophyceae). *PLoS ONE* **2011**, *6*, e24302. [CrossRef]
138. Chiu, L.; Ho, S.-H.; Shimada, R.; Ren, N.-Q.; Ozawa, T. Rapid in Vivo Lipid/Carbohydrate Quantification of Single Microalgal Cell by Raman Spectral Imaging to Reveal Salinity-Induced Starch-to-Lipid Shift. *Biotechnol. Biofuels* **2017**, *10*, 9. [CrossRef]
139. Mishra, K.B.; Vítek, P.; Mishra, A.; Hájek, J.; Barták, M. Chlorophyll a Fluorescence and Raman Spectroscopy Can Monitor Activation/Deactivation of Photosynthesis and Carotenoids in Antarctic Lichens. *Spectrochim. Acta Part A Mol. Biomol. Spectrosc.* **2020**, *239*, 118458. [CrossRef]
140. Mirsafavi, R.; Moskovits, M.; Meinhart, C. Detection and Classification of Fentanyl and Its Precursors by Surface-Enhanced Raman Spectroscopy. *Analyst* **2020**, *145*, 3440–3446. [CrossRef]

cells

MDPI

Review

Sustainable Agriculture through Multidisciplinary Seed Nanopriming: Prospects of Opportunities and Challenges

Amruta Shelar [1], Ajay Vikram Singh [2,*], Romi Singh Maharjan [2], Peter Laux [2], Andreas Luch [2], Donato Gemmati [3], Veronica Tisato [3], Shubham Pratap Singh [4], Maria Fernanda Santilli [5], Akanksha Shelar [6], Manohar Chaskar [7] and Rajendra Patil [8,*]

1 Department of Technology, Savitribai Phule Pune University, Pune 411007, India; amrutavijaykumarshelar@gmail.com
2 Department of Chemical and Product Safety, German Federal Institute for Risk Assessment (BfR), Max-Dohrn-Strasse 8-10, 10589 Berlin, Germany; Romi-Singh.Maharjan@bfr.bund.de (R.S.M.); Peter.Laux@bfr.bund.de (P.L.); Andreas.Luch@bfr.bund.de (A.L.)
3 Department of Translational Medicine, University of Ferrara, 44121 Ferrara, Italy; cet@unife.it (D.G.); veronica.tisato@unife.it (V.T.)
4 Faculty of Informatics, Otto von Guericke University, 39106 Magdeburg, Germany; shubhamp.singh@yahoo.com
5 The Nanoinformatics Innovation Centre, La Plata, Buenos Aires C1062, Argentina; mfsantilli@gmail.com
6 Department of Microbiology, Savitribai Phule Pune University, Pune 411007, India; akankshavijaykumar2@gmail.com
7 Ramkrishna More Arts, Commerce and Science College, Pune 411044, India; dean.st@unipune.ac.in
8 Department of Biotechnology, Savitribai Phule Pune University, Pune 411007, India
* Correspondence: Ajay-Vikram.Singh@bfr.bund.de (A.V.S.); rpatil@unipune.ac.in (R.P.)

check for updates

Citation: Shelar, A.; Singh, A.V.; Maharjan, R.S.; Laux, P.; Luch, A.; Gemmati, D.; Tisato, V.; Singh, S.P.; Santilli, M.F.; Shelar, A.; et al. Sustainable Agriculture through Multidisciplinary Seed Nanopriming: Prospects of Opportunities and Challenges. *Cells* **2021**, *10*, 2428. https://doi.org/10.3390/cells10092428

Academic Editors: Suleyman Allakhverdiev, Alexander G. Ivanov and Marian Brestic

Received: 1 August 2021
Accepted: 12 September 2021
Published: 15 September 2021

Abstract: The global community decided in 2015 to improve people's lives by 2030 by setting 17 global goals for sustainable development. The second goal of this community was to end hunger. Plant seeds are an essential input in agriculture; however, during their developmental stages, seeds can be negatively affected by environmental stresses, which can adversely affect seed vigor, seedling establishment, and crop production. Seeds resistant to high salinity, droughts and climate change can result in higher crop yield. The major findings suggested in this review refer nanopriming as an emerging seed technology towards sustainable food amid growing demand with the increasing world population. This novel growing technology could influence the crop yield and ensure the quality and safety of seeds, in a sustainable way. When nanoprimed seeds are germinated, they undergo a series of synergistic events as a result of enhanced metabolism: modulating biochemical signaling pathways, trigger hormone secretion, reduce reactive oxygen species leading to improved disease resistance. In addition to providing an overview of the challenges and limitations of seed nanopriming technology, this review also describes some of the emerging nano-seed priming methods for sustainable agriculture, and other technological developments using cold plasma technology and machine learning.

Keywords: seed priming; nanotechnology; germination; seed resistance; sustainability; cold plasma technology; machine learning

1. Introduction

World food production must increase 50% by 2050 to meet the needs of 9 billion people [1]. The growing food demand and rapidly changing climatic conditions across the world motivates us to look for technological solutions that can provide food security for the future generations. Seed is a primary requirement for crop production, which carries the genetic potential of the variations and determines the ultimate productivity. Therefore, seed production is always the basic pre-requisite of any food security undertaking. A resilient and climate hardened seed can give the maximum output to the farmers. To increase the seed vigor and crop production, different chemical-based fertilizers and pesticides are used extensively in agriculture. In light of leaching, degradation, hydrolysis, and pollution associated with conventional chemical-based practices, they are being discouraged [2].There is an urgent need to develop a sustainable technology that can contribute to the green revolution to address these growing concerns and to restore the damage caused to the ecosystem [3]. Seed priming is an innovative sustainable seed technology to increase the seed vigor and crop production without harming the ecosystem [4–7]. As mentioned in Figure 1, priming of seeds dates back to Theophrastus, (371–287 B.C.) who, observed during an investigation that soaking cucumber seeds in water causes their germination to be faster and more uniform than unprimed seeds (enquiry into plants, book VII, I.6). A similar preparation of cucumber seeds in honey and water for seed germination was reported by the Roman naturalist Gaius Plinius Secundus (23–79 A.D.) in Gaius' Encyclopedia (Gaius 1949–1954). The French botanist Oliver de Serres, in 1539–1619, found that soaking seeds in manure water for 2 days and drying them prior to sowing was an effective cure for poor crop growth. The osmo-priming process was tested on lettuce and cress seeds in seawater by Charles Darwin, who observed that the primed seeds germination was higher than that of non-primed seeds. By identifying the critical parameters of seed treatment, Ells (1963) presented the modern concept of seed priming. His experiments with nutrient solutions showed a high germination rate. Khodakovskaya et al. published one of the first studies to demonstrate the potential for nanomaterials to affect seed germination [8]. The latest innovative technique with potential application in seed priming is seed nanopriming, an important emerging seed technology that blends seed priming science with updated nanotechnology [9,10]. A lot of attention has recently been brought to the development and optimization of nanomaterials for application in agriculture, including improved growth, plant protection, and overall performance. Agricultural nanomaterials are still in a juvenile state in terms of their application to sustainable agriculture [11]. Nanomaterials will be applied to multiple functions in agriculture as the understanding of nanotechnology increases. It has been noticed that most agricultural lands are affected by abiotic stress factors such as salinity and drought, which is limiting plant distribution in the habitat [12]. In order to combat abiotic stress factors, engineered nanoparticles have been found to be a promising alternative [13]. Recent studies have shown that nanomaterials can significantly impact plant metabolism, genes expression, and antioxidant enzyme activity [14]. Abiotic stress can be improved by nanoparticles such as halloysite, cerium oxide, chitosan-selenium and titanium dioxide by enabling their antioxidant system to perform better [15–18]. The use of nanoparticles as seed priming agents has demonstrated encouraging results in the field of crop productivity and seed germination [19]. By using nanotechnology, it is possible to release priming agents at specific sites and in controlled ways, which is revolutionary for ecosystems, crop improvement, animal health, and pesticide use. Nanopriming is an ongoing effort to create nano-agrochemicals for releasing specific nutrients in a controlled manner, which maintains soil fertility. Using high quality seed detection technologies makes it more likely to identify a variety of better crops since seed quality is an important factor in crop production. An increasing number of effective methods are needed that can detect nanoprimed seed quality in a non-destructive, objective manner as quickly as possible. The presented review describes the potential application of engineered nanomaterials in seed nanopriming for sustainable agriculture and machine learning technology for detection and classification of nanoprimed seeds to improve crop production.

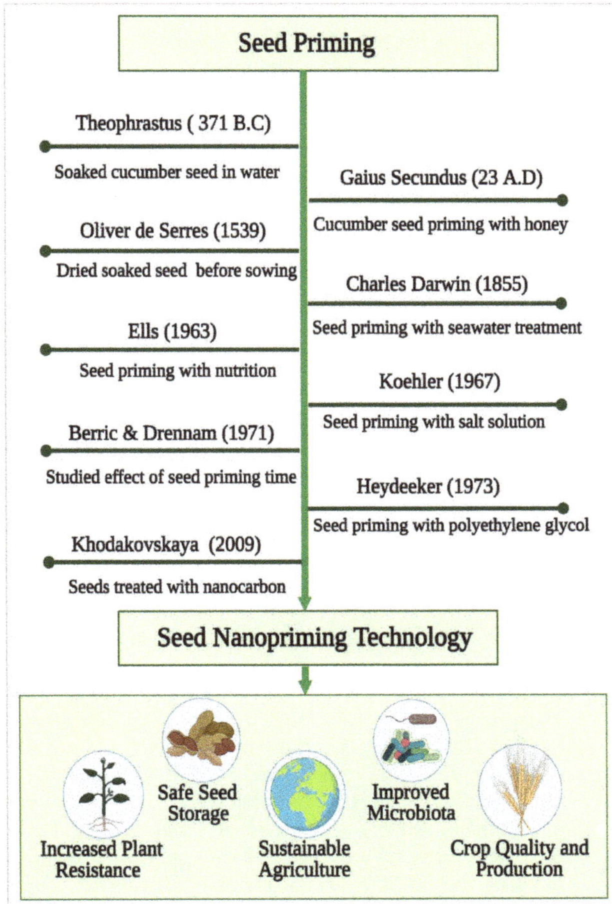

Figure 1. A timeline of seed priming development towards nanopriming technology to improve seed germination. For sustainable food production, nanopriming improves the ability to withstand environmental stressors, improves seed storage, and boosts microbiota.

2. Seed Nanopriming Technology

Seeds are miniature plants which, when fertilized, hatch into ovules containing both a germinating embryo and enough food to grow. A variety of treatments can be applied to a seed before planting in order to enhance its quality and potential yield. The process of priming seeds, in its broadest sense, involves soaking them in a solution where enough water is absorbed so that germination of the seed becomes possible, but not enough to allow the radicle to protrude through the seed coat [20]. Using this technique, seeds are advanced to an equal stage of germination to facilitate emergence from seed quickly and uniformly [21]. Thus, priming of seeds is a very important practice in order to ensure seed productivity [22]. Seed priming methods include hydro-priming, halo-priming, osmo-priming, hormonal priming, solid matrix priming, and bio-priming to stimulate seeds, encourage germination, and reduce environmental stress. Among the most well-known and cost-effective pre-sowing seed priming methods, hydro-priming involves treating seeds with water before sowing. The seeds are soaked in water, then re-dried to their original moisture content [23]. The process of halo priming involves treating seeds with inorganic solutions such as sodium chloride, potassium nitrate, calcium chloride, and calcium sulphate to improve germination. A known application of halopriming during the germination, seedling emergence, emergence of plants, and growth of plants is well known [24]. In osmo-priming, seeds are soaked in different concentrations of osmotic solutions. Different osmotic solutions are used depending on the species, including sugar, polyethylene glycol, glycerol, sorbitol, and mannitol and then air drying is followed by sowing. Solutions containing low water potential facilitate seed imbibitions and, therefore, early stages of germination occur [25]. Hormonal-priming involves the use of hormone solutions to prime seeds. During hormonal priming, plant growth regulators are used to imbibe seeds that have direct effects on seed metabolism. Hormo-priming uses various regulators such as abscisic acid, salicylic acid, ascorbic acid, cytokinins, auxins, gibberellins, kinetin, ethylene, and polyamines [26]. In solid matrix priming (SMP or Matri-conditioning), seeds are mixed in known proportions with a solid material and water. Water-absorbing seeds will reach a point of equilibrium, precisely at which priming will take place. A thorough washing and drying follow the separated seeds from the matrix. A hydrated, metabolically active seed can then be achieved, which is an important germination step [27]. As an ecological approach, bio-priming involves seed inoculation with beneficial organisms to protect seed from disease. Seed treatment with this new trend involves hydrating seeds with beneficial microorganisms and improving germination procedures. In terms of disease management, biopriming provides better results than pelleting or coating [28]. Nanopriming, a technique that involves seeds soaking in nanomaterials, has been shown to facilitate germination and growth by allowing nanoparticles to penetrate the seed coat and increase water uptake. When compared with unprimed seeds or seeds treated with other priming agents, nanopriming improves storage, quality, seedling emergence, yields, and tolerance to environmental stress [7]. In addition, nanopriming of seeds can help prevent diseases caused by pathogens present in seeds. A high percentage of nanomaterials are retained on the seed surface as coatings when nanoparticles are absorbed, and a tortuous pathway prevents uncontrolled water uptake while reducing gas permeability, which leads to more stable seed storage. A nanopriming process can alter the metabolism of seeds and signalling pathways, thereby influencing their germination and growth. In addition to affecting almost every existing scientific field, nanoparticles can gain significant impact on agricultural sustainability by being introduced to the agricultural domain [7]. Studies have demonstrated that nanoparticle application can stimulate germination and growth of plants in numerous ways. Nanoparticles are effective because of their small size and unique physio-chemical properties, which make them an ideal seed priming agent [21]. Nanoparticles are molecular or atomic aggregates that have a measured dimension between 1 nm and 100 nm and which may produce significantly higher chemical and physical properties than typical bulk materials [29]. It is important to note that nanomaterials have a wide range of physio-chemical properties depending on the shape, size, surface area, surface/volume ratio, chemical behavior, particle charge, production method, coating, and so forth (as shown in Figure 2A). The unique features of nanomaterials, such as their high surface to mass ratio, enables them to enhance catalysis and deliver materials of interest, as well as adsorb substances of interest. The nanopriming process triggers a special metabolic reaction that is naturally triggered during the early stages of germination, as shown in Figure 2B. It increases seed germination, by modulating the metabolism of seed, which leads to enhanced water uptake, starch hydrolysis rate, cell wall loosening, endosperm weakening, rapid embryo growth, and rapid root-shoot development. The nanopriming method improves the emergence of seedlings, their growth, production, and quality [30]. The cellular and molecular mechanisms of plant interaction with the environment are also modulated by seed nanopriming. The goal of using nanoparticles in agriculture and natural ecosystems is to increase the performance and sustainability of plants and soil by using less of the input parameters defined above [31,32]. It is important to include different factors that affect priming success, such as the amount of water in the soil, the kind of priming treatment, the amount of time the seeds are exposed to the treatment, and the conditions in which the seeds are stored. Numerous studies have demonstrated the ability of nanoparticles to penetrate seed coats and enhance water uptake over time, which facilitates germination and flowering [33,34]. As seed pretreatment agents, several metal-metal oxide nanoparticles and carbon-based nanoparticles have been applied to enhance germination and seedling growth of some crops and to strengthen their stress tolerance [35–37]. Seed priming, in which seeds are re-dried to their original moisture content before planting, has been used in only a few studies, but it is not widely used. In this case, seed nanopriming would provide a different mechanism to that of pre-sowing seed treatment without drying. Additionally, comprehensive studies have not been conducted on the physiological and molecular mechanisms of nanopriming on seed germination, thus there are many unanswered questions, particularly concerning the mechanism behind nanoparticles-induced seed germination. A summary of the types of nanoparticles that are used for seed priming is shown in Table 1, with their potential effects as stimulants or protective against biotic and abiotic stress.

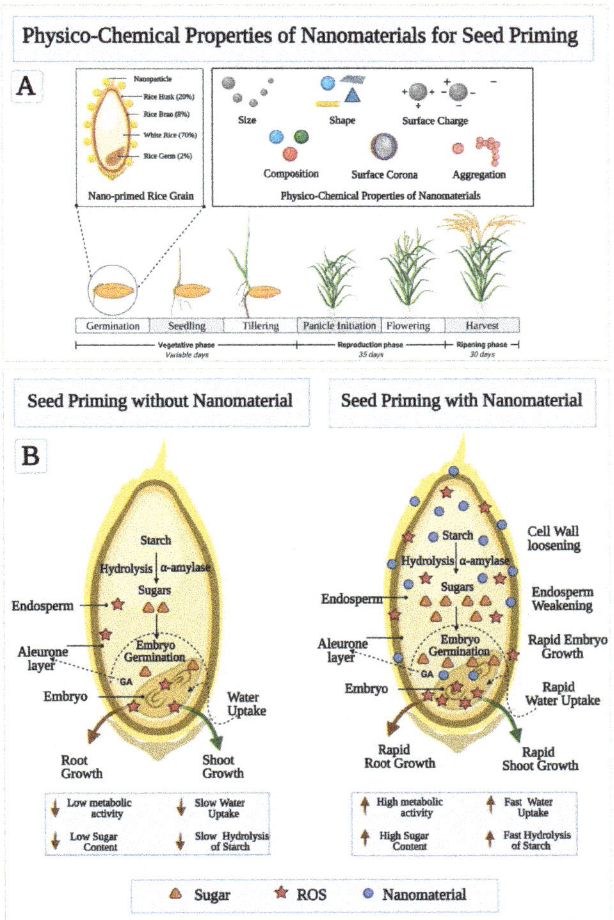

Figure 2. Physicochemical properties of nanomaterials influencing seed priming and its proposed mechanism. (**A**) The nanomaterial's size, shape, and surface chemistry make them a versatile candidate to modulate the seed priming phenomena via size-mediated diffusions and relocating to specific regions of the seed to activate synergistic mechanisms resulting in germination. (**B**) The versatile catalytic function of NMs selectively play role in promoting enzymatic and biochemical pathways to promote root/shoot growth.

Table 1. An overview of nanoparticles used for seed priming, their physico-chemical properties, and the main effects on some species of seeds when evaluated against biotic and abiotic stress [7].

Type of Nanomaterials	Size of Nanomaterials	Concentrations of Nanomterials	Seed Species	Results	References
Chitosan nanoparticles containing zinc	387.7 ± 4 nm	0.01, 0.04, 0.08, 0.12, and 0.16% *w/v*	Maize seeds (*Zea mays* L.)	Improved seed and seedling vigor and biotic resistance	[38]
Chitosan nanoparticles containing copper	374.3 ± 8.2 nm	0.01, 0.04, 0.08, 0.12, and 0.16% *w/v*	Maize seeds (*Zea mays* L.)	Improved seed and seedling vigor	[39]
Chitosan loaded with gibberellic acid	450 ± 10 nm	0.05, 0.005, and 0.0005 mg/mL	Tomato (*Solanum lycopersicum* var. *cerasiforme*)	Improved seed vigor and plant morphology with increased biomass	[40]
Lignin nanoparticles loaded with gibberellic acid	200–250 nm	0.5, 1, and 1.5 mg/mL	Arugula (*Eruca visicaria* (L.) Cav. subsp. *sativa*), tomato (*Solanum lycopersicum* L. cv. Ciliegino), and chickpea (*Cicer arietinum* L.)	Improved seed and seedlings vigor	[41]
Cobalt and molybdenum oxide nanoparticles	60–80 nm	1 L/40 kg of seeds	Soybean seeds (*Glycine max* (L). Merr.)	Improved seed vigor and plant morphology with increased biomass	[42]
Multi-walled carbon nanotubes	13–14 nm	70, 80, and 90 µg/mL	Wheat (*Triticum aestivum* L.)	Improved seed vigor and plant morphology	[43]
Silver nanoparticles	141.3 ± 0.78 nm	31.3 µg/mL	Watermelon (*Citrullus lanatus* (Thunb.) Matsum. and Nakai)	Improved seed vigor and plant morphology	[44]
Iron nanoparticles	~80 nm	25, 50, 100, 200, 300, 400, 500, and 1000 µg/mL	Wheat (*Triticum aestivum* L.)	Improved seed vigor and plant morphology	[45]
Zinc nanoparticles	21.3 nm	20, 40, and 60 mg/L.	Lupin (*Lupinis termis* L.).	Increased salinity resistance and biochemical activity	[46]
Zinc nanoparticles	20 nm	1, 10, 100, 1000, and 5000 mg/L	Common bean (*Phaseolus vulgaris* L.)	Increased biomass	[47]
Copper nanoparticles	25, 40, and 80 nm	1, 10, 100, and 1000 mg/L	Common bean (*Phaseolus vulgaris* L.)	Increased seed vigor and biomass	[48]
Iron (II) sulfide aqua nanoparticles	6–20 nm	30 µg/mL	Rice (*Oryza sativa* L.)	Improved seed vigor and disease resistance	[49]
Manganese (III) oxide nanoparticles	50 nm	0.1, 0.5, and 1 mg/mL	Jalapeño (*Capsicum annuum* L.)	Increased salinity resistance and antioxidant enzymes	[50]
Chitosan/tripoly phosphate nanoparticles	259.4 ± 4.7 nm	1–100 µg/mL	Wheat (*Triticum aestivum* L.)	Improved plant morphologyand upregulation of plant growth regulator	[51]
Silicon nanoparticles	90 nm	300, 600, 900, and 1200 mg/L	Wheat (*Triticum aestivum* L.)	Increased biomass and biochemical activity, reduced cadmium uptake	[52]
Silver nanoparticles	6–26 nm	10 and 20 mg/mL	Rice seeds (*Oriza sativa* L. cv. KDML 105)	Upregulation of aquaporin gene expression, improved seed and seedlings vigor	[35]
Iron oxide nanoparticles	<50 nm	10, 50, 100, and 500 mg/L	Sorghum (*Sorghum bicolor* (L.) Moench)	Increased biochemical activity and biomass, improved water content in leaves	[53]
Iron nanoparticles	19–30 nm	20, 40, 80, and 160 mg/L	Watermelon (*Citrullus lanatus* (Thunb.) Matsum and Nakay varieties).	Improved plant morphology, reduced phytotoxicity	[54]
Zinc, titanium, and silver	ZnO, TiO2, A g 35–40, 100, 85 nm, resp.	750, 1000, and 1250 mg/kg	Chilli (*Capsicum annuum* L.)	Improved seed vigor, increased disease resistance	[55]

3. Molecular Targets to Seed Nanopriming

An analysis of the studies shows that the benefits of nanomaterial in promoting seed germination include (a) developing nanopores in the seed coat, (b) introducing reactive oxygen species (ROS) to the seed, and (c) using the nanocatalyst to boost enzyme activity at the starch-degrading site (Schematic Figure 2B) [35]. Although the exact mechanism behind seed nanopriming is not clearly understood, it appears that nanoparticles can induce physiological effects on seed germination. A nanomaterial that penetrates the seed coat and creates small pores can lead to increased water uptake and upregulation of aquaporin gene expression. Nano-pretreated seeds accumulate more ROS during germination than unprimed seed and other priming treatments, suggesting that significant ROS is needed to stimulate seed germination as shown schematically in Figure 2B. A study by Kibinza et al. demonstrated that osmo-priming sunflower seeds increases expression of a gene encoding catalase, suggesting that catalase may be a key enzyme responsible for the recovery of vigor in older seeds [56]. The enzyme superoxide dismutase scavenges $O_2\bullet-$ and converts it into H_2O_2, and the enzyme catalase transforms H_2O_2 into water. A significant increase in antioxidant enzyme amounts could be the result of nanoparticle-mediated reduction of ROS. Due to the relationship between reactive oxygen species and antioxidant enzymes, seeds that are nanoprimed tend to contain a lot of antioxidant enzymes through nanoparticle-mediated ROS mitigation. Furthermore, aquaporin and ROS-mediated interactions have been found to influence seed development [57]. The activation of seed germination could be achieved through aquaporins and ROS. Several studies suggest that aquaporins facilitate the transfer of H_2O_2 and ROS through biological membranes as well as water uptake [58]. As a result of the nanopriming process, nanopores are formed, allowing rapid influx of water into the seeds, and aquaporin genes are activated. For seed growth and germination to occur, seed antioxidant systems must regulate ROS in order to trigger oxidative signaling molecules to function. The nanopriming process leads to increased soluble sugar levels. Sugar concentrations in the cells can reduce osmotic potential, which reduces the water potential. In this way, the difference (gradient) between the water potential outside and inside the tissues increases, facilitating water movement into the seeds. Nanoprimed seeds have increased soluble sugars content, resulting in the increase of amylase activity, which leads to an increase in water uptake in the seeds due to the change in internal osmotic potential due to the increase of soluble sugars (solutes). Studies have demonstrated that ROS, including OH, contribute to radiation growth, cellular reorganization, and endosperm and testicular wasting [59]. Nanoparticles can generate ROS by triggering the formation of •OH. A theory suggests that during the short period of time (i.e., 24 h) spent soaking seeds in nanopriming solutions, the OH produced by bound nanoparticles would facilitate a process of loosened cell walls, which would stimulate the growth of seedlings [60]. As a nanocatalyst, nanoparticles absorbed from the seeds may be crucial in accelerating the reaction rate of starch hydrolysis catalyzed by α-amylase. (Figure 2B) [35]. A recent study showed that nanoparticles can act as nanocatalysts, and that soluble starch can be rapidly digested with nanosilver [61]. As reported by Mahakham et al. [35] the aquaporin gene (PIP1;1 and PIP2;1) expressed higher levels in rice seeds primed with silver nanoparticles. In conjunction with triggering the expression of aquaporin genes, aquaporins (water channels) are trans membrane proteins that facilitate the movement of water and metabolic gases across biological membranes, and regulate water homeostasis. Nanopriming of rice seeds is an efficient method for boosting their starch metabolism. The hydrolytic enzyme α-amylase converts reserved carbohydrates into soluble sugars during seed germination, which keeps the respiration metabolism active until sufficient photosynthesis occurs. A study conducted by Jing et al. [62] examined the transcriptomes of cotton seeds treated with poly(acrylic acid)-coated cerium oxide nanoparticles (PNCs) and then exposed to saline stress. Compared to unexposed control groups, ROS are accumulated under conditions of salinity stress. When salinity stress is applied to nanoprimed seed, the expression of 13 genes related to ROS signaling pathways and 10 genes related to ion homeostasis was detected. Plants produce antioxidant enzymes to counteract oxidative damage, including peroxidases (POD), glutathione S-transferases (GST), and peroxiredoxins (PRX). Comparing seedlings under salt stress, POD and GST were significantly up regulated, while PRX was down regulated by PNC priming. A PNC priming also increases magnesium levels in cotton roots, as well as genes (CAD1 and TPS) with functions related to terpene synthase. In response to salt stress, PNC seed priming regulates cotton seedling development through signaling ion and antioxidant pathways. Among the oomycete species investigated in the study, Siddaiah et al. [63] found that chitosan nanoparticles showed antimicrobial activity against *Sclerospora graminicola*, which causes downy mildew. By priming seeds with chitosan nanoparticles, plants display improved immunity, with increased gene expression of alkaline peroxidase, polyphenoloxidase, and phenylalanine ammonia lyase. The nanochitosan seed priming over expression involves PR1 and PR5, which are involved in the salicylic acid pathways. Using zinc oxide nanoparticles to combat environmental stress factors was studied by Chaudhary et al. [38]. In response to ZnO nanopriming, miR156 and miR159 expression increased, which plays a significant role in helping plants resist biotic and abiotic stresses. A study by Ye et al. [50] demonstrated that manganese nanoparticles improved seedling growth under salinity conditions. Nanopriming with manganese upregulated SOD (superoxide dismutase), which offers protection against ROS damage and prevents phytotoxicity. It is possible that nanoparticles are present within the seed to account for this behavior. Nanoparticles appear to enhance starch hydrolysis, but the exact mechanism is still unclear [64]. Nevertheless, further research is warranted to determine the exact mechanism.

4. Effects of Seed Nanopriming under Abiotic and Biotic Stresses

Abiotic and biotic stresses cause damage to seed growth and eventually lead to economic losses. In the wake of global warming and climate change, seeds are subject to an increased number of biotic and abiotic stress combinations, which negatively affect their growth and yield [65]. The simultaneous presence of abiotic stress factors such as drought, flood, salinity, heavy mineral contamination, cold and heat has been shown to severely deter seed germination (Figure 3A). The biotic stress that is exhibited in Figure 3B includes a variety of plant pathogens including bacteria, fungi, viruses, nematodes, insects, and others. As a result of pathogen infection, changes in plant physiology often result in reduced biomass, early flowering, decreased seed set, accumulation of protective metabolites, and many other changes [66]. Researchers have reported that a variety of nanomaterials reduce biotic and abiotic stress and improve seed germination [67,68].

4.1. Abiotic Stress Ameliorating Nanomaterials

By seed priming with cerium oxide nanoparticles, An et al. studied the molecular mechanisms behind plant salinity stress tolerance. Poly (acrylic acid)-coated cerium oxide nanoparticles (PNC) show morphological, physiological, biochemical, and transcriptomic effects on cotton seedling priming under salinity stress, according to An et al. An advantage of PNC nanoparticle seed priming is that it can be used to increase a crop tolerance to stress during the early seedling stage in a sustainable, practical, and scalable way. These results showed promising molecular mechanisms that could be synergistically operated to enhance plant salt tolerance [62]. According to a recent study published by Baz et al., nanopriming with water-soluble carbon nanoparticles (CNPs) significantly increases seed vigor and seedling growth of lettuce under salinity stress. Under high temperatures and salinity stress, CNPs have been shown to significantly promote seed germination. A CNP-assisted nanopriming treatment enhanced lateral root growth but slightly inhibited the elongation of primary roots, resulting in a balanced accumulation of chlorophyll in high salinity stress [69]. The effects of manganese (III) oxide nanoparticles (MnNPs) on the salinity stress of *capsicum annum L* were studied by Ye et al. It showed that MnNPs can penetrate seed coats and form corona complexes. The results of the study demonstrate that MnNPs can modulate biochemical interactions in seeds that exhibit salinity tolerance in order to promote sustainable agriculture [50]. The study showed that nano-silica primed zea mays seeds had a higher rate of germination and a higher seedling vigor index. The priming process helps to increase the activity of antioxidant enzymes, which can suppress lipid peroxidation by suppressing the production of ROS under salinity stress. Additionally, priming reduces abscisic acid content while increasing gibberellin content. As a result of this hormonal balance, the hydrolysis enzymes (amylase and lipase) are activated. It was found that nano-silica priming enhanced the metabolic activity of maize seeds when exposed to salinity [70]. Khan et al. investigated the effect of nano ZnO and nano Fe on cadmium accumulation in wheat. Nanoprimed seeds show increased length of spikes, shoots, and roots, as well as increased grain size. Superoxide dismutase activity and electrolyte leakage were reduced in nanoprimed seeds. Nanoparticles in seeds increase wheat biomass and nutrient content as well as decreasing Cd toxicity overall [71]. The role of bulk and nanosized SiO_2 particles in fenugreek germination has been investigated by R. Ivani et al. A study concluded nanosized particles of SiO_2 protect fenugreek seeds from salt stress and improve growth attributes [72]. A study by S. Hojjat et al. examined the effects of silver nanoparticles on the germination of fenugreek seeds under salinity conditions. AgNPs improved salinity tolerance in fenugreek seedlings. This may increase various plant defense mechanisms that reduce salt stress [73]. In a study by Konate et al., magnetic (Fe_3O_4) nanoparticles were shown to mitigate heavy metals uptake and toxicity in wheat seedlings. Physiological mechanisms were investigated in wheat seedlings to determine how magnetic nanoparticles (nano-Fe_3O_4) mitigated the toxic effects of heavy metals (Pb, Zn, Cd, and Cu) [74].

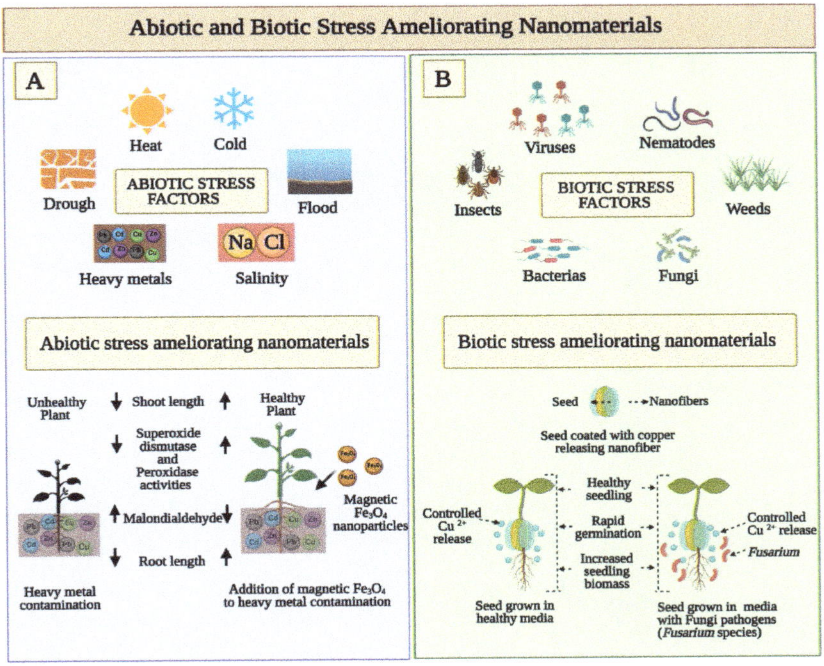

Figure 3. Effect of nanopriming agents to improve seed germination in abiotic and biotic stress conditions. (**A**) Environmental conditions in combination with soil quality may pose stress to different phases of seed germination affecting biochemical pathways involved in the root-shoot development. The metal chelators and biocatalytic NPs could play a rescue role mitigating those abiotic conditions, giving a healthy plant development. (**B**) Seed nanopriming with broad spectrum NMs with known microbicidal and anti-parasitic effect improve seed germination in biotic stress conditions boosting overall crop yield.

4.2. Biotic Stress Ameliorating Nanomaterials

As a bioprotectant against rice borne pathogens *P. grisea* and *X. oryzae*, Sathayabama and Muthukumar investigated the antimicrobial activity of chitosan guar nanoparticles (CGNP). The treated rice showed no symptoms of blast disease. Rice plants with CGNP-applied antimicrobial protection displayed an enhanced rate of seed germination and growth [75]. T. Xu et al. demonstrated enhanced seedling development and agrochemical delivery by biodegradable, biopolymer-based nanofiber seed coatings. The germination and subsequent growth of nanofiber-coated tomato and lettuce seeds were studied in a greenhouse, with and without a fungal pathogen (*Fusarium* species) as shown in Figure 2B. Based on recent observations, it appears to be possible to use nanomaterials to deliver active ingredients at precise locations. Nanofiber seed coatings provide precise agrochemical delivery and significantly improve seedling germination and biomass in comparison to conventional film coating techniques used in the industry due to their nanofibrous structure and controlled release kinetics [76]. The effects of cinnamaldehyde encapsulated mesoporous silica nanoparticles (MSNP) on seed borne diseases were studied by Bravo Cadena et al. Based on the results of this study, it is clear that MSNP significantly enhanced the antimicrobial activity of plant products, which allows the use of volatile biocides such as essential oils at very low concentrations to prevent microbial diseases in crop plants [77]. Choudhary et al. evaluated the effect of zinc chitosan nanoparticles as a seed priming agent and foliar application to maize plants. An in vitro study showed that seed nanopriming with Zn-chitosan nanoparticles enhanced seed germination and inhibited fungal growth. These results showed that Zn-chitosan nanoparticles have strong fungicidal activity, are an effective micronutrient fortifier, and could stimulate maize crop growth [38].

5. Effect of Combined Seed Priming Treatment of Nanomaterials with Microbes

Seed germination and seedling growth characteristics of several plant species have been enhanced by nanopriming with a variety of nanomaterials [9,44]. Several studies on the interaction between nanomaterials and plant-benefitting bacteria, called Plant Growth-Promoting Rhizobacteria (PGPR), have been identified as a novel approach in sustainable agriculture [78]. Through the application of both PGPR and nanomaterials, plants may be able to resist environmental stress through several physiological and morphological mechanisms, such as improved root system nutrient uptake and induced expression of cellular antioxidative enzymes [79]. By using silicon nanoparticles as a seed priming agent, Mahakham et al. studied the role of silicon nanoparticles in the cell growth process. This study examined the effect of silicon nanoparticle-mediated seed priming on *Pseudomonas* species enhancing growth, physiology, and antioxidant metabolism in *Melissa officinalis* L. According to Figure 4, seed priming, pre-sowing, and seedling inoculation with bio-elicitor improves plant growth and phytochemical constituents. Combined with silicon nanoparticles, *P. fluorescens* and *P. putida* significantly increased plant biomass indices, the relative water content of leaves, photosynthetic pigment values, essential oil yield, and all major components (except thymol). It has been found that seed priming with nano-silicon particles combined with inoculation with pseudomonas strains boosts the number of both primary and secondary metabolites in lemon balm plants [35]. Shcherbakova et al. demonstrated that pre-seed treatments with microbial inoculants and molybdenum (Mo) nanoparticles. The goal of this study was to determine the effects of pre-seed treatment with microbial inoculants and Mo nanoparticles on the composition of root exudates of chickpea plants and the diversity of microbes in the rhizosphere. Shcherbakova et al. examined the activity of enzymes involved in the antioxidant protection system and the formation of plant-microbial systems induced by nodules and rhizosphere bacteria as well as Mo nanoparticles before sowing seeds [80]. De Gregorio et al. investigated the application of beneficial rhizobacteria immobilized in nanofibers to bioinoculants soybean seeds. Their findings showed that inoculating seeds with PGPR provides soil with a significant number of beneficial microorganisms. A nanofiber-immobilized rhizobacteria-coated soybean seed was also evaluated for its influence on bacterial survival during seed storage and on characteristics of germination and plant growth. Seed coating with *P. agglomerans* improved germination, size, and weight of the roots. Moreover, seed coating with *Bacillus californianus* increased leaf number and dry weight. In view of this, the technique applied in the current study to prime seeds with nanofiber-immobilized PGPR could be regarded as a promising eco-friendly strategy to enhance soybean production by using microbial inoculants [81]. According to Taran et al., colloidal solutions of metals for micronutrients enhance plant health by enhancing resistance to exposure to unfavorable environmental conditions and enhancing plant nutrition by increasing the penetration of nanoscale elements in the cell walls. Combining seed treatment with colloidal solution of Mo nanoparticles and microbial preparation results in a four-fold increase in nodule formation compared to control plants [82].

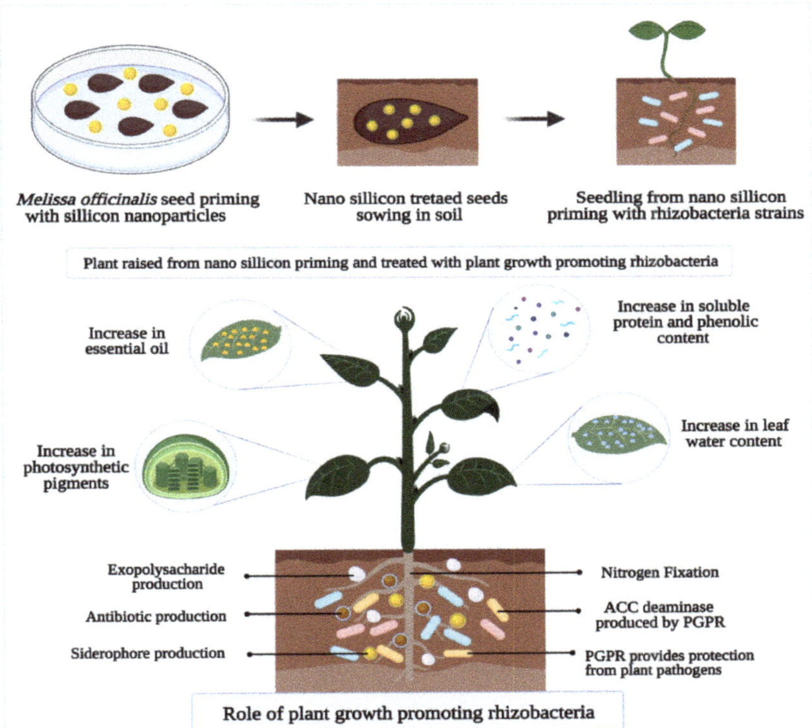

Figure 4. The plant growth-promoting rhizobacteria (PGPR) combined seed nanopriming treatment with a selection of microbicidal nanomaterials without influencing the PGPR spares beneficial microbes of the soil and plant tissues. This promotes N_2-fixation improves harvest quality and yield via selectively increasing threshold towards biotic stresses.

6. Effect of Combined Seed Priming Treatment of Nanoparticles and Cold Plasma Technology

The combined potential benefits of working with cold plasma and metal-based nanoparticles seed priming are rarely studied. The reactive oxygen and nitrogen species (ROS and RNS) are important signaling molecules that are involved in seed signaling pathways, cellular physiology, gene expression, differentiation, and growth [83]. Seed treatment with cold plasma and nanomaterials is an eco-agricultural high-tech technique that can increase crop yields [84,85]. The non-ionizing low-level radiation activates the vitality of a seed without causing gene mutation, so there is no genetic risk. A new path to increasing grain yield is provided by the application of cold plasma and nanotechnology in agriculture, since it is a fast, cheap, green, and riskless method [86]. Using cold plasma could help increase germination rates and peroxidase activities in seeds [87]. Figure 5 illustrates that both UV and biologically active agents (nitrogen and oxygen active species) produce molecular changes in seeds which then influence food composition [88]. In a study, Abedi et al. reported that seed priming with cold plasma enhanced early growth and flowering while protecting *Cichorium intybus* with selenium nanoparticles. This study provides a better understanding of the potential advantages of cold plasma in improving early growth, protection, and production. Figure 6 indicate that plasma with nano-selenium is capable of improving plant tolerance to stress conditions via activation of plant defense systems, especially antioxidants. As a result, plasma priming combined with nano-Selenium at a low dose is an effective approach to promote plant growth, biochemistry, and protection [89].

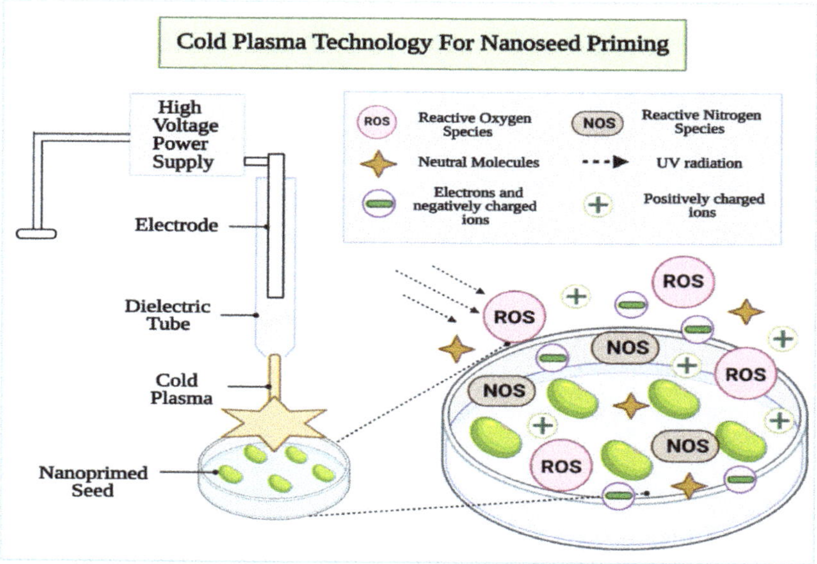

Figure 5. The cold plasma (dielectric barrier discharge) mediated seed priming improve seed vigor via synergistical activation of plant defense machinery against ROS, supporting redox homeostasis in plant. Bases on cold vs. non-thermal plasma used in seed priming, the energetic electrons, charged particle, and reactive species are produced, which dissolve tough seed coat promoting water uptake and drought yield in the crop.

Using non-thermal plasma, A. Babajani et al. studied selenium and zinc oxide nanoparticle reactions of seed priming. Medicinal plant *Melissa officinalis* has shown to respond differently to zinc oxide (nZnO) and selenium (nSe) nanoparticles when primed with cold plasma. The primed seeds with plasma were cultured in nutrient solution modulated with nSe and nZnO. As a result of plasma priming, the growth-related characteristics (stem length, root length, and width of the leaf) as well as biomass accumulation were improved, as well as the toxicity signs of nSe were attenuated [88]. The study described differential physiology and expression of phenylalanine ammonia lyase (PAL) and universal stress protein (USP) in endangered species published by Moghanloo et al. (2019). Treatments with plasma and nano silicon (nSi) simultaneously had the highest expression rates of the gene for phenylalanine ammonia lyase. In contrast, the plasma treatments did not make a significant difference in the expression of USP gene, but nSi-treated seedlings exhibited higher levels of USP expression. After plasma and nSi were applied, leaf thickness and vascular tissue development were reinforced (xylem and phloem). In the present study, nSi and plasma as potential antagonists of phytotoxicity are described, which may be used as a theoretical base for possible commercialization [90]. Cold plasma has been reported to reduce the toxicity signs of nano zinc oxide in capsicum cayenne by modifying growth, differentiation, and physiology, according to A. Iranbakhsh et al. In this study, the systemically applied cold plasma solution and zinc oxide nanoparticles (nZnO) were evaluated within plant (*Capsicum annuum*) under in vitro and in pot conditions using a functional scientific device and metal-based nanoparticles. Eliciting peroxidase enzyme activity in both culture media was achieved by treating either with plasma or nZnO on both roots, and it was found that plasma activity and/or nZnO activity of phenylalanine ammonia-lyase were significantly greater. Plant growth was most influenced by soaking seeds with nanomaterials before plasma treatment in the pot experiment [91].

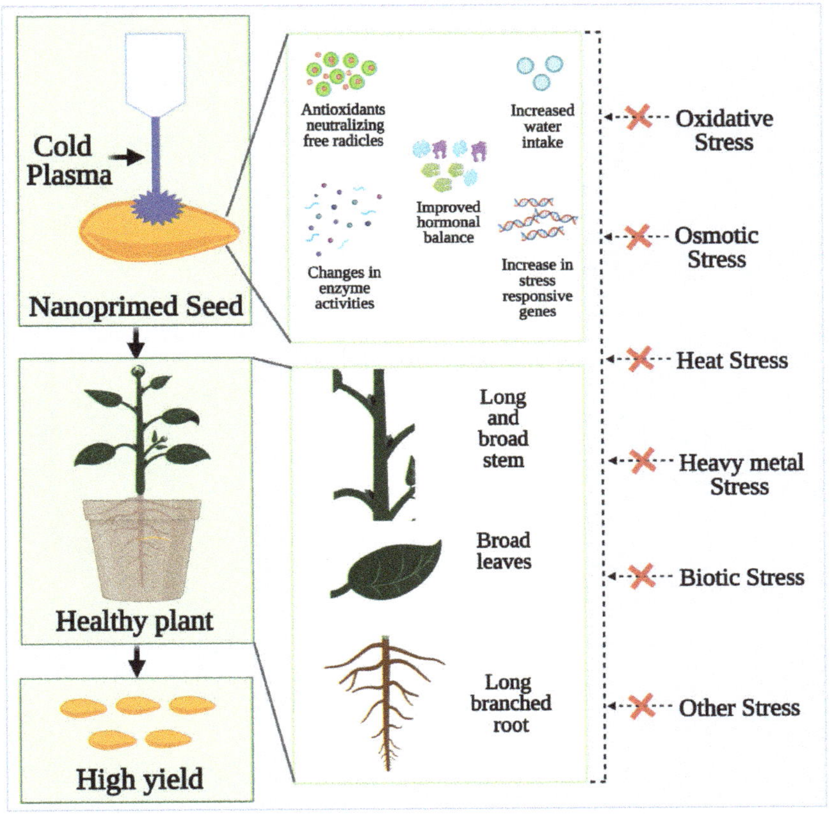

Figure 6. As an alternative to ecotoxic chemical treatments, non-thermal and cold plasma-based seed nanopriming has advantages in managing environmental stressors towards improving seed vigor and enhancing seed germination. The enhanced antioxidant system and activated defense response towards improved physiological processes; seed scarification and pathogen inactivation via plasma affects seed surface environment ameliorating improved features at molecular level and leave/root/shoot proportions.

7. Artificial Intelligence and Machine Learning Technology for Nanoprimed Seed Diagnostics

Seeds with better seed germination and seedling emergence rates can ensure reliable emergence under a variety of agricultural conditions and are therefore instrumental in ensuring yield potential and uniformity. As of today, germination scoring is typically done by human observation, and therefore, has a limited frequency, scale, and accuracy.

In order to handle this bottleneck, many attempts have been made to automate both seed diagnostics and associated phenotypic analysis, with more than one research-based solution such as germinator, phenoSeeder, and the MultiSense tool as a result [92]. As shown in Figure 7B, by using these softwares, post seed nanopriming diagnosis can be done to predict nanoprimed seeds viability and phenotypic analysis.

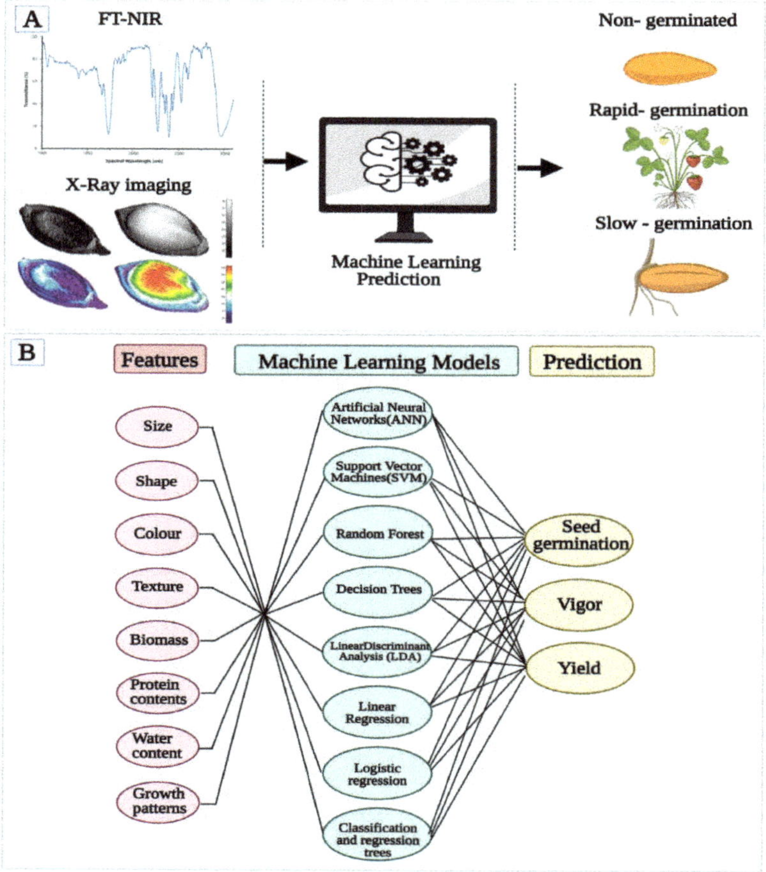

Figure 7. (**A**) Artificial intelligence and machine learning may assist with infrared spectroscopy and X-ray image data mining and curation. AI and ML may further complement via optimizing seed priming technology based on chemianalytic profiling for heat and moisture content assisting in breaking seed dormancy. (**B**) Post nanopriming, AI and ML tools can be used to predict seed germination, vigor, and yield of nanoprimed seeds.

Recently, advanced computer vision and machine learning methods have been applied to germination assays; among them, the Rice Seed Germination Evaluation System which uses artificial neural networks to gauge the germination status of Thai rice species [93]. Numerous computer vision and machine learning that combined analytic methods have been developed to automate phenotypic analyses of organs as diverse as leaves, roots, and reproductive organs [94,95]. Nanoprimed seed germination can be measured dynamically and objectively based on the color, texture, morphology, and growth pattern of a seed, thus enabling new biological findings about seed physiology. In addition, the automation of seed germination scoring represents an excellent opportunity to start standardizing seed science research. The quality of seeds and their vigor can be accessed through digital analysis, but also biological experiments under varying conditions can be quantitatively compared to increase the confidence of research results. A study by Genze et al. presented accurate germination detection, prediction and quality assessment based on machine learning [96]. Several attempts have been made to automate seed testing in order to reduce the number of error-prone manual steps required in this process. Various analyses of seeds have recently been carried out using modern image analysis techniques since they are easily automatized and can provide unbiased and quantitative measurements which are error-free [97]. Modern convolutional networks were used in the present study to detect individual seeds and to distinguish germinated from non-germinated seeds with higher precision. By using artificial intelligence to measure germination potential, the evaluation process could be sped up. Compared to manual methods and conventional methods, it has a lower error rate and a higher performance. This allows it to provide more accurate germination indices and assess the quality of seeds. An approach for determining seed quality was developed using FT-NIR spectroscopy and X-ray imaging data by A D Medeiros et al. [98]. The results shown in Figure 7A illustrate how FT-NIR and X-ray imaging can be used in conjunction with machine learning algorithms to improve seed germination and vigor prediction. This study examines the use of FT-NIR spectroscopy in conjunction with X-ray imaging in the prediction of seed quality traits (germination and vigor). Seed germination capacity can be accurately measured (85% accuracy) using the proposed approach. A machine learning algorithm developed using both NIR spectra and X-ray images can easily, efficiently, and accurately predict seed germination. A D Medeiros et al. developed a system for classifying soybean seedlings based on quality data [98]. In this study, it was proposed to use traditional machine learning methods for determining the size, shape, and physiological potential of soybean seeds and seedlings. The models were developed using free and low-cost software based on images of soy seeds and seedlings. Using the developed model for seed and seedling classifications, 94% of seed and seedlings were accurately classified. By using interactive and traditional machine learning models, high precision was demonstrated in the models developed, which were able to classify soybean seeds by their appearance, as well as beans and their seedlings by vigor quickly and non-subjectively. A deep learning architecture is an artificial neural network (ANN) that is used to find patterns in data or model complex relationships between inputs and outputs. Classifiers are based on multilayer neural networks for identifying wheat grain (seed) irregularities grown in a variety of agricultural environmental zones. The scientific and technological developments of agricultural enterprises enable them to generate a large amount of data. Automated analysis methods are needed to process this data size. It is a very useful way to support agricultural center experts using advanced computing techniques [99,100]. Combined with real-time processing of data, it helps farmers take correct decisions concerning harvesting crops, planting, and fertilization. The following Table 2 displays some examples of seed analysis software with their features for possible use to diagnose nanoprimed seeds.

Table 2. Software that processes images simultaneously and produces accurate analyses of seedling germination and establishment. Using artificial neural networks, advanced computer-vision and machine-learning techniques have been applied to germination assays to assess the germination status of various nanoprimed seed.

Seed Analysis Software	Type of Seed Studied	Feature	Remark	Reference
Germinator	*Arabidopsis thaliana* seed	An indicator of germination	An automated high-throughput evaluation procedure for germination	[101]
Seed Vigor Imaging System	Soybean and Corn Seed	Used to calculate the length of seeds	Physiological differences between seed lots could be identified	[102]
SeedUSoon	-	Tracking of different mutations	It provides a visual representation of the genetic link between related seed batche	[103]
SmartGrain	*Arabidopsis thaliana* seed	Recognize varieties of seeds	Understanding the genes and mechanisms of grain or seed size and shape	[104]
SeedExtractor	Rice seeds	Measures seed size, shape (including length, width, circle and color	Crop yield traits can be determined	[105]
SeedCounter	Wheat grain	The use of a mobile device can be used to estimate the number and size of grains	It is possible to improve the accuracy of seed analysis.	[106]

8. Limitations of Nanopriming Techniques and Future Prospects

It is recognized that despite remarkable progress, scientists do not have a complete understanding of how these nanomaterials can affect the macro- and micro-environments of seeds. It is alarming that there is still a lack of basic understanding regarding the possible health and safety effects of engineered nanomaterials on both human and non-human receptors considering the actual and projected levels of exposure [107]. A general rule concerning seed nanopriming does not exist, and there is no clear trend regarding priming responses depending on the taxonomic position of the species. As a result of some nanopriming treatments, there is a possibility of contamination of the medium with fungi and bacteria, which may greatly hinder subsequent seed germination [7]. A seed that has been nanoprimed is dried back to its original moisture content, but this process is done faster than the dehydration of mature seeds. Several researchers have hypothesized that brutal desiccation procedures alter the effects of nanopriming [7]. As a consequence, nanoprimed seed material can be less stable, and higher maintenance costs for seed companies and farmers are consequently incurred. In some cases, repeated nanopriming treatments can partly prevent seed viability losses, whereas in others, such losses are permanent and cannot be reversed [108]. Possibly requiring an additional treatment may be both an extra cost and a source of variability because germination potential may not be fully restored.

Nanopriming can be a potential future target for sustainable agriculture with the current global supply versus demand trend. A fortification program consists of adding vitamins and minerals to processed foods as a public health measure to improve the nutritional quality of the food supply. The goal is to enhance the nutrient intake by the population [52]. As a result, biofortification reduces the runoff of fertilizers and other agrochemicals and their inputs into the environment. As a result of seed biofortification, stands have been established and plant production, yields, nutrient content, and water utilization have increased as well as plant tolerance to biotic and abiotic stresses have increased [109]. The technique has been found to be a simple, practical, and cost-effective approach to improving the quality of seed and crops in resource-constrained regions. Compared to conventional seed priming, seed priming using engineered nanomaterial produced higher germination rates at equivalent or lower life cycle embodied energy. Nanotechnology is also an exceptionally potent way to deliver nutrients effectively due to its ability to deliver a wide range of engineered nanomaterials in a more efficient manner than before [34,110]. By introducing nano-bio-fortification into seeds, less fertilizer potentially may be consumed. This results in the need to use less water and less resources to cultivate the same volume of nutrient-rich food. Nano-bio fortification of seeds could also assist in shortening the growth cycle of plants (e.g., requiring less time for them to germinate and to mature), allowing the land to grow crops faster in the future. In similar ways to other agronomic methods, the success of seed nano-biofortification will depend on how well it is adapted to the setting, including the type of soil, the type of crop and the climate, and the location [111]. The advantages of nanotechnology need to be compared to other biofortification procedures that may provide long-term and cost-effective solutions, including plant breeding and CRISPR/transgenic technology [111]. In addition to assessing the overall environmental implications, it is important to determine if nano-based solutions are practical compared to conventional practices [112]. Nanomaterials continue to be manufactured in larger scales, which lead to further decreases in production costs. It is highly likely that nanotechnology will integrate with seed biofortification practices in the future, especially as highly specialized and tunable nanomaterials emerge. Nano-enabled seed biofortification is an important topic that merits intense scrutiny and greater attention given its potential benefits and the increasing global food insecurity that we will face in the coming decades.

Author Contributions: Conceptualization, A.S. (Amruta Shelar), A.V.S. and R.P.; data curation, A.S. (Amruta Shelar), A.S. (Akanksha Shelar), S.P.S. and A.V.S.; writing—original draft preparation, A.S. (Amruta Shelar) and A.V.S. and R.P.; writing—review and editing, A.V.S., M.F.S., R.S.M. and V.T., P.L., A.L.; graphic design and visualization, A.S. (Amruta Shelar), R.S.M., M.F.S. and A.V.S.; supervision, R.P., A.V.S. and D.G.; project administration, A.V.S., R.P., M.C.; funding acquisition, A.V.S. & R.P. All authors have read and agreed to the published version of the manuscript.

Funding: This research received no external funding.

Institutional Review Board Statement: Not applicable.

Informed Consent Statement: Not applicable.

Data Availability Statement: Data available in a publicly accessible repository.

Acknowledgments: This work was supported by BfR SFP 1322-725; BfR SFP 1322-735 (A.V.S.).

Conflicts of Interest: The authors declare no competing financial interest.

References

1. Fróna, D.; Szenderák, J.; Harangi-Rákos, M. The Challenge of Feeding the World. *Sustainability* **2019**, *11*, 5816. [CrossRef]
2. Umesha, S.; Manukumar, H.M.G.; Chandrasekhar, B. Chapter 3—Sustainable Agriculture and Food Security. In *Biotechnology for Sustainable Agriculture*; Singh, R.L., Mondal, S., Eds.; Woodhead Publishing: Cambridge, UK, 2018; pp. 67–91. [CrossRef]
3. Durán-Lara, E.F.; Valderrama, A.; Marican, A. Natural Organic Compounds for Application in Organic Farming. *Agriculture* **2020**, *10*, 41. [CrossRef]
4. Paparella, S.; Araújo, S.S.; Rossi, G.; Wijayasinghe, M.; Carbonera, D.; Balestrazzi, A. Seed priming: State of the art and new perspectives. *Plant Cell Rep.* **2015**, *34*, 1281–1293. [CrossRef] [PubMed]
5. Waqas, M.; Korres, N.E.; Khan, M.D.; Nizami, A.-S.; Deeba, F.; Ali, I.; Hussain, H. Advances in the Concept and Methods of Seed Priming. In *Priming and Pretreatment of Seeds and Seedlings: Implication in Plant Stress Tolerance and Enhancing Productivity in Crop Plants*; Hasanuzzaman, M., Fotopoulos, V., Eds.; Springer Singapore: Singapore, 2019; pp. 11–41. [CrossRef]
6. Marthandan, V.; Geetha, R.; Kumutha, K.; Renganathan, V.G.; Karthikeyan, A.; Ramalingam, J. Seed Priming: A Feasible Strategy to Enhance Drought Tolerance in Crop Plants. *Int. J. Mol. Sci.* **2020**, *21*, 8258. [CrossRef]
7. do Espirito Santo Pereira, A.; Caixeta Oliveira, H.; Fernandes Fraceto, L.; Santaella, C. Nanotechnology Potential in Seed Priming for Sustainable Agriculture. *Nanomaterials* **2021**, *11*, 267. [CrossRef]
8. Khodakovskaya, M.; Dervishi, E.; Mahmood, M.; Xu, Y.; Li, Z.; Watanabe, F.; Biris, A.S. Carbon Nanotubes Are Able To Penetrate Plant Seed Coat and Dramatically Affect Seed Germination and Plant Growth. *ACS Nano* **2009**, *3*, 3221–3227. [CrossRef] [PubMed]
9. Abbasi Khalaki, M.; Moameri, M.; Asgari Lajayer, B.; Astatkie, T. Influence of nano-priming on seed germination and plant growth of forage and medicinal plants. *Plant Growth Regul.* **2021**, *93*, 13–28. [CrossRef]

10. Shang, Y.; Hasan, M.K.; Ahammed, G.J.; Li, M.; Yin, H.; Zhou, J. Applications of Nanotechnology in Plant Growth and Crop Protection: A Review. *Molecules* **2019**, *24*, 2558. [CrossRef]
11. Ioannou, A.; Gohari, G.; Papaphilippou, P.; Panahi Rad, S.; Akbari, A.; Dadpour, M.; Krasia-Christoforou, T.; Fotopoulos, V. Advanced nanomaterials in agriculture under a changing climate: The way to The future? *Environ. Exp. Bot.* **2020**, *176*. [CrossRef]
12. Gohari, G.; Panahirad, S.; Sadeghi, M.; Akbari, A.; Zareei, E.; Zahedi, S.M.; Bahrami, M.K.; Fotopoulos, V. Putrescine-functionalized carbon quantum dot (put-CQD) nanoparticles effectively prime grapevine (*Vitis vinifera* cv. 'Sultana') against salt stress. *BMC Plant Biol.* **2021**, *21*, 120. [CrossRef]
13. Gohari, G.; Zareei, E.; Rostami, H.; Panahirad, S.; Kulak, M.; Farhadi, H.; Amini, M.; Martinez-Ballesta, M.d.C.; Fotopoulos, V. Protective effects of cerium oxide nanoparticles in grapevine (*Vitis vinifera* L.) cv. Flame Seedless under salt stress conditions. *Ecotoxicol. Environ. Saf.* **2021**, *220*, 112402. [CrossRef]
14. Azimi, F.; Oraei, M.; Gohari, G.; Panahirad, S.; Farmarzi, A. Chitosan-selenium nanoparticles (Cs–Se NPs) modulate the photosynthesis parameters, antioxidant enzymes activities and essential oils in *Dracocephalum moldavica* L. under cadmium toxicity stress. *Plant Physiol. Biochem.* **2021**, *167*, 257–268. [CrossRef]
15. Masoudniaragh, A.; Oraei, M.; Gohari, G.; Akbari, A.; Faramarzi, A. Using halloysite nanotubes as carrier for proline to alleviate salt stress effects in sweet basil (*Ocimum basilicum* L.). *Sci. Hortic.* **2021**, *285*, 110202. [CrossRef]
16. Sheikhalipour, M.; Esmaielpour, B.; Behnamian, M.; Gohari, G.; Giglou, M.T.; Vachova, P.; Rastogi, A.; Bresti, M.; Skalicky, M.J.N. Chitosan Selenium Nanoparticle (C" Se NP) Foliar Spray Alleviates Salt Stress in Bitter Melon. *Nanomaterials* **2021**, *11*, 684. [CrossRef]
17. Sheikhalipour, M.; Esmaielpour, B.; Gohari, G.; Haghighi, M.; Jafari, H.; Farhadi, H.; Kulak, M.; Kalisz, A. Salt Stress Mitigation via the Foliar Application of Chitosan-Functionalized Selenium and Anatase Titanium Dioxide Nanoparticles in Stevia (*Stevia rebaudiana* Bertoni). *Molecules* **2021**, *26*, 4090. [CrossRef] [PubMed]
18. Mohammadi, M.H.Z.; Panahirad, S.; Navai, A.; Bahrami, M.K.; Kulak, M.; Gohari, G. Cerium oxide nanoparticles (CeO2-NPs) improve growth parameters and antioxidant defense system in Moldavian Balm (*Dracocephalum moldavica* L.) under salinity stress. *Plant Stress* **2021**, *1*, 100006. [CrossRef]
19. Sanzari, I.; Leone, A.; Ambrosone, A. Nanotechnology in Plant Science: To Make a Long Story Short. *Front. Bioeng. Biotechnol.* **2019**, *7*, 120. [CrossRef]
20. Kaur, R.; Chandra, J.; Keshavkant, S. Nanotechnology: An efficient approach for rejuvenation of aged seeds. *Physiol. Mol. Biol. Plants* **2021**, *27*, 399–415. [CrossRef]
21. Mittal, D.; Kaur, G.; Singh, P.; Yadav, K.; Ali, S.A. Nanoparticle-Based Sustainable Agriculture and Food Science: Recent Advances and Future Outlook. *Front. Bioeng. Biotechnol.* **2020**, *2*, 9954. [CrossRef]
22. Duhan, J.S.; Kumar, R.; Kumar, N.; Kaur, P.; Nehra, K.; Duhan, S. Nanotechnology: The new perspective in precision agriculture. *Biotechnol. Rep.* **2017**, *15*, 11–23. [CrossRef]
23. Damalas, C.A.; Koutroubas, S.D.; Fotiadis, S. Hydro-Priming Effects on Seed Germination and Field Performance of Faba Bean in Spring Sowing. *Agriculture* **2019**, *9*, 201. [CrossRef]
24. Patade, V.Y.; Bhargava, S.; Suprasanna, P. Halopriming imparts tolerance to salt and PEG induced drought stress in sugarcane. *Agric. Ecosyst. Environ.* **2009**, *134*, 24–28. [CrossRef]
25. Chen, K.; Arora, R. Dynamics of the antioxidant system during seed osmopriming, post-priming germination, and seedling establishment in Spinach (*Spinacia oleracea*). *Plant Sci. Int. J. Exp. Plant Biol.* **2011**, *180*, 212–220. [CrossRef]
26. Moori, S.; Ahmadi-Lahijani, M.J. Hormopriming instigates defense mechanisms in Thyme (*Thymus vulgaris* L.) seeds under cadmium stress. *J. Appl. Res. Med. Aromat. Plants* **2020**, *19*, 100268. [CrossRef]
27. Sen, S.K.; Chouhan, D.; Das, D.; Ghosh, R.; Mandal, P. Improvisation of salinity stress response in mung bean through solid matrix priming with normal and nano-sized chitosan. *Int. J. Biol. Macromol.* **2020**, *145*, 108–123. [CrossRef]
28. Rakshit, A.; Sunita, K.; Pal, S.; Singh, A.; Singh, H. *Bio-Priming Mediated Nutrient Use Efficiency of Crop Species*; Springer: New Delhi, India, 2015; pp. 181–191. [CrossRef]
29. Nasrollahzadeh, M.; Sajadi, S.M.; Sajjadi, M.; Issaabadi, Z. Chapter 1—An Introduction to Nanotechnology. In *Interface Science and Technology*; Nasrollahzadeh, M., Sajadi, S.M., Sajjadi, M., Issaabadi, Z., Atarod, M., Eds.; Elsevier: Amsterdam, The Netherlands, 2019; Volume 28, pp. 1–27. [CrossRef]
30. García-Gómez, C.; Fernández, M.D. Chapter Four—Impacts of metal oxide nanoparticles on seed germination, plant growth and development. In *Comprehensive Analytical Chemistry*; Verma, S.K., Das, A.K., Eds.; Elsevier: Amsterdam, The Netherlands, 2019; Volume 84, pp. 75–124. [CrossRef]
31. Prasad, R.; Bhattacharyya, A.; Nguyen, Q.D. Nanotechnology in Sustainable Agriculture: Recent Developments, Challenges, and Perspectives. *Front. Microbiol.* **2017**, *8*, 1014. [CrossRef]
32. Chen, H. Metal based nanoparticles in agricultural system: Behavior, transport, and interaction with plants. *Chem. Speciat. Bioavailab.* **2018**, *30*, 123–134. [CrossRef]
33. Bandala, E.R.; Berli, M. Nanomaterials: New Agrotechnology Tools to Improve Soil Quality? In *Agricultural Nanobiotechnology: Modern Agriculture for a Sustainable Future*; López-Valdez, F., Fernández-Luqueño, F., Eds.; Springer International Publishing: Cham, The Netherlands, 2018; pp. 127–140. [CrossRef]
34. Lowry, G.V.; Avellan, A.; Gilbertson, L.M. Opportunities and challenges for nanotechnology in the agri-tech revolution. *Nat. Nanotechnol.* **2019**, *14*, 517–522. [CrossRef] [PubMed]

35. Mahakham, W.; Sarmah, A.K.; Maensiri, S.; Theerakulpisut, P. Nanopriming technology for enhancing germination and starch metabolism of aged rice seeds using phytosynthesized silver nanoparticles. *Sci. Rep.* **2017**, *7*, 8263. [CrossRef] [PubMed]
36. Vera-Reyes, I.; Vázquez-Núñez, E.; Lira-Saldivar, R.H.; Méndez-Argüello, B. Effects of Nanoparticles on Germination, Growth, and Plant Crop Development. In *Agricultural Nanobiotechnology: Modern Agriculture for a Sustainable Future*; López-Valdez, F., Fernández-Luqueño, F., Eds.; Springer International Publishing: Berlin/Heidelberg, Germany, 2018; pp. 77–110. [CrossRef]
37. Szőllősi, R.; Molnár, Á.; Kondak, S.; Kolbert, Z. Dual Effect of Nanomaterials on Germination and Seedling Growth: Stimulation vs. Phytotoxicity. *Plants* **2020**, *9*, 1745. [CrossRef]
38. Choudhary, R.C.; Kumaraswamy, R.V.; Kumari, S.; Sharma, S.S.; Pal, A.; Raliya, R.; Biswas, P.; Saharan, V. Zinc encapsulated chitosan nanoparticle to promote maize crop yield. *Int. J. Biol. Macromol.* **2019**, *127*, 126–135. [CrossRef]
39. Saharan, V.; Kumaraswamy, R.V.; Choudhary, R.C.; Kumari, S.; Pal, A.; Raliya, R.; Biswas, P. Cu-Chitosan Nanoparticle Mediated Sustainable Approach To Enhance Seedling Growth in Maize by Mobilizing Reserved Food. *J. Agric. Food Chem.* **2016**, *64*, 6148–6155. [CrossRef] [PubMed]
40. Pereira, A.d.E.S.; Oliveira, H.C.; Fraceto, L.F. Polymeric nanoparticles as an alternative for application of gibberellic acid in sustainable agriculture: A field study. *Sci. Rep.* **2019**, *9*, 7135. [CrossRef] [PubMed]
41. Falsini, S.; Clemente, I.; Papini, A.; Tani, C.; Schiff, S.; Salvatici, M.C.; Petruccelli, R.; Benelli, C.; Giordano, C.; Gonnelli, C.; et al. When Sustainable Nanochemistry Meets Agriculture: Lignin Nanocapsules for Bioactive Compound Delivery to Plantlets. *ACS Sustain. Chem. Eng.* **2019**, *7*, 19935–19942. [CrossRef]
42. Chau, N.H.; Doan, Q.H.; Chu, T.H.; Nguyen, T.T.; Dao Trong, H.; Ngo, Q.B. Effects of Different Nanoscale Microelement-Containing Formulations for Presowing Seed Treatment on Growth of Soybean Seedlings. *J. Chem.* **2019**, *2019*, 8060316. [CrossRef]
43. Joshi, A.; Kaur, S.; Dharamvir, K.; Nayyar, H.; Verma, G. Multi-walled carbon nanotubes applied through seed-priming influence early germination, root hair, growth and yield of bread wheat (*Triticum aestivum* L.). *J. Sci. Food Agric.* **2018**, *98*, 3148–3160. [CrossRef]
44. Acharya, P.; Jayaprakasha, G.K.; Crosby, K.M.; Jifon, J.L.; Patil, B.S. Nanoparticle-Mediated Seed Priming Improves Germination, Growth, Yield, and Quality of Watermelons (*Citrullus lanatus*) at multi-locations in Texas. *Sci. Rep.* **2020**, *10*, 5037. [CrossRef]
45. Sundaria, N.; Singh, M.; Upreti, P.; Chauhan, R.; Jaiswal, J.P.; Kumar, A. Seed Priming with Iron Oxide Nanoparticles Triggers Iron Acquisition and Biofortification in Wheat (*Triticum aestivum* L.) Grains. *J. Plant. Growth Regul.* **2019**, *38*, 122–131. [CrossRef]
46. Abdel Latef, A.A.H.; Abu Alhmad, M.F.; Abdelfattah, K.E. The Possible Roles of Priming with ZnO Nanoparticles in Mitigation of Salinity Stress in Lupine (*Lupinus termis*) Plants. *J. Plant Growth Regul.* **2017**, *36*, 60–70. [CrossRef]
47. Savassa, S.M.; Duran, N.M.; Rodrigues, E.S.; de Almeida, E.; van Gestel, C.A.M.; Bompadre, T.F.V.; de Carvalho, H.W.P. Effects of ZnO Nanoparticles on *Phaseolus vulgaris* Germination and Seedling Development Determined by X-ray Spectroscopy. *ACS Appl. Nano Mater.* **2018**, *1*, 6414–6426. [CrossRef]
48. Duran, N.M.; Savassa, S.M.; Lima, R.G.d.; de Almeida, E.; Linhares, F.S.; van Gestel, C.A.M.; Pereira de Carvalho, H.W. X-ray Spectroscopy Uncovering the Effects of Cu Based Nanoparticle Concentration and Structure on *Phaseolus vulgaris* Germination and Seedling Development. *J. Agric. Food Chem.* **2017**, *65*, 7874–7884. [CrossRef] [PubMed]
49. Ahuja, R.; Sidhu, A.; Bala, A. Synthesis and evaluation of iron(ii) sulfide aqua nanoparticles (FeS-NPs) against *Fusarium verticillioides* causing sheath rot and seed discoloration of rice. *Eur. J. Plant Pathol.* **2019**, *155*, 163–171. [CrossRef]
50. Ye, Y.; Cota-Ruiz, K.; Hernández-Viezcas, J.A.; Valdés, C.; Medina-Velo, I.A.; Turley, R.S.; Peralta-Videa, J.R.; Gardea-Torresdey, J.L. Manganese Nanoparticles Control Salinity-Modulated Molecular Responses in *Capsicum annuum* L. through Priming: A Sustainable Approach for Agriculture. *ACS Sustain. Chem. Eng.* **2020**, *8*, 1427–1436. [CrossRef]
51. Li, R.; He, J.; Xie, H.; Wang, W.; Bose, S.K.; Sun, Y.; Hu, J.; Yin, H. Effects of chitosan nanoparticles on seed germination and seedling growth of wheat (*Triticum aestivum* L.). *Int. J. Biol. Macromol.* **2019**, *126*, 91–100. [CrossRef]
52. Hussain, A.; Rizwan, M.; Ali, Q.; Ali, S. Seed priming with silicon nanoparticles improved the biomass and yield while reduced the oxidative stress and cadmium concentration in wheat grains. *Environ. Sci. Pollut. Res. Int.* **2019**, *26*, 7579–7588. [CrossRef]
53. Maswada, H.F.; Djanaguiraman, M.; Prasad, P.V.V. Seed treatment with nano-iron (III) oxide enhances germination, seeding growth and salinity tolerance of sorghum. *J. Agron. Crop. Sci.* **2018**, *204*, 577–587. [CrossRef]
54. Kasote, D.M.; Lee, J.H.J.; Jayaprakasha, G.K.; Patil, B.S. Seed Priming with Iron Oxide Nanoparticles Modulate Antioxidant Potential and Defense-Linked Hormones in Watermelon Seedlings. *ACS Sustain. Chem. Eng.* **2019**, *7*, 5142–5151. [CrossRef]
55. Gnanasekaran, D.k.; Raja, K.; Natarajan, N.; Govindaraju, K.; Subramanian, K. Invigouration treatment of metal and metal oxide nanoparticles for improving the seed quality of aged chilli seeds (*Capsicum annum* L.). *Mater. Chem. Phys.* **2019**, *242*, 122492. [CrossRef]
56. Kibinza, S.; Bazin, J.; Bailly, C.; Farrant, J.; Corbineau, F.; El-Maarouf-Bouteau, H. Catalase is a key enzyme in seed recovery from ageing during priming. *Plant Sci. Int. J. Exp. Plant Biol.* **2011**, *181*, 309–315. [CrossRef]
57. Maurel, C.; Boursiac, Y.; Luu, D.-T.; Santoni, V.; Shahzad, Z.; Verdoucq, L. Aquaporins in Plants. *Physiol. Rev.* **2015**, *95*, 1321–1358. [CrossRef]
58. Bienert, G.P.; Chaumont, F. Aquaporin-facilitated transmembrane diffusion of hydrogen peroxide. *Biochim. Biophys. Acta* **2014**, *1840*, 1596–1604. [CrossRef]
59. Müller, K.; Linkies, A.; Vreeburg, R.A.M.; Fry, S.C.; Krieger-Liszkay, A.; Leubner-Metzger, G. In Vivo Cell Wall Loosening by Hydroxyl Radicals during Cress Seed Germination and Elongation Growth. *Plant Physiol.* **2009**, *150*, 1855. [CrossRef] [PubMed]
60. Gomes, M.P.; Garcia, Q.S. Reactive oxygen species and seed germination. *Biologia* **2013**, *68*, 351–357. [CrossRef]

61. Ernest, V.; Shiny, P.J.; Mukherjee, A.; Chandrasekaran, N. Silver nanoparticles: A potential nanocatalyst for the rapid degradation of starch hydrolysis by α-amylase. *Carbohydr. Res.* **2012**, *352*, 60–64. [CrossRef]
62. An, J.; Hu, P.; Li, F.; Wu, H.; Shen, Y.; White, J.C.; Tian, X.; Li, Z.; Giraldo, J.P. Emerging investigator series: Molecular mechanisms of plant salinity stress tolerance improvement by seed priming with cerium oxide nanoparticles. *Environ. Sci. Nano* **2020**, *7*, 2214–2228. [CrossRef]
63. Siddaiah, C.N.; Prasanth, K.V.H.; Satyanarayana, N.R.; Mudili, V.; Gupta, V.K.; Kalagatur, N.K.; Satyavati, T.; Dai, X.-F.; Chen, J.-Y.; Mocan, A.; et al. Chitosan nanoparticles having higher degree of acetylation induce resistance against pearl millet downy mildew through nitric oxide generation. *Sci. Rep.* **2018**, *8*, 2485. [CrossRef] [PubMed]
64. Misson, M.; Zhang, H.; Jin, B. Nanobiocatalyst advancements and bioprocessing applications. *J. R. Soc. Interface* **2015**, *12*, 20140891. [CrossRef]
65. Raza, A.; Razzaq, A.; Mehmood, S.S.; Zou, X.; Zhang, X.; Lv, Y.; Xu, J. Impact of Climate Change on Crops Adaptation and Strategies to Tackle Its Outcome: A Review. *Plants* **2019**, *8*, 34. [CrossRef]
66. Chojak-Koźniewska, J.; Kuźniak, E.; Zimny, J. The Effects of Combined Abiotic and Pathogen Stress in Plants: Insights From Salinity and *Pseudomonas syringae* pv lachrymans Interaction in Cucumber. *Front. Plant Sci.* **2018**, *9*, 1691. [CrossRef]
67. Aslani, F.; Bagheri, S.; Muhd Julkapli, N.; Juraimi, A.S.; Hashemi, F.S.G.; Baghdadi, A. Effects of Engineered Nanomaterials on Plants Growth: An Overview. *Sci. World J.* **2014**, *2014*, 641759. [CrossRef]
68. Ansari, M.H.D.; Lavhale, S.; Kalunke, R.M.; Srivastava, P.L.; Pandit, V.; Gade, S.; Yadav, S.; Laux, P.; Luch, A.; Gemmati, D.; et al. Recent Advances in Plant Nanobionics and Nanobiosensors for Toxicology Applications. *Curr. Nanosci.* **2020**, *16*, 27–41. [CrossRef]
69. Baz, H.; Creech, M.; Chen, J.; Gong, H.; Bradford, K.; Huo, H. Water-Soluble Carbon Nanoparticles Improve Seed Germination and Post-Germination Growth of Lettuce under Salinity Stress. *Plants* **2020**, *10*, 1192.
70. Naguib, D.M.; Abdalla, H. Metabolic Status during Germination of Nano Silica Primed Zea mays Seeds under Salinity Stress. *J. Crop Sci. Biotechnol.* **2019**, *22*, 415–423. [CrossRef]
71. Khan, Z.S.; Rizwan, M.; Hafeez, M.; Ali, S.; Javed, M.R.; Adrees, M. The accumulation of cadmium in wheat (*Triticum aestivum*) as influenced by zinc oxide nanoparticles and soil moisture conditions. *Environ. Sci. Pollut. Res. Int.* **2019**, *26*, 19859–19870. [CrossRef]
72. Ivani, R.; Sanaei Nejad, S.H.; Ghahraman, B.; Astaraei, A.R.; Feizi, H. Role of bulk and Nanosized SiO(2) to overcome salt stress during Fenugreek germination (*Trigonella foenum- graceum* L.). *Plant Signal. Behav.* **2018**, *13*, e1044190. [CrossRef]
73. Hojjat, S.S.; Kamyab, M. The effect of silver nanoparticle on Fenugreek seed germination under salinity levels. *Russ. Agric. Sci.* **2017**, *43*, 61–65. [CrossRef]
74. Konate, A.; He, X.; Zhang, Z.; Ma, Y.; Zhang, P.; Alugongo, G.M.; Rui, Y. Magnetic (Fe3O4) Nanoparticles Reduce Heavy Metals Uptake and Mitigate Their Toxicity in Wheat Seedling. *Sustainability* **2017**, *9*, 790. [CrossRef]
75. Sathiyabama, M.; Muthukumar, S. Chitosan guar nanoparticle preparation and its in vitro antimicrobial activity towards phytopathogens of rice. *Int. J. Biol. Macromol.* **2020**, *153*, 297–304. [CrossRef]
76. Xu, T.; Ma, C.; Aytac, Z.; Hu, X.; Ng, K.W.; White, J.C.; Demokritou, P. Enhancing Agrichemical Delivery and Seedling Development with Biodegradable, Tunable, Biopolymer-Based Nanofiber Seed Coatings. *ACS Sustain. Chem. Eng.* **2020**, *8*, 9537–9548. [CrossRef]
77. Bravo Cadena, M.; Preston, G.M.; Van der Hoorn, R.A.L.; Flanagan, N.A.; Townley, H.E.; Thompson, I.P. Enhancing cinnamon essential oil activity by nanoparticle encapsulation to control seed pathogens. *Ind. Crop. Prod.* **2018**, *124*, 755–764. [CrossRef]
78. Vejan, P.; Abdullah, R.; Khadiran, T.; Ismail, S.; Nasrulhaq Boyce, A. Role of Plant Growth Promoting Rhizobacteria in Agricultural Sustainability-A Review. *Molecules* **2016**, *21*, 573. [CrossRef] [PubMed]
79. Kumari, B.; Mallick, M.A.; Solanki, M.K.; Hora, A.; Mani, M. 8—Applying nanotechnology to bacteria: An emerging technology for sustainable agriculture. In *Role of Plant Growth Promoting Microorganisms in Sustainable Agriculture and Nanotechnology*; Kumar, A., Singh, A.K., Choudhary, K.K., Eds.; Woodhead Publishing: Cambridge, UK, 2019; pp. 121–143. [CrossRef]
80. Shcherbakova, E.; Shcherbakov, A.; Andronov, E.; Gonchar, L.; Kalenska, S.; Chebotar, V. Combined pre-seed treatment with microbial inoculants and Mo nanoparticles changes composition of root exudates and rhizosphere microbiome structure of chickpea (*Cicer arietinum* L.) plants. *Symbiosis* **2017**, *73*, 57–69. [CrossRef]
81. De Gregorio, P.R.; Michavila, G.; Ricciardi Muller, L.; de Souza Borges, C.; Pomares, M.F.; Saccol de Sá, E.L.; Pereira, C.; Vincent, P.A. Beneficial rhizobacteria immobilized in nanofibers for potential application as soybean seed bioinoculants. *PLoS ONE* **2017**, *12*, e0176930. [CrossRef] [PubMed]
82. Taran, N.Y.; Gonchar, O.M.; Lopatko, K.G.; Batsmanova, L.M.; Patyka, M.V.; Volkogon, M.V. The effect of colloidal solution of molybdenum nanoparticles on the microbial composition in rhizosphere of *Cicer arietinum* L. *Nanoscale Res. Lett.* **2014**, *9*, 289. [CrossRef]
83. Jiang, J.; He, X.; Li, L.; Li, J.; Shao, H.; Xu, Q.; Ye, R.; Dong, Y. Effect of Cold Plasma Treatment on Seed Germination and Growth of Wheat. *Plasma Sci. Technol.* **2013**, *16*, 54. [CrossRef]
84. Jiang, J.; Lu, Y.; Li, J.; Li, L.; He, X.; Shao, H.; Dong, Y. Effect of Seed Treatment by Cold Plasma on the Resistance of Tomato to *Ralstonia solanacearum* (Bacterial Wilt). *PLoS ONE* **2014**, *9*, e97753. [CrossRef]
85. Volin, J.; Denes, F.S.; Young, R.A.; Park, S.M.T. Modification of seed germination performance through cold plasma chemistry technology. *Crop Sci.* **2000**, *40*, 1706–1718. [CrossRef]

86. Mildažienė, V.; Aleknavičiūtė, V.; Žūkienė, R.; Paužaitė, G.; Naučienė, Z.; Filatova, I.; Lyushkevich, V.; Haimi, P.; Tamošiūnė, I.; Baniulis, D. Treatment of Common Sunflower (*Helianthus annus* L.) Seeds with Radio-frequency Electromagnetic Field and Cold Plasma Induces Changes in Seed Phytohormone Balance, Seedling Development and Leaf Protein Expression. *Sci. Rep.* **2019**, *9*, 6437. [CrossRef]
87. Ling, L.; Jiafeng, J.; Jiangang, L.; Minchong, S.; Xin, H.; Hanliang, S.; Yuanhua, D. Effects of cold plasma treatment on seed germination and seedling growth of soybean. *Sci. Rep.* **2014**, *4*, 5859. [CrossRef]
88. Babajani, A.; Iranbakhsh, A.; Oraghi Ardebili, Z.; Eslami, B. Seed Priming with Non-thermal Plasma Modified Plant Reactions to Selenium or Zinc Oxide Nanoparticles: Cold Plasma as a Novel Emerging Tool for Plant Science. *Plasma Chem. Plasma Process.* **2019**, *39*, 21–34. [CrossRef]
89. Abedi, S.; Iranbakhsh, A.; Oraghi Ardebili, Z.; Ebadi, M. Seed priming with cold plasma improved early growth, flowering, and protection of Cichorium intybus against selenium nanoparticle. *J. Theor. Appl. Phys.* **2020**, *14*, 113–119. [CrossRef]
90. Moghanloo, M.; Iranbakhsh, A.; Ebadi, M.; Oraghi Ardebili, Z. Differential physiology and expression of phenylalanine ammonia lyase (PAL) and universal stress protein (USP) in the endangered species *Astragalus fridae* following seed priming with cold plasma and manipulation of culture medium with silica nanoparticles. *3 Biotech* **2019**, *9*, 288. [CrossRef] [PubMed]
91. Iranbakhsh, A.; Oraghi Ardebili, Z.; Oraghi Ardebili, N.; Ghoranneviss, M.; Safari, N. Cold plasma relieved toxicity signs of nano zinc oxide in *Capsicum annuum* cayenne via modifying growth, differentiation, and physiology. *Acta Physiol. Plant.* **2018**, *40*, 154. [CrossRef]
92. Colmer, J.; O'Neill, C.M.; Wells, R.; Bostrom, A.; Reynolds, D.; Websdale, D.; Shiralagi, G.; Lu, W.; Lou, Q.; Le Cornu, T.; et al. SeedGerm: A cost-effective phenotyping platform for automated seed imaging and machine-learning based phenotypic analysis of crop seed germination. *New Phytol.* **2020**, *228*, 778–793. [CrossRef]
93. Nguyen, T.; Hoang, V.-N.; Le, T.; Tran, T.-H.; Hai, V. A Vision Based Method for Automatic Evaluation of Germination Rate of Rice Seeds. In Proceedings of the 1st International Conference on Multimedia Analysis and Pattern Recognition (MAPR), Ho Chi Minh, Vietnam, 5–6 April 2018; pp. 1–6. [CrossRef]
94. Zhou, J.; Applegate, C.; Alonso, A.; Reynolds, D.; Orford, S.; Mackiewicz, M.; Griffiths, S.; Penfield, S.; Pullen, N. Leaf-GP: An open and automated software application for measuring growth phenotypes for arabidopsis and wheat. *Plant Methods* **2017**, *13*, 117. [CrossRef]
95. Falk, K.G.; Jubery, T.Z.; Mirnezami, S.V.; Parmley, K.A.; Sarkar, S.; Singh, A.; Ganapathysubramanian, B.; Singh, A.K. Computer vision and machine learning enabled soybean root phenotyping pipeline. *Plant Methods* **2020**, *16*, 5. [CrossRef] [PubMed]
96. Genze, N.; Bharti, R.; Grieb, M.; Schultheiss, S.J.; Grimm, D.G. Accurate machine learning-based germination detection, prediction and quality assessment of three grain crops. *Plant Methods* **2020**, *16*, 157. [CrossRef]
97. Dell'Aquila, A. Digital Imaging Information Technology Applied to Seed Germination Testing: A Review. *Agron. Sustain. Dev.* **2009**, *29*, 213–221. [CrossRef]
98. Dantas de Medeiros, A.; Silva, L.; Ribeiro, J.P.; Ferreira, K.; Fim Rosas, J.; Santos, A.; Barboza da Silva, C. Machine Learning for Seed Quality Classification: An Advanced Approach Using Merger Data from FT-NIR Spectroscopy and X-ray Imaging. *Sensors* **2020**, *20*, 4319. [CrossRef]
99. Singh, A.V.; Chandrasekar, V.; Janapareddy, P.; Mathews, D.E.; Laux, P.; Luch, A.; Yang, Y.; Garcia-Canibano, B.; Balakrishnan, S.; Abinahed, J.; et al. Emerging Application of Nanorobotics and Artificial Intelligence To Cross the BBB: Advances in Design, Controlled Maneuvering, and Targeting of the Barriers. *ACS Chem. Neurosci.* **2021**, *12*, 1835–1853. [CrossRef]
100. Singh, A.V.; Maharjan, R.S.; Jungnickel, H.; Romanowski, H.; Hachenberger, Y.U.; Reichardt, P.; Bierkandt, F.; Siewert, K.; Gadicherla, A.; Laux, P.; et al. Evaluating Particle Emissions and Toxicity of 3D Pen Printed Filaments with Metal Nanoparticles As Additives: In Vitro and in Silico Discriminant Function Analysis. *ACS Sustain. Chem. Eng.* **2021**, *9*, 11724–11737. [CrossRef]
101. Joosen, R.V.L.; Kodde, J.; Willems, L.A.J.; Ligterink, W.; van der Plas, L.H.W.; Hilhorst, H.W.M. GERMINATOR: A software package for high-throughput scoring and curve fitting of Arabidopsis seed germination. *Plant J.* **2010**, *62*, 148–159. [CrossRef] [PubMed]
102. Hoffmaster, A.F.; Xu, L.; Fujimura, K.; Bennett, M.A.; Evans, A.F.; McDonald, M.B. The Ohio State University Seed Vigor Imaging System (SVIS) for Soybean and Corn Seedlings. *Seed Technol.* **2005**, *27*, 7–24.
103. Charavay, C.; Segard, S.; Pochon, N.; Nussaume, L.; Javot, H. SeedUSoon: A New Software Program to Improve Seed Stock Management and Plant Line Exchanges between Research Laboratories. *Front. Plant Sci.* **2017**, *8*. [CrossRef]
104. Tanabata, T.; Shibaya, T.; Hori, K.; Ebana, K.; Yano, M. SmartGrain: High-throughput phenotyping software for measuring seed shape through image analysis. *Plant Physiol* **2012**, *160*, 1871–1880. [CrossRef]
105. Zhu, F.; Paul, P.; Hussain, W.; Wallman, K.; Dhatt, B.K.; Sandhu, J.; Irvin, L.; Morota, G.; Yu, H.; Walia, H. SeedExtractor: An Open-Source GUI for Seed Image Analysis. *Plant Sci.* **2021**, *11*, 1546. [CrossRef]
106. Komyshev, E.; Genaev, M.; Afonnikov, D. Evaluation of the SeedCounter, A Mobile Application for Grain Phenotyping. *Front. Plant Sci.* **2017**, *7*, 1990. [CrossRef]
107. Marmiroli, N.; White, J.C. Editorial: Nanotoxicology and Environmental Risk Assessment of Engineered Nanomaterials (ENMs) in Plants. *Front. Plant Sci.* **2016**, *7*, 1370. [CrossRef] [PubMed]
108. Wang, W.; He, A.; Peng, S.; Huang, J.; Cui, K.; Nie, L. The Effect of Storage Condition and Duration on the Deterioration of Primed Rice Seeds. *Plant Sci.* **2018**, *9*, 172. [CrossRef]

109. Farooq, M.; Usman, M.; Nadeem, F.; Rehman, H.; Wahid, A.; Basra, S.; Siddique, K. Seed priming in field crops—Potential benefits, adoption and challenges. *Crop Pasture Sci.* **2019**, *70*, 731–771. [CrossRef]
110. Kah, M.; Tufenkji, N.; White, J.C. Nano-enabled strategies to enhance crop nutrition and protection. *Nat. Nanotechnol.* **2019**, *14*, 532–540. [CrossRef] [PubMed]
111. Garg, M.; Sharma, N.; Sharma, S.; Kapoor, P.; Kumar, A.; Chunduri, V.; Arora, P. Biofortified Crops Generated by Breeding, Agronomy, and Transgenic Approaches Are Improving Lives of Millions of People around the World. *Front. Nutr.* **2018**, *5*, 12. [CrossRef] [PubMed]
112. Gilbertson, L.M.; Pourzahedi, L.; Laughton, S.; Gao, X.; Zimmerman, J.B.; Theis, T.L.; Westerhoff, P.; Lowry, G.V. Guiding the design space for nanotechnology to advance sustainable crop production. *Nat. Nanotechnol.* **2020**, *15*, 801–810. [CrossRef] [PubMed]

cells

MDPI

Review

Insights into the Interactions among Roots, Rhizosphere, and Rhizobacteria for Improving Plant Growth and Tolerance to Abiotic Stresses: A Review

Naeem Khan [1,*], Shahid Ali [2], Muhammad Adnan Shahid [3], Adnan Mustafa [4], R. Z. Sayyed [5] and José Alfredo Curá [6]

1 Department of Agronomy, Institute of Food and Agricultural Sciences, University of Florida, Gainesville, FL 32611, USA
2 College of Life Sciences, Northeast Forestry University, Harbin 150040, China; shahidsafi926@gmail.com
3 Department of Agriculture, Nutrition and Food Systems, University of New Hampshire, Durham, NH 03824, USA; muhammad.shahid@unh.edu
4 Biology Center CAS, SoWa RI, Na Sadkach 7, 370-05 České Budějovice, Czech Republic; adnanmustafa780@gmail.com
5 Department of Microbiology, P.S.G.V.P. Mandal's, Arts, Science, and Commerce College, Shahada 425409, India; sayyedrz@gmail.com
6 Facultad de Agronomía, Universidad de Buenos Aires, Av. San Martín 4453, Ciudad Autónoma de Buenos Aires C1417DSE, Argentina; acura@agro.uba.ar
* Correspondence: naeemkhan@ufl.edu or naeemkhan001@gmail.com

Citation: Khan, N.; Ali, S.; Shahid, M.A.; Mustafa, A.; Sayyed, R.Z.; Curá, J.A. Insights into the Interactions among Roots, Rhizosphere, and Rhizobacteria for Improving Plant Growth and Tolerance to Abiotic Stresses: A Review. *Cells* **2021**, *10*, 1551. https://doi.org/10.3390/cells10061551

Academic Editors: Suleyman Allakhverdiev, Alexander G. Ivanov and Marian Brestic

Received: 18 April 2021
Accepted: 17 June 2021
Published: 19 June 2021

Abstract: Abiotic stresses, such as drought, salinity, heavy metals, variations in temperature, and ultraviolet (UV) radiation, are antagonistic to plant growth and development, resulting in an overall decrease in plant yield. These stresses have direct effects on the rhizosphere, thus severely affect the root growth, and thereby affecting the overall plant growth, health, and productivity. However, the growth-promoting rhizobacteria that colonize the rhizosphere/endorhizosphere protect the roots from the adverse effects of abiotic stress and facilitate plant growth by various direct and indirect mechanisms. In the rhizosphere, plants are constantly interacting with thousands of these microorganisms, yet it is not very clear when and how these complex root, rhizosphere, and rhizobacteria interactions occur under abiotic stresses. Therefore, the present review attempts to focus on root–rhizosphere and rhizobacterial interactions under stresses, how roots respond to these interactions, and the role of rhizobacteria under these stresses. Further, the review focuses on the underlying mechanisms employed by rhizobacteria for improving root architecture and plant tolerance to abiotic stresses.

Keywords: root; rhizosphere; rhizobacteria; root morphology; abiotic stresses

1. Introduction

Stress is any environmental factor that can adversely affect plant growth and development and decrease the final yield. All the major abiotic stresses lead to the major declines in the yield of globally important crop plants. Drought stress affects leaf expansion, stem elongating, root proliferation, and reduces water and nutrient uptake [1]. Water stress for a prolonged duration also declines leaf water potential and stomatal closing and opening, delays flowering, and decreases seed number and size [2–4]. Salt stress is the most stubborn among all abiotic stresses and has prolonged deleterious effects on glycophytes [5,6]. Many plants cannot endure salt concentrations of more than 200 mM [7], as high salinity reduces the rate of seed germination, the establishment of seedlings, vegetative growth, and increases the osmotic pressure, with ion toxicity ultimately leading to oxidative damage [8–10]. Heat stress affects overall plant morphology, physiology, and biochemistry, leading to stunted plant growth and a reduction in plant biomass and productivity [11,12]. Similarly, heavy metals have direct and indirect effects on plant growth, significantly reduce plant growth, and disturb various physiological and molecular mechanisms of plants, resulting in chlorosis, senescence, and an inhibition of chlorophyll and photosynthesis [13–15]. The above-mentioned stresses are collectively known as abiotic stresses, which are a pretesting threat to crop growth and agriculture and responsible for great crop yield loss and future food safety [16–18].

Plant roots interconnect with a specific group of soil microorganisms that inhabit the root vicinity, known as the rhizosphere. The rhizosphere is considered one of the most composite ecosystems on earth, containing millions of microbial cells—but the number changes according to plant genotype and growth stages [19]. In the rhizosphere, plant roots secrete various compounds that act as a chemical attractant for soil microorganisms [20]. These root exudates also modify the physico-chemical properties of the soil and thus, adjust the structure of soil microbial communities in the close vicinity of the root surface [21,22]. Rhizobacteria that inhabit the rhizosphere alleviate the influences of abiotic stresses on plants through a number of different mechanisms, which include alterations in phytohormone levels, metabolic adjustments, antioxidant defenses, bacterial exopolysaccharides (EPS), and protecting and improving the root growth. These microorganisms can modulate the expression of plant metabolites and improve their photosynthetic, carbohydrate, and protein content, thus improving the yield-related traits under stress [23]. They improve plant growth by enhancing nutrient and water uptake from the soil even under stressful environments [24]. They also improve the phosphate and nitrate reductase activities under water-stressed conditions [25] and limit the Na^+ accumulation under salt stress [26]. Furthermore, they indirectly promote plant growth by decreasing the damaging effects of pathogenic organisms by producing antagonistic substances (Figure 1) [27]. The PGPR *Burkholderia phytofirmans* improved the photosynthesis and defense responses of *Arabidopsis thaliana* to pathogenic attack [28], whereas Pseudomonas putida improved the systemic resistance and priming of wheat plants to pathogen attack [29].

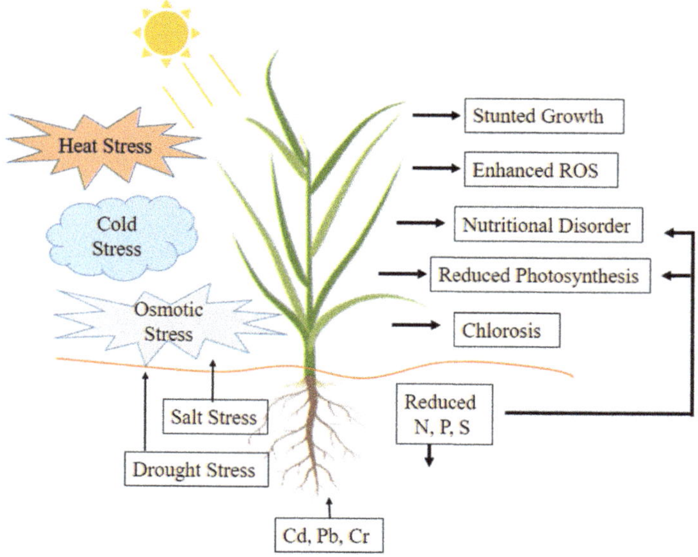

Figure 1. Adverse effects of abiotic stresses on root and shoot growth. Abiotic stresses adversely affect root growth, which results in an overall decrease in plant growth due to an extreme deficiency of water and nutrients in the aboveground parts of the plant.

In this review, we summarize and discuss the current understanding of root–rhizosphere and rhizobacterial interactions to abiotic stresses. We first summarize the impacts of abiotic stresses on overall plant growth and yield. We then elucidate the role of rhizobacteria under abiotic stresses and evaluate the strategies of rhizobacteria for improving root growth and plant tolerance mechanisms.

2. The Root–Rhizosphere and Rhizobacterial Alliance

Root-associated rhizobacterial communities play an important role in the maintenance of plant health under abiotic stresses. Plant–microbial associations happen at the rhizosphere. The rhizosphere consists of both beneficial and pathogenic microorganisms. This rhizobacterial community of the rhizosphere changes with changes in soil properties [29]. Interactions between rhizobacteria in the rhizosphere have intuitive effects (i.e., improvement of the soil nutrient content, remediation of HMs, minimization of soil disturbances and root growth and soil immune responses) on soil health and improve the nutritional status of the soil, which is important for better plant growth [30]. A surfeit of these interactions between the roots, rhizosphere, and rhizobial microbes also improves root growth and proliferation, which play a critical role in the exchange of resources between the shoots and the soil environment [31,32]. Rhizobacteria also benefit crop production by reducing the dependency on chemical fertilizers to attain high production yields.

This rich rhizosphere–rhizobacterial interaction defends root exudates, which consist of various organic compounds that attract the microorganisms towards the root vicinity [33,34]. Root exudation arbitrates plant–microbe interactions by root colonization and the promotion of root growth. As Neal et al. [35] reported, there is an increase in the removal of benzoxazinone from the maize rhizosphere due to the presence of *P. putida*. Root exudates contain a wide array of chemical constituents, including amino acids, peptides, sugars, enzymes, vitamins, organic acids, and different types of primary and secondary metabolites [36,37]. The microbial soil diversity depends on the type and composition of root exudate, which supports the growth of useful microorganisms that can assist in plant health and their productivity, while, in other cases, some root exudates also prevent the growth of harmful microbes [38–42]. The proteome data also provide evidence on the biological process that occurs in the rhizosphere, as Baysal et al. [43] carried out a proteomic approach for studying the control of soil-borne pathogens with the help of *Bacillus* species. Bona et al. [44] used the metaproteome approach for studying the microbial communities and their activities in the rhizosphere. They reported the rhizosphere proteome of *Vitis vinifera*, where they found that the bacterial species belonging to the *Bacillus*, *Pseudomonas*, *Bradyrhizobium*, *Streptomyces*, and *Bulkhorderia* genuses are more active in protein expression and their rhizospheres have more metabolic processes and regulation.

The root exudate strigolactone is also an important signaling molecule that regulates primary root and root hair length. They are present in root exudates of monocotyledonous and dicotyledonous plants and are involved in mutualistic interactions with arbuscular mycorrhizal fungi in the rhizosphere [45]. Strigolactones induce hyphal branching near the host plant, which enhances the chances of interactions between the host plant roots and fungi [46]. They may also play an important role in legume–rhizobia symbiosis [47,48].

Among all these root exudates, the most important are organic acids, which not only act as a source of energy for microbial–cellular metabolism but also act as intermediary compounds in bio-geochemical cyclic reactions taking place in the rhizosphere [49,50]. The low-carbon molecules of root exudates act as the originator for the biosynthesis of rhizobacterial-produced phytohormones, whereas tryptophan (Trp) present in the root exudates acts as a precursor for the production of indole-3-acetic acid (IAA), and is mostly present in the root tip region [51]. In addition, the precursor for ethylene, aminocyclopropane-1-carboxylic acid (ACC), also oozes out from plants and can be utilized as a source of nitrogen and carbon by rhizobacteria [52–54]. Other compounds identified as flavonoids are released by the roots of leguminous plants, persuading the transcription of rhizobia Nod factors (NF). Nod factors are responsible for the formation of root hairs and also play an important role in nodule initiation [55–58].

The root–rhizosphere and rhizobacterial interactions also influence plant responses to environmental stresses [59–61]. These rhizobacterial species are reported to impart abiotic stress lenience by up or down-regulating the stress-responsive genes, such as S-adenosyl-methionine synthetase, ascorbate peroxidase, and heat shock proteins [62–65]. The root–rhizobial association also prompts resistance against root herbivores and guards the roots against a number of diseases [66]. The effects of rhizobacteria on the growth of root hairs and root system architecture were inspected on seedlings grown in vitro in upright agar plates comprising a mineral medium inoculated with a 10^8 cfu per mL of rhizobacteria. The results of the experiment showed significant positive effects of the inoculated rhizobacteria on root hairs and architecture under in vitro conditions [67–70].

The role of root–rhizosphere and rhizobacterial interactions is essential for plant growth promotion, nutrient acquisition, and yield quality [71]. It is apparent that mutual communications occur among plants, soil, and microorganisms, and all such interactions are intricate and should be accounted for useful outcomes in terms of plant growth and soil health (Figure 2) [72–76].

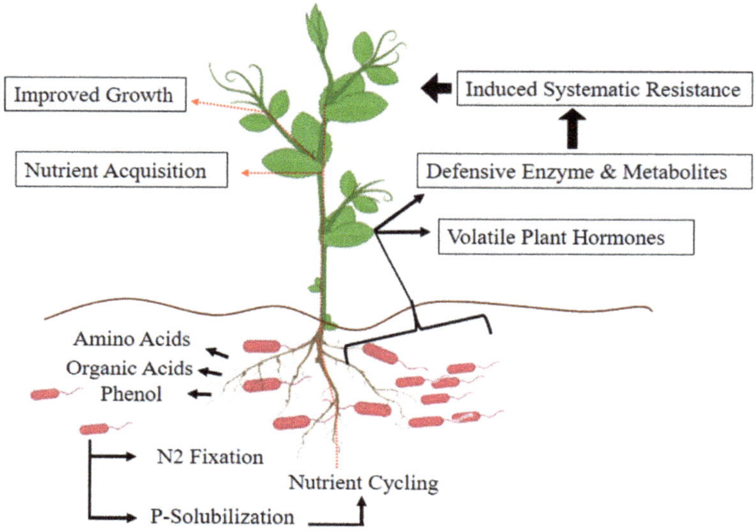

Figure 2. Mechanisms employed by rhizobacteria for increases in plant growth and tolerance to abiotic stresses. Rhizobacteria improve nutrient content and nutrient cycling and help plants to withstand harsh environmental conditions.

3. Effects of Abiotic Stresses on Root Growth and Rhizosphere

Abiotic stresses adversely affect plant growth and development as well as the overall growth and morphology of roots, which not only affects the crop quality but also the final yield. An increase in carbon dioxide (CO_2) levels results in global climate change, consisting of a rise in temperatures and disturbing weather patterns that significantly affect the plant rhizosphere [77–80].

Plant roots are the major organs responsible for nutrient and water acquisition and maintaining normal plant growth and yield [81,82]. However, abiotic stresses result in poor root growth, which results in decreased water and nutrient uptake. Drought stress has more severe effects on plant roots than any other stress and significantly reduces its biomass [83,84]. Salinity causes ion toxicity due to an excess of Na^+ and Cl^-, which also damages root growth and development [85,86]. High temperatures adversely affect the root architecture and the roots' interactions with the surrounding microorganisms (Figure 1), whereas a decrease in temperature at the root zone adversely affects the process of nodulation and N-fixation [87,88]. Similarly, the presence of heavy metals (HMs) in the rhizosphere has toxic effects on root growth (Figure 1) [89]. Among the HMs, lead (Pb) is the most widespread, causing inhibition of cell division in the root tip and rapid inhibition of root growth [90]. The presence of a high concentration of cadmium (Cd) in the rhizosphere causes visible injuries to the root and shoot, browning of the root tips, and chlorosis in plant shoots [91–93]. Similarly, chromium (Cr) toxicity also causes chlorosis in newly developing leaves, and injury to roots [94,95].

Role of Rhizobacteria under Abiotic Stresses

Rhizosphere microorganisms, mainly beneficial bacteria, can increase plant performance under stressful conditions and, consequently, improve soil health and enhance root growth and plant yield [96]. Rhizobacteria either exert a direct stimulation on root and overall plant growth by fixing nitrogen, the production of plant hormones, and sequestering iron and solubilizing phosphate [97–99]. The microbial-produced phytohormones promote root growth and alter root architecture, triggering an increase in root surface area [100]. This is considered one of the basic mechanisms employed by root-associated bacteria for the increases in nutrient uptake. In the rhizosphere, plant–rhizobacteria interactions assist plants through the induction of systemic resistance against pathogens and 1-aminocyclopropane 1-carboxylic acid (ACC)-deaminase activity. Such stimuli of rhizobacteria can benefit plant defense against pathogens and can also improve overall plant yield (Figure 2; Table 1) [101–105].

Table 1. List of rhizobacteria species responsible for abiotic stress tolerance in common crop plants.

Crop	Stress	Rhizobacteria	References
Helianthus annuus	Drought	*Achromobacter xylosoxidans* (SF2) *Bacillus pumilus* (SF3 and SF4)	Castillo et al. [106]
Oryza sativa	Drought	*Azospirillum brasilense* Az.39	Ruíz-Sanches et al. [107]
Vigna radiata	Drought	*Pseudomonas fluorescens* strain Pf1 *Bacillus subtilis* EPB5, EPB22 and EPB31	Saravanakumar et al. [108]
Cucurbita pepo	Drought	*Bacillus circulans ML2, Bacillus megaterium* ML3	El-Meihy [109]
Zea mays	Drought	*Klebsiella variicola* F2, *Pseudomonas fluorescens* YX2 *Raoultella planticola* YL2	Gou et al. [110]
Arachis hypogea	Salinity	*B. licheniformis* K11	Lim et al. [111]
Phaseolus vulgaris	Salinity	*Aneurinibacillus aneurinilyticus, Paenibacillus sp.*	Gupta and Pandey [112]
Steva rebaundia	Salinity	*Steptomyces spp.*	Tolba et al. [113]
Abelmoschus esculentus	Salinity	*Enterobacter sp.*	Habib et al. [114]
Lycopersicon esculentum	Heavy metal	*Pseudomonas aeruginosa, Burkholderia gladioli*	Khana et al. [115]
Triticum aestivum	Heavy metal	*Bacillus siamensis*	Awan et al. [116]
Brassica nigra	Heavy metal	*Bacillus cereus*	Akhtar et al. [117]
Pisum sativum	Heavy metal	*V. paradoxus* 5C-2	Belimov et al. [118]
Solanum nigrum	Heavy metal	*Bacillus* genus	He et al. [119]
Mentha piperita	Heavy metal	*Alcalegenes faecalis, B. amyloliquefaciens*	Zafar-ul-Haye et al. [120]
Triticum aestivum	Heat	*Pseudomonas brassicacearum, Bacillus thuringiensis, Bacillus subtilis*	Ashraf et al. [121]
Triticum aestivum	Heat	*Bacillus velezensis* 5113	Abde El-Daim [122]
Lycopersicon esculentum	*Heat*	*Bacillus cereus*	Khan et al. [123]
Solanum tuberosum	*Salt/Drought/HMs*	*Bacillus pumilus DH 11, Bacillus firmus 40*	Gururani et al. [124]

It has also been reported by Marulanda et al. [125] that rhizobacterial-inoculated plants exhibit significant increases in plant growth and yield, as well as in drought tolerance to semi-arid and arid environments. The application of *Phyllobacterium brassicacearum* strain STM196 to *Arabidopsis thaliana* improved its resistance to moderate water deficit stress by modulating the rate of transpiration and delaying maturity [126]. The inoculation of plants with rhizobacteria helps in the mitigation of the deleterious effects of various stresses by assisting them in the acquisition of less available nutrients and by increasing the levels of plant growth regulators [12]. Microorganisms with the capability to persist under severe environmental conditions are more active at vindicating the undesirable impacts of drought on crop plants [127]. Niu et al. [128] reported that drought-tolerant strains are capable of producing exopolysaccharides (EPS), which stimulate seed germination and seedling growth under drought stress. Among all PGPR strains, *Pseudomonas fluorescens* has the highest capability of producing EPS and ACC deaminase. Recently, Batool et al. [129] reported the effects of rhizobacteria in reducing the impacts of drought and maintaining the higher growth and physio-chemical properties of the plants. They noted a higher growth rate and leaf area and an increase in dry matter production in inoculated plants. The isolated PGPR-HAS31 maintained higher chlorophyll, photosynthesis, soluble proteins, sugars, and enzymatic activities in relation to uninoculated plants.

Kumar et al. [130] studied the effects of salt-tolerant (ST) bacterial strains. Their results exposed the inoculation of paddy plants with the rhizobacteria *Pseudomonas aeruginosa*, and *Lysinibacillus* sp. boosted the seedlings' growth under salinity stress. *Pseudomonas aeruginosa* exhibited more profound effects than other species and significantly improved the rate of seed germination and the lengths of shoots and roots. Shultana et al. [131] measured the effects of rhizobacterial strains isolated from the saline rice field in Malaysia on the growth and yield of rice plants. Their results revealed significant useful effects of bacterial inoculation on the rate of transpiration, photosynthesis, and stomatal conductance, which also resulted in a higher increase in yield. The most frequently used halotolerant rhizobacteria are *Azotobacter*, *Acinetobacter*, Bacillus sp., *Pseudomonas* sp., *Rhizobium* sp., and *Serratia* sp., which employ different mechanisms, including N-fixation, P-solubilization, and siderophore formation [132–135]. Similarly, many different reports revealed that halotolerant microbes significantly enhanced the growth of many crops under both normal and saline conditions [136,137]. Various ST rhizobacterial species improve the salt tolerance in plants by the production of different types of osmolytes and antioxidant enzymes and synthesizing ACC deaminase [138–140].

Temperature is another important variable that influences plant root growth. Fluctuations in the temperature of the root zone also alter shoot growth responses by inducing changes in the temperature of the shoot apical meristems [77]. It has adverse effects on the plasma membrane, photosynthesis, phytohormones, and enzyme activity. However, the rhizosphere microbes have the ability to mitigate the adverse effects of high temperature stress. They protect their membranes and nucleic acids under such conditions and contribute to normal plant growth. Some microorganisms are better in the production of biofilm and can help plants to tolerate high salt and temperature stress [141]. Moreover, *B. subtilis* Co1-6 and *P. polymyxa* Mc5Re-14 showed better production of the bioactive secondary metabolite apigenin-7-*O*-glucoside, whereas inoculation with *Pseudomonas* sp. strain AKM-P6 and *P. putida* strain AKM-P7 enhanced the tolerance of sorghum and wheat seedlings to high temperature stress due to the synthesis of high-molecular-weight proteins, and also improved the levels of cellular metabolites [142,143]. Zhu et al. [144] observed positive physiological effects of the arbuscular mycorrhizal fungus *Glomus etunicatum* on *Zea mays* plants at a range of different temperatures (5–40 °C) when compared with uninoculated plants. Similarly, Pedranzani et al. [145] showed an increase in antioxidants in the shoots and roots of *Digitaria eriantha* and a reduction in cellular lipid peroxidation in plants inoculated with the arbuscular mycorrhizal fungus *Rhizophagus irregularis* under cold stress (4 °C).

Soil microbes maintain an efficient flow of water and nutrients to plants under heat stress conditions [146], whereas thermotolerant phosphate solubilizing bacteria act as biofertilizers in agriculture and are involved in the biogeochemical cycling of phosphorus [147]. One of the common mechanisms inked by rhizosphere microorganisms for reducing the effects of heat stress in plants is the induction of osmoprotectants and heat shock proteins (HSP). The modulation of the levels of phytohormones, secondary metabolites, and the production of ROS are some of the important mechanisms employed by rhizosphere microorganisms for combating the adverse effects of heat stress. Kang et al. [148] reported an increase in the content of GA and ABA and the reduction in the content of jasmonate and salicylate in pepper plants inoculated with a gibberellin-producing PGPR. This alteration in the content of phytohormone/plant growth regulators resulted in an increase in plant growth under low temperature stress conditions. Issa et al. [149] reported that bacterium *Burkholderia phytofirmans* significantly enhanced the production of phenolics, proline, and starch under heat stress and was able to protect the tissues of tomato against heat. Rodriguiz et al. [150] also reported *Curvularia protuberate*-induced heat stress tolerance in tomatoes. Gram-positive microorganisms possess heat-resistant spores that are used in the formulation of stable and dry powder products [151].

Rhizobacteria are known to affect the movement and accessibility of HMs by releasing various chelating agents or by the process of acidification, phosphate solubilization, and redox reaction and thus, enhance the phytoremediation of HMs [152]. The aptitude of microorganisms to degrade pollutants largely depends on the pH, temperature, and moisture content of the environment in which they live [153]. Microorganisms can also cleanse metals by valence conversion, volatilization, or extracellular chemical precipitation [40]. However, the co-inoculation of *Bacillus subtilis* was found to be more effective against the remediation of HMs than a single inoculation. Some bacterial species produce iron-chelating substances called siderophores, which enhance the mobility and reduce the bioavailability of metals [154]. Sulfate-reducing bacteria have the ability to convert sulfate to hydrogen sulfate, which then reacts with heavy metals and converts them to insoluble forms of metal sulfides [155]. The oxalate crystals produced by mycorrhizal fungi are also known to immobilize and detoxify HM [156].

Tiwari et al. [157] reported that plant-associated bacteria reduce the accumulation of various metals in plant tissues and also assist in reducing metal availability in the soil through a number of different mechanisms. The practice of rhizobacteria in combination with plants is estimated to deliver high efficacy for phytoremediation [158–160]. Khanna et al. [161] also reported the role of *Pseudomonas aeruginosa* and *Burkholderia gladioli* in the alleviation of Cd stress (0.4 mM) in the 10-day old seedlings of *L. esculentum*. They revealed the adverse effects of Cd stress on root and shoot growth and on plant biomass. However, the bacterial-inoculated plants showed improved plant growth and resistance to Cd toxicity. The usage of these beneficial rhizobacteria is reflected as one of the most capable approaches for harmless crop management practices in HM-contaminated soils. Plant–microbe interactions help in adapting plants to metalliferous environments and increase microbial-assisted metal tolerance.

4. Strategies of Rhizobacteria for Improving Root Architecture under Stresses

Rhizobacteria efficiently colonize the roots of crop plants and enhance their growth by a number of different direct and indirect mechanisms. The alteration of root system architecture by root-associated rhizobacteria involves the assembly of phytohormones, for example, auxin, gibberellins, and other signaling molecules that lead to greater lateral root branching and growth of root hairs. As these rhizobacteria attach to the plant root surface, they convert root exudates into phytohormones [162–166]. The composition of the root exudates change along with the plant development; hence, the rhizo-microbiome alignment varies consequently [167–169]. They also show the antagonistic activities against the phytopathogenic microorganisms by producing siderophores, enzymes, the synthesis of antibiotics, antifungal compounds, and essential nutrients, thus improving the root architecture under all these stressful conditions [170–172].

The root colonization pattern of rhizobacteria like *Bacillus* and *Pseudomonas* has been well-described in numerous plants, including tomato [173], cucumber [174], *Arabidopsis thaliana* [175], wheat [176], and grape [177]. Erturk et al. [178] studied the effects of various strains of *Bacillus* on rooting and root growth in kiwifruit. The highest rooting ratios (47.5%) were obtained as a result of Bacillus RC03 and *Bacillus simplex* RC19 treatments. The inoculation of wheat and maize plants with these bacterial species also delayed the onset of the drought symptoms on plants. Both of the applied rhizobacterial species were synergistic to root branching and length, compared to the control. *Enterobacter* sp. demonstrated greater effects on root branching, length, and diameter when compared to the control (Figure 3) [179–182].

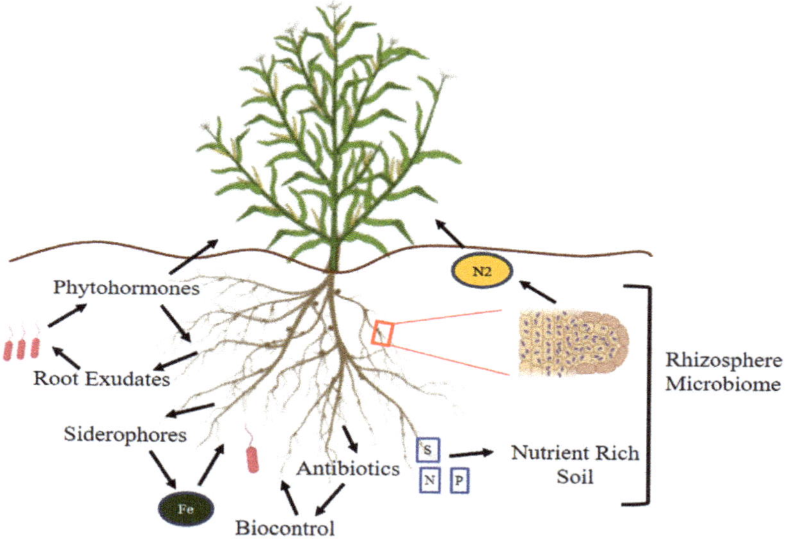

Figure 3. Strategies of PGPR for improving root architecture and overall plant growth under abiotic stresses. These microorganisms form a rhizosheath around the roots and produce antibiotics and biocontrol agents, thus protecting the roots from the adverse effects of environmental stresses.

5. Stress Responsive Metabolites Mediated by Rhizobacteria

Plants experience diverse abiotic stresses throughout their life cycle that need to be handled in order to survive. Abiotic stress lenience has been studied in relation to rhizobacteria in order to offer a biological understanding of the alteration and persistence of rhizobacteria under such stresses [183,184]. Several stress tolerance mechanisms have been considered for rhizobacterial-arbitrated stress tolerance in plants. It has been reported that rhizobacterial inoculation alters the metabolic expression in plants under stress and helps in activating stress-responsive genes and metabolites [185]. The potential of rhizobacteria producing stress-related metabolites is gaining importance these days. They also have the ability to modulate the transcriptional machinery responsible for stress tolerance in plants. Their involvement in the upregulation of ABA-signaling cascades that lead to the expression of *TaWRKY* and *TaMYB* has been reported previously [68,186]. Many genetic studies have been carried out in plants grown under abiotic stresses to characterize the bacterial-mediated changes in plants at genetic and metabolic levels [187,188]. Previously, the genetic studies of drought stress tolerance were categorized by means of molecular and genetic approaches in pepper plants [184,189–194].

A large number of secondary metabolites, such as compatible solutes and volatile organic compounds (VOCs), have been reported to be from halotolerant rhizobacteria that are crucial for bettering the adverse effects of salinity stress in plants and their associated partners [195]. Halotolerant rhizobacteria employ key mechanisms for stress tolerance, which include osmotic adjustments at a cellular level, regulation in ionic transportation, and maintaining homeostasis by reducing the toxic effects of sodium (Na^+) and chlorine (Cl^-) ions [196]. Moreover, these microbes synthesize different types of volatile compounds and antifungal or antibacterial metabolites, for example, sugar, betaines, amino acids, and polyols, which help plants to withstand harsh environmental conditions [197–200]. Some of the halotolerant bacteria can endure stress caused by high salinity due to their innate ability to accrue some of the vital compatible osmolytes essential for retaining intracellular osmotic homeostasis that benefit them to persist under high salinity, and they are also liable for supporting plant growth and survival under such stresses (Figure 4) [201–205].

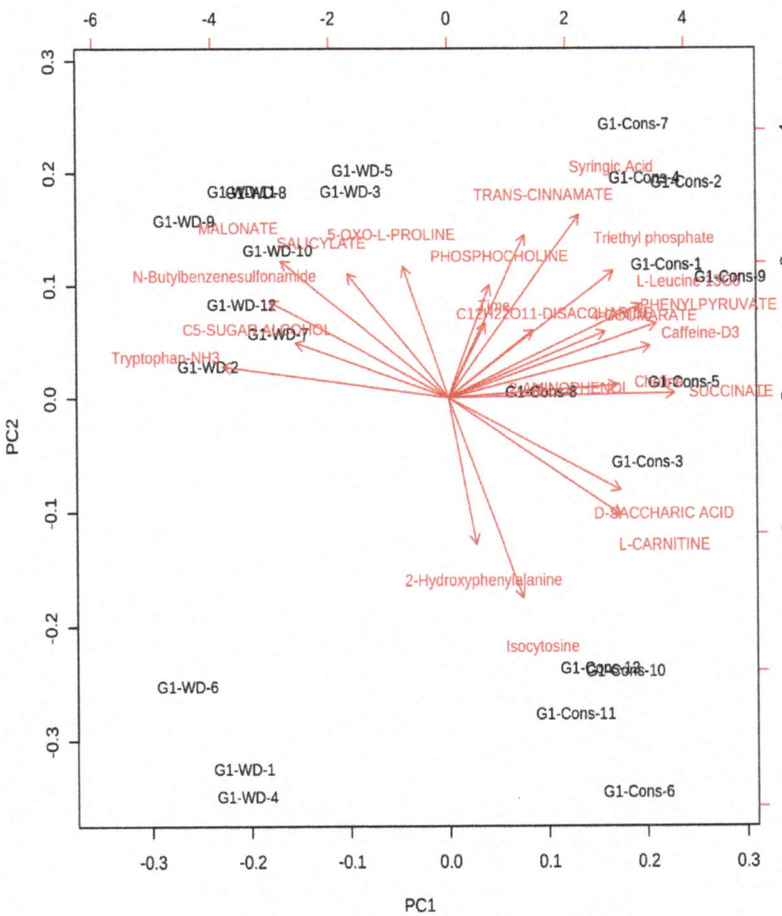

Figure 4. A PCA-based biplot showing the associations among different metabolites induced by plant growth-promoting rhizobacteria (PGPR) in chickpea leaves grown under stress conditions. The figure was generated by uploading data files to the MetaboAnalyst 3.0 server (http://www.metaboanalyst.ca accessed on 10 February 2021). We selected normalization by sum, log transformation, and auto-scaling for the analysis.

The overproduction of reactive oxygen species (ROS) under stress conditions alters redox states, causes damage to DNA, proteins, and membrane fluidity, and lastly, causes cell death [189]. Various growth-promoting rhizobacterial species are described to endure oxidative stress with the support of antioxidant enzyme activity. Sandhya et al. [206] reported an increase in the activity of ascorbate peroxidases (APX) in *Enterobacter* inoculated tomato seedlings grown under high saline conditions. Higher catalase (CAT) and superoxide dismutase activities were recorded in bacterial inoculated gladiolus plants when compared to the control [207,208].

Endophytic bacteria are capable of synthesizing nitrogenase enzymes under HM stress and destitute nitrogen conditions by giving ample nitrogen to connected plants. Doty et al. [209] also isolated endophytic genera (*Acinetobacter, Burkholderia, Rahnella,* and *Sphingomonas*) from *Populus trichocarpa* and *Salix sitchensis*, with the ability to synthesize nitrogenase enzymes, and were capable to fix atmospheric nitrogen [210]. The production of citric acid, gluconic acid, and oxalic acid by rhizobacteria plays an effective role in the mobilization of heavy metals, thus protecting the plant roots from the lethal effects of HMs [211–213]. Paredes-Páliz et al. [214] selected biofilm-forming rhizobacteria based on their ability for metal tolerance and applied them to *Spartina densiflora*. The inoculated plants were then allowed to grow for four months and then harvested. The frozen harvested plant parts were used for the determination of enzyme assays and gene expression. They noted increases in the activity of SOD, CAT, and APOX and a 2-fold increase of thiobarbituric acid reactive substances (TBARs) that resulted in membrane and cell damage. However, the created oxidative stress index (OSI) was significantly decreased (>50%) upon inoculation.

Changes in gene expression in relation to ethylene biosynthesis have been reported in rhizobacterial-inoculated plants grown under abiotic stresses [215–218]. The existence of ethylene (ET) is vital for normal plant growth and fruit ripening, but under stress conditions, the production of ethylene significantly increases, which has negative effects on seed germination and root growth [219,220]. However, ACC deaminase-containing rhizobacteria can hydrolyze ACC, the precursor of ET, thus decreasing the extra ethylene production under stress and saving plants from its inhibitory effects [221–223]. Beneficial rhizobacteria enhance the synthesis of proline in abiotically stressed plants. The most important proline synthesizing rhizobacteria are *Burkholderia* [224], *Arthrobacter*, and *Bacillus* [225].

6. Conclusions

Soil microbiomes and especially rhizobacteria possess different mechanisms by which they improve soil health, root growth, and the tolerance of plants to various abiotic stresses. The ability of these bacteria to survive under abiotic stresses makes them a brilliant candidate for sustainable agriculture. They improve root access to nutrients and water and improve their translocation to the above-ground parts of the plants, leading to overall improvements in plant growth and yield. These bacterial strains mitigate the adverse effects of abiotic stress by producing different types of metabolites, including phytohormone, exopolysaccharides, siderophores, antioxidant enzymes, and volatile compounds. Improvements in plant tolerance to abiotic stresses will result in increased yields and production of crops, even under stressful environments. This can be achieved via the search, selection, and engineering of rhizobacterial species capable of resistance to abiotic stresses.

Author Contributions: Conceptualization, N.K.; formal analysis, N.K. and S.A.; project administration, N.K.; software, N.K. and S.A.; methodology, M.A.S. and A.M.; validation, M.A.S. and R.Z.S.; data curation, A.M. and R.Z.S.; resources, N.K., S.A. and J.A.C.; supervision, N.K.; funding acquisition, J.A.C.; verified the numerical results, A.M., R.Z.S., S.A. and M.A.S.; writing review.; N.K.; editing, J.A.C., M.A.S., A.M., S.A. and R.Z.S. All authors have read and agreed to the published version of the manuscript.

Funding: This study received no external funding. The APC was funded by the UBACyT Project 20020170100080BA.

Institutional Review Board Statement: Not applicable.

Informed Consent Statement: Not applicable.

Data Availability Statement: All the data generated or analyzed during this study are included in this article.

Conflicts of Interest: The authors declare no conflict of interest.

References

1. Farooq, M.; Wahid, A.; Kobayashi, N.; Fujita, D.B.S.M.A.; Basra, S.M.A. Plant drought stress: Effects, mechanisms and management. In *Sustainable Agriculture*; Springer: Dordrecht, The Netherlands, 2009; pp. 153–188. [CrossRef]
2. Meena, K.K.; Sorty, A.M.; Bitla, U.M.; Choudhary, K.; Gupta, P.; Pareek, A.; Singh, D.P.; Prabha, R.; Sahu, P.K.; Gupta, V.K.; et al. Abiotic stress responses and microbe-mediated mitigation in plants: The omics strategies. *Front. Plant Sci.* **2017**, *8*, 172. [CrossRef]
3. Li, G.; Zhao, H.; Liu, Z.; Wang, H.; Xu, B.; Guo, X. The wisdom of honeybee defenses against environmental stresses. *Front. Microbiol.* **2018**, *9*, 722. [CrossRef] [PubMed]
4. Xu, K.; Lee, Y.S.; Li, J.; Li, C. Resistance mechanisms and reprogramming of microorganisms for efficient biorefinery under multiple environmental stresses. *Synth. Syst. Biotechnol.* **2019**, *4*, 92–98. [CrossRef] [PubMed]
5. Negrão, S.; Schmöckel, S.M.; Tester, M. Evaluating physiological responses of plants to salinity stress. *Ann. Bot.* **2017**, *119*, 1–11. [CrossRef] [PubMed]
6. Egamberdieva, D.; Lugtenberg, B. Use of plant growth-promoting rhizobacteria to alleviate salinity stress in plants. In *Use of Microbes for the Alleviation of Soil Stresses*; Springer: New York, NY, USA, 2014; Volume 1, pp. 73–96.
7. Timmusk, S.; Timmusk, K.; Behers, L. Rhizobacterial plant drought stress tolerance enhancement: Towards sustainable water resource management and food security. *J. Food Secur.* **2013**, *1*, 6–9.
8. Kaushal, M.; Wani, S.P. Rhizobacterial-plant interactions: Strategies ensuring plant growth promotion under drought and salinity stress. *Agric. Ecosyst. Environ.* **2016**, *231*, 68–78. [CrossRef]
9. Kumari, B.; Mallick, M.A.; Solanki, M.K.; Solanki, A.C.; Hora, A.; Guo, W. Plant growth promoting rhizobacteria (PGPR): Modern prospects for sustainable agriculture. In *Plant Health under Biotic Stress*; Springer: Singapore, 2019; pp. 109–127.
10. Subiramani, S.; Ramalingam, S.; Muthu, T.; Nile, S.H.; Venkidasamy, B. Development of abiotic stress tolerance in crops by plant growth-promoting rhizobacteria (PGPR). In *Phyto-Microbiome in Stress Regulation*; Springer: Singapore, 2020; pp. 125–145.
11. Barnawal, D.; Bharti, N.; Pandey, S.S.; Pandey, A.; Chanotiya, C.S.; Kalra, A. Plant growth-promoting rhizobacteria enhance wheat salt and drought stress tolerance by altering endogenous phytohormone levels and TaCTR1/TaDREB2 expression. *Physiol. Plant.* **2017**, *161*, 502–514. [CrossRef]
12. Morcillo, R.J.; Manzanera, M. The Effects of Plant-Associated Bacterial Exopolysaccharides on Plant Abiotic Stress Tolerance. *Metabolites* **2021**, *11*, 337. [CrossRef]
13. Van Loon, L.C. Plant responses to plant growth-promoting rhizobacteria. In *New Perspectives and Approaches in Plant Growth-Promoting Rhizobacteria Research*; Springer: Dordrecht, The Netherlands, 2007; pp. 243–254.
14. Bhat, M.A.; Kumar, V.; Bhat, M.A.; Wani, I.A.; Dar, F.L.; Farooq, I.; Bhatti, F.; Koser, R.; Rahman, S.; Jan, A.T. Mechanistic insights of the interaction of plant growth-promoting rhizobacteria (PGPR) with plant roots toward enhancing plant productivity by alleviating salinity stress. *Front. Microbiol.* **2020**, *11*, 1952. [CrossRef]
15. Goswami, D.; Thakker, J.N.; Dhandhukia, P.C. Portraying mechanics of plant growth promoting rhizobacteria (PGPR): A review. *Cogent Food Agric.* **2016**, *2*, 1127500. [CrossRef]
16. Bhattacharyya, P.N.; Jha, D.K. Plant growth-promoting rhizobacteria (PGPR): Emergence in agriculture. *World J. Microbiol. Biotechnol.* **2012**, *28*, 1327–1350. [CrossRef]
17. Zhu, J.K. Abiotic stress signaling and responses in plants. *Cell* **2016**, *167*, 313–324. [CrossRef]
18. Kosová, K.; Vítámvás, P.; Urban, M.O.; Prášil, I.T.; Renaut, J. Plant abiotic stress proteomics: The major factors determining alterations in cellular proteome. *Front. Plant Sci.* **2018**, *9*, 122. [CrossRef] [PubMed]
19. Singhal, P.; Jan, A.T.; Azam, M.; Haq, Q.M.R. Plant abiotic stress: A prospective strategy of exploiting promoters as alternative to overcome the escalating burden. *Front. Life Sci.* **2016**, *9*, 52–63. [CrossRef]
20. Pandey, P.; Irulappan, V.; Bagavathiannan, M.V.; Senthil-Kumar, M. Impact of combined abiotic and biotic stresses on plant growth and avenues for crop improvement by exploiting physio-morphological traits. *Front. Plant Sci.* **2017**, *8*, 537. [CrossRef] [PubMed]
21. Bechtold, U.; Field, B. Molecular Mechanisms Controlling Plant Growth during Abiotic Stress. *J. Exp. Bot.* **2018**, *69*, 2753–2758. [CrossRef] [PubMed]
22. Yang, J.; Kloepper, J.W.; Ryu, C.M. Rhizosphere bacteria help plants tolerate abiotic stress. *Trends Plant Sci.* **2009**, *14*, 1–4. [CrossRef] [PubMed]
23. Ismail, M.A.; Amin, M.A.; Eid, A.M.; Hassan, S.E.D.; Mahgoub, H.A.; Lashin, I.; Abdelwahab, A.T.; Azab, E.; Gobouri, A.A.; Elkelish, A.; et al. Comparative Study between Exogenously Applied Plant Growth Hormones versus Metabolites of Microbial Endophytes as Plant Growth-Promoting for *Phaseolus vulgaris* L. *Cells* **2021**, *10*, 1059. [CrossRef]
24. Ahkami, A.H.; White, R.A., III; Handakumbura, P.P.; Jansson, C. Rhizosphere engineering: Enhancing sustainable plant ecosystem productivity. *Rhizosphere* **2017**, *3*, 233–243. [CrossRef]

25. Kohler, J.; Hernández, J.A.; Caravaca, F.; Roldán, A. Plant-growth-promoting rhizobacteria and arbuscular mycorrhizal fungi modify alleviation biochemical mechanisms in water-stressed plants. *Funct. Plant Biol.* **2008**, *35*, 141–151. [CrossRef]
26. Wang, Q.; Dodd, I.C.; Belimov, A.A.; Jiang, F. Rhizosphere bacteria containing 1-aminocyclopropane-1-carboxylate deaminase increase growth and photosynthesis of pea plants under salt stress by limiting Na+ accumulation. *Funct. Plant Biol.* **2016**, *43*, 161–172. [CrossRef]
27. Sivasakthi, S.; Usharani, G.; Saranraj, P. Biocontrol potentiality of plant growth promoting bacteria (PGPR)-Pseudomonas fluorescens and Bacillus subtilis: A review. *Afr. J. Agric. Res.* **2014**, *9*, 1265–1277.
28. Su, F.; Villaume, S.; Rabenoelina, F.; Crouzet, J.; Clément, C.; Vaillant-Gaveau, N.; Dhondt-Cordelier, S. Different *Arabidopsis thaliana* photosynthetic and defense responses to hemibiotrophic pathogen induced by local or distal inoculation of Burkholderia phytofirmans. *Photosynth. Res.* **2017**, *134*, 201–214. [CrossRef] [PubMed]
29. Pérez-de-Luque, A.; Tille, S.; Johnson, I.; Pascual-Pardo, D.; Ton, J.; Cameron, D.D. The interactive effects of arbuscular mycorrhiza and plant growth-promoting rhizobacteria synergistically enhance host plant defences against pathogens. *Sci. Rep.* **2017**, *7*, 1–10. [CrossRef] [PubMed]
30. Badri, D.V.; Weir, T.L.; Van der Lelie, D.; Vivanco, J.M. Rhizosphere chemical dialogues: Plant–microbe interactions. *Curr. Opin. Biotechnol.* **2009**, *20*, 642–650. [CrossRef] [PubMed]
31. Zhang, R.; Vivanco, J.M.; Shen, Q. The unseen rhizosphere root–soil–microbe interactions for crop production. *Curr. Opin. Microbiol.* **2017**, *37*, 8–14. [CrossRef]
32. Traxler, M.F.; Kolter, R. Natural products in soil microbe interactions and evolution. *Nat. Prod. Rep.* **2015**, *32*, 956–970. [CrossRef]
33. el Zahar Haichar, F.; Santaella, C.; Heulin, T.; Achouak, W. Root exudates mediated interactions belowground. *Soil Biol. Biochem.* **2014**, *77*, 69–80. [CrossRef]
34. Semchenko, M.; Saar, S.; Lepik, A. Plant root exudates mediate neighbour recognition and trigger complex behavioural changes. *New Phytol.* **2014**, *204*, 631–637. [CrossRef]
35. Neal, A.L.; Ahmad, S.; Gordon-Weeks, R.; Ton, J. Benzoxazinoids in root exudates of maize attract Pseudomonas putida to the rhizosphere. *PLoS ONE* **2012**, *7*, e35498. [CrossRef] [PubMed]
36. Basiliko, N.; Stewart, H.; Roulet, N.T.; Moore, T.R. Do root exudates enhance peat decomposition? *Geomicrobiol. J.* **2012**, *29*, 374–378. [CrossRef]
37. Korenblum, E.; Dong, Y.; Szymanski, J.; Panda, S.; Jozwiak, A.; Massalha, H.; Meir, S.; Rogachev, I.; Aharoni, A. Rhizosphere microbiome mediates systemic root metabolite exudation by root-to-root signaling. *Proc. Natl. Acad. Sci. USA* **2020**, *117*, 3874–3883. [CrossRef]
38. Berlanas, C.; Berbegal, M.; Elena, G.; Laidani, M.; Cibriain, J.F.; Sagües, A.; Gramaje, D. The fungal and bacterial rhizosphere microbiome associated with grapevine rootstock genotypes in mature and young vineyards. *Front. Microbiol.* **2019**, *10*, 1142. [CrossRef] [PubMed]
39. Raklami, A.; Bechtaoui, N.; Tahiri, A.I.; Anli, M.; Meddich, A.; Oufdou, K. Use of rhizobacteria and mycorrhizae consortium in the open field as a strategy for improving crop nutrition, productivity and soil fertility. *Front. Microbiol.* **2019**, *10*, 1106. [CrossRef] [PubMed]
40. Dilnashin, H.; Birla, H.; Hoat, T.X.; Singh, H.B.; Singh, S.P.; Keswani, C. Applications of agriculturally important microorganisms for sustainable crop production. In *Molecular Aspects of Plant Beneficial Microbes in Agriculture*; Academic Press: Cambridge, MA, USA, 2020; pp. 403–415.
41. Akiyama, K.; Hayashi, H. Strigolactones: Chemical signals for fungal symbionts and parasitic weeds in plant roots. *Ann. Bot.* **2006**, *97*, 925–931. [CrossRef] [PubMed]
42. Ahemad, M.; Kibret, M. Mechanisms and applications of plant growth promoting rhizobacteria: Current perspective. *J. King Saud Univ. Sci.* **2014**, *26*, 1–20. [CrossRef]
43. Baysal, Ö.; Lai, D.; Xu, H.H.; Siragusa, M.; Çalışkan, M.; Carimi, F.; Da Silva, J.A.T.; Tör, M. A proteomic approach provides new insights into the control of soil-borne plant pathogens by Bacillus species. *PLoS ONE* **2013**, *8*, e53182. [CrossRef]
44. Bona, E.; Massa, N.; Novello, G.; Boatti, L.; Cesaro, P.; Todeschini, V.; Magnelli, V.; Manfredi, M.; Marengo, E.; Mignone, F.; et al. Metaproteomic characterization of the *Vitis vinifera* rhizosphere. *FEMS Microbiol. Ecol.* **2019**, *95*, fiy204. [CrossRef]
45. Breuillin, F.; Schramm, J.; Hajirezaei, M.; Ahkami, A.; Favre, P.; Druege, U.; Hause, B.; Bucher, M.; Kretzschmar, T.; Bossolini, E.; et al. Phosphate systemically inhibits development of arbuscular mycorrhiza in *Petunia hybrida* and represses genes involved in mycorrhizal functioning. *Plant J.* **2010**, *64*, 1002–1017. [CrossRef]
46. De Cuyper, C.; Fromentin, J.; Yocgo, R.E.; De Keyser, A.; Guillotin, B.; Kunert, K.; Boyer, F.D.; Goormachtig, S. From lateral root density to nodule number, the strigolactone analogue GR24 shapes the root architecture of *Medicago truncatula*. *J. Exp. Bot.* **2015**, *66*, 137–146. [CrossRef]
47. Peláez-Vico, M.A.; Bernabéu-Roda, L.; Kohlen, W.; Soto, M.J.; López-Ráez, J.A. Strigolactones in the Rhizobium-legume symbiosis: Stimulatory effect on bacterial surface motility and down-regulation of their levels in nodulated plants. *Plant Sci.* **2016**, *245*, 119–127. [CrossRef]
48. Yang, J.L.; Fan, W.; Zheng, S.J. Mechanisms and regulation of aluminum-induced secretion of organic acid anions from plant roots. *J. Zhejiang Univ. Sci. B* **2019**, *20*, 513–527. [CrossRef] [PubMed]
49. Yang, L.T.; Qi, Y.P.; Jiang, H.X.; Chen, L.S. Roles of organic acid anion secretion in aluminium tolerance of higher plants. *BioMed Res. Int.* **2013**, *2013*, 173682. [CrossRef] [PubMed]

50. Wu, D.; Zhao, M.; Shen, S.; Fu, Y.; Sasaki, T.; Yamamoto, Y.; Wei, W.; Shen, H. Al-induced secretion of organic acid, gene expression and root elongation in soybean roots. *Acta Physiol. Plant.* **2013**, *35*, 223–232. [CrossRef]
51. Li, G.X.; Wu, X.Q.; Ye, J.R.; Yang, H.C. Characteristics of Organic Acid Secretion Associated with the Interaction between Burkholderia multivorans WS-FJ9 and Poplar Root System. *BioMed. Res. Int.* **2018**, *2018*, 9619724. [CrossRef] [PubMed]
52. Xiang, G.; Ma, W.; Gao, S.; Jin, Z.; Yue, Q.; Yao, Y. Transcriptomic and phosphoproteomic profiling and metabolite analyses reveal the mechanism of NaHCO 3-induced organic acid secretion in grapevine roots. *BMC Plant Biol.* **2019**, *19*, 1–15. [CrossRef] [PubMed]
53. Pini, F.; East, A.K.; Appia-Ayme, C.; Tomek, J.; Karunakaran, R.; Mendoza-Suárez, M.; Edwards, A.; Terpolilli, J.J.; Roworth, J.; Downie, J.A.; et al. Bacterial biosensors for in vivo spatiotemporal mapping of root secretion. *Plant Physiol.* **2017**, *174*, 1289–1306. [CrossRef]
54. Ziegler, J.; Schmidt, S.; Chutia, R.; Müller, J.; Böttcher, C.; Strehmel, N.; Scheel, D.; Abel, S. Non-targeted profiling of semi-polar metabolites in Arabidopsis root exudates uncovers a role for coumarin secretion and lignification during the local response to phosphate limitation. *J. Exp. Bot.* **2016**, *67*, 1421–1432. [CrossRef]
55. Sugiyama, A. The soybean rhizosphere: Metabolites, microbes, and beyond—A review. *J. Adv. Res.* **2019**, *19*, 67–73. [CrossRef]
56. Clemens, S.; Weber, M. The essential role of coumarin secretion for Fe acquisition from alkaline soil. *Plant Signal. Behav.* **2016**, *11*, e1114197. [CrossRef]
57. Chen, Y.T.; Wang, Y.; Yeh, K.C. Role of root exudates in metal acquisition and tolerance. *Curr. Opin. Plant Biol.* **2017**, *39*, 66–72. [CrossRef]
58. Pii, Y.; Mimmo, T.; Tomasi, N.; Terzano, R.; Cesco, S.; Crecchio, C. Microbial interactions in the rhizosphere: Beneficial influences of plant growth-promoting rhizobacteria on nutrient acquisition process. A review. *Biol. Fertil. Soils* **2015**, *51*, 403–415. [CrossRef]
59. Nadeem, S.M.; Ahmad, M.; Zahir, Z.A.; Javaid, A.; Ashraf, M. The role of mycorrhizae and plant growth promoting rhizobacteria (PGPR) in improving crop productivity under stressful environments. *Biotechnol. Adv.* **2014**, *32*, 429–448. [CrossRef] [PubMed]
60. Khan, N.; Ali, S.; Zandi, P.; Mehmood, A.; Ullah, S.; Ikram, M.; ISMAIL, M.A.S.; BABAR, M. Role of sugars, amino acids and organic acids in improving plant abiotic stress tolerance. *Pak. J. Bot.* **2020**, *52*, 355–363. [CrossRef]
61. Chaparro, J.M.; Sheflin, A.M.; Manter, D.K.; Vivanco, J.M. Manipulating the soil microbiome to increase soil health and plant fertility. *Biol. Fertil. Soils* **2012**, *48*, 489–499. [CrossRef]
62. Hashem, A.; Tabassum, B.; Abd_Allah, E.F. Bacillus subtilis: A plant-growth promoting rhizobacterium that also impacts biotic stress. *Saudi J. Biol. Sci.* **2019**, *26*, 1291–1297. [CrossRef] [PubMed]
63. Bharti, N.; Pandey, S.S.; Barnawal, D.; Patel, V.K.; Kalra, A. Plant growth promoting rhizobacteria Dietzia natronolimnaea modulates the expression of stress responsive genes providing protection of wheat from salinity stress. *Sci. Rep.* **2016**, *6*, 1–16. [CrossRef] [PubMed]
64. Jatan, R.; Chauhan, P.S.; Lata, C. Pseudomonas putida modulates the expression of miRNAs and their target genes in response to drought and salt stresses in chickpea (*Cicer arietinum* L.). *Genomics* **2019**, *111*, 509–519. [CrossRef]
65. Gontia-Mishra, I.; Sapre, S.; Sharma, A.; Tiwari, S. Amelioration of drought tolerance in wheat by the interaction of plant growth-promoting rhizobacteria. *Plant Biol.* **2016**, *18*, 992–1000. [CrossRef]
66. Maheshwari, D.K.; Dheeman, S.; Agarwal, M. Phytohormone-producing PGPR for sustainable agriculture. In *Bacterial Metabolites in Sustainable Agroecosystem*; Springer: Cham, Switzerland, 2015; pp. 159–182.
67. Prieto, P.; Schilirò, E.; Maldonado-González, M.M.; Valderrama, R.; Barroso-Albarracín, J.B.; Mercado-Blanco, J. Root hairs play a key role in the endophytic colonization of olive roots by *Pseudomonas* spp. with biocontrol activity. *Microb. Ecol.* **2011**, *62*, 435–445. [CrossRef]
68. Vacheron, J.; Desbrosses, G.; Bouffaud, M.L.; Touraine, B.; Moënne-Loccoz, Y.; Muller, D.; Legendre, L.; Wisniewski-Dyé, F.; Prigent-Combaret, C. Plant growth-promoting rhizobacteria and root system functioning. *Front. Plant Sci.* **2013**, *4*, 356. [CrossRef]
69. Bishnoi, U. PGPR interaction: An ecofriendly approach promoting the sustainable agriculture system. *Adv. Bot. Res.* **2015**, *75*, 81–113.
70. Reddy, P.P. Potential role of PGPR in agriculture. In *Plant Growth Promoting Rhizobacteria for Horticultural Crop Protection*; Springer: New Delhi, India, 2014; pp. 17–34.
71. Rahimi, S.; Talebi, M.; Baninasab, B.; Gholami, M.; Zarei, M.; Shariatmadari, H. The role of plant growth-promoting rhizobacteria (PGPR) in improving iron acquisition by altering physiological and molecular responses in quince seedlings. *Plant Physiol. Biochem.* **2020**, *155*, 406–415. [CrossRef] [PubMed]
72. Kumar, A.; Maurya, B.R.; Raghuwanshi, R. Isolation and characterization of PGPR and their effect on growth, yield and nutrient content in wheat (*Triticum aestivum* L.). *Biocatal. Agric. Biotechnol.* **2014**, *3*, 121–128. [CrossRef]
73. Etesami, H.; Adl, S.M. Plant growth-promoting rhizobacteria (PGPR) and their action mechanisms in availability of nutrients to plants. In *Phyto-Microbiome in Stress Regulation*; Springer: Singapore, 2020; pp. 147–203. [CrossRef]
74. Anbi, A.A.; Mirshekari, B.; Eivazi, A.; Yarnia, M.; Behrouzyar, E.K. PGPRs affected photosynthetic capacity and nutrient uptake in different Salvia species. *J. Plant Nutr.* **2020**, *43*, 108–121. [CrossRef]
75. Danish, S.; Zafar-ul-Hye, M. Co-application of ACC-deaminase producing PGPR and timber-waste biochar improves pigments formation, growth and yield of wheat under drought stress. *Sci. Rep.* **2019**, *9*, 1–13. [CrossRef]
76. Wang, D.; Gao, Y.; Li, M.; Sturrock, C.J.; Gregory, A.S.; Zhang, X. Change in hydraulic properties of the rhizosphere of maize under different abiotic stresses. *Plant Soil* **2020**, *452*, 615–626. [CrossRef]

77. Saleem, M.; Law, A.D.; Sahib, M.R.; Pervaiz, Z.H.; Zhang, Q. Impact of root system architecture on rhizosphere and root microbiome. *Rhizosphere* **2018**, *6*, 47–51. [CrossRef]
78. Khan, N.; Zandi, P.; Ali, S.; Mehmood, A.; Adnan Shahid, M.; Yang, J. Impact of salicylic acid and PGPR on the drought tolerance and phytoremediation potential of *Helianthus annus*. *Front. Microbiol.* **2018**, *9*, 2507. [CrossRef]
79. Vescio, R.; Malacrinò, A.; Bennett, A.E.; Sorgonà, A. Single and combined abiotic stressors affect maize rhizosphere bacterial microbiota. *Rhizosphere* **2021**, *17*, 100318. [CrossRef]
80. Yadav, A.N. Agriculturally important microbiomes: Biodiversity and multifarious PGP attributes for amelioration of diverse abiotic stresses in crops for sustainable agriculture. *Biomed. J. Sci. Tech. Res.* **2017**, *1*, 861–864.
81. Qu, Q.; Zhang, Z.; Peijnenburg, W.J.G.M.; Liu, W.; Lu, T.; Hu, B.; Chen, J.; Chen, J.; Lin, Z.; Qian, H. Rhizosphere microbiome assembly and its impact on plant growth. *J. Agric. Food Chem.* **2020**, *68*, 5024–5038. [CrossRef]
82. Pérez-Jaramillo, J.E.; Mendes, R.; Raaijmakers, J.M. Impact of plant domestication on rhizosphere microbiome assembly and functions. *Plant Mol. Biol.* **2016**, *90*, 635–644. [CrossRef]
83. Vives-Peris, V.; de Ollas, C.; Gómez-Cadenas, A.; Pérez-Clemente, R.M. Root exudates: From plant to rhizosphere and beyond. *Plant Cell Rep.* **2020**, *39*, 3–17. [CrossRef] [PubMed]
84. Timmusk, S.; Abd El-Daim, I.A.; Copolovici, L.; Tanilas, T.; Kännaste, A.; Behers, L.; Nevo, E.; Seisenbaeva, G.; Stenström, E.; Niinemets, Ü. Drought-tolerance of wheat improved by rhizosphere bacteria from harsh environments: Enhanced biomass production and reduced emissions of stress volatiles. *PLoS ONE* **2014**, *9*, e96086. [CrossRef] [PubMed]
85. Ali, M.A.; Naveed, M.; Mustafa, A.; Abbas, A. The good, the bad, and the ugly of rhizosphere microbiome. In *Probiotics and Plant Health*; Springer: Singapore, 2017; pp. 253–290.
86. Zerrouk, I.Z.; Benchabane, M.; Khelifi, L.; Yokawa, K.; Ludwig-Müller, J.; Baluska, F. A Pseudomonas strain isolated from date-palm rhizospheres improves root growth and promotes root formation in maize exposed to salt and aluminum stress. *J. Plant Physiol.* **2016**, *191*, 111–119. [CrossRef] [PubMed]
87. Nihorimbere, V.; Ongena, M.; Smargiassi, M.; Thonart, P. Beneficial effect of the rhizosphere microbial community for plant growth and health. *Biotechnol. Agron. Société Environ.* **2011**, *15*, 327–337.
88. Grover, M.; Ali, S.Z.; Sandhya, V.; Rasul, A.; Venkateswarlu, B. Role of microorganisms in adaptation of agriculture crops to abiotic stresses. *World J. Microbiol. Biotechnol.* **2011**, *27*, 1231–1240. [CrossRef]
89. Zia, R.; Nawaz, M.S.; Siddique, M.J.; Hakim, S.; Imran, A. Plant survival under drought stress: Implications, adaptive responses, and integrated rhizosphere management strategy for stress mitigation. *Microbiol. Res.* **2020**, *242*, 126626. [CrossRef] [PubMed]
90. Dessaux, Y.; Grandclément, C.; Faure, D. Engineering the rhizosphere. *Trends Plant Sci.* **2016**, *21*, 266–278. [CrossRef] [PubMed]
91. Sharma, S.; Chandra, D.; Sharma, A.K. Rhizosphere Plant–Microbe Interactions under Abiotic Stress. In *Rhizosphere Biology: Interactions between Microbes and Plants*; Springer: Singapore, 2021; pp. 195–216.
92. Mommer, L.; Hinsinger, P.; Prigent-Combaret, C.; Visser, E.J. Advances in the rhizosphere: Stretching the interface of life. *Plant Soil* **2016**, *407*, 1–8. [CrossRef]
93. Li, S.; Fu, Q.; Chen, L.; Huang, W.; Yu, D. *Arabidopsis thaliana* WRKY25, WRKY26, and WRKY33 coordinate induction of plant thermotolerance. *Planta* **2011**, *233*, 1237–1252. [CrossRef] [PubMed]
94. Asseng, S.; Foster, I.A.N.; Turner, N.C. The impact of temperature variability on wheat yields. *Glob. Chang. Biol.* **2011**, *17*, 997–1012. [CrossRef]
95. Boo, H.O.; Heo, B.G.; Gorinstein, S.; Chon, S.U. Positive effects of temperature and growth conditions on enzymatic and antioxidant status in lettuce plants. *Plant Sci.* **2011**, *181*, 479–484. [CrossRef]
96. Asati, A.; Pichhode, M.; Nikhil, K. Effect of heavy metals on plants: An overview. *Int. J. Appl. Innov. Eng. Manag.* **2016**, *5*, 56–66.
97. Halušková, L.U.; Valentovičová, K.; Huttová, J.; Mistrík, I.; Tamás, L. Effect of heavy metals on root growth and peroxidase activity in barley root tip. *Acta Physiol. Plant.* **2010**, *32*, 59. [CrossRef]
98. Pavel, V.L.; Sobariu, D.L.; Diaconu, M.; Stătescu, F.; Gavrilescu, M. Effects of heavy metals on *Lepidium sativum* germination and growth. *Environ. Eng. Manag. J. (EEMJ)* **2013**, *12*, 727–733. [CrossRef]
99. Samardakiewicz, S.; Woźny, A. Cell division in Lemna minor roots treated with lead. *Aquat. Bot.* **2005**, *83*, 289–295. [CrossRef]
100. Prasad, M.N.V. (Ed.) *Heavy Metal Stress in Plants: From Biomolecules to Ecosystems*; Springer Science & Business Media: Berlin, Germany, 2013.
101. Rahman, Z.; Singh, V.P. The relative impact of toxic heavy metals (THMs)(arsenic (As), cadmium (Cd), chromium (Cr)(VI), mercury (Hg), and lead (Pb)) on the total environment: An overview. *Environ. Monit. Assess.* **2019**, *191*, 1–21. [CrossRef] [PubMed]
102. Nagajyoti, P.C.; Lee, K.D.; Sreekanth, T.V.M. Heavy metals, occurrence and toxicity for plants: A review. *Environ. Chem. Lett.* **2010**, *8*, 199–216. [CrossRef]
103. Jaishankar, M.; Tseten, T.; Anbalagan, N.; Mathew, B.B.; Beeregowda, K.N. Toxicity, mechanism and health effects of some heavy metals. *Interdiscip. Toxicol.* **2014**, *7*, 60. [CrossRef]
104. Nazir, R.; Khan, M.; Masab, M.; Rehman, H.U.; Rauf, N.U.; Shahab, S.; Ameer, N.; Sajed, M.; Ullah, M.; Rafeeq, M.; et al. Accumulation of heavy metals (Ni, Cu, Cd, Cr, Pb, Zn, Fe) in the soil, water and plants and analysis of physico-chemical parameters of soil and water collected from Tanda Dam Kohat. *J. Pharm. Sci. Res.* **2015**, *7*, 89.
105. Benáková, M.; Ahmadi, H.; Dučaiová, Z.; Tylová, E.; Clemens, S.; Tůma, J. Effects of Cd and Zn on physiological and anatomical properties of hydroponically grown *Brassica napus* plants. *Environ. Sci. Pollut. Res.* **2017**, *24*, 20705–20716. [CrossRef] [PubMed]

106. Castillo-Lorenzo, E.; Pritchard, H.W.; Finch-Savage, W.E.; Seal, C.E. Comparison of seed and seedling functional traits in native Helianthus species and the crop *H. annuus* (sunflower). *Plant Biol.* **2019**, *21*, 533–543. [CrossRef] [PubMed]
107. Ruíz-Sánchez, M.; Armada, E.; Muñoz, Y.; de Salamone, I.E.G.; Aroca, R.; Ruíz-Lozano, J.M.; Azcón, R. Azospirillum and arbuscular mycorrhizal colonization enhance rice growth and physiological traits under well-watered and drought conditions. *J. Plant Physiol.* **2011**, *168*, 1031–1037. [CrossRef]
108. Saravanakumar, D.; Kavino, M.; Raguchander, T.; Subbian, P.; Samiyappan, R. Plant growth promoting bacteria enhance water stress resistance in green gram plants. *Acta Physiol. Plant.* **2011**, *33*, 203–209. [CrossRef]
109. El-Meihy, R.M. Evaluation of pgpr as osmoprotective agents for squash (*Cucurbita pepo* L.) growth under drought stress. *Middle East J.* **2016**, *5*, 583–595.
110. Gou, W.; Tian, L.; Ruan, Z.; Zheng, P.E.N.G.; Chen, F.U.C.A.I.; Zhang, L.; Cui, Z.; Zheng, P.; Li, Z.; Gao, M.; et al. Accumulation of choline and glycinebetaine and drought stress tolerance induced in maize (*Zea mays*) by three plant growth promoting rhizobacteria (PGPR) strains. *Pak. J. Bot.* **2015**, *47*, 581–586.
111. Lim, J.H.; Ahn, C.H.; Jeong, H.Y.; Kim, Y.H.; Kim, S.D. Genetic monitoring of multi-functional plant growth promoting rhizobacteria Bacillus subtilis AH18 and Bacillus licheniformis K11 by multiplex and real-time polymerase chain reaction in a pepper farming field. *J. Korean Soc. Appl. Biol. Chem.* **2011**, *54*, 221–228. [CrossRef]
112. Gupta, S.; Pandey, S. ACC deaminase producing bacteria with multifarious plant growth promoting traits alleviates salinity stress in French bean (*Phaseolus vulgaris*) plants. *Front. Microbiol.* **2019**, *10*, 1506. [CrossRef]
113. Tolba, S.T.; Ibrahim, M.; Amer, E.A.; Ahmed, D.A. First insights into salt tolerance improvement of Stevia by plant growth-promoting Streptomyces species. *Arch. Microbiol.* **2019**, *201*, 1295–1306. [CrossRef] [PubMed]
114. Habib, S.H.; Kausar, H.; Saud, H.M. Plant growth-promoting rhizobacteria enhance salinity stress tolerance in okra through ROS-scavenging enzymes. *BioMed. Res. Int.* **2016**, *2016*, 6284547. [CrossRef] [PubMed]
115. Ke, T.; Guo, G.; Liu, J.; Zhang, C.; Tao, Y.; Wang, P.; Xu, Y.; Chen, L. Improvement of the Cu and Cd phytostabilization efficiency of perennial ryegrass through the inoculation of three metal-resistant PGPR strains. *Environ. Pollut.* **2021**, *271*, 116314. [CrossRef] [PubMed]
116. Awan, S.A.; Ilyas, N.; Khan, I.; Raza, M.A.; Rehman, A.U.; Rizwan, M.; Rastogi, A.; Tariq, R.; Brestic, M. Bacillus siamensis Reduces Cadmium Accumulation and Improves Growth and Antioxidant Defense System in Two Wheat (*Triticum aestivum* L.) Varieties. *Plants* **2020**, *9*, 878. [CrossRef] [PubMed]
117. Akhtar, N.; Ilyas, N.; Yasmin, H.; Sayyed, R.Z.; Hasnain, Z.; A Elsayed, E.; El Enshasy, H.A. Role of *Bacillus cereus* in Improving the Growth and Phytoextractability of *Brassica nigra* (L.) K. Koch in Chromium Contaminated Soil. *Molecules* **2021**, *26*, 1569. [CrossRef] [PubMed]
118. Belimov, A.A.; Safronova, V.I.; Tsyganov, V.E.; Borisov, A.Y.; Kozhemyakov, A.P.; Stepanok, V.V.; Martenson, A.M.; Gianinazzi-Pearson, V.; Tikhonovich, I.A. Genetic variability in tolerance to cadmium and accumulation of heavy metals in pea (*Pisum sativum* L.). *Euphytica* **2003**, *131*, 25–35. [CrossRef]
119. He, X.; Xu, M.; Wei, Q.; Tang, M.; Guan, L.; Lou, L.; Xu, X.; Hu, Z.; Chen, Y.; Shen, Z.; et al. Promotion of growth and phytoextraction of cadmium and lead in Solanum nigrum L. mediated by plant-growth-promoting rhizobacteria. *Ecotoxicol. Environ. Saf.* **2020**, *205*, 111333. [CrossRef]
120. Zafar-ul-Hye, M.; Tahzeeb-ul-Hassan, M.; Wahid, A.; Danish, S.; Khan, M.J.; Fahad, S.; Brtnicky, M.; Hussain, G.S.; Battaglia, M.L.; Datta, R. Compost mixed fruits and vegetable waste biochar with ACC deaminase rhizobacteria can minimize lead stress in mint plants. *Sci. Rep.* **2021**, *11*, 1–20. [CrossRef]
121. Ashraf, A.; Bano, A.; Ali, S.A. Characterisation of plant growth-promoting rhizobacteria from rhizosphere soil of heat-stressed and unstressed wheat and their use as bio-inoculant. *Plant Biol.* **2019**, *21*, 762–769. [CrossRef]
122. Abd El-Daim, I.A.; Bejai, S.; Meijer, J. Bacillus velezensis 5113 induced metabolic and molecular reprogramming during abiotic stress tolerance in wheat. *Sci. Rep.* **2019**, *9*, 1–18. [CrossRef]
123. Khan, M.A.; Asaf, S.; Khan, A.L.; Jan, R.; Kang, S.M.; Kim, K.M.; Lee, I.J. Extending thermotolerance to tomato seedlings by inoculation with SA1 isolate of *Bacillus cereus* and comparison with exogenous humic acid application. *PLoS ONE* **2020**, *15*, e0232228. [CrossRef]
124. Gururani, M.A.; Upadhyaya, C.P.; Baskar, V.; Venkatesh, J.; Nookaraju, A.; Park, S.W. Plant growth-promoting rhizobacteria enhance abiotic stress tolerance in Solanum tuberosum through inducing changes in the expression of ROS-scavenging enzymes and improved photosynthetic performance. *J. Plant Growth Regul.* **2013**, *32*, 245–258. [CrossRef]
125. Marulanda, A.; Azcón, R.; Chaumont, F.; Ruiz-Lozano, J.M.; Aroca, R. Regulation of plasma membrane aquaporins by inoculation with a Bacillus megaterium strain in maize (*Zea mays* L.) plants under unstressed and salt-stressed conditions. *Planta* **2010**, *232*, 533–543. [CrossRef] [PubMed]
126. Khan, N.; Bano, A. Rhizobacteria and abiotic stress management. In *Plant Growth Promoting Rhizobacteria for Sustainable Stress Management*; Springer: Singapore, 2019; pp. 65–80.
127. Ghosh, P.K.; De, T.K.; Maiti, T.K. *Role of ACC Deaminase as a Stress Ameliorating Enzyme of Plant Growth-Promoting Rhizobacteria Useful in Stress Agriculture: A Review. Role of Rhizospheric Microbes in Soil*; Springer: Singapore, 2018; pp. 57–106. [CrossRef]
128. Niu, X.; Song, L.; Xiao, Y.; Ge, W. Drought-tolerant plant growth-promoting rhizobacteria associated with foxtail millet in a semi-arid agroecosystem and their potential in alleviating drought stress. *Front. Microbiol.* **2018**, *8*, 2580. [CrossRef] [PubMed]

129. Batool, T.; Ali, S.; Seleiman, M.F.; Naveed, N.H.; Ali, A.; Ahmed, K.; Abid, M.; Rizwan, M.; Shahid, M.R.; Alotaibi, M.; et al. Plant growth promoting rhizobacteria alleviates drought stress in potato in response to suppressive oxidative stress and antioxidant enzymes activities. *Sci. Rep.* **2020**, *10*, 1–19. [CrossRef] [PubMed]
130. Kumar, A.; Patel, J.S.; Meena, V.S.; Srivastava, R. Recent advances of PGPR based approaches for stress tolerance in plants for sustainable agriculture. *Biocatal. Agric. Biotechnol.* **2019**, *20*, 101271. [CrossRef]
131. Shultana, R.; Tan Kee Zuan, A.; Yusop, M.R.; Mohd Saud, H.; Ayanda, A.F. Effect of salt-tolerant bacterial inoculations on rice seedlings differing in salt-tolerance under saline soil conditions. *Agronomy* **2020**, *10*, 1030. [CrossRef]
132. Kechid, M.; Desbrosses, G.; Rokhsi, W.; Varoquaux, F.; Djekoun, A.; Touraine, B. The NRT 2.5 and NRT 2.6 genes are involved in growth promotion of Arabidopsis by the plant growth-promoting rhizobacterium (PGPR) strain *Phyllobacterium brassicacearum* STM 196. *New Phytol.* **2013**, *198*, 514–524. [CrossRef]
133. Bresson, J.; Vasseur, F.; Dauzat, M.; Labadie, M.; Varoquaux, F.; Touraine, B.; Vile, D. Interact to survive: *Phyllobacterium brassicacearum* improves Arabidopsis tolerance to severe water deficit and growth recovery. *PLoS ONE* **2014**, *9*, e107607. [CrossRef]
134. Galland, M.; Gamet, L.; Varoquaux, F.; Touraine, B.; Touraine, B.; Desbrosses, G. The ethylene pathway contributes to root hair elongation induced by the beneficial bacteria *Phyllobacterium brassicacearum* STM196. *Plant Sci.* **2012**, *190*, 74–81. [CrossRef]
135. Islam, F.; Yasmeen, T.; Ali, Q.; Ali, S.; Arif, M.S.; Hussain, S.; Rizvi, H. Influence of Pseudomonas aeruginosa as PGPR on oxidative stress tolerance in wheat under Zn stress. *Ecotoxicol. Environ. Saf.* **2014**, *104*, 285–293. [CrossRef]
136. Sultana, S.; Paul, S.C.; Parveen, S.; Alam, S.; Rahman, N.; Jannat, B.; Hoque, S.; Rahman, M.T.; Karim, M.M. Isolation and identification of salt-tolerant plant growth-promoting rhizobacteria and its application for rice cultivation under salt stress. *Can. J. Microbiol.* **2019**. [CrossRef]
137. Rajput, L.U.B.N.A.; Imran, A.; Mubeen, F.; Hafeez, F.Y. Salt-tolerant PGPR strain Planococcus rifietoensis promotes the growth and yield of wheat (*Triticum aestivum* L.) cultivated in saline soil. *Pak. J. Bot.* **2013**, *45*, 1955–1962.
138. Damodaran, T.; Sah, V.; Rai, R.B.; Sharma, D.K.; Mishra, V.K.; Jha, S.K.; Kannan, R. Isolation of salt tolerant endophytic and rhizospheric bacteria by natural selection and screening for promising plant growth-promoting rhizobacteria (PGPR) and growth vigour in tomato under sodic environment. *Afr. J. Microbiol. Res.* **2013**, *7*, 5082–5089.
139. Vimal, S.R.; Singh, J.S. Salt tolerant PGPR and FYM application in saline soil paddy agriculture sustainability. *Clim. Chang. Environ. Sustain.* **2019**, *7*, 61–71. [CrossRef]
140. Nawaz, A.; Shahbaz, M.; Asadullah, A.I.; Marghoob, M.U.; Imtiaz, M.; Mubeen, F. Potential of salt tolerant PGPR in growth and yield augmentation of wheat (*Triticum aestivum* L.) under saline conditions. *Front. Microbiol.* **2020**, *11*, 2019. [CrossRef]
141. Bal, H.B.; Nayak, L.; Das, S.; Adhya, T.K. Isolation of ACC deaminase producing PGPR from rice rhizosphere and evaluating their plant growth promoting activity under salt stress. *Plant Soil* **2013**, *366*, 93–105. [CrossRef]
142. Egamberdieva, D.; Wirth, S.; Bellingrath-Kimura, S.D.; Mishra, J.; Arora, N.K. Salt-tolerant plant growth promoting rhizobacteria for enhancing crop productivity of saline soils. *Front. Microbiol.* **2019**, *10*, 2791. [CrossRef]
143. Silambarasan, S.; Logeswari, P.; Cornejo, P.; Kannan, V.R. Role of plant growth–promoting rhizobacterial consortium in improving the *Vigna radiata* growth and alleviation of aluminum and drought stresses. *Environ. Sci. Pollut. Res.* **2019**, *26*, 27647–27659. [CrossRef]
144. Khan, M.A.; Asaf, S.; Khan, A.L.; Adhikari, A.; Jan, R.; Ali, S.; Imran, M.; Kim, K.M.; Lee, I.J. Halotolerant rhizobacterial strains mitigate the adverse effects of NaCl stress in soybean seedlings. *BioMed Res. Int.* **2019**, *2019*, 9530963. [CrossRef]
145. Zhu, X.; Song, F.; Xu, H. Influence of arbuscular mycorrhiza on lipid peroxidation and antioxidant enzyme activity of maize plants under temperature stress. *Mycorrhiza* **2010**, *20*, 325–332. [CrossRef]
146. Li, L.; Ye, Y.; Pan, L.; Zhu, Y.; Zheng, S.; Lin, Y. The induction of trehalose and glycerol in Saccharomyces cerevisiae in response to various stresses. *Biochem. Biophys. Res. Commun.* **2009**, *387*, 778–783. [CrossRef]
147. Paulucci, N.S.; Gallarato, L.A.; Reguera, Y.B.; Vicario, J.C.; Cesari, A.B.; de Lema, M.B.G.; Dardanelli, M.S. Arachis hypogaea PGPR isolated from Argentine soil modifies its lipids components in response to temperature and salinity. *Microbiol. Res.* **2015**, *173*, 1–9. [CrossRef]
148. Kang, C.H.; So, J.S. Heavy metal and antibiotic resistance of ureolytic bacteria and their immobilization of heavy metals. *Ecol. Eng.* **2016**, *97*, 304–312. [CrossRef]
149. Issa, A.; Esmaeel, Q.; Sanchez, L.; Courteaux, B.; Guise, J.F.; Gibon, Y.; Ballias, P.; Clément, C.; Jacquard, C.; Vaillant-Gaveau, N.; et al. Impacts of Paraburkholderia phytofirmans strain PsJN on tomato (*Lycopersicon esculentum* L.) under high temperature. *Front. Plant Sci.* **2018**, *9*, 1397. [CrossRef] [PubMed]
150. Rodriguez, R.J.; Henson, J.; Van Volkenburgh, E.; Hoy, M.; Wright, L.; Beckwith, F.; Kim, Y.O.; Redman, R.S. Stress tolerance in plants via habitat-adapted symbiosis. *ISME J.* **2008**, *2*, 404–416. [CrossRef] [PubMed]
151. Ali, S.Z.; Sandhya, V.; Grover, M.; Kishore, N.; Rao, L.V.; Venkateswarlu, B. Pseudomonas sp. strain AKM-P6 enhances tolerance of sorghum seedlings to elevated temperatures. *Biol. Fertil. Soils* **2009**, *46*, 45–55. [CrossRef]
152. Ali, S.Z.; Sandhya, V.; Grover, M.; Linga, V.R.; Bandi, V. Effect of inoculation with a thermotolerant plant growth promoting Pseudomonas putida strain AKMP7 on growth of wheat (*Triticum* spp.) under heat stress. *J. Plant Interact.* **2011**, *6*, 239–246. [CrossRef]
153. Chang, C.H.; Yang, S.S. Thermo-tolerant phosphate-solubilizing microbes for multi-functional biofertilizer preparation. *Bioresour. Technol.* **2009**, *100*, 1648–1658. [CrossRef] [PubMed]

154. Desoky, E.S.M.; Merwad, A.R.M.; Semida, W.M.; Ibrahim, S.A.; El-Saadony, M.T.; Rady, M.M. Heavy metals-resistant bacteria (HM-RB): Potential bioremediators of heavy metals-stressed *Spinacia oleracea* plant. *Ecotox. Environ. Safety* **2020**, *198*, 110685. [CrossRef]
155. Ullah, S.; Ashraf, M.; Asghar, H.N.; Iqbal, Z.; Ali, R. Review Plant growth promoting rhizobacteria-mediated amelioration of drought in crop plants. *Soil Environ.* **2019**, *38*, 1–20. [CrossRef]
156. Ghosh, D.; Gupta, A.; Mohapatra, S. A comparative analysis of exopolysaccharide and phytohormone secretions by four drought-tolerant rhizobacterial strains and their impact on osmotic-stress mitigation in *Arabidopsis thaliana*. *World J. Microbiol. Biotechnol.* **2019**, *35*, 1–15. [CrossRef]
157. Tiwari, S.; Muthamilarasan, M.; Lata, C. Genome-wide identification and expression analysis of Arabidopsis GRAM-domain containing gene family in response to abiotic stresses and PGPR treatment. *J. Biotechnol.* **2021**, *325*, 7–14. [CrossRef] [PubMed]
158. Merdy, P.; Gharbi, L.T.; Lucas, Y. Pb, Cu and Cr interactions with soil: Sorption experiments and modelling. *Colloids Surf. A Physicochem. Eng. Asp.* **2009**, *347*, 192–199. [CrossRef]
159. Kang, S.M.; Shahzad, R.; Khan, M.A.; Hasnain, Z.; Lee, K.E.; Park, H.S.; Kim, L.R.; Lee, I.J. Ameliorative effect of indole-3-acetic acid-and siderophore-producing *Leclercia adecarboxylata* MO1 on cucumber plants under zinc stress. *J. Plant Interact.* **2021**, *16*, 30–41. [CrossRef]
160. Javaherdashti, R. Impact of sulphate-reducing bacteria on the performance of engineering materials. *Appl. Microbiol. Biotechnol.* **2011**, *91*, 1507–1517. [CrossRef]
161. Khanna, K.; Jamwal, V.L.; Gandhi, S.G.; Ohri, P.; Bhardwaj, R. Metal resistant PGPR lowered Cd uptake and expression of metal transporter genes with improved growth and photosynthetic pigments in *Lycopersicon esculentum* under metal toxicity. *Sci. Rep.* **2019**, *9*, 1–14. [CrossRef]
162. Gadd, G.M.; Bahri-Esfahani, J.; Li, Q.; Rhee, Y.J.; Wei, Z.; Fomina, M.; Liang, X. Oxalate production by fungi: Significance in geomycology, biodeterioration and bioremediation. *Fungal Biol. Rev.* **2014**, *28*, 36–55. [CrossRef]
163. Khan, N.; Ali, S.; Tariq, H.; Latif, S.; Yasmin, H.; Mehmood, A.; Shahid, M.A. Water Conservation and Plant Survival Strategies of Rhizobacteria under Drought Stress. *Agronomy* **2020**, *10*, 1683. [CrossRef]
164. Etesami, H.; Maheshwari, D.K. Use of plant growth promoting rhizobacteria (PGPRs) with multiple plant growth promoting traits in stress agriculture: Action mechanisms and future prospects. *Ecotoxicol. Environ. Saf.* **2018**, *156*, 225–246. [CrossRef]
165. Arora, N.K.; Fatima, T.; Mishra, J.; Mishra, I.; Verma, S.; Verma, R.; Verma, M.; Bhattacharya, A.; Verma, P.; Mishra, P.; et al. Halo-tolerant plant growth promoting rhizobacteria for improving productivity and remediation of saline soils. *J. Adv. Res.* **2020**, *26*, 69–82. [CrossRef]
166. Khan, N.; Bano, A. Exopolysaccharide producing rhizobacteria and their impact on growth and drought tolerance of wheat grown under rainfed conditions. *PLoS ONE* **2019**, *14*, e0222302. [CrossRef]
167. Kumar, K.; Amaresan, N.; Madhuri, K. Alleviation of the adverse effect of salinity stress by inoculation of plant growth promoting rhizobacteria isolated from hot humid tropical climate. *Ecol. Eng.* **2017**, *102*, 361–366. [CrossRef]
168. ALKahtani, M.D.; Fouda, A.; Attia, K.A.; Al-Otaibi, F.; Eid, A.M.; Ewais, E.E.D.; Hijri, M.; St-Arnaud, M.; Hassan, S.E.D.; Khan, N.; et al. Isolation and characterization of plant growth promoting endophytic bacteria from desert plants and their application as bioinoculants for sustainable agriculture. *Agronomy* **2020**, *10*, 1325. [CrossRef]
169. Tiwari, S.; Lata, C. Heavy metal stress, signaling, and tolerance due to plant-associated microbes: An overview. *Front. Plant Sci.* **2018**, *9*, 452. [CrossRef]
170. He, Z.L.; Yang, X.E. Role of soil rhizobacteria in phytoremediation of heavy metal contaminated soils. *J. Zhejiang Univ. Sci. B* **2007**, *8*, 192–207.
171. Moreira, H.; Pereira, S.I.; Marques, A.P.; Rangel, A.O.; Castro, P.M. Selection of metal resistant plant growth promoting rhizobacteria for the growth and metal accumulation of energy maize in a mine soil—Effect of the inoculum size. *Geoderma* **2016**, *278*, 1–11. [CrossRef]
172. Hartman, K.; Tringe, S.G. Interactions between plants and soil shaping the root microbiome under abiotic stress. *Biochem. J.* **2019**, *476*, 2705–2724. [CrossRef]
173. Chen, Y.; Palta, J.A.; Wu, P.; Siddique, K.H. Crop root systems and rhizosphere interactions. *Plant Soil* **2019**, *439*, 1–5. [CrossRef]
174. Naylor, D.; Coleman-Derr, D. Drought stress and root-associated bacterial communities. *Front. Plant Sci.* **2018**, *8*, 2223. [CrossRef] [PubMed]
175. Liang, J.G.; Tao, R.X.; Hao, Z.N.; Wang, L.; Zhang, X. Induction of resistance in cucumber against seedling damping-off by plant growth-promoting rhizobacteria (PGPR) Bacillus megaterium strain L8. *Afr. J. Biotechnol.* **2011**, *10*, 6920–6927.
176. Rahmoune, B.; Morsli, A.; Khelifi-Slaoui, M.; Khelifi, L.; Strueh, A.; Erban, A.; Kopka, J.; Prell, J.; van Dongen, J.T. Isolation and characterization of three new PGPR and their effects on the growth of Arabidopsis and Datura plants. *J. Plant Interact.* **2017**, *12*, 1–6. [CrossRef]
177. Turan, M.; Gulluce, M.; Cakmakci, R.; Oztas, T.; Sahin, F.; Gilkes, R.J.; Prakongkep, N. The effect of PGPR strain on wheat yield and quality parameters. In Proceedings of the 19th World Congress of Soil Science: Soil Solutions for a Changing World, Brisbane, Australia, 1–6 August 2010; pp. 209–212.
178. Erturk, Y.; Ercisli, S.; Haznedar, A.; Cakmakci, R. Effects of plant growth promoting rhizobacteria (PGPR) on rooting and root growth of kiwifruit (*Actinidia deliciosa*) stem cuttings. *Biol. Res.* **2010**, *43*, 91–98. [CrossRef]

179. Curá, J.A.; Franz, D.R.; Filosofía, J.E.; Balestrasse, K.B.; Burgueño, L.E. Inoculation with *Azospirillum* sp.; *Herbaspirillum* sp. bacteria increases the tolerance of maize to drought stress. *Microorganisms* **2017**, *5*, 41. [CrossRef]
180. Almaghrabi, O.A.; Massoud, S.I.; Abdelmoneim, T.S. Influence of inoculation with plant growth promoting rhizobacteria (PGPR) on tomato plant growth and nematode reproduction under greenhouse conditions. *Saudi J. Biol. Sci.* **2013**, *20*, 57–61. [CrossRef]
181. Jones, P.; Garcia, B.J.; Furches, A.; Tuskan, G.A.; Jacobson, D. Plant host-associated mechanisms for microbial selection. *Front. Plant Sci.* **2019**, *10*, 862. [CrossRef] [PubMed]
182. De-la-Peña, C.; Loyola-Vargas, V.M. Biotic interactions in the rhizosphere: A diverse cooperative enterprise for plant productivity. *Plant Physiol.* **2014**, *166*, 701–719. [CrossRef] [PubMed]
183. De la Fuente Canto, C.; Simonin, M.; King, E.; Moulin, L.; Bennett, M.J.; Castrillo, G.; Laplaze, L. An extended root phenotype: The rhizosphere, its formation and impacts on plant fitness. *Plant J.* **2020**, *103*, 951–964. [CrossRef] [PubMed]
184. Jochum, M.D.; McWilliams, K.L.; Borrego, E.J.; Kolomiets, M.V.; Niu, G.; Pierson, E.A.; Jo, Y.K. Bioprospecting plant growth-promoting rhizobacteria that mitigate drought stress in grasses. *Front. Microbiol.* **2019**, *10*, 2106. [CrossRef]
185. Mishra, J.; Fatima, T.; Arora, N.K. Role of secondary metabolites from plant growth-promoting rhizobacteria in combating salinity stress. In *Plant Microbiome: Stress Response*; Springer: Singapore, 2018; pp. 127–163.
186. Gamez, R.; Cardinale, M.; Montes, M.; Ramirez, S.; Schnell, S.; Rodriguez, F. Screening, plant growth promotion and root colonization pattern of two rhizobacteria (*Pseudomonas fluorescens* Ps006 and *Bacillus amyloliquefaciens* Bs006) on banana cv. Williams (*Musa acuminata* Colla). *Microbiol. Res.* **2019**, *220*, 12–20. [CrossRef]
187. Kousar, B.; Bano, A.; Khan, N. PGPR modulation of secondary metabolites in tomato infested with *Spodoptera litura*. *Agronomy* **2020**, *10*, 778. [CrossRef]
188. Vílchez, J.I.; Yang, Y.; He, D.; Zi, H.; Peng, L.; Lv, S.; Kaushal, R.; Wang, W.; Huang, W.; Liu, R.; et al. DNA demethylases are required for myo-inositol-mediated mutualism between plants and beneficial rhizobacteria. *Nat. Plants* **2020**, *6*, 983–995. [CrossRef] [PubMed]
189. Zhou, D.; Huang, X.F.; Chaparro, J.M.; Badri, D.V.; Manter, D.K.; Vivanco, J.M.; Guo, J. Root and bacterial secretions regulate the interaction between plants and PGPR leading to distinct plant growth promotion effects. *Plant Soil* **2016**, *401*, 259–272. [CrossRef]
190. Vurukonda, S.S.K.P.; Vardharajula, S.; Shrivastava, M.; SkZ, A. Enhancement of drought stress tolerance in crops by plant growth promoting rhizobacteria. *Microbiol. Res.* **2016**, *184*, 13–24. [CrossRef] [PubMed]
191. Naseem, H.; Ahsan, M.; Shahid, M.A.; Khan, N. Exopolysaccharides producing rhizobacteria and their role in plant growth and drought tolerance. *J. Basic Microbiol.* **2018**, *58*, 1009–1022. [CrossRef] [PubMed]
192. Singh, B.N.; Hidangmayum, A.; Singh, A.; Shera, S.S.; Dwivedi, P. *Secondary Metabolites of Plant Growth Promoting Rhizomicroorganisms*; Springer: Berlin, Germany, 2019.
193. Bakka, K.; Challabathula, D. Amelioration of Salt Stress Tolerance in Plants by Plant Growth-Promoting Rhizobacteria: Insights from "Omics" Approaches. In *Plant Microbe Symbiosis*; Springer: Cham, Switzerland, 2020; pp. 303–330.
194. Lim, J.H.; Kim, S.D. Induction of drought stress resistance by multi-functional PGPR Bacillus licheniformis K11 in pepper. *Plant Pathol. J.* **2013**, *29*, 201. [CrossRef] [PubMed]
195. Abbas, R.; Rasul, S.; Aslam, K.; Baber, M.; Shahid, M.; Mubeen, F.; Naqqash, T. Halotolerant PGPR: A hope for cultivation of saline soils. *J. King Saud Univ. Sci.* **2019**, *31*, 1195–1201. [CrossRef]
196. Upadhyay, S.K.; Singh, D.P. Effect of salt-tolerant plant growth-promoting rhizobacteria on wheat plants and soil health in a saline environment. *Plant Biol.* **2015**, *17*, 288–293. [CrossRef] [PubMed]
197. Kumar, A.; Verma, J.P. Does plant—Microbe interaction confer stress tolerance in plants: A review? *Microbiol. Res.* **2018**, *207*, 41–52. [CrossRef] [PubMed]
198. Li, H.; Qiu, Y.; Yao, T.; Ma, Y.; Zhang, H.; Yang, X. Effects of PGPR microbial inoculants on the growth and soil properties of Avena sativa, Medicago sativa, and *Cucumis sativus* seedlings. *Soil Tillage Res.* **2020**, *199*, 104577. [CrossRef]
199. Khan, M.N.N.; Ahmad, Z.; Ghafoor, A. Genetic diversity and disease response of rust in bread wheat collected from Waziristan Agency, Pakistan. *Int. J. Biodivers. Conserv.* **2011**, *3*, 10–18.
200. Dimkpa, C.; Weinand, T.; Asch, F. Plant–rhizobacteria interactions alleviate abiotic stress conditions. *Plant Cell Environ.* **2009**, *32*, 1682–1694. [CrossRef] [PubMed]
201. Pare, P.W.; Farag, M.A.; Krishnamachari, V.; Zhang, H.; Ryu, C.M.; Kloepper, J.W. Elicitors and priming agents initiate plant defense responses. *Photosynth. Res.* **2005**, *85*, 149–159. [CrossRef] [PubMed]
202. Yu, P.; Hochholdinger, F. The role of host genetic signatures on root–microbe interactions in the rhizosphere and endosphere. *Front. Plant Sci.* **2018**, *9*, 1896. [CrossRef] [PubMed]
203. Barea, J.M.; Pozo, M.J.; Azcon, R.; Azcon-Aguilar, C. Microbial co-operation in the rhizosphere. *J. Exp. Bot.* **2005**, *56*, 1761–1778. [CrossRef] [PubMed]
204. Nanjundappa, A.; Bagyaraj, D.J.; Saxena, A.K.; Kumar, M.; Chakdar, H. Interaction between arbuscular mycorrhizal fungi and *Bacillus* spp. in soil enhancing growth of crop plants. *Fungal Biol. Biotechnol.* **2019**, *6*, 1–10. [CrossRef]
205. Ivanov, V.B.; Bystrova, E.I.; Seregin, I.V. Comparative impacts of heavy metals on root growth as related to their specificity and selectivity. *Russ. J. Plant Physiol.* **2003**, *50*, 398–406. [CrossRef]
206. Sandhya, V.S.K.Z.; Ali, S.Z.; Grover, M.; Reddy, G.; Venkateswarlu, B. Effect of plant growth promoting *Pseudomonas* spp. on compatible solutes, antioxidant status and plant growth of maize under drought stress. *Plant Growth Regul.* **2010**, *62*, 21–30. [CrossRef]

207. Misra, J.; Pandey, V.; Singh, N. Effects of some heavy metals on root growth of germinating seeds of *Vicia faba*. *J. Environ. Sci. Health Part A* **1994**, *29*, 2229–2234.
208. Luo, H.; Xu, H.; Chu, C.; He, F.; Fang, S. High temperature can change root system architecture and intensify root interactions of plant seedlings. *Front. Plant Sci.* **2020**, *11*, 160. [CrossRef]
209. Doty, S.L.; Oakley, B.; Xin, G.; Kang, J.W.; Singleton, G.; Khan, Z.; Vajzovic, A.; Staley, J.T. Diazotrophic endophytes of native black cottonwood and willow. *Symbiosis* **2009**, *47*, 23–33. [CrossRef]
210. Santos, F.; Peñaflor, M.F.G.; Paré, P.W.; Sanches, P.A.; Kamiya, A.C.; Tonelli, M.; Nardi, C.; Bento, J.M.S. A novel interaction between plant-beneficial rhizobacteria and roots: Colonization induces corn resistance against the root herbivore *Diabrotica speciosa*. *PLoS ONE* **2014**, *9*, e113280. [CrossRef]
211. Desbrosses, G.; Contesto, C.; Varoquaux, F.; Galland, M.; Touraine, B. PGPR-Arabidopsis interactions is a useful system to study signaling pathways involved in plant developmental control. *Plant Signal. Behav.* **2009**, *4*, 319–321. [CrossRef]
212. Hassan, M.K.; McInroy, J.A.; Kloepper, J.W. The interactions of rhizodeposits with plant growth-promoting rhizobacteria in the rhizosphere: A review. *Agriculture* **2019**, *9*, 142. [CrossRef]
213. Rosier, A.; Medeiros, F.H.; Bais, H.P. Defining plant growth promoting rhizobacteria molecular and biochemical networks in beneficial plant-microbe interactions. *Plant Soil* **2018**, *428*, 35–55. [CrossRef]
214. Paredes-Páliz, K.; Rodríguez-Vázquez, R.; Duarte, B.; Caviedes, M.A.; Mateos-Naranjo, E.; Redondo-Gómez, S.; Caçador, M.I.; Rodríguez-Llorente, I.D.; Pajuelo, E. Investigating the mechanisms underlying phytoprotection by plant growth-promoting rhizobacteria in *Spartina densiflora* under metal stress. *Plant Biol.* **2018**, *20*, 497–506. [CrossRef]
215. Mhlongo, M.I.; Piater, L.A.; Madala, N.E.; Labuschagne, N.; Dubery, I.A. The chemistry of plant–microbe interactions in the rhizosphere and the potential for metabolomics to reveal signaling related to defense priming and induced systemic resistance. *Front. Plant Sci.* **2018**, *9*, 112. [CrossRef]
216. Igiehon, N.O.; Babalola, O.O. Below-ground-above-ground plant-microbial interactions: Focusing on soybean, rhizobacteria and mycorrhizal fungi. *Open Microbiol. J.* **2018**, *12*, 261. [CrossRef] [PubMed]
217. Parmar, N.; Dufresne, J. Beneficial interactions of plant growth promoting rhizosphere microorganisms. In *Bioaugmentation, Biostimulation and Biocontrol*; Springer: Berlin/Heidelberg, Germany, 2011; pp. 27–42.
218. Castro-Sowinski, S.; Herschkovitz, Y.; Okon, Y.; Jurkevitch, E. Effects of inoculation with plant growth-promoting rhizobacteria on resident rhizosphere microorganisms. *FEMS Microbiol. Lett.* **2007**, *276*, 1–11. [CrossRef] [PubMed]
219. Liu, F.C.; Xing, S.J.; Ma, H.L.; Du, Z.Y.; Ma, B.Y. Effects of inoculating plant growth-promoting rhizobacteria on the biological characteristics of walnut (*Juglans regia*) rhizosphere soil under drought condition. *Ying Yong Sheng Tai Xue Bao J. Appl. Ecol.* **2014**, *25*, 1475–1482.
220. Majeed, A.; Abbasi, M.K.; Hameed, S.; Imran, A.; Rahim, N. Isolation and characterization of plant growth-promoting rhizobacteria from wheat rhizosphere and their effect on plant growth promotion. *Front. Microbiol.* **2015**, *6*, 198. [CrossRef]
221. Singh, S.; Parihar, P.; Singh, R.; Singh, V.P.; Prasad, S.M. Heavy metal tolerance in plants: Role of transcriptomics, proteomics, metabolomics, and ionomics. *Front. Plant Sci.* **2016**, *6*, 1143. [CrossRef]
222. Yadav, S.K. Heavy metals toxicity in plants: An overview on the role of glutathione and phytochelatins in heavy metal stress tolerance of plants. *South Afr. J. Bot.* **2010**, *76*, 167–179. [CrossRef]
223. Fahr, M.; Laplaze, L.; Bendaou, N.; Hocher, V.; El Mzibri, M.; Bogusz, D.; Smouni, A. Effect of lead on root growth. *Front. Plant Sci.* **2013**, *4*, 175. [CrossRef] [PubMed]
224. Chibuike, G.U.; Obiora, S.C. Heavy metal polluted soils: Effect on plants and bioremediation methods. *Appl. Environ. Soil Sci.* **2014**, *2014*, 752708. [CrossRef]
225. Ahmed, S.; Choudhury, A.R.; Chatterjee, P.; Samaddar, S.; Kim, K.; Jeon, S.; Sa, T. The role of plant growth-promoting rhizobacteria to modulate proline biosynthesis in plants for salt stress alleviation. In *Plant Growth Promoting Rhizobacteria for Sustainable Stress Management*; Springer: Singapore, 2019; pp. 1–20.

MDPI
St. Alban-Anlage 66
4052 Basel
Switzerland
www.mdpi.com

Cells Editorial Office
E-mail: cells@mdpi.com
www.mdpi.com/journal/cells

www.ingramcontent.com/pod-product-compliance
Lightning Source LLC
Chambersburg PA
CBHW061757210926
51850CB00006B/576
* 9 7 8 3 0 3 6 5 9 1 4 3 8 *